T0262912

EVAPOTRANSPIRATION

Principles and Applications for
Water Management

EVAPOTRANSPIRATION

Principles and Applications for Water Management

Edited by

Megh R. Goyal, PhD, PE, Editor-in-Chief
and Eric W. Harmsen, PhD, PE, Co-editor

Apple Academic Press

TORONTO   NEW JERSEY

Apple Academic Press Inc.	Apple Academic Press Inc.
3333 Mistwell Crescent	9 Spinnaker Way
Oakville, ON L6L 0A2	Waretown, NJ 08758
Canada	USA

©2014 by Apple Academic Press, Inc.

First issued in paperback 2021

Exclusive worldwide distribution by CRC Press, a member of Taylor & Francis Group
No claim to original U.S. Government works

ISBN 13: 978-1-77463-286-4 (pbk)
ISBN 13: 978-1-926895-58-1 (hbk)

Library of Congress Control Number: 2013948146

Library and Archives Canada Cataloguing in Publication

Evapotranspiration: principles and applications for water management/edited by Megh R. Goyal, PhD, PE, and Eric W. Harmsen, PhD, PE.

Includes bibliographical references and index.
ISBN 978-1-926895-58-1
1. Evapotranspiration--Measurement. 2. Water--Management. I. Goyal, Megh Raj, editor of compilation II. Harmsen, Eric W., editor of compilation

QC915.5.E93 2013	551.57'2	C2013-905616-5

Apple Academic Press also publishes its books in a variety of electronic formats. Some content that appears in print may not be available in electronic format. For information about Apple Academic Press products, visit our website at **www.appleacademicpress.com** and the CRC Press website at **www.crcpress.com**

CONTENTS

Contents

LIST OF CONTRIBUTORS

Abdurrahman Ali Alazba
Professor in Water Research, College of Agriculture and Food Sciences, Kingdom of Saudi Arabia, PO Box: 2460/11451, Riyadh, Saudi Arabia

M. T. Amin
Assistant Professor in Water Research, College of Agriculture and Food Sciences, King Saud University, Kingdom of Saudi Arabia, PO Box: 2460/11451, Riyadh, Saudi Arabia.

Carmen L. Arcelay
Former graduate research assistant, University of Puerto Rico, Mayaguez, PR, USA
Email: cl_arcelay@hotmail.com

B. Keith Bellingham
P.H., Certified Professional Hydrologist, and Soil Scientist , Stevens Water Monitoring Systems Inc., 12067 NE Glenn Widing Drive Suite 106, Portland-Oregon-97220-USA. Phone: (800) 452-5272, http://www.stevenswater.com/contact.aspx
Email: kbellingham@stevenswater.com

Melvin J. Cardona-Soto
Graduate research assistant, University of Puerto Rico, Mayaguez Campus.

Kenneth L. Clark
Research Forester, Silas Little Experiment Station, Northern Research Station, United States Department of Agriculture Forest Service, 11 Campus Blvd., Suite 200 Newtown Square, PA 19073 Phone: 609-894-0325, Email: kennethclark@fs.fed.us

Joel T. Colon
Former graduate research assistant, University of Puerto Rico, Mayaguez, PR,
Email: joel.j.colon@cbp.dhs.gov

R. Dominguez
Institute of Engineering, National Autonomous University of Mexico, University Avenue 3000, 04510 Coyoacan, Mexico D.F., Mexico.

Michael D. Dukes
Associate Professor, Department of Agricultural and Biological Engineering, University of Florida, Gainesville, FL., U.S.A., Email: mddukes@ufl.edu

Mohammad N. Elnesr
Assistant Professor, Shaikh Mohammad Alamoudi Chair for Water Research, College of Agriculture and Food Sciences, King Saud University, Kingdom of Saudi Arabia, PO Box: 2460/11451, Riyadh, Saudi Arabia. E-mail: melnesr@ksu.edu.sa or drnesr@gmail.com

Antonio González-Pérez
Undergraduate Research Assistant, Department of Marine Science and Director of the Caribbean Atmospheric Research Center, University of Puerto Rico, Mayagüez, PR 00681

F. J. González
Institute of Engineering, National Autonomous University of Mexico, University Avenue 3000, 04510 Coyoacan, Mexico D.F., Mexico. Email: jramosh@iingen.unam.mx

Jorge E. Gonzalez

Professor, Department of Mechanical Engineering, City College of New York.
Email: gonzalez@me.ccny.cuny.edu

Megh R. Goyal

P.E., Retired Professor in Agricultural and Biomedical Engineering, University of Puerto Rico – Mayaguez
Campus; and Senior Technical Editor-in-Chief in Agriculture Sciences and Biomedical Engineering, Apple
Academic Press Inc., PO Box 86, Rincon, PR 00677, USA. Email: goyalmegh@gmail.com

Eric W. Harmsen

Professor, Department of Agricultural and Biosystems Engineering, University of Puerto Rico, Mayagüez,
PR 00681, Email: eric.harmsen@upr.edu

Marvin E. Jensen

Retired Research Program Leader at USDA-ARS and Irrigation Consultant. 1207 Spring Wood Drive, Fort
Collins, Colorado 80525, USA. Email: mjensen419@aol.com

Xinhua Jia

Assistant Professor, Department of Agricultural and Biosystems Engineering, North Dakota State University, Fargo, ND, U.S.A. Email: Xinhua.Jia@ndsu.edu

Preeyaphorn Kosa

Assistant Professor [Water Resources Engineering], School of Civil Engineering [Institute of Engineering],
Suranaree University of Technology, 111 University Avenue, Muang District, Nakhon Ratchasima, 30000,
Thailand. Email: kosa@sut.ac.th

L. Marrufo

Institute of Engineering, National Autonomous University of Mexico, University Avenue 3000, 04510
Coyoacan, Mexico D.F., Mexico.

Ariel Mercado-Vargas

Graduate Research Assistant, Department of Computer and Electrical Engineering, UPRM;

G. Miller

Professor in Soil Science, North Carolina State University, Raleigh, NC, 27695

Norman L. Miller

Climate Science Department, Earth Sciences Division, Berkeley National Laboratory.
Email: NLMiller@lbl.gov

Robert S. Nicholson

Chief, Environmental Studies Program, New Jersey Water Science Center, United States Geological Survey,
810 Bear Tavern Road, West Trenton, NJ, 08628. Phone: 609-771-3925, USA
Email: rnichol@usgs.gov

Luis R. Pérez-Alegía

Professor, Department of Agricultural and Biosystems Engineering, University of Puerto Rico, P.O. Box
9030, Mayagüez, PR 00681, U.S.A. Email: luperez@uprm.edu

Timothy G. Porch

Genetics Researcher, Tropical Agricultural Research Station, USDA-ARS, Mayagüez - Puerto Rico, USA.
Email: Timothy.Porch@ars.usda.gov

G. S. Rajput

Principal Scientist, Jawaharlal Nehru Agricultural University, Jabalpur, Madhya Pradesh, India-482 004

Victor H. Ramirez-Builes

Agroclimatology and Crop Science Researcher, National Coffee Research Center (Cenicafe), Chinchina
(Caldas, Colombia). Email: victor.ramirez@cafedecolombia.com

Judith Ramos-Hernandez
Chemical Engineer, Institute of Engineering, National Autonomous University of Mexico, University Avenue 3000, 04510 Coyoacan, Mexico D.F., Mexico. Email: jramosh@iingen.unam.mx

Dionel C. Rodríguez
Former undergraduate research assistant, University of Puerto Rico, Mayaguez, PR

Alejandra Rojas-Gonzalez
Graduate Students, University of Puerto Rico, Mayaguez Campus

Nicole J. Schlegel
Earth and Planetary Science Department, University of California, Berkeley. Email: Schlegel@EPS.Berkeley.edu

S. F. Shih
P.E., Late Professor of Hydrology at University of Florida, Institute of Food and Agricultural Sciences, Gainesville, Florida

P. S. Shirgure
Senior Scientist, National Research Center for Citrus, Indian Council of Agricultural Research, Amravati Road, Nagpur, India, 440010. Email: shirgure@gmail.com

T. Sinclair
University of Florida, Gainesville, FL 32611.

A. K. Srivastava
Principal Scientist, Jawaharlal Nehru Agricultural University, Jabalpur, Madhya Pradesh, India – 482 004

David M. Sumner
Hydrologist, United States Geological Survey, Florida Water Science Center, 12703 Research Parkway, Orlando, FL, 32826, USA. Phone: 407-803-5518, Email: dmsumner@usgs.gov

Ramon E. Vasquez
P.E., Former Dean, College of Engineering and Professor, Department of Computer and Electrical Engineering, University of Puerto Rico, Mayaguez, PR, Email: reve@ece.uprm.edu

Benjamin G. Wherley
Assistant Professor, Turfgrass Physiology and Ecology, Dept. of Soil and Crop Sciences, Texas A&M University, College Station, TX 77843-2474. Email: b-wherley@tamu.edu

Amos Winter
Professor, Department of Marine Science and Director of the Caribbean Atmospheric Research Center, University of Puerto Rico, Mayagüez, PR 00681

LIST OF ABBREVIATIONS

°C	degree celsius
AC-FT	acre foot
ACTMO	Agricultural Chemical Transport Model (USDA)
AET	actual evapotranspiration
AHPS	Advanced Hydrologic Prediction Service
AI	artificial intelligence
ANN	Artificial Neural Network
ARS	Agricultural Research Service
ASABE	American Society of Agricultural and Biological Engineers
ASCE	American Society of Civil Engineers
ASCS	Agriculture Stabilization and Conservation Service
ATP	adenosine triphosphate
AWRA	American Water Resources Association
AWS	Agro-Climatic Weather Stations
AWWA	American Water Works Association
B-C	Blaney-Criddle
BOR	Bureau of Reclamation
BP	back propagation
BPNNs	back propagation neural networks
CFS	cubic feet per second
CFSM	water depth, cubic feet per second per square mile
CGDD	cumulative growing degree days
CLENS	Computing Loading Estimates from Non-point Sources
CO_2	carbon dioxide
CU	consumptive use
CWSI	crop water stress index
DOY	day of the year
EP	evolutionary programming
EPA	Environmental Protection Agency
EPAN	pan evaporation
ET	evapotranspiration
FAO	Food and Agricultural Organization, Rome
FC	field capacity
FWS	Fish and Wildlife Service
GA	genetic algorithm
GM	general model
GUI	graphical user interface
HRG	Hargreaves
IASWC	Indian Association of Soil and Water Conservationists

ICAR	Indian Council of Agriculture Research
ICRAF	International Council for Research in Agroforestry
IIM	International Irrigation Management
IR	infrared
ISAE	Indian Society of Agricultural Engineers
ISC	Indian Society of Citriculture
KSA	Kingdom of Saudi Arabia
LAI	leaf area index
MAD	maximum allowable depletion
MAE	mean absolute error
mg/l	milligrams per liter
mg/d	million gallons per day
MLP	multi-layer perceptron
MLR	multiple linear regression
MSE	mean squared error
MSL	mean sea level
NAE	National Academy of Engineers
NARR	North American Regional Reanalysis
NAS	National Academy of Sciences
NAWDEX	National Water Data Exchange
NBS	National Biological Survey (US Dept. of the Interior)
NCEP	National Centers for Environmental Prediction
NCPC	National Capital Planning Commission
NEPA	National Environmental Policy Act
NET	net value
NEWS	Northeastern US Water Supply Study
NIWR	National Institutes of Water Resources
NMD	Nagpur model development
NOAA	National Oceanographic and Atmospheric Administration
NPS	non-point source
NRCS	National Resources Conservation Service
NURP	Nationwide Urban Runoff Program (EPA)
NWSRFS	National Weather Service River Forecasting System
PAR	photosynthetically active radiation
PDP	parallel distributed processing
PE	processing elements
PET	potential evapotranspiration
pH	acidity/alkalinity measurement scale
PLC	programmable logic controller
PM	Penman-Monteith
PME	Presidency of Meteorology and Environment
PPB	one part per billion
PPM	one part per million
PPT	one part per trillion
PTF	pedotransfer functions

PVC	poly vinyl chloride
PWP	permanent wilting point
RBF	radial basis function
RBFNNs	radial basis function neural networks
RH	relative humidity
RMAX	maximum relative humidity
RMIN	minimum relative humidity
RMSE	root mean squared error
RS	solar radiation
SAC-SMA	Sacramento soil moisture accounting
SCRR	simple conceptual rainfall-runoff
SCS-BC	SCS Blaney-Criddle
SDWA	Safe Drinking Water Act
SEB	surface energy balance techniques
SEBAL	surface energy balance model for land
SF	sap flow
SNHR	sunshine hours
SRB	Sonora River Basin
SSE	sum squared error
SWB	soil water balance
SWCB	State Water Control Board
SWMM	Storm Water Management Model
SWP	soil water potential
TDR	time-domain reflectometry
TE	transpiration efficiency
TEW	total evaporable water
TMAX	maximum temperature
TMDL	total maximum daily loads
TMIN	minimum temperature
TR	temperature range
TUE	transpiration use efficiency
USBR	US Bureau of Reclamation
USDA	US Department of Agriculture
USDA-SCS	US Department of Agriculture-Soil Conservation Service
USDAHL	US Department of Agriculture Hydrograph Laboratory Model
USGS	United States Geological Survey (Department of Interior)
USNPS	US National Park Service (Department of the Interior)
VITA	Volunteers in Technical Assistance
VPD	vapor pressure deficit
VWC	volumetric water content
WATBAL	water balance
WDSD	Water Data Sources Directory
WEC	World Environment Center

WED	World Environment Day
WFI	World Forest Institute
WFP	World Food Program, United Nations
WISP	wind speed
WMO	World Meteorological Organization
WRI	World Resources Institute
WRSIC	Water Resources Scientific Information Center
WS	Weather Stations
WSEE	weighed standard error of estimate
μg/g	micrograms per gram
μg/L	micrograms per liter

LIST OF SYMBOLS

B	eddy-covariance method, in W/m^2
C_p	specific heat capacity of air, in $J/(g \cdot °C)$
C_s	specific heat of dry soil
C_w	specific heat of water
d	zero displacement height (m)
D_r	root zone depletion (mm)
E	evapotranspiration rate, in $g/(m^2 \cdot s)$
e	vapor pressure, in kPa
EB	covariance method, in W/m^2
E_p	pan evaporation
E_s	saturation vapor pressure, in kPa
ET	evapotranspiration rate, in mm/year
ETa	reference ET, in the same water evaporation units as Ra
f	fractional vegetative cover
f_c	daily fraction of the soil surface
f_w	soil surface wetted
G	soil heat flux at land surface, in W/m^2
g	acceleration due to gravity, $9.81 \ m.s^{-2}$
h	canopy height, in m
H	sensible heat flux, in W/m^2
HBR	sensible heat flux as estimated by the Bowen ratio energy-budget variant of the eddy
I	identity matrix
J	Jacbian matrix (J_{ij})
k	monthly CU coefficient
K_o	extinction coefficient of hygrometer for oxygen, in $m^3/(g \cdot cm)$
K_p	Pan coefficient
K_{RS}	an empirical coefficient usually estimated as 0.16 and 0.19 for inland and coastal areas, respectively. No units.
K_w	extinction coefficient of hygrometer for water, in $m^3/(g \cdot cm)$
N^I	number of input nodes
N^{TR}	number of training sample
P	atmospheric pressure (kPa)
p	mean monthly percent of annual daytime hours
Pa	atmospheric pressure, 101.3 Pa at mean sea level.
q	specific humidity, g water/g moist air
r_a	aerodynamic resistance
Ra	extraterrestrial radiation, in the same water evaporation units as ETa
r_c	canopy resistance

R_d	gas constant for dry air, equal to 0.28704 J/°C/g
R_n	net radiation, in W/m²
R_s	incoming solar radiation on land surface, in the same water evaporation units as ETa
S	change in storage of energy in the biomass, air, and any standing water, in W/m²
T	air temperature (°F)
t	mean monthly air temperature (°F)
T_a	air temperature, in °C
T_s	surface temperature
U	lateral wind speed along coordinate x-direction, in m/s
u_z	wind speed at height z
v	lateral wind speed along coordinate y-direction, in m/s
w	wind speed along coordinate z-direction, in m/s
W	water content of soil
x	one of two orthogonal coordinate directions within a plane parallel to canopy surface
y	one of two orthogonal coordinate directions within a plane parallel to canopy surface
z	coordinate direction perpendicular to canopy surface
Z_h	height of humidity measurements (m)
z_m	roughness length of canopy for momentum, in m
Z_m	height of wind measurements (m)
z_s	height of sensors above land surface, in m
Z_t	rooting depth (m)

Greek Symbols

α	Priestley-Taylor coefficient, dimensionless
γ	psychometric constant, in kPa/°C
Δ	slope of the saturation vapor pressure curve, in kPa/°C
Δt	time interval between measurement (sec)
ΔT_s	soil temperature interval between measurement
ε	ratio molecular weight of water to molecular weight of air
ε_a	atmospheric emissivity
ε_o	surface emissivity
η	angle of rotation about the z-axis to align u into the x-direction on the x–y plane, in radians
θ	angle of rotation in the y-direction to align w along the z-direction, in radians
θ_{FC}	water content at field capacity
q_s	soil volumetric water content
θ_{WP}	water content at wilting point
λ	latent heat of vaporization, in J/g
λE	latent heat flux, in W/m²

λE_{BR}	latent heat flux as estimated by the Bowen ratio energy-budget variant of the eddy
λE_{RE}	latent heat flux as estimated by the residual energy-budget variant
μ	ratio of molecular weight of air to molecular weight of water
ρ	air density, in g/m³
σ	ratio of vapor density (ρ_v) to air density (ρ)
ϵ	small quantity

PREFACE 1

Due to increased agricultural production, irrigated land has increased in the arid and subhumid zones around the world. Agriculture has started to compete for the water use with industries, municipalities and other sectors. This increasing demand along with increments in water and energy costs have made it necessary to develop new technologies for the adequate management of water. The intelligent use of water for crops requires understanding of evapotranspiration processes.

Evapotranspiration (ET) is a combination of two processes: evaporation and transpiration. Evaporation is a physical process that involves conversion of liquid water into water vapor and then into the atmosphere. Evaporation of water into the atmosphere occurs on the surface of rivers, lakes, soils and vegetation. Transpiration is a physical process that involves flow of liquid water from the soil (root zone) through the trunk, branches and surface of leaves through the stomates. An energy gradient is created during the evaporation of water, which causes the water movement into and out of the plant stomates. In the majority of green plants, stomates remain open during the day and stay closed during the night. If the soil is too dry, the stomates will remain closed during the day in order to slow down the transpiration.

Evaporation, transpiration and ET processes are important for estimating crop water requirements and for irrigation scheduling. To determine crop water requirements, it is necessary to estimate ET by on-site measurements or by using meteorological data. On-site measurements are very costly and are mostly employed to calibrate ET methods using climatological data. There are a number of proposed mathematical equations that require meteorological data and are used to estimate the ET for periods of one day or more. The generalized Penman-Monteith (PM) method is a physically based method. The simplest methods require only data about average temperature of air, length of the day and the crop. Other equations require daily radiation data, temperature, vapor pressure and wind velocity. After comparing the PM method with other 20 methods worldwide, the FAO has stated that the PM method can be used at all locations and that it is better to use the PM method (estimating the missing data) than to use simpler method that ignore variables.

Potential ET is the ET from a well-watered crop, which completely covers the surface. Meteorological processes determine the ET of a crop. Closing of stomates and reduction in transpiration are usually important only under drought or under stress conditions of a plant. The ET depends on four factors: (1) climate, (2) vegetation, (3) water availability in the soil and (4) behavior of stomates. Vegetation affects the ET in various ways. It affects the ability of the soil surface to reflect light. The vegetation affects the amount of energy absorbed by the soil surface. Soil properties, including soil moisture, also affect the amount of energy that flows through the soil. The height and density of vegetation influence efficiency of the turbulent heat interchange and the water vapor of the foliage.

The mission of this compendium is to serve as a textbook or a reference manual for graduate and undergraduate students of agricultural sciences and engineering. This book will be a valuable reference for professionals and technicians who work with water management for agriculture and forestry; and will be beneficial to students, hydrologists, engineers, meteorologists, water managers and others.

There are several books on ET, such as, *ET in the Soil-Plant-Atmosphere System* by Viliam Novák, Springer; *ET* by Jesse Russell et al., by Book on Demand Ltd; *Evaporation and ET: Measurements and Estimations* by Wossenu Abtew et al., Springer; *ET Covers for Landfills and Waste Sites* by Victor L. Hauser, CRC Press; *The ASCE Standardized Reference ET Equation* by Richard G. Allen et al., American Society of Civil Engineers; *ET: Webster's Timeline History by Icon Group* International; *Evaluating ET for Six Sites in Benton, Spokane, and Yakima Counties, Washington* by Stewart A. Tomlinson, University of Michigan Library; *ET and Irrigation Water Requirements ASCE Manual No 70* by M. E. Jensen et al., ASCE; *Crop Evapotranspiration: Guidelines for Computing Crop Water Requirements* (FAO Report 56); *Water Requirements for Irrigation and the Environment* by Marinus G. Bos et al., Springer; *Evapotranspiration* by Leszek Labedzki, InTech Open Access.

This compendium complements all books on evapotranspiration that are currently available on the market, and our intention is not to replace anyone of these. This book on evapotranspiration is unique because it is complete and simple, a one-stop manual, with worldwide applicability to water management in agriculture. Its coverage of the field of ET includes historical review; basic principles and applications; how to generate missing climatic data; research results using remotely sensed climatic data; research studies from Colombia, India, Mexico, Puerto Rico, Saudi Arabia, Thailand, Trinidad and U.S.A.; studies related to agronomical crops and forest trees in arid, humid, semiarid, and tropical climates; and methods and techniques that can be easily applied to other locations (not included in this book). This book offers basic principles, knowledge and techniques of evapotranspiration to water management in agriculture and forestry, which are necessary to understand before designing/developing and evaluating an agricultural water management system. This book is a must for those interested in water resources planning and management, namely, researchers, scientists, educators and students.

This book, *Evapotranspiration: Principles and Applications for Water Management*, includes 30 chapters that are presented in three parts—Part I: Principles, Part II: Applications and Part III: Water Management in the Tropics. Book chapters include Historical evolution of ET methods by M. E. Jensen; Water vapor flux models for agriculture by V. H. Ramirez and E. W. Harmsen; Direct measurement of transpiration by V. H. Ramirez; Design of lysimeter for turfgrass water use by B. Wherely, T. Sinclair, M. Dukes and G. Miller; ET: Meteorological methods by M. R. Goyal; Evaporation estimations with neural networks by P. Shirgure and G. S. Rajput; Pan evaporation modeling: Indian agriculture by P. Shirgure and G. S. Rajput; ET for cypress and pine forests: Florida, USA by D. M. Sumner; ET for pinelands in New Jersey, USA by D. M. Sumner, R. S. Nicholson and K. L. Clark; Water management in citrus by P. Shirgure, A. K. Srivastava and S. Singh; Vegetation water demand and basin water availability in Mexico by J. Ramos, F. J. Gonzalez, L. Marrufo and R. Dominguez; Turfgrass deficit irrigation practices for water conservation by B. Wherley; ET for Saudi Arabia: Modified Hargreaves model by M. N. Elnesr; ET with distant weather stations: Saudi Arabia by M. N. Elnesrr; Actual ET using LANDSAT 5™ in Thailand by P. Kosa; Sensor based irrigation scheduling by K. Bellingham; Snow budgeting and water resources by K. Bellingham; Historical overview of ET in Puerto Rico by E. W. Harmsen; Reference ET for Colombia by V. H. Ramírez; Water management for agronomic crops in Trinidad by M. R. Goyal; Crop water stress index for common beans by V. H. Ramírez, E.W. Harmsen and T. Porch; Temperature versus elevation relationships: ET by M. R. Goyal; Generation of missing climatic data: Puerto Rico by M. R. Goyal; Estimation of pan evaporation coefficients by E. W. Harmsen, A. Gonzaléz and A. Winter; Daily ET estimations using satellite remote sensing by E. W. Harmsen, J. Mecikalski, M. J. Cardona, A. Rojas and R. Vasquez; Vapor flux measurements by E. W. Harmsen, V. H. Ramirez, M. D. Dukes, X. Jia, L. R. Pérez, and R. Vasquez; Climate change impacts on agricultural water resources: 2090 by E. W. Harmsen, N. L. Miller, N. J. Schlegel and J. E. Gonzalez; ET using satellite remote sensing for the tropical climate by E. W. Harmsen, J. Mecikalski, A. Mercado and P. Tosado; Water management for sweet peppers by E. W. Harmsen, J. Colón, C. L. Arcelay and D. Cádiz; Web based irrigation scheduling by E. W. Harmsen; Appendix; and Subject Index.

The contributions by all cooperating authors to this edition has been most valuable in the compilation of this compendium. Their names are mentioned in each chapter. This book would not have been written without the valuable cooperation of these investigators, many of whom are renowned scientists who have worked in the field of evapotranspiration throughout their professional careers.

Dr. Eric W. Harmsen and I will like to thank editorial staff, Sandy Jones Sickels, Vice President, and Ashish Kumar, Publisher and President, at Apple Academic Press, Inc., for making every effort to publish the book when the diminishing water resources is a major issue worldwide. Special thanks are due to the AAP Production Staff, Apple Academic Press, Inc., for typesetting the entire manuscript and for the quality production of this book. We request the reader to offer us your constructive suggestions that may help to improve the next edition. The reader can order a copy of this book for the library, the institute or for a gift from CRC Press [Taylor and Francis Group], 6000 Broken Sound Parkway, NW Suite 300, Boca Raton, FL, 33487, USA; Tel.: 800-272-7737.

Eric W. Harmsen joins me to express thanks to our families for their understanding and collaboration during the preparation of this book. With our whole heart and best affection, we dedicate this book to our mothers, Daya W. Goyal and June Rose Harmsen, whose hopes were that their sons would add their drop to the ocean of service to the world of humanity.

— **Megh R. Goyal, PhD, PE, Editor-in-Chief**

PREFACE 2

Knowledge of evapotranspiration (ET) is critically important for efficient management of water resources. ET is often an overlooked component of the hydrologic cycle, unlike rainfall or surface water, which are perceptible to the human eye. Nevertheless, this invisible flux of water vapor is relentless in its ability to remove water from the soil, and in many cases lead to devastating drought. Only by understanding its complex nature can we hope to manage our water resources in a wise manner.

During the last century great advances have been made in our understanding of the process of evapotranspiration. Initially, the methods for estimating ET were simple, yet effective, as for example nonweighing lysimeters and pan evaporation tanks. This was followed by a rich period of development of meteorological methods, some empirically based, others physically based. This period also included the development and increased use of weighing lysimeters in research.

Two important milestones were reached with the publication of the United Nations Food and Agriculture Organization (FAO) 1977 and 1998 documents on estimating crop water requirements, representing concrete examples of the global promotion of increasing agricultural water use efficiency. The more recent FAO publication resulted in the worldwide adoption of the Penman-Monteith (PM) methodology for estimating crop ET, and enumerated data and knowledge gaps that scientists have been addressing since the date of that publication. Recent studies have led to the widespread adoption of the eddy covariance and satellite remote sensing methods; while the former method has remained a research instrument because of its relatively high cost and complexity, the latter has evolved into a practical tool which farmers can use on a daily basis via web-based applications (e.g., *see* Chapter 30).

Despite the advances, many farmers, especially in developing countries, are not aware of crop water requirements and are consequently over- or under-applying irrigation. This is a serious problem leading to loss in crop yields, wasting of resources (water, energy, chemicals) and in many cases contamination of the environment. During an era of climate change, over-population, and extreme poverty, such inefficiency is morally unacceptable.

Initially I got interested in the phenomenon referred to as ET during my graduate work at Michigan State University. This of course was coupled with the need to explain to my family and friends that ET did not mean to Extraterrestrial, but rather that it meant evapotranspiration, after which their eyes glazed over and the conversation quickly changed to another subject. During my PhD at the University of Wisconsin-Madison, a postdoc at North Carolina State University, and eight years in private environmental consulting, I focused on topics like groundwater flow and transport and vadose zone processes. In the last 13 years, I have had the privilege of being able to pursue various studies related to ET in the tropics, with the goal of promoting agricultural water conservation, especially within the island nations of the Caribbean. This region is rich with problems to be solved, and God willing, I will spend the remainder of my career working on the solution to some of these problems.

I am indebted to my wife, Rhea Harmsen-Howard, who has been my unwavering support since we initiated our undergraduate studies together so many years ago. I would also like to acknowledge the NOAA-CREST (grant NA06OAR4810162) and USDA Hatch (H-402) projects, which provided partial financial support for my participation on this book project. It is my hope that this book may serve as a stepping-stone for further development in the field and for the promotion of water conservation during the twenty-first century.

— **Eric W. Harmsen, PhD, PE, Co-editor**

FOREWORD

Megh R. Goyal, PhD, PE, and Eric W. Harmsen, PhD, PE, have chosen to write a textbook on a very important topic in soil and water conservation engineering. *Evapotranspiration: Principles and Applications for Water Management* presents principles, procedures and applications as they apply to the more critical components of water management, concentrating on the engineering approaches to understanding the water management problems in agriculture. The chapters provide comprehensive coverage of evaporation and transpiration processes in the hydrological cycle. The book should serve as a text for a course or as a valuable reference with practical applications of evapotranspiration processes that will be useful to students, instructors, researchers and device designers alike. Engineering scientists, and students should find this book equally valuable. I am honored to write the foreword for this valuable and unique compendium.

I want to share with the readers of this compendium the revised/edited version of my article: "Basics of Evapotranspiration", <http://www.stevenswater.com/articles/etbasics.aspx>. Evapotranspiration (ET) is a combination of direct evaporation and plant transpiration. ET represents the loss of water from the Earth's surface. ET is usually expressed as a rate such as inches per day or milliliters per day. Knowledge of ET is important for irrigation scheduling but it is also an important factor for other land use applications, such as septic tank drain fields, urban planning, water shed budgeting, and climate and weather models. ET can be used as a historical tool but usually it is predicted or used in a forecast to help irrigators optimize irrigation.

Factors affecting ET are: (1) *Current weather conditions:* Conditions include wind speed, air temperature, relative humidity, and sunlight. The effect weather has on ET rate is rather intuitive. Warm, sunny, and dry weather with a lot of wind will have greater ET rates; (2) *Plant type:* The ET rates between different plant species can vary greatly. For example, needle leaf trees such as pine trees will have a much greater ET than a deciduous tree such as an oak tree. Even though the pine needle can be very small, the needles have much more surface area than deciduous leaves allowing for more transpiration; (3) *Soil chemistry (Soil conditions):* The chemical makeup of soil is also a major factor affecting ET rates. At the molecular level, most clays will have a chemical affinity (attraction) for water. Clay's chemical affinity for water results from a planner geometry and the charge distribution at the molecular level. The chemical affinity for water is much less for sands and silts. The chemical interaction between the clay and the water impedes both the evaporation and the plant transpiration. The evaporation rate of a soil that is mostly sand, on the other hand, will be close to what the evaporation rate would be out of a pan. For example, if the soil moisture is 20% by volume, the ET rate in sand would be very high and very low in clay under the same weather conditions. On the other hand, if the soil moisture is 30% by volume, the sandy soil will dry out quicker than the clay-rich soil and the clay-rich soil will have a higher ET rate over a longer period of time; (4) *Soil salinity (Soil conditions):* Another factor that affects the plant transpiration is the salt content of soil. Plant transpiration is the movement of water from the roots and the subsequent loss of water vapor from the stomata in the leaves. The primary driving force of the movement of water from the roots to the leaves is osmoses. Osmosis is the diffusion or movement of water that is driven by a salt concentration gradient. Increasing soil salinity will decrease plant transpiration because the salt concentration gradient will diminish, thus affecting the overall ET; (5) *Geographical locations:* Elevation, longitude, latitude and time of year for the location for which ET data are needed. In general, ET rates increase toward the equator (Fig. 1) and will be higher in summer than in winter. Figure 1 shows the areas on Earth where ET is the highest. In areas where ET is greater than precipitation, there will be very little recharge to the aquifer and a net upward movement of water. This net upward movement of water followed by ET will cause salts and minerals to accumulate near the surface. This explains why soils in arid lands tend to have a high pH and a higher salinity. Conversely, if precipitation is greater than ET, minerals and nutrients will be leached out of the soil, and there will be a shallow aquifer. Highly leached soils will develop more clay loam textured soils and will typically have a lower base saturation.

ET Methods: The simplest method for approximating ET is the pan evaporation method. Measuring the rate of evaporation of water in a pan provides a quick and reasonable approximation of ET with little or no cost. While convenient, the pan evaporation method does not take into account the contribution of plant transpiration, and it makes the assumption that water evaporates out of the soil at the same rate it would out of a pan. Developed in the late 1940s, the Penman-Monteith method provides a correlation between ET and the energy the earth's surface receives. One of the problems with the previous ET models such as pan evaporation methods does not take into account the variability of ET from one kind of crop to the next. The Penman-Monteith method became a widely accepted ET method because a reference ET could be obtained from weather data. A reference ET is the calculated ET for a standard grass of a standard height. The ET for a specific crop could then be calculated by simply multiplying the reference ET by a crop coefficient. With the development of better and more reliable weather instruments, farm agencies could publish reference ET values along with the crop coefficients so the ET could be determined for the entire crop in a particular region. Despite the advances in weather measurements in the 1950s, the Penman-Monteith method had a few drawbacks. The reference ET calculation was mathematically tedious and few computers in the past had the computational power to calculate it and few mathematicians were able to calculate with a pencil and paper. In the 1980s, the Priestly Taylor method was introduced as an alternative to Penman-Monteith method. Based on the same heat budget principles as the Penman-Monteith method, the Priestly Taylor method relies on several assumptions and simplifies the calculation of reference ET. However, with advances in computers in the 1990s the Priestly Taylor method was short lived and the Penman-Monteith method became standard method for calculating ET by the scientific community and organizations such as the American Society of Civil Engineers and the Food and Agriculture Organization (FAO) of the United Nations. While much emphasis has been placed on calculating ET from weather calculation modules such and the Penman-Monteith method, a previously overlooked method for calculating ET using Darcy's Law and soil hydrology has been slow to emerge. ET rates can also be measured with soil moisture sensors by looking at the water balance in the soil. For example, the amount of water recharging the water table can be measured with a deeply place soil moisture sensor while changes in soil moisture can be monitored with soil sensors throughout the profile. By examining changes in soil moisture through the soil profile over a period of time, an ET rate can be calculated.

Evapotranspiration Measurement Networks: An ET station consists of weather sensors, a data logger, a solar panel to provide electric power and radio telemetry. The radio telemetry communicates the weather data in real time back to a computer where a computer program calculates the reference ET. The typical sensors included on an ET station are an anemometer for wind speed and direction, a relative humidity sensor, air temperature sensor, pyronometer for solar radiation measurements, a barometer, and rain gauge.

Federal agencies such and the US Bureau of Reclamation, The US Department of Agriculture National Resources Conservation Service (NRCS) and state soil water conversation districts provide daily ET rates along with crop coefficients for local crops for network of ET stations. One of the most comprehensive ET networks in the western US is the AgriMet network provided by the US Bureau of Reclamation. In the Midwestern US, many states provide ET rates from state-funded Mesonets.

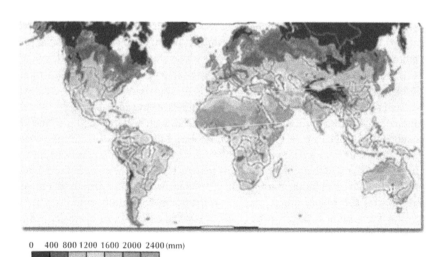

0   400 800 1200 1600 2000 2400 (mm)

FIGURE 1   Regional annual evapotranspiration rates. ET rate increase toward the equator (United Nations World Atlas of Desertification, 1997).

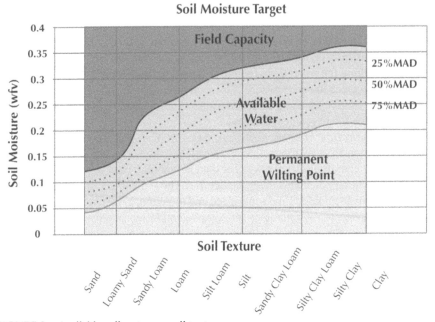

FIGURE 2   Available soil water vs. soil texture.

ET and Irrigation Management: Perhaps the most important application for ET rates is irrigation scheduling. The weekly irrigation schedule can be calculated from forecasted ET rates, sprinkler efficiencies, root zone depth, and the available water capacity of the soil. With ET irrigation scheduling, an irrigator will know the approximate amount of water the crop will need for any given week. While ET-based irrigation scheduling has shown to save water and to optimize crop yields, ET-based irrigation scheduling is based on many assumptions to give an approximation of the actual water requirements of the crops. Microclimates and highly varying soil conditions introduce a lot of error in weather station based ET calculation, which in turn will increase error in the scheduling. Because of this uncertainty, crop advisors and irrigation scheduling calculators will manipulate factors in the ET calculation so that the schedule will overirrigate by 10 to 20%. In other words, if there is a 10 to 20% error in the ET-based irrigation schedule calculation, adding 10 to 20% more water will ensure that the crop will not be stressed. The ultimate goal for an irrigation scheduling is to keep the soil moisture at a level that is best for the crops. ET-based calculations basically estimate what the soil moisture would be based on weather data. Figure 1 indicates regional annual evapotranspiration rates. ET rates increase toward the equator. Figure 2 shows the range of plant available water based on soil texture and illustrates the changes in field capacity and permanent wilting point, as the soil becomes clay rich.

Basically, with almost any crop, the soil moisture needs to be maintained above permanent wilting point and in general stay below field capacity. Permanent wilting point is a condition where the soil moisture is at low level where a plant can't uptake any water. Field capacity is the amount of water that can be held in soil before gravity will begin to drain the soil. Deep infiltration and aquifer recharge will occur if the soil moisture stays above field capacity.

In precision agriculture, the most accurate irrigation schedules use regional ET forecasts in combination with soil moisture sensors to ensure that the soil moisture values stay at a level that is best for the crop health and yields. The ET forecasts help the irrigator create a weekly irrigation schedule, while soil moisture data takes into account microclimates and soil conditions and fine-tune the weekly schedule on a daily basis. With the growing demands on agricultural products and the growing importance to protect and restore natural aquatic habitats, water conservation is more important than ever. Proper irrigation management is a key to ensure healthy, high quality crops while protecting valuable water resources. Knowledge of ET is critical for environmental and economic best irrigation management practices.

In addition to ET models to help manage crops, products such as the Stevens Hydra Probe or SAM (Stevens Agricultural Monitoring) system can further help define exactly how much water is needed and when it should be applied. Combining actual soil moisture measurements with a local ET forecast can give the user an incredibly detailed look at actual soil moisture information, leading to better managed crops with increased yield and reduced waste.

Professors Goyal and Harmsen are well qualified to present this important book. I have learned a great deal while browsing this book. I recommend Apple Academic Press Inc., for publishing more books on water management.

B. Keith Bellingham
Certified Professional Hydrologist
Soil Scientist and Geochemist
Stevens Water Monitoring Systems Inc,
<http://www.stevenswater.com/contact.aspx>
12067 NE Glenn Widing Drive Suite 106
Portland, Oregon 97220
Phone: United States (800) 452-5272; (503) 445-8000
Fax: (503) 445-8001
E-mail: kbellingham@stevenswater.com

B. Keith Bellingham, with 20 years of experience in the water resources industry, holds MS in Environmental Science and Engineering from the Oregon Health and Science University at the Oregon Graduate Institute; a BS in Chemistry from the University of Cincinnati. He is also a sensor technology advisor for Safe Harvest, providing integrated hydrological and meteorological monitoring instrumentation, and information systems that optimize water resource management and enhance forecasting.

WARNING/DISCLAIMER

PLEASE READ CAREFULLY

The editors, the contributing authors, the publisher and the printing company have made every effort to make this book as complete and as accurate as possible. However, there still may be grammatical errors or mistakes in the content or typography. Therefore, the contents in this book should be considered as a general guide and not a complete solution to address any specific situation in water management. For example, all water management systems are not same and are different.

The editors, the contributing authors, the publisher and the printing company shall have neither liability nor responsibility to any person, any organization or entity with respect to any loss or damage caused, or alleged to have caused, directly or indirectly, by information or advice contained in this book. Therefore, the purchaser/reader must assume full responsibility for the use of the book or the information therein.

The mentioning of commercial brands and trade names are only for technical purposes. It does not mean that a particular product is endorsed over to another product or equipment not mentioned. Editors, cooperating authors, educational institutions, and the publisher Apple Academic Press Inc. do not have any preference for a particular product.

All weblinks that are mentioned in this book were active on June 30, 2013. The editors, the contributing authors, the publisher and the printing company shall have neither liability nor responsibility, if anyone of the weblinks is inactive at the time of reading of this book.

ABOUT THE EDITOR-IN-CHIEF

Megh R. Goyal received his BSc degree in Agricultural Engineering in 1971 from Punjab Agricultural University, Ludhiana, India; his MSc degree in 1977 and PhD degree in 1979 from the Ohio State University, Columbus; his Master of Divinity degree in 2001 from Puerto Rico Evangelical Seminary, Hato Rey – Puerto Rico.

Since 1971, he has worked as Soil Conservation Inspector; Research Assistant at Haryana Agricultural University and the Ohio State University; and Research Agricultural Engineer at Agricultural Experiment Station of UPRM. At present, he is a Retired Professor in Agricultural and Biomedical Engineering in the College of Engineering at University of Puerto Rico – Mayaguez Campus; and Senior Acquisitions Editor in Agriculture and Biomedical Engineering for Apple Academic Press Inc.

He was the first agricultural engineer to receive a professional license in Agricultural Engineering in 1986 from the College of Engineers and Surveyors of Puerto Rico. On September 16, 2005, he was proclaimed as "Father of Irrigation Engineering in Puerto Rico for the 20th Century" by the ASABE – Puerto Rico Section, for his pioneer work on microirrigation, evapotranspiration, agroclimatology, and soil and water engineering. During his professional career of 42 years, he has received awards such as: Scientist of the Year, Blue Ribbon Extension Award, Research Paper Award, Nolan Mitchell Young Extension Worker Award, Agricultural Engineer of the Year, Citations by Mayors of Juana Diaz and Ponce, Membership Grand Prize for ASAE Campaign, Felix Castro Rodriguez Academic Excellence, Rashtrya Ratan Award and Bharat Excellence Award and Gold Medal, Domingo Marrero Navarro Prize, Adopted son of Moca, Irrigation Protagonist of UPRM, Man of Drip Irrigation by Mayor of Municipalities of Mayaguez/ Caguas/ Ponce and Senate/Secretary of Agric. of ELA – Puerto Rico. He has authored more than 200 journal articles and textbooks including: *Elements of Agroclimatology (Spanish)* by UNISARC, Colombia, two *Bibliographies on Drip Irrigation*, *Biofluid Dynamics of Human Body Systems* by Apple Academic Press Inc., *Biomechanics of Artificial Organs and Prostheses* by Apple Academic Press Inc., and *Management of Drip/Trickle or Micro Irrigation* by Apple Academic Press Inc. Readers may contact him at: <goyalmegh@gmail.com>

ABOUT THE CO-EDITOR

 Dr. Eric W. Harmsen received his BSc and MSc degrees in Agricultural Engineering from Michigan State University and his PhD degree from the University of Wisconsin. He holds a Professional Engineer License. Currently he is a professor in the Department of Agricultural and Biosystems Engineering, University of Puerto Rico-Mayaguez Campus. He teaches courses on agricultural hydrology, agroclimatology and irrigation. His research interests include measurement and modeling all components of the hydrologic cycle; remote sensing of water and energy balance in the tropics; agroclimatology. Some of Dr. Harmsen's water management related publications and presentation can be found at the following link: http://pragwater.com/selected-publications-and-presentations/

PART I: PRINCIPLES

CHAPTER 1

HISTORICAL EVOLUTION OF EVAPOTRANSPIRATION METHODS[1]

MARVIN E. JENSEN

CONTENTS

[1] I thank the organizers of Evapotranspiration Workshop on 10th March, 2010 by Colorado State University and USDA – ARS for giving me an opportunity to express my personal involvement in the ET research. I ask the readers of this chapter to bear with me, as I have over emphasized my involvement, it is because I have been associated with the development of ET estimating methods for the past 50 years.

1.1 INTRODUCTION

This chapter is a condensation of a more detailed paper that Rick Allen and I prepared in 2000 [53] for the presentation at the Fourth National Irrigation Symposium, ASAE in Phoenix, Arizona. This chapter is also an edited version of my original paper that was prepared for a workshop on Evapotranspiration. This chapter emphasizes my involvement or association with my colleagues in the development and dissemination of new technology for estimating evapotranspiration [ET] in the United States of America [USA]. I reviewed current ET literature and older documents relating to the development of early evapotranspiration estimating methods in the USA. In this chapter, I have included some history of the development of the "ASCE Manual 70 Evapotranspiration and Water Requirements [54]," "FAO-56 Crop Evapotranspiration [4]," development of programs to calculate ET using satellite data [5, 9, 10], the new "ASCE Standardized Reference ET Equation [8]," a proposal for a one-step approach to estimate ET, and an update on the second edition of ASCE Manual 70. More detailed but less personnel-oriented, progress in measuring and modeling ET in agriculture can be found in a recent review article by Farahani et al. [34].

1.2 EARLY STUDIES

With the rapid development of irrigation in the western USA after about 1850, efforts to reduce water losses from both beneficial and nonbeneficial vegetation became more important. Measured water deliveries varied widely and deliveries often greatly exceeded consumptive use. This problem was recognized early on as Buffum [24] stated that overirrigation was the first and most serious mistake made by early settlers. Most early methods that were developed for estimating evapotranspiration (ET) or consumptive use (CU) for irrigated areas were for seasonal values based on observed or measured water deliveries. Air temperature was the main weather variable that was used. Solar radiation was not considered directly as a separate variable. For monthly values, crop stage of growth effects became important. As developers of empirical estimating methods adjusted and modified their temperature based methods, the methods tended to become more complex. A measured change in soil moisture content over periods of seven or more days was the main source of measured ET data before and during the 1950s. Reliable published data were scarce, especially for measured ET rates by stage of crop growth.

1.3 MY INVOLVEMENT

My involvement in methods of estimating ET began in 1960, when I evaluated monthly coefficients for the Blaney-Criddle equation. In the late 1950s, the United States Department of Agriculture-Soil Conservation Service [USDA-SCS] asked the USDA - Agricultural Research Service (USDA-ARS) to develop monthly coefficients for the Blaney-Criddle (B-C) method. Howard Haise and Harry Blaney had requested USDA-ARS researchers to use their available data to calculate monthly coefficients for the B-C method. A stack of data sheets was collected and given to me when I arrived in Fort Collins in 1959. Before this request, the data had not been analyzed, because Harry Blaney went to Israel for a year and Howard Haise went to University of

California, Davis – California for a six-month sabbatical leave. I was asked to work on the data during their absence. I started reviewing the numbers and the results were highly variable. Researchers with little or no prior experience with the B-C method had calculated monthly coefficients. Many did not have the experience to judge good input data from bad data such as drainage that may have occurred following rains. Also, some time periods were too short to be reliable. The plotted monthly coefficients were widely scattered. I revised the questionnaire that was sent to ARS researchers by Howard Haise, as he was well known to ARS researchers at that time. As part of this task, I reviewed current literature and many early publications on development of irrigation and ET estimating methods. These reviews were never published, but served as a good personal reference and the main source of information for a later 1962 ET workshop report [55].

The main difference in the new questionnaire that I prepared was that we asked for basic soil water and weather data, mainly air temperature and precipitation along with supporting data and not the B-C coefficients that they had computed. We established criteria for reviewing datasets before making our own calculations. Where possible, we estimated solar radiation for each measurement period, usually from cloud cover at one or more nearby weather stations. We ended up with about 1,000 ET rates for periods of seven or more days. When searching for full crop cover ET data, we found only about 100 datasets represented reasonably reliable values. The general equation [56] summarizing the ET rates and solar radiation (Rs) data from the full cover datasets was developed:

$$ET = ((0.014 \times T) - 0.37) \times (Rs) \qquad (1)$$

where, T is an air temperature in °F, and Rs is solar radiation in mm/day or inches/day. A tabulated summary of the results was presented at an Consumptive Use (CU) workshop that was organized by ARS-SCS in March 1972 and later at the annual meeting of the Soil Sci. Society of America [80]. The workshop report contained: Charts of the ratio of ET to solar radiation for seven crops versus percent of the growing season and between cuttingsof alfalfa grown in lysimeters at Reno, Nevada; a summary of solar radiation relationships; and tabulated weekly mean solar radiation, mean air temperature and cloud cover for 20 locations in the western USA. Copies of this workshop report were sent to others involved in estimating ET who did not attend the workshop such as Jerry Christiansen at Utah State University and David Robb at U.S. Bureau of Reclamation (USBR) in Denver – Colorado.

The main results of my research were published in 1963 [56]. The main purpose of this paper was to encourage engineers, soil scientists and agronomists to begin thinking about radiant energy as the primary source of energy for evaporation instead of air temperature, which had been the practice for decades. Using our 1962 report, Dave Robb [75] with the USBR in Denver, Colorado developed a set of nine crop coefficient curves for use with the Jensen–Haise Eq. (1).

1.4 EARLY RESEARCH: IRRIGATION AND CONSUMPTIVE USE

Many books on irrigation had been written on irrigation in England, France and Italy from 1846 to 1888 along with reports on irrigation in California [25]. In 1897, joint efforts were started for irrigation research by USDA – Experiment Stations under the supervision of the Office of Experiment Stations [82]. These investigations were continued for the next 55 years under various USDA departments, but were transferred to the Soil Conservation Service (USDA-SCS) in 1939 [59]. In 1902, the Bureau of Plant Industry was established in the USDA and detailed studies of transpiration were conducted by this organization in the early 1900s.

From about 1890 to 1920, the term duty of water was used to describe the water use in irrigation with no standard definition [48]. Duty of water was sometimes reported as the number of acres that could be irrigated by a constant flow of water such as 1 cubic foot per second or as depth of water applied. Water measured at farm turnouts was referred to as net duty of water. Examples of early studies were those by: Widstoe [93, 94] in Utah from 1902 to 1911 on 14 crops; Harris [41, 42] who summarized 17 years of study in the Cache Valley of Utah; Lewis [61] who conducted studies near Twin Falls – Idaho from 1914 to 1916; Hemphill [44] who summarized studies in the Cache La Poudre river valley of northern Colorado; Israelson and Winsor [47, 48] who made duty of water studies in the Sevier River valley in Utah from 1914 to 1918; and Crandall [28] who worked in the Snake River area near Twin Falls, Idaho in 1917 and 1918. An excellent summary of early seasonal CU studies was presented in a progress report of the Duty of Water Committee in the Irrigation Division of the American Society of Civil Engineers (ASCE). It was presented in 1927 by O.W. Israelson and later published [7].

L. J. Briggs, a biophysicist, and H. L. Shantz, a plant physiologist, conducted highly significant studies of transpiration in eastern Colorado [19–22]. Briggs and Shantz recognized that solar radiation was the primary cause of the cyclic change of environmental factors [23]. They developed hourly transpiration prediction equations using: the vertical component of solar radiation and temperature rise, solar radiation, and vapor saturation deficit. They also recognized the significance of advected energy. A summary of the Briggs and Shantz 1910–1917studies was later published by Shantz and Piemeizel [76]. Widstoe [92] began studying the influence of various factors affecting evaporation and transpiration in 1902. Harris and Robinson [42] conducted similar studies from 1912 through 1916.Widstoe and McLaughlin [94] concluded that temperature is the most important factor than the sunshine and relative humidity.

Other studies conducted during the 1900–1920 period were related with the factors causing and controlling water loss due to irrigation. During the next two decades emphasis was on the development of procedures to estimate seasonal CU using available climatic data.

1.5   EARLY ESTIMATING METHODS: EVAPOTRANSPIRATION

In 1920, the U.S. Bureau of Reclamation (USBR) began studying the relationships between CU and temperature [62]. Hedke [43] proposed a procedure to the ASCE Duty of Water Committee in 1924 for estimating valley CU based on heat available. Heat available was estimated using degree-days. Radiant energy was not considered directly. The ASCE committee concluded in 1927 that there was an urgent need for a relatively simple method of estimating CU [7]. In the 1920s, Harry Blaney began measuring CU of crops based on soil samples. He worked on crops grown along the Pecos

River for the Division of Irrigation and Water Conservation under the USDA-SCS. His first procedure for estimating seasonal and annual CU used mean temperature, percent of annual daylight hours and average humidity [17]. From 1937 to 1940, Lowry with the National Resources Planning Board, and Johnson with the USBR, developed a procedure for estimating: Seasonal CU using maximum temperature above 32°F during the growing season, and annual inflow minus outflow data for irrigation projects [62]. Thornthwaite [84] correlated mean monthly temperature with ET based on eastern river basin water balance studies and developed an equation for potential evapotranspiration which was widely used for years. Thornthwaite recognized the limitations of his equation pointing out the lack of understanding of why potential ET at a given temperature is not the same everywhere. Because of its simplicity, the equation was applied everywhere and, in general, underestimated ET in arid areas. All of these early methods were based on correlations of measured or estimated CU data with various available or calculated climatic data. The resulting equations were relatively simple because computers were not available—only slide rules and perhaps hand-operated adding machines. This may be difficult for young people to comprehend today since they are very dependent on personal computers.

Numerous other reports and publications were prepared between 1920 and 1945 by various state and federal agencies. Many of them dealt with investigations of water requirements for specific areas and measured farm deliveries and not on techniques for estimating CU or ET. Bibliographies of publications on seasonal CU can be found in books such as Israelsen [47] and Houk [45].

1.5.1 BLANEY-CRIDDLE METHOD

In the 1950s and 1960s, the most widely known empirical ET estimating method used in the USA was the Blaney-Criddle (BC) method. The procedure was first proposed by Blaney and Morin in 1942 [17]. It was modified later by Blaney and Criddle [12–15]. The Eq. (2) was well-known older engineers:

$$U = KF = \sum k \times f \qquad (2)$$

where, U = estimated CU (or ET) in inches; F = the sum of monthly CU factors, f, for the period ($f = t \times p/100$); t = mean monthly air temperature in °F; p = mean monthly percent of annual daytime hours (daytime is defined as the period between sunrise and sunset); K = empirical CU coefficient (irrigation season or growing period); and k = monthly CU coefficient.

For long-time periods mean air temperature was considered to be a good measure of solar radiation [15]. Phelan [73] developed a procedure for adjusting monthly k values as a function of air temperature, which later became part of SCS publication on the BC method [86]. Criddle also developed a table of daily peak ET rates as a function of depth of water to be replaced during irrigation. Hargreaves [38] developed a procedure similar to the BC method for transferring CU data to other areas of the globe. Christiansen and Hargreaves developed a series of regression equations for estimating monthly grass ET based on pan evaporation, air temperature and humidity data [26, 27]. They initiated efforts to reduce weather data requirements to only air temperature, calculated extraterrestrial radiation and the difference between maximum and minimum air temperature to predict the effects of relative humidity and cloudiness. A culmination of these efforts was the well-known 1985 Hargreaves equation for grass reference ET [39, 40].

A summary of early studies and the BC method was published in a USDA technical bulletin in cooperation with the Office of Utah State Engineer [15]. Criddle was Utah State Engineer at that time. This publication presented the BC equation in English and metric units and updated records of measured CU by crops, percent of daytime hours of the year for latitudes of 0 to 65°N. and 0 to 50°S, monthly CU factors by states, suggested monthly crop coefficients (k) for selected locations, monthly CU factors (f) and average precipitation in various foreign countries, and a summary of BC method applications that were made by various consultants such as Claude Fly in Afghanistan, Tipton and Kalmbach in Egypt and West Pakistan. The BC method is still used in some states because historical water records such as water rights have been based on this method.

Prior to about 1960 and U.S. engineers were usually taught only the Blaney-Criddle method of estimating CU or ET. Today, students and young engineers generally have had a fairly broad training in modern methods of estimating ET and ET-climate relationships.

It is unfortunate that Blaney and Criddle did not select extraterrestrial solar radiation as an index of solar energy instead used percent of daytime hours. Daytime hours from sunshine tables of Marvin [65] do not adequately account for effects of solar angle, especially in higher latitudes as illustrated in Fig. 1. "Smithsonian Meteorological Tables" would have had an equation for calculating total daily solar radiation at the top of the atmosphere and total daily solar radiation for selected dates during the year at that time.

1.5.2 TRANSITION METHODS IN THE USA

In the 1960s, estimating ET methods in the USA began to change from methods based primarily on mean air temperature to methods considering both temperature and solar radiation. Several of these methods are listed below:

Alfalfa reference ET	Jensen and Haise [56]	$ET = ((0.014 \times T - 0.37)) \times (R_s)$ (3)
Grass reference ET	Hargreaves and Samani [40] and Hargreaves et al. [39]	$ET = (0.0023) \times R_a \times (T + 17.8) \times (TD)^{0.50}$ (4)
Grass reference ET, Florida	Stephens and Stewart	$ET = (0.0082 \times T_f - 0.19) \times (R_s)$ (5)

where, ET = Potential evapotranspiration in mm/day, T is a mean air temperature in ° C, R_s is solar radiation in mm per day or inches per day, R_a = Extraterrestrial radiation, mm/ day, and TD is the difference between maximum and minimum daily air temperature in° C.

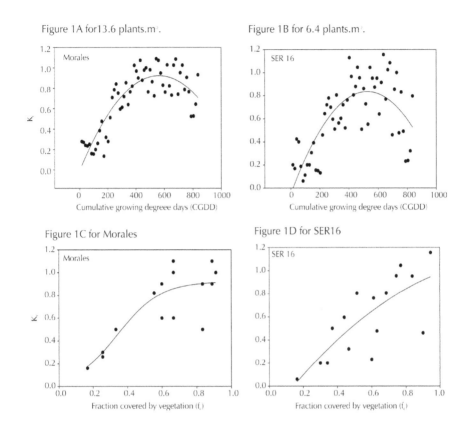

FIGURE 1   Change in relative solar radiation vs. change in relative percent daylight hours.

1.5.3   THEORETICAL METHODS

The Bowen ratio is the ratio of temperature to vapor pressure gradients, $\Delta t/\Delta e$, [18 26]. Bowen ratio and energy balance concepts were not incorporated at an early date into methods for estimating ET as they were for estimating evaporation from water surfaces. Examples of early work on evaporation from water using the Bowen ratio were those of Cummings and Richardson [31], McEwen [96], Richardson [97], Cummings [29], Kennedy and Kennedy [58] and Cummings [30].

In contrast to development of largely empirical methods in the USA, Penman [67] in the United Kingdom took a basic approach and related ET to energy balance and rates of sensible heat and water vapor transfer. Penman's work was based on the physics of the processes, and it laid the foundation for current ET estimating methodology using standard weather measurements of solar radiation, air temperature, humidity, and wind speed. The Penman equation [69–71] stands out as the most commonly applied physics-based equation. Penman [69] used the Bowen ratio principle in developing the well-known equation for estimating evaporation from a water surface along with reduction coefficients for grass. Later, a surface resistance term was added [66, 72, 74]. Howell and Evett [46] provided an excellent summary of the Penman equation and its evolution to the development of a standardized equation by the American Society of Civil Engineers. The modern combination equation applied to standardized surfaces is currently referred to as the Penman-Monteith Eq. (PM). It represents the state-of-the-art in estimating hourly and daily ET. When applied to standardized surfaces, it is now calledthe Standardized Reference ET Equation [8].

Other methods of estimating and measuring ET range from eddy covariance and energy balance using Bowen ratio or sensible heat flux based on surface temperature, radiosonde measurements of complete boundary layer profiles of temperature, humidity and energy balance estimates based on satellite imagery.

A **radiosonde** (*Sonde* is French and German for *probe*) is a unit for use in things such as weather balloons that measures various atmospheric parameters and transmits them to a fixed receiver. Radiosondes may operate at a radio frequency of 403 MHz or 1680 MHz and both types may be adjusted slightly higher or lower as required. A **rawinsonde** is a radiosonde that is designed to only measure wind speed and direction. Colloquially, rawinsondes are usually referred to as radiosondes. Modern radiosondes measure or calculate the following variables: Pressure, Altitude, Geographical position (Latitude/ Longitude), Temperature, Relative humidity, Wind (both wind speed and wind direction), Cosmic ray readings at high altitude. Radiosondes measuring ozone concentration are known as ozonesondes [<http://en.wikipedia.org/wiki/Radiosonde>].

1.5.4   OTHER METHODS

During the 1950s and early 1960s, many other equations were proposed, but they have not been widely used in the USA. Some of these are Halkias et al. [37], Ledbedevich [60], Romanov [98], Makkink [63] and Vitkevich [90]. In 1957, Makkink published a formula for estimating potential ET based on solar radiation and air temperature [74] that is still used in Western Europe. Makkink used the energy-weighting term of the Penman equation, solar radiation and a small negative constant. The Makkink equation formed the basis of the subsequent FAO Radiation method that was included in FAO Irrigation and Drainage Paper No. 24 on crop water requirements by Doorenbos and Pruitt [33, 34]. Turc developed a formula in 1960, which was later modified [85]. It was based on mean air temperature and solar radiation for 10-day periods. Rijtema [74] proposed the Turc formula for individual crops using crop factors and length of growing season. Olivier [68] in England developed a procedure for estimating average monthly CU for planning new projects where climatic data were limited. The Olivier's equation used average monthly wet-bulb depression and a factor based on clear sky solar radiation values.

In 1968, I described the process of using the rate of ET from a well-watered crop with an aero-dynamically rough surface like alfalfa with 30–50 cm of growth as a measure of potential ET or Eo [50]. ET for a given crop could be related to Eo using a coefficient, now commonly known as a crop coefficient:

$$ET = [K_c] \times [E_o] \qquad (6)$$

where, K_c is a dimensionless crop coefficient similar to that proposed by van Wijk and de Vries [89] representing the combined effects of resistance to water movement from the soil to the evaporating surfaces, resistance to diffusion of water vapor from the evaporating surfacesthrough the laminar boundary layer, resistance to turbulent transfer to the free atmosphere, and relative amount of radiant energy available as compared to the reference crop. At that time, methods other than those based on air temperature were not well known. In order to facilitate the understanding of the of the "[$E_o \times K_c$] process," illustrating the change in the Kc as crop cover develops enabled users to visualize how the coefficient changed from a value near 0.15 for bare soil to 1.0 at full cover.

Routine use of computers for estimating ET was in its infancy. Estimates of alfalfa reference ET andETr, that were used in our first computerized irrigation scheduling program [51] were calculated using the Penman method with alfalfa wind speed coefficients developed at Kimberly, ID [57]. Since we did not have a computer at Kimberly – Idaho, we used a time-share computer located in Phoenix – Arizona connected via the telephone system with a teletype paper tape reader and printer at Kimberly. In 1970, copies of a FORTRAN computer program that Ben Pratt and I had written for estimating daily ET and scheduling irrigations using the Penman equation was widely distributed following an informal workshop at Kimberly – Idaho in 1970.

The method of using reference ET and crop coefficients has been widely used for nearly a half century. In general, it has been relatively robust. For example, in 2004, Ivan Walter and I estimated ET for the Imperial Irrigation District in California using this approach. Our estimateof ET for agricultural land for CY 1998 was 2% higher than that of a SEBAL estimate for entire district; and 5% higher for agricultural crops for water year 1998 [83] of 2.01 million ac-ft or 2.5 km^3. Whether the [$ET_{ref} \times K_c$] method, also known as the two-step method, will be replaced by the direct PM method, or one-step method, in the near future is uncertain.

Various methods of measuring ET and methods of estimating ET were summarized during a conference held in Chicago following the ASAE winter meeting on Dec. 5–6, 1966. Leading researchers from many different organizations and disciplines presented papers on the current state-of-the-art of estimating ET in the early 1960s. I was the conference chairman and program committee members represented ASCE, American Meteorological Society, Soil Sci. Society of America, International Commission of Irrigation and Drainage, and several Canadian organizations. W.O. Pruitt assisted in editing the proceedings [49].

Unfortunately, we not did get anyone to specifically discuss the Penman method. I tried to get Howard Penman to attend and present a paper, but he declined after several letters and a phone call. Leo Fritschen discussed the energy balance method, C.H.M. van Bavel discussed combination methods [87, 88], and Champ Tanner discussed a comparison of energy balance and mass transfer methods for measuring ET [81]. The conference proceedings contained pictures of the authors who were leaders in their field at the time and session chairmen.

1.6 DISSEMINATING AND ADOPTING NEW ESTIMATING TECHNOLOGY FOR EVAPOTRANSPIRATION

Disseminating and adopting advances in ET estimating technology has not been rapid as compared with many other fields because dissemination-adaption involves many disciplines such as agricultural engineering, hydrology, water science, meteorology, soils, and plants. Dissemination and adaption of new technology did not occur rapidly as it did in single disciplines. The users of the technology, a half-century ago, were mainlyengineers.

Penman had a comprehensive understanding of the physical processes involved in ET. He presented an introductory paper at the informal meeting on physics that was held in The Netherlands in September 1955 [70]. He concluded with the sentence: "Though the physicist still has some problems he can solve by himself, much of his future contribution to understanding evaporation in agriculture must be in collaboration with the biologist and the soil scientist."

In my 1960 review of old and current literature, it was clear that new advances in estimating ET had to depart from the traditional use of temperature as the primary input variable. The Penman equation had been developed, but was not in general use by engineers designing and managing irrigation projects. Generally, the Penman equation was thought to be too complicated for use by engineers given the status of computational tools and weather data commonly collected at that time. In developing countries, it is still considered complex.

During the 1966 meeting of the Irrigation and Drainage Division of ASCE in Las Vegas, Nevada, I was asked to chair the ASCE committee on Irrigation Water Requirements, formerly the Committee on CU of Crops and Native Vegetation. The committee was charged with developing a manual on CU. The committee had been chaired by Harry Blaney, but it was not making much progress on this task. A few papers that had been prepared were reproductions of old temperature based methods of estimating ET. My one condition on accepting chairmanship was that I could bring in nonASCE members into the committee such as Bill Pruitt. Bill, who was with the University of California at Davis and was not an ASCE member then. Bill had been measuring ET using weighing lysimeters and collecting associated basic meteorological data. The committee had control of the Corresponding and Non-member advisors. Control members were: Robert Burman, University of Wyoming; Harlan Collins, SCS; Albert Gibbs, USBR; Marvin E. Jensen, ARS-USDA; and Arnold Johnson, USGS. Harry Blaney remained on the committee, but never attended any of our meetings. We made progress and produced a report that was the start of the ASCE manual on ET. In 1973, we prepared as a camera-ready document at Kimberly – Idaho that was printed as the ASCE report Consumptive Use and Irrigation Requirements [52] and widely disseminated in the U.S. and in other countries like China.

I left the ET committee for three years when I became a member of the executive committee of the ASCE-I&D Division in October 1974. Members of the ET committee continued to work on the manual. Several members served as ET committee chairman between October 1973 and 1986. In 1986, a subcommittee of the ET committee was formed. Its members were Rick Allen, Ron Blatchley, Bob Burman, Marvin E. Jensen, Eldon Johns, Jack Stone and Jim Wright. I was designated as chairman. We were charged with the task of preparing the CU or ET manual. The first draft, which had both English and metric units, was completed for review by one or several ASCE committees in 1988. One reviewer suggested that we use only metric units. This was a good advice, but delayed the manuscript another year as changes were made in the manuscript. The manual was published in 1990 as ASCE Manual No. 70 [54]. The ASCE-PM equation from ASCE Manual 70 has been widely accepted for standardized calculations such as in the Natural Resources Conservation Service National Engineering Handbook [64].

Allen et al. [2] prepared a paper for publication in the Agron. J. to disseminate new information to soil scientists and agronomists. In 1990, the FAO needed to update its 1975–1977 publication on crop water requirements (FAO-24). It organized an expert's consultation on the revision of FAO-24 in Rome on 28–30 May 1990 [79]. There were 10 participants from seven countries: John Monteith from the International Crops Research Institute for Semi-arid Tropics, D. Rijks from WMO, and several participants from FAO. USA participants were Rick Allen, Marvin E. Jensen, and Bill Pruitt. At this conference several manuscript copies of the ASCE Manual 90 were made available and became a key reference. The revision of FAO-24 with major contributions by Rick Allen resulted in the well-known FAO-56 publication on crop evapotranspiration by Allen et al. [4].

In the Netherlands, R.A. Feddes had been working on models of ET, plant root systems and soil water extraction [35]. In 1998, Bastiaanssen et al., including Feddes, published a paper describing a surface energy balance model for land (SEBAL) using satellite-based imagery [9, 10]. In the USA, Allen et al., developed a high resolution-mapping model with internalized calibration using principles and similar techniques as used in the SEBAL model [5]. It uses a near-surface temperature gradient, dT, that is indexed to the radiometric surface as in SEBAL. The METRIC technique [5] uses the SEBAL technique for estimating dT.METRIC also uses weather-based reference ET to establish energy balance conditions for a cold pixel and is internally calibrated at two extreme conditions (wet and dry) using local available weather data. The auto-calibration is done for each image using alfalfa-based reference ET. Both of these models are now being used in the U.S. to map ET at a high resolution over large areas.

In 2000, the Irrigation Association and landscape industry requested the ASCE Irrigation Water Requirements committee, now renamed Committee on Evapotranspiration in Irrigation and Hydrology, to recommend a single procedure for estimating reference ET for use in the USA. This request in part resulted in an ASCE task committee of the ET Committee consisting of engineers and scientists from around the USA. They agreed on a single equation for estimating reference crop ET. The equation is a simplification of the ASCE-PM equation. The request was made to help standardize the basis for the myriad of landscape (i.e., crop) coefficients that have been developed since the late 1980s. The task committee suggested applying the PM equation to both a tall reference crop like alfalfa and short reference crop like clipped grass by changing several coefficients in order to support usage in both agricultural and landscape industries. A reduced form of the PM equation was adopted for both reference types, with the grass form being the same as in the FAO-56 publication. This was done to promote agreement in usage between the USA and other countries. Sets of coefficients were presented in a table for estimating daily or hourly reference ET for the short and tall references. Details of that equation and its development were presented in separate papers at the 2000 ASAE National Irrigation Conference in Phoenix, Arizona [91]. The various forms and applications of the PM and Penman equations, as well as commonly used empirical equations, were implemented in REF-ET software (Allen, 2000: Personal communication) that was available for free downloading from: [<http://www.kimberly.uidaho.edu/ref-et> Current Version: 3.1.08, updated January, 2012]

Rick Snyder also has several reference ET programs (monthly, daily, and hourly) for Excel spreadsheets on the web site: http://biomet.ucdavis.edu/evapotranspiration.html.

A summary of the reference ET methodology and tables of mean and basal crop coefficients were published as a major chapter in the 2nd edition of the ASABE book Design and Operation of Farm Irrigation Systems [6]. It contains a great deal of detail on factors controlling ET and on estimating ET. There are numerous recent publications on many different crop coefficients written by authors in the USA and other countries. Before using these coefficients, the users need to carefully review how the calculations were made. FAO-56 crop coefficients have been refined incorporating the fraction of ground cover and plant height [3]. Others have used remote sensing with ground-based or aircraft-based cameras, and satellite images to measure the normalized difference vegetation index [NDVI] and then related NDVI to ETref-based crop coefficients. The use of the reference ET and crop coefficient method is expected to continue because of the extensive collection of available crop coefficients.

Some scientists are suggesting a more direct approach to estimating crop water requirements such a one-step method or direct PM [66]. Shuttleworth derived a Penman-Monteith based, one-step estimation equation that allows for different aerodynamic characteristics of crops in all conditions of atmospheric aridity to estimate crop ET from any crop of a specified height using standard 2-m climate data [77]. Not everyone agrees that the concept can adequately account for surface soil water conditions. Shuttleworth called for field studies to address the problem of effective values for surface resistance for different crops equivalent to that for crop coefficients. Shuttleworth and Wallace summarized a detailed study that was conducted in Australia using the one-step approach [78]. I am not aware of any specific applications that have been made in the USA. In the second edition of Manual 70, we have a chapter on "Direct Penman-Monteith method."

The first edition of Manual 70 is out of print. In 2000, the ET committee decided to have a technical committee prepare a second edition. Marvin E. Jensen and Rick Allen were designated cochairmen. Other members were Terry Howell, Derrel Martin, Rick Snyder, and Ivan Walter. The second edition has been completely restructured. We are near having a final draft ready for review by ASCE committees. I hope that this can be completed this year.

Major progress on developing improved methods of estimating ET was made during the past third of a century because of the efforts of many Europe and U.S.

individuals. Some are R. Feddes and W. Bastiaannsen in the Netherlands; L. Pereira in Portugal; and M. Smith with FAO who led the FAO effort. Many individuals in the U.S. were major contributors to ET development technology. Some of these are Rick Allen, Terry Howell, Bill Pruitt, Joe Ritchie, Rick Snyder and Jim Wright. Bill Pruitt measured ET in weighing lysimeters near Davis – California along with detailed weather data for many years. His technical guidance and data were very valuable in the development of new technology. Jim Wright measured ET from various crops using a weighing lysimeter at Kimberly, ID over an eight year period and he developed the concept of the basal crop coefficient representing conditions when soil evaporation is minimal and most of the ET is transpiration [95]. Wright's measured data were also used to refine net radiation and crop coefficients.

1.7 SUMMARY

This chapter summarizes a century of progress in the development of modern methodology for accurately estimating daily and hourly evapotranspiration. Why did this process take so long? Evaporation from soil and plant services is a complex process involving plants, soils, local weather data like wind speed and humidity, solar and long-wave radiation. Its developments involved many disciplines. It has only been 200 years since hydrologic principles were first understood and described by Dalton so perhaps a century is not that unreasonable.

Most of the progress in developing new methodology in the U.S. was made during the last third of a century. Many scientists and engineers were involved in the evolution of ET-estimating technology. Scientists and engineers in Europe have also been instrumental in advancing the technology. In this chapter, I tried to highlight major contributions of people many of whom I knew personally or at least I had met them briefly. My involvement in the process of development better ET estimating technology and association with other engineers and scientists was a learning experience. It started me on a very rewarding career path. The experience that I gained working leading scientists and engineers over the past half century has been very rewarding personally. If I have emphasized my involvement too much it is because I have been associated with the development of ET estimating methods for the past 50 years. More detailed progress on measuring and modeling ET can be found in a review paper by Farahani et al. [34].

KEYWORDS

- Blaney
- Blaney-Criddle
- Bowen ratio
- Colorado
- consumptive use
- crop coefficient

- **crop water requirement**
- **duty of water**
- **energy balance**
- **evaporation**
- **evapotranspiration**
- **FAO**
- **ground cover**
- **irrigation**
- **irrigation engineering**
- **meteorology**
- **Monteith**
- **Penman – Monteith**
- **Penman**
- **satellite- based imagery**
- **SEBAL**
- **soil moisture**
- **solar angle**
- **solar radiation**
- **transpiration**
- **USA**
- **USDA–ARS**
- **USDA–SCS**
- **water requirement**

REFERENCES

1. Allen, Richard G. Estimating crop coefficients from fraction of ground cover and height. *Irrig. Sci.* **2009**, *28,* 17–34.
2. Allen, R. G.; Jensen, M. E.; Wright, J. L.; Burman, R. D.;Operational estimates of evapotranspiration. *Agron. J.* **1989**, *81(4),* 650–662.
3. Allen, R. G.; and Pereira, L. S.; Estimating crop coefficients from fraction of ground cover and height. *Irrig. Sci.* **2009**, *28,* 17–34.
4. Allen, R. G.; Pereira, L. S.; Raes, D.; Smith, M.; **1998,** Crop Evapotranspiration: Guidelines for Computing Crop Water Requirements. United Nations Food and Agriculture Organization, Irrigation and Drainage Paper No. 56, Rome, Italy. 300 pages.
5. Allen, R. G.; Tassumi, M.; Trezzo, R.; **2007,** Satellite-based energy balance for mapping evapotranspiration with internalized calibration (METRIC)—Model. J. Irrig. and Drain. Engr.; *133(4),* 380–406.
6. Allen, R.G.; Wright, J. L.; Pruitt, W. O.; Pereira, L. S.; Jensen, M. E.; **2007,** Water requirements. Chap. 8, pp. 208–288, In: *Design and Operation of Farm Irrigation Systems.* Second ed.; 875. Am. Soc. Agr. Biol. Engrs, St. Joseph, MI.

7. Anonymous, Consumptive Use of Water in Irrigation. Prog. Rep.; Duty of Water Comm.; Irrig. Div.; Trans. of ASCE, **1930**, *94*, 1349–1399.

8. ASCE-EWRI, **2005**, *The ASCE Standardized Reference Evapotranspiration Equation*. Edited by Allen, R. G.; Walter, I. A.; Elliott, R. L.; Howell, T. A.; D.; Itenfisu, Jensen, M. E.; Snyder, R. L.; Am. Soc. Civ. Engrs.; 69 pages with appendices A to F Index.

9. Bastiaanssen, W. G. M.; Mensenti, M.; Feddes, R. A.; Holtslag, A. A. M.; A remote sensing surface energy balance algorithm for land (SEBAL), part 1, Formulation. J. Hydrology, **1998**, *212*, 198–212

10. Bastiaanssen, W. G. M.; Pelgrum, H.; Wang, J.; Ma, Y.; Moreno, J.; Roerink, G. J. Surface energy balance algorithm for land (SEBAL), part 2, Validation. J. Hydrology, **1998**, *212*, 213–229.

11. Blaney, H. F.; Definitions, methods, and research data: consumptive use of water. A symposium, Trans. Am. Soc. Civ. Engr.; **1952**, *117*, 849–973.

12. Blaney, H. F.; Criddle, W. D.; **1945**, Determining water requirements in irrigated areas from climatological data. Processed, 17 p.

13. Blaney, H. F.; Criddle, W. D.; **1945**, A method of estimating water requirements in irrigated areas from climatological data. **1945**, (mimeo).

14. Blaney, H. F.; Criddle, W. D.; **1950**, Determining water requirements in irrigated areas from climatological and irrigation data. USDA-SCS-TP-96 Report, 50 pages.

15. Blaney, H. F.; Criddle, W. D.; **1962**, Determining Consumptive Use and Water Requirements. USDA, Technical Bull. No. **1275**, 63 pages.

16. Blaney, H. F.; Rich, L. R.; Criddle, W. D.; Consumptive use of water. Trans. Am. Soc. Civ. Engr.; **1952**, *117*, 948–1023.

17. Blaney, H. F.; Morin, K. V.; Evaporation and consumptive use of water formulas. Am. Geophys. Union Trans.; **1942**, *1*, 76–82.

18. Bowen, I. S.; The ratio of heat losses by conduction and by evaporation from any water surface. *Physics Review*, **1926**, *27*, 779–787.

19. Briggs, L. J. Shantz, H. L.; The water requirements of plants: I.; Investigations in the Great Plains in 1910, and **1911**, USDA *Bur. Plant Indr. Bull.* **1913**, *284*, 49 pp.

20. Briggs, L. J. Shantz, H. L.; Relative water requirements of plants. *J. Agr. Res.;* **1914**, *III(1)*, 11–64.

21. Briggs, L. J. Shantz, H. L.; Hourly transpiration rate on clear days as determined by cyclic environmental factors. J. Agr. Res.; **1916a**, *5(14)*, 583–648.

22. Briggs, L. J. Shantz, H. L.; Daily transpiration during the normal growth period and its correlation with the weather. J. Agr. Res.; **1916b**, *7(4)*, 155–212.

23. Briggs, L. J. Shantz, H. L.; A comparison of the hourly transpiration rate of atmometers and free water surfaces with the transpiration rate of Medicago sativa. J. Agr. Res.; **1917**, *9(9)*, 279–292.

24. Buffum, B. C.; **1892**, Irrigation and duty of water. Wyo. Agr. Exp. Bull. No. 8.

25. Carpenter, L. G.; **1890**, Section of meteorology and irrigation engineering. 3rd Ann. Report of ColoradoExp. Stn.; Fort Collins.

26. Christiansen, J. E. Pan evaporation and evapotranspiration from climatic data. J. Irrig. Drain. Div.; **1968**, *94*, 243–256.

27. Christiansen, J. E.; Hargreaves, G. H.; Irrigation requirements from evaporation. Trans. Int. Comm. Irrig. Drain.; **1969**, *3(23):* 569–23.596.

28. Crandall, L.; **1918**, Report of use of water on Twin Falls North Side (unpublished data).

29. Cummings, N. W.; **1936**, Evaporation from water surfaces. Am. Geophys. Union Trans.; Part 2, 507–509.

30. Cummings, N. W.; The evaporation-energy equations and their practical application. Am. Geophys. Union Trans.; **1940**, *21*, 512–522.

31. Cummings, N. W.; Richardson, B.; Evaporation from lakes. Physics. Rev.; **1927**, *30(4)*, 527–534.

32. Doorenbos, J. Pruitt, W. O.; Guidelines for predicting crop water requirements. FAO Irrig. Drain. Paper No. 24, FAO, Rome, **1975**, 179 pp.

33. Doorenbos, J. Pruitt, W. O.; Guidelines for predicting crop water requirements. FAO Irrig. Drain. Paper No. 24 (Revized), FAO, Rome, **1977**, 179 pp.
34. Farahani, H. J. Howell, T. A.; Shuttleworth W. J.; Bausch, W. C.; Evapotranspiration: Progress in measurement and modeling in agriculture. Trans. of the ASBE, **2007**, *50(5),* 1627–1638.
35. Feddes, R. A.; Menenti, M.; Kabat P.; Bastiaanssen, W. G. M.; Is large scale modeling of unsaturated flow with areal average evaporation and surface soil moisture as estimated from remote sensing feasible? J. Hydrol.; **1993**, *143,* 125–152.
36. Fritschen, L. J. Energy balance method. **1966.** 34–37, In: Jensen, M. E. (ed), *Proceedings onEvapotranspiration and its Role in Water resources Management.* Am. Soc. Agr. Engr.; Chicago, IL, December 5–6, 68 pp.
37. Halkias, N. A.; Veihmeyer, F. J.; Hendrickson, A. H.; Determining water needs for crops from climatic data. Hilgardia, **1955**, *24(9).*
38. Hargreaves, G. H.; Irrigation requirements based on climatic data. Proc. Am. Soc. Civ. Engr.; Irrig. Drain. Div.; 82(IR3): Paper 1105, **1956**, 1–10.
39. Hargreaves, G. L.; G. H.; Hargreaves and Riley, J. P.; Agricultural benefits for Senegal river basin. J. Irrig. Drain. Engr.; Am. Soc. Civ. Engr.; **1985**, *111(2),* 113–124.
40. Hargreaves, G. H.; Samani, Z. A.; Reference crop evapotranspiration from temperature. Applied Eng. in Agr. Am. Soc. Agr. Engr, **1985**, *1(2),* 96–99.
41. Harris, F. S.; **1920**, The duty of water in the Cache Valley Utah. Utah Agr. Exp. Stn. Bull. Number 173.
42. Harris, F. S.; Robinson, J. S.; Factors affecting the evaporation of moisture from soil. J. Agr. Res.; **1916**, *7(10).*
43. Hedke, C. R.; **1924**, Consumptive use of water by crops. New Mexico State Engr. Office, July.
44. Hemphill, R. G.; **1922**, Irrigation in Northern Colorado. USDA Bull. Number 1026.
45. Houk, I. E. **1951**, Irrigation Engineering, 1, Agricultural and Hydrological Phases. John Wiley & Sons, Inc.; New York.
46. Howell, T. A.; Evett, S. R.; **2002**, The Penman-Monteith method. <www.cprl.ars.usda.gov>
47. Israelsen, O. W. **1950**, Irrigation Principles and Practices. 2nd ed.; John Wiley & Sons, Inc.; NewYork, 405.
48. Israelsen, O. W.; Winsor, L. M.; **1922**, The net duty of water in the Sevier Valley. Utah. UtahAgr Exp. Stn. Bull. 182.
49. Jensen, M. E. (ed), Proceedings on Evapotranspiration and its Role in Water resources Management. Am.Soc. Agr. Engr.; Chicago, Dec. 5–6, **1966**, 68 pp.
50. Jensen, M. E. Water consumption by agricultural plants. In: T. T.; Kozlowski (ed.), *WaterDeficits and Plant Growth*: Academic Press, New York. **1968**, *2,* 1–22.
51. Jensen, M. E. Scheduling irrigations with computers. J. Soil and Water Conservation, **1969**, *24(5),* 193–195.
52. Jensen, M. E. (ed), Consumptive use of Water and Irrigation Water Requirements. Rept.Tech. Com. on Irrig. Water Req.; Irrig. and Drain. Div.; Am. Soc. Civ. Engr, **1974**, 229 pp.
53. Jensen, M. E.; Allen, R. G.; Evolution of practical ET estimating methods. In Evans, R. R.; B. L.; Benham and T. P.; Trooien (eds), National Irrigation Symposium, Am. Soc. Agr. Engr.; St. Joseph, MI. **2000**, 52–65.
54. Jensen, M. E.; Burman R. D.; Allen R. G.; (eds), Evapotranspiration and Irrigation Water Requirements, Am. Soc. Civ. Engr. Manuals and Repts. Eng. Practice No. 70, ISBN 0-87262-763-2, **1990**, 360 pp.
55. Jensen, M. E.; H R Haise, **1962**, Estimating Evapotranspiration from Solar Radiation. Prel. Rept. for discussion at an ARS-SCS Consumptive Use Workshop, Phoenix, AZ, Mar. 6–8.
56. Jensen, M. E.; Haise, H. R.; **1963**, Estimating evapotranspiration from solar radiation. Proc. J. Irrig. Drain. Div.; Am. Soc. Civ. Engr.; 89(IR4): 15–41, Closure, 91(IR1): 203–205.
57. Jensen, M. E.; D. Robb, C. N.; Franzoy, C. E.; Scheduling irrigations using climate-crop-soil data. J. Irrig. Drain. Div.; Am. Soc. Civ. Engr.; **1970**, *96,* 25–28.
58. Kennedy, R. E.; Kennedy, R. W.; Evaporation computed by the energy-equation. Am. Geophys. Union Trans. **1936**, *17,* 426–430.

59. Knoblauch, H. C.; Law, E. N.; Mayer, W. P.; and others. **1962,** State Experiment Stations. USDA Misc. Publ. 904.

60. Ledbedevich, N. F. **1956,** Water regime in peat and swamp soils in the Belo Russian and crop yields. Belo Russian, S. S. R.; 6 (translated from Russian), USDC PST Cat. 489, 1961.

61. Lewis, M. R. **1919,** Experiments on the proper time and amount of irrigation. Twin Falls Exp. Stn.; 1914, 1915, and 1916.

62. Lowry, R. L.; Jr. and Johnson, A. F.; **1942,** Consumptive use of water for agriculture. Trans. Am. Soc. Civ. Engr.; *107,* 1243–1302.

63. Makkink, G. F. Testing the Penman formula by means of lysimeters. J. Inst. Water Eng.; **1957,** *11(3),* 277–278.

64. Martin, D. L.; J. Gilley, **1993,** Irrigation Water Requirements. Chapter *2,*Part 623. In: *National Engineering Handbook.* USDA – Soil Conservation Service. 284 pages.

65. Marvin, C. F. **1905,** Sunshine Tables. U. S.; Weather Bur. Bull. *805,* (Reprinted 1944). McEwen, G. F. **1930,** "Results of evaporation studies conducted at the Scripps Institute of Oceanography and the California Institute of Technology." Bull. the Scripps Inst. of Oceanography, Techn. Series, *2(11),* 401–415.

66. Monteith, J. L.; **1965,** Evaporation and the environment. p. 205–234, In: *The State and Movement of Water in Living Organisms.* XIXth Symposium. Soc. for Exp. Biol.; Swansea, Cambridge Univ. Press.

67. Monteith, J. L.; **1986,** Howard Latimer Penman, 10 April **1909,** to 13 October 1*984,* Biographical Memoirs of Fellows of the Royal Society, 32 (Dec. 1986): 379–404. Stable URL: <http: // www.jstor.org/stable/770117>

68. Olivier, H.; **1961,** Irrigation and Climate. Edward Arnold Publ.; LTD.; London.

69. Penman, H. L.; Natural evaporation from open water, bare soil and grass. Proc. Roy. Soc. London, Series, A.; **1948,** *193,* 120–146

70. Penman, H. L.; Evaporation: An introductory survey. Neth. J. Agr. Res.; **1956,** *4(1),* 9–29.

71. Penman, H. L.; **1963,** Vegetation and Hydrology. Tech. Common. *53,*Commonwealth Bureau of Soils, Harpenden, England.

72. Penman, H. L.; and Long, I. F.; Weather in wheat: An essay in micrometeorology. Qtrly. J. Roy. Meteorol. Soc.; **1960,** *86,* 16–50.

73. Phelan, J. T. **1962,** Estimating monthly "k" values for the Blaney-Criddle formula. ARS-SCS Workshop onConsumptive Use, Mar. 6–8 (mimeo).

74. Rijtema, P. E.; **1965,** Analysis of actual evapotranspiration. Agr. Res. Rep. No. 69. Centre for Agr. Publ. and Doc.; Wageningen, the Netherlands, 111 pp. Rijtema, P. E.; **1958,** Calculation methods of potential evapotranspiration. Rept. on the Conf. on Supplemental Irrigation, Comm. V1 Intl. Soc. Soil Sci. Copenhagen, June 30–July 4.

75. Robb, D. C. N. **1966,** Consumptive use estimates from solar radiation and air temperature. Proc. Methods for Estimating Evapotranspiration, ASCE Irrig. Drain. Spec. Conf.; Las Vegas, NV, 169–191.

76. Shantz, H. L.; Piemeizel, L. N.; The water requirement of plants at Akron, Colorado. J. Agr. Res.; **1927,** *34(12),* 1093–1190.

77. Shuttleworth, W. J. Towards one-step estimation of crop water requirements. Trans. of the ASABE, **2006,** 925–935.

78. Shuttleworth, W. J. Wallace, J. S.; Calculating the water requirements of irrigated crops in Australia using the Matt-Shuttleworth approach. Trans. of the ASBE, **2009,** *56(6),* 1895–1906.

79. Smith, M.; Allen, R.; Monteith, J. Perrier, A.; Pereira, L.; Segeren, A. Report of the expert consultation on procedures for revision of FAO guidelines for prediction of crop water requirements. UN-FAO, Rome, Italy, **1991,** 54 p.

80. Soil Sci. Society of America meeting, Aug. 20–24, **1962,** Ithaca, NY.

81. Tanner, C. B. **1966,** Comparison of energy balance and mass transport methods for measuring evaporation. 45–48, IN Jensen (ed). **1966,** Proc. Evapotranspiration and its Role in Water resources Management. Am. Soc. Agr. Engr.; Chicago, Dec. 5–6, 68 p.

82. Teele, R. P. **1904,** Irrigation and drainage investigations of the Office of the Experiment Stations, USDA.

83. Thoreson, Bryan, **2009,** Personal information. Davids Engineering.

84. Thornthwaite, C. W. An approach toward a rational classification of climate. Geographical Review, **1948,** *38,* 55–94.

85. Turc, L. Evaluation des besoins en eau d=irrigation, evapotranspiration potentialle, formule climatique simplifice et mize a jour. (English: Estimation of irrigation water requirements, potential evapotranspiration: A simple climatic formula evolved up todate). Ann. Agron.; **1961,** *12,* 13–49.

86. USDA, Soil Conservation Service. Irrigation Water Requirements. Tech. Release No. 21, (rev.), **1970,** 92 pages.

87. Van Bavel, C. H. M. Potential evaporation: The combination concept and its experimental verification. Water Resources Res.; **1966a,** *2(3),* 455–467.

88. Van Bavel, C. H. M. Combination (Penman type) methods. *48,*IN Jensen (ed). 1966, Proc. Evapotranspiration and its Role in Water resources Management. Am. Soc. Agr. Engr.; Chicago, Dec. 5–6, **1966b,** 68 p.

89. Van Wijk, W. R.; de Vries, D. A. Evapotranspiration. Neth. J. Agr. Sci. **1954,** *2,* 105–119.

90. Vitkevich, **1958,** Determining evaporation from the soil surface. USDC PST Cat. 310, (Trans. from Russian).

91. Walter, I. A.; Allen, R. G.; Elliott, R.; Jensen, M. E.; Itenfisu, D.; Mecham, B.; Howell, T. A.; Snyder, R.; Brown, P.; Eching, S.; Spofford, T.; Hattendorf, M.; Cuenca, R. H.; Wright, J. L.; Martin, D.; **2000,** ASCE's standardized reference evapotranspiration equation. Proc. Fourth Nat'l. Irrig. Symp.; ASAE, Phoenix, AZ.

92. Widstoe, J. A. Irrigation investigations: Factors influencing evaporation and transpiration. Utah Agr. Exp. Bull. **1909,** *105,* 64 p.

93. Widstoe, J. A. The production of dry matter with different quantities of water. Utah Agr. Exp. Stn. Bull.; **1912,** *116,* 64 p.

94. Widstoe, J. A.; McLaughlin. 1912, The movement of water in irrigated soils. Utah Agr. Exp. Stn. Bull. 115.

95. Wright, J. L. New evapotranspiration crop coefficients. J. Irrig. Drain. Div.; **1982,** *108(IR1):* 57–74.

96. McEwen, G. F.; Results of evaporation studies conducted at the Scripps Institute of Oceanography and the California Institute of Technology. Bull. of the Scripps Inst. of Oceanography, Technical Series, **1930,** *2(11),* 401–415.

97. Richardson, B.; Evaporation as a function of insolation. Trans. Am. Soc. Civil Engr.; **1931,** *99,* 996–1019.

98. Romanoff, V. V. **1956,** Evaporation computations by simplified thermal balance method. Trudy CGI.

CHAPTER 2

WATER VAPOR FLUX MODELS FOR AGRICULTURE[1]

VICTOR H. RAMIREZ, and ERIC W. HARMSEN

CONTENTS

[1] Printed with permission and modified from: V. H. Ramirez and E. W. Harmsen, 2011, Water Vapor Flux in Agroecosystems: Methods and Models Review. Chapter 1, pp. 3–48. In: *Evapotranspiration* by Leszek Labedzki (Editor), ISBN 978-953-307-251-7, InTech. Available from: http://www.intechopen.com/books/evapotranspiration/turfgrass-growth-quality-and-reflective-heat-load-in-response-to-deficit-irrigation-practices.

2.1   INTRODUCTION

The water vapor flux in agroecosystems is the second largest component in the hydrologycal cycle. Water vapor flux or evapotranspiration (ET) from the vegetation to the atmosphere is a widely studied variable throughout the world. ET is important for determining the water requirements for the crops, climatic characterization, and for water management. The estimation of ET from vegetated areas is a basic tool to compute water balances and to estimate water availability and requirements. In the last 60 years, several methods and models to measure the water flux in agroecosystems have been developed. The aim of this chapter is to provide a literature review on the subject, and provide an overview of methods and model developed which are widely used to estimate and/or measure ET in agroecosystems.

Evapotranspiration constitutes an important component of the water fluxes of our hydrosphere and atmosphere [21], and is a widely studied variable throughout the world, due to it applicability in various disciplines, such as hydrology, climatology, and agricultural science. Pereira et al. [80] has reported that the estimation of ET from vegetated areas is a basic tool for computing water balances and to estimate water availability and requirements for plants. Measurement of ET is needed for many applications in agriculture, hydrology and meteorology [102]. ET is a major component of the hydrologic water budget, but one of the least understood [120]. ET permits the return of water to the atmosphere and induces the formation of clouds, as part of a never-ending cycle. ET also permits the movement of water and nutrients within the plant; water moving from the soil into the root hairs, and then to the plant leaves.

ET is a complicated process because it is the product the different processes, such as evaporation of water from the soil, and water intercepted by the canopy, and transpiration from plant leaves. Physiological, soil and climatic variables are involved in these processes. Symons in 1867 described evaporation as "...*the most desperate art of the desperate science of meteorology*" [69]. The first vapor flux measurements were initiated by Thornthwaite and Holzman in 1930s, but that works was interrupted by World War II [69]. In the late 1940s Penman [78] published the paper "*Natural Evaporation from open Water, Bare Soil and Grass*" in which he combined a thermodynamic equation for the surface heat balance and an aerodynamic equation for vapor transfer. The "Penman equation" is one of the most widely used equations in the world. The equation was later modified by Monteith [67, 68] and is widely known as the "*The Penman-Monteith Model.*" It is also necessary to introduce a review of the work of Bowen, who in 1926 [11] published the relationship between the sensible and latent heat fluxes, which is known as the "*Bowen ratio.*" Measurement of the water vapor flux became a common practice by means of the "*Bowen ratio energy balance method*" [106]. Allen et al. [5] classified the factors that affect the ET into three groups:

 a. **Weather parameters**, such as radiation, air temperature, humidity and wind speed: The evapotranspirational component of the atmosphere is expressed by the reference crop evapotranspiration (ET_o) as the Penman-Monteith (FAO-56), or using direct measurements of pan evaporation data [22], or using other empirical equations;
 b. **Crop factors** such as the crop type, variety and developmental stage should be considered when assessing the ET from crops grown in large, well-man-

agement fields. Differences in resistance to transpiration, crop height, crop roughness, reflection, ground cover and crop rooting characteristics result in different ET levels in different types of crops under identical environmental conditions. Crop ET under standard conditions (ET_c) refers to excellent management and environmental conditions, and achieves full production under given climatic conditions (Eq. (2)); and

c. **Management and environmental conditions** (ET_{cadj}). Factors such soil salinity, poor land fertility, limited applications of fertilizers, the presence of hard or impermeable soil horizons, the absence the control of dizease and pest and poor soil management may limit the crop development *etc.*, and reduce the ET, (ETc_{adj} Eq. (3)).

One of the most common and fairly reliable techniques for estimating ET_0 is using evaporation pan data when adjustments are made for the pan environment [31] using the pan evaporation and the pan coefficient (K_p).

$$ET_0 = K_p \cdot E_p \tag{1}$$

where, E_p is the pan evaporation (mmday^{-1}), and K_p is the pan coefficient, and depends on location. It is important to know or calculate pan coefficient before calculating the ET_0. Allen et al., [5] gave a methodology to calculate it. K_p is essentially a correction factor that depends on the prevailing upwind fetch distance, average daily wind speed, and relative humidity conditions associated with the siting of the evaporation pan [22].

2.2 CROP WATER FLUX USING SINGLE CROP COEFFICIENTS: THE FAO APPROACH

The FAO approach is well known as the *Two steps method*, which is very useful for single crops and when "references" conditions are available (i.e., no crop water stress). In this case, crop evapotranspiration (ET_c) can be estimated using Eq. (2) [5, 22]:

$$ET_c = K_c \cdot ET_0 \tag{2}$$

where, K_c is the coefficient expressing the ratio of between the crop and reference ET for a grass surface. The crop coefficient can be expressed as a single coefficient, or it can be split into two factors, one describing the affect of evaporation and the other the affect of transpiration. As soil evaporation may fluctuate daily, as a result of rainfall and/or irrigation, the single crop coefficient expresses [5] only the time-average (multiday) effects of crop ET, and has been considered within four distinct stages of growth (*see* FAO #56 by Allen et al. [5]). When stress conditions exist, the effects can be accounted for by a crop water stress coefficient (K_s) as follows:

$$ET_{cadj} = K_s \cdot K_c \cdot ET_0 \tag{3}$$

2.2.1 CROP COEFFICIENTS

Although a number of ET_c estimation techniques are available, the crop coefficient (K_c) approach has emerged as the most widely used method for irrigation scheduling [45]. As ET is not only a function of the climatic factors, the crop coefficients can include conditions related to the crop development (K_c), and nonstandard conditions (K_s). The

K_c is the application to two concepts [5]: a. Crop transpiration represented by the basal crop coefficient (K_{cb}); and *b.* The soil evaporation, K_e, is calculated using Eq. (4):

$$K_c = K_{cb} + K_e \qquad (4)$$

K_c is an empirical ratio between ET_c and ET_o over grass or alfalfa, based on historically measurements. The curve for K_c is constructed for an entire crop growing season, and which attempts to relate the daily water use rate of the specific crop to that of the reference crop [45]. The United Nations Food and Agriculture Paper FAO #56 by Allen et al. [5] provided detailed instructions for calculating these coefficients. For limited soil water conditions, the fractional reduction of K_c by K_s depends on the crop, soil water content, and magnitude of the atmospheric evaporative demand [22].

The value for K_c equals K_{cb} for conditions: the soil surface layer is dry (i.e., when $K_e = 0$) and the soil water within the root zone is adequate to sustain the full transpiration (nonstressed conditions, i.e., $K_s = 1$). When the available soil water of the root zone becomes low enough to limit potential ET_c, the value of the K_s coefficient is less than 1 [5, 44, 45].

The soil evaporation coefficient accounts for the evaporation component of ET_c when the soil surface is wet, following irrigation or rainfall [5,45]. When the available soil water of the root zone become low enough, crop water stress can occur and reduce ET_c. In the FAO-56 procedures, the effects of water stress are accounted for by multiplying K_{cb} (or K_c) by the water stress coefficient (K_s):

$$K_c \cdot K_s = (K_{cb} \cdot K_s + K_e) = ET_c/ET_o \qquad (5)$$

where, $K_s < 1$ when the available soil water is insufficient for the full ET_c and $K_s = 1$ when there is no soil water limitation on ET_c. Thus, to determine K_s, the available soil water within the crop zone for each day needs to be measured or calculated using a soil water balance approach [45].

The estimation of K_e using the FAO-56 method, requires the use of the soil field capacity (FC), the permanent wilting point (PWP), total evaporable water (TEW), the fraction of the soil surface wetted (f_w) during each irrigation or rain, and the daily fraction of the soil surface shaded by vegetation (f_c), or conversely the unshaded fraction ($1-f_c$). Hunsaker et al., [45] reported an exponential relation between $1-f_c$ and height to the Alfalfa crop.

The measurement of K_e and K_{cb} can be made by performing a *daily water balance,* and use of the following equations from FAO Paper 56 [5].

$$ET_c = (K_{cb} + K_e) \, ET_o \qquad (6)$$

$$K_{cb} = (ET_c/ET_o) - K_e \qquad (7)$$

The soil evaporation (E) can be calculated using the Eqs. (8) and (9):

$$E = K_e \, ET_o \text{ and} \qquad (8)$$

$$K_e = E/ET_o \qquad (9)$$

The soil evaporation (E) can be measured using the water balance Eq. (10):

$$E = D_{e,i-1} - \left(P_i - RO_i\right) - \frac{I_i}{f_w} + \frac{f_i}{f_{ew}} + Tew_{,i} + DP_{e,i} \tag{10}$$

where, $D_{e,\,i-1}$ is the cumulative depth of evaporation following complete wetting from the exposed and wetted fraction of the topsoil at the end of day $i-1$ (mm), P_i is the precipitation on day i (mm); RO_i is precipitation runoff from the surface on day i (mm), I_i is the irrigation depth on day i that infiltrates into the soil (mm), E_i is evaporation on day i (i.e., $E_i = K_e/ET_o$) (mm), $T_{ew,\,i}$ is depth of transpiration from the exposed and wetted fraction of the soil surface layer on day i (mm), f_w is fraction of soil surface wetted by irrigation (0.01 to 1), and f_{ew} is the exposed and wetted soil fraction (0.001 to 1).

The ratio of reference evaporation to reference transpiration depends on the development stage of the leaf canopy expressed as "δ" the dimensionless fraction of incident beam radiation that penetrates the canopy [15] mentioned by Zhang et al. [124].

$$\delta = \exp(-K.LAI) \tag{11}$$

where, K is the dimensionless canopy extinction coefficient, and therefore, evaporation and transpiration can be calculated using Eqs. (12) and (13):

$$E_0 = \delta.ET_0 \tag{12}$$

$$T = (1-\delta).ET \tag{13}$$

Hunsaker, [44] found that ET_c in cotton was higher when the crop was submitted to high depth of irrigation (820–811 mm) that when have low depth of irrigation level (747–750 mm), similar to the $K_{cb}K_s$ curves, obtaining higher values than the treatment with high frequency (i.e., $K_{cb}K_s = 1.5$, 90 days after planting) than the low frequency (i.e., $K_{cb}K_s = 1.4$, 90 days after planting).

2.2.2 LIMITATIONS IN THE USE OF K_C

Katerji and Rana, [52] reviewed recent literature related to K_c and found differences of $\pm 40\%$ between K_c values reported in the FAO-56 paper [5] and the values experimentally obtained, especially in the mid growth stage. According to the authors, these large differences are attributable to the complexity of the coefficient K_c, which actually integrate several factors: aerodynamic factors linked to the height of the crop, biological factors linked to the growth and senescence of the surfaces leaves, physical factors linked to evaporation from the soil, physical factors linked to the response of the stomata to the vapor pressure deficit and agronomic factors linked to crop management (distance between rows, using mulch, irrigation system, etc.). For this reason K_c values needs to be evaluated for local conditions

The variation in crop development rates between location and year have been expressed as correlations between crop coefficients and indices such as the thermal base index, ground cover, days after emergence or planting, and growth rate (i.e., Wright and Jensen, [122]; Hunsaker, [44]; Brown et al., [14]; Nasab et al., [72]; Hanson and May [34]; Madeiros et al., [61]; and Ramírez, [87]). The K_c is well related with the growing degree grades-GDD and with the fraction of the soil cover by vegetation (f_c)

(Fig. 1), and depends on the genotype and plant densities [87]. The Eqs. (15) and (16) are for common bean genotype Morales with 13.6 plants.m^{-2}. The Eqs. (17) and (18) are for common bean genotype SER 16, with 6.4 plants.m^{-2}.

$$K_c = -3 \times 10^{-6} CGDD^2 + 0.053; R^2 = 0.76; \; P<0.0000 \tag{15}$$

$$K_c = -1.4019 f_c^2 + 2.5652 f_c - 0.2449; R^2 = 0.70; p < 0.0003 \tag{16}$$

$$K_c = -3x10^{-6} CGDD^2 + 0.0034 CGDD - 0.0515; R^2 = 0.60; p < 0.0001 \tag{17}$$

$$K_c = -0.6726 f_c^2 + 1.90086 f_c - 0.2560; R^2 = 0.60; p < 0.0032 \tag{18}$$

2.2.3   WATER STRESS COEFFICIENT (K_s)

The soil water stress coefficient, K_s, is mainly estimated by its relationship to the average soil moisture content or matric potential in a soil layer, and it can usually be estimated by an empirical formula based in soil water content or relative soil water available content [124].

The K_s is an important coefficient because it indicates the sensitivity of the crop to water deficit conditions, for example corn grain yield is especially sensitive to moisture stress during tasselling and continuing through grain fill. Roygard et al., [94] observed that depletion of soil water to the wilting point for 1 or 2 days during tasselling or pollenization reduced yield by 22%. Six to eight days of stress reduced yield by 50%. Allen et al. [5], presented the following methodology for estimating K_s:

$$K_s = \frac{TAW - Dr}{TAW - RAW} = \frac{TAW - Dr}{(1 - p)TAW} \tag{19}$$

where, TAW is total available water and refers to the capacity of the soil to retain water available for plants (mm); Dr is root zone depletion (mm); RAW is the readily available soil water in the root zone (mm), p is the fraction of TAW that the crop can extract from the root zone without suffering water stress.

$$TAW = 1000(\theta_{FC} - \theta_{WP})Z_t \tag{20}$$

where, θ_{FC} is the water content at field capacity (m^3.m^{-3}), θ_{WP} is the water content at wilting point (m^3.m^{-3}), and Z_t is the rooting depth (m).

$$RAW = pTAW \tag{21}$$

Allen et al. [5] give values for different crops (FAO #56. p. 163) [5]. Roygard et al. [94] and Zhang et al. [124], reported that K_s is a logarithmic function of soil water availability (Aw), and can be estimated as follows:

$$K_s = \ln(Aw + 1) / \ln(101) \tag{22}$$

where, Aw is calculate according to the Eq. (23):

$$Aw = 100 \left(\frac{\theta_a - \theta_{wp}}{\theta_{FC} - \theta_{wp}} \right) \tag{23}$$

where, θ_a is average soil water content in the layers of the root zone depth. An example of the relationships between K_s and available soil water changes, estimated as a root zone depletion, is presented by Ramirez [87] The root zone depletion (D_r), can be calculated using the water balance Eq. (24):

$$D_{r,i} = D_{r,i-1} - (P - RO)_i - I_i + ET_{c,i} + DP_i \tag{24}$$

where, $D_{r,i}$ is the root zone depletion at the end of day i; $D_{r,i-1}$ is water content in the root zone at the end of the previous day, $i-1$; $(P-RO)_i$ is the difference between precipitation and surface runoff on day i; I_i is the irrigation depth on day i; $ET_{c,i}$ is the crop ET on day i and DP_i is the water loss from the root zone by deep percolation on day i; all the units are in mm.

The root zone depletion associated with a $K_s = 1.0$ (i.e., no water stress), was up to 10 mm for a root depth between 0 to 20 cm, and up to 15 mm for a root depth of 0 to 40 cm in common beans. Fifty percent of the transpiration reduction was reached for $D_r = 22$ mm and 25 mm for the common bean genotype Morales and genotype SER 16, respectively. Transpiration ceased completely ($K_s = 0$) when Dr = 37 mm and 46 mm, respectively, for Morales and SER 16 [87].

2.3   DIRECT WATER VAPOR FLUX MEASUREMENT: LYSIMETERS

The word '*lysimeter*' is derived from the Greek root 'lysis,' which means dissolution or movement, and 'metron,' which means to measure [41]. Lysimeters are tanks filled with soil in which crops are grown under natural conditions to measure the amount of water lost by evaporation and transpiration [5]. A lysimeter is the method of determining ET directly. The lysimeter are tanks buried in the ground to measure the percolation of water through the soil. Lysimeter are the most dependable means of directly measuring the ET rate, but their installation must meet four requirements for the data to be representative of field conditions [19]:

Requirement 1: The lysimeter itself should be fairly large and deep to reduce the boundary effect and to ovoid restricting root development. For short crops, the lysimeter should be at least one cubic meter in volume. For tall crops, the size of the lysimeter should be much larger.

Requirement 2: The physical conditions within the lysimeter must be comparable to those outside. The soil should not be loosened to such a degree that the root ramification and water movement within the lysimeter are greatly facilitated. If the lysimeter is unclosed on the bottom, precaution must be taken to avoid the persistence of a water table and presence of an abnormal thermal regime. To ensure proper drainage, the bottom of an isolated soil column will often require the artificial application of a moisture suction, equivalent to that present at the same depth in the natural soil [20].

Requirement 3: The lysimeter will not be representative of the surrounding area if the crop in the lysimeter is taller, shorter, denser, or thinner, or if the lysimeter is on the periphery of no-cropped area. The effective area of the lysimeter is defined as the ratio of the lysimeter ET per unit area of the surrounding field. The values of this ratio, other than unity, are caused by the in homogeneity of the surface. The maintenance of uniform crop height and density is not an easy task in a tall crop, spaced in rows. If the surface is indeed inhomogeneous, there is no adequate way to estimate the effective area from tank area overlap corrections or plant counts.

Requirement 4: Each lysimeter should have a "guard-ring" area around it maintained under the same crop and moisture conditions in order to minimize the clothesline effect. In arid climates, Thornthwaite in 1954, suggested that a "guard-ring" area of ten acres may or may not be large enough. Where several lysimeters are installed in the same field, the "guard-ring" radius may have to be about ten times the lysimeter separation [19].

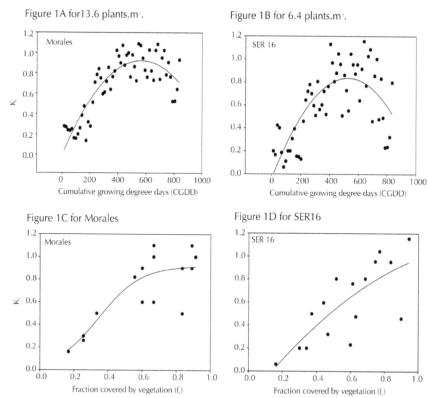

FIGURE 1   Crop coefficients (K_c) as related to cumulative growing degree days (CGDD) and fraction covered by vegetation (f_c) for two common bean genotypes: (**a**) Morales CGDD vs K_c, (**b**) SER 16 CGDD vs K_c (**c**) Morales f_c vs K_c (**d**) SER 16 f_c vs K_c. The curves were fitted from growth periods V1 to R9 (Data from: Ramirez, 2007). (These data were obtained under the project sponsored by NOAA-CREST (NA17AE1625), NASA-EPSCoR (NCC5–595), USDA-TSTAR-100, and University of Puerto Rico Agricultural Experiment Station, Mayaguez, USA).

Lysimeters surrounded by sidewalks or gravel will not provide reliable data, nor will lysimeters planted to a tall crops if it is surrounded by short grass, or planted to grass and surrounded by a tall crop. Differences in growth and maturity between the lysimeters plants and surrounding plants can result in significant differences in measured ET in and outside the lysimeter [4]. The lysimeters are classified basically in two types: Weighing and Non-weighing.

2.3.1   NON-WEIGHING LYSIMETERS (DRAINAGE LYSIMETERS)

These operate on the principle that ET is equal to the amount of rainfall and irrigation water added to the system, minus percolation, runoff and soil moisture changes. Since the percolation is a slow process, the drainage lysimeters is accurate only for long periods for which the water content at the beginning exactly equals that at end. The length of such a period varies with the rainfall regime, frequency and amount of irrigation water application, depth of the lysimeters, water movement, and the like. Therefore, records of drainage lysimeters should be presented only in terms of a long-period more than one day [19], and they are not useful for estimating hourly ET.

Allen et al. [4] discusses two types the nonweighing lysimeters: a) *non weighing constant water-table type*, which provides reliable data in areas where a high water table normally exists and where the water table level is maintained essentially at the same level inside as outside the lysimeters; b) *Non-weighing percolation type*, in which changes in water stored in the soil are determined by sampling or neutron methods or other soil humidity sensors like TDR, and the rainfall and percolation are measured.

2.3.1.1   GENERAL PRINCIPLES OF A DRAINAGE LYSIMETER

Provisions are made at the bottom of the lysimeter container to collect and measure volumetrically the deep percolation. Precipitation is measured by rain gauge(s). *Evapotranspiration* is considered as the difference among *water applied, water drainage and soil water change* [108,123].

When filling-in a lysimeter, the soil dug out from the pit of a lysimeter is replaced in the container, special precautions are needed to return the soil to its original status by restoring the correct soil profile and compacting the soil layers to the original density. It is desirable to have a similar soil state inside the lysimeter relative to the outside. However if the roots are well developed and nutrients are available, and as long as the water supply to the roots is unrestricted, dissimilar soil will not give significant variation in water use and yield, provided other conditions are similar [123].

Although disturbed soil in filled-in lysimeters does not pose serious problems in ET measurement, the soil can affect plant growth. Breaking up the soil, will change soil structure, aeration, and soil moisture retention characteristics. The lysimeters should provide a normal rooting profile. It should be large enough to lender the effect of the rim insignificant. It can give relatively large errors in the ET measurement if the container is small. However, the greater the lysimeters area, the more costly and complicated the installation and operation becomes [123].

Installation and walls: The wall can be different materials: reinforced concrete, polyester reinforced with steel, fiberglass or plastic. The installation proceeds in the following steps: Excavation (e.g. 1 m×1 m×1.2 m) in the experimental site. Each layer of soil (e.g., 0–30 cm, 30–60 cm and 60–100 cm) is separated. Once the excavation it completed, the lysimeter is placed in the excavated hole with 4 wooden boards outside. Before repacking the soil layers, make a V-shaped slope at the bottom and place a 25 mm inside diameter perforated PVC pipe (horizontal). There should be a screen material placed around the perforated pipe to ovoid the soil particles from entering the pipe. Connect an access tube (25 mm PVC), approximately 1 m long (vertical). Cover the horizontal pipe with fine gravel approximately 3–5 cm thick. Fill the container with the excavated soil where each layer is repacked inside the lysimeter to match the original vertical soil state [123].

2.3.2 WEIGHING LYSIMETER

A weighing Lysimeter is capable of measuring ET for periods as short as ten minutes. Thus, it can provide much more additional information than a drainage lysimeter can. Problems such as diurnal pattern of ET, the phenomenon of midday wilt, the short-term variation of energy partitioning, and the relationship between transpiration and soil moisture tension, can be investigated only by studying the records obtained from a weighing lysimeter [5, 19, 60, 76, 92, 101, 105, 117].

Weighing lysimeters make direct measurements of water loss from a growing crop and the soil surface around a crop and thus, provide basic data to validate other water vapor flux prediction methods [23, 59, 85, 116]. The basic concept of this type of lysimeter is that it measures the difference between two mass values, the mass change is then converted into ET (mm) [47, 62].

During periods without rainfall, irrigation and drainage, the ET rate is computed as indicated by Howell [41], as:

$$ET = \left\lfloor A_l \left[(M_i - M_{i-1}) / A_l \right] / A_f \right\rfloor / T_i \qquad (25)$$

where, ET is in units of (mm.h^{-1} or Kg.m^2) for time interval i; M is the lysimeter soil mass, (Kg); A_l is lysimeter inner tank surface area (m^2); A_f is lysimeter foliage area (mid wall-air gap area) (m^2); T is the time period (h). The ratio A_f / A_l is the correction factor for the lysimeter effective area. This correction factor assumes the outside and inside vegetation foliage overlap evenly on all of the sides or edges. If there is no overlap, as occurs in short grass, the $A_f / A_l = 1.0$ [41].

Weighing lysimeters provide the most accurate data for short time periods, and can be determined accurately over periods as short as one hour with a mechanical scale, load cell system, or floating lysimeters [4]. Some weighing lysimeters use a weighing mechanism consisting of scales operating on a lever and pendulum principle [62]. However, some difficulties are very common like: electronic data logger replacement, data logger repair, load cell replacement, multiflexor installation etc. [62].

The measurement control in these lysimeters are important because of the following issues: a) recalibration requirements, b) measurement drift (e.g., slope drift, variance drift), c) instrument problems (e.g., localized nonlinearity of load cell, load cell damage, data logger damage), d) human error (e.g., incorrectly recording data during calibration) and e) confidence in measurement results [62].

A load cell is a transducer that coverts a load acting on it into an analog electrical signal. The electrical signal is proportional to the load and the relationship is determined through calibration, employing linear regressions models (mV/V/mm water), and it is used to determine mass changes of a lysimeter over the period interest (e.g., day, hour, etc.).

The lysimeter characteristics can be different, for example: Malone et al., [62] built a lysimeter of the following form: 8.1 m² in surface area and 2.4 m depth, the lysimeter is constructed without disturbing the soil profile and the underlying fracture bedrock. The soil monolith is supported by a scale frame that includes a 200:1 lever system and a counterweight for the deadweight of the soil monolith. The gap between the soil in the lysimeters and the adjacent soil is between 5.1 cm and 7.0 cm except at the bottom slope where the runoff trough is located, this same author has given instructions for achieving a good calibration for this type of lysimeter.

Tyagi et al., [114] in wheat and sorghum used two rectangular tanks, an inner and outer tank, constructed from 5-mm welded steel plates. The dimensions of the inner tank were $1.985 \times 1.985 \times 1.985$ m³ and those of the outer tank were $2.015 \times 2.015 \times 2.015$ m³. The lysimeters were situated in the center of a 20-ha field. The size ratio of the outer tank to the inner tank is 1.03, so the error due to wall thickness is minimal. The effective area for crop ET was 4 m². The height of the lysimeter rim was maintained near ground level to minimize the boundary layer effect in and around the lysimeter. The lysimeter tank was suspended on the outer tank by four load cells. The load cells were made out of the steel shear beam type with 40,000-kg design load capacity. The total suspended mass of the lysimeter including tank, soil, and water was about 14,000 kg. This provided a safety factor of 2.85. The high safety factor was provided to allow replacement of a load cell without the danger of overloading and also to account for shock loading. A drainage assembly connected with a vertical stand and gravel bedding to facilitate pumping of drainage water was provided. The standpipe also can be used to raize the water table in the lysimeter.

To calculate the ET using Lysimeter, we need to employ the soil water balance (SWB) equation:

$$ET = R + I - P - Rff \pm \Delta SM \qquad (26)$$

where, R is the rain, I is the irrigation, Rff is the runoff, and $\pm \Delta SM$ soil moisture changes, all in mm. The size of the Lysimeter is an important element to be considered in water vapor flux studies with this method. For example, Dugas et al. [23] evaluated small square lysimeters (<1.0 m²) and reported significant differences in the ET estimations, basically associated with the differences in the leaf area index (LAI) inside the lysimeters, which differed among lysimeter, this problem can be addressed using LAI corrections.

2.3.3   CALIBRATION OF THE WEIGHING LYSIMETER

Seyfried et al., [98] made a weighing lysimeter calibration by placing known weights on the lysimeters and then recording the resultant pressure changes. The weights used in that study were as follows: 19.9 kg for supportive blocks placed on the lysimeter, 43.4 kg for the tank which contained the weights, and then 20-four 22.7 kg sacks of rock added in four-sack increments. The weight of each sack corresponded to about a 13 mm addition of water; so that weight increments were equivalent to ~52 mm and the total range was ~360 mm of water. Measurements were made both as weight was added and removed.

The main arguments against the use of weighing lysimeters for monitoring water balance parameters and measuring solute transport parameters in the soil and unsaturated zone has been the discussion of potential sources of error, such as the well known oasis effect, preferential flow paths at the walls of the lysimeter cylinders due to an insufficient fit of soil monoliths inside the lysimeters, or the influence of the lower boundary conditions on the outflow rates [25].

2.4 THE MICROMETEOROLOGICAL METHODS

For many agricultural applications, micrometeorological methods are preferred since they are generally nonintrusive, can be applied on a semicontinuous basis, and provide information about the vertical fluxes that are occurring on scales ranging from tens of meter to several kilometers, depending the roughness of the surface, the height of the instrumentation, and the stability of the atmosphere surface layer. Meyers and Baldocchi [65] have separated micrometeorological methods into four categories: 1) eddy covariance, 2) flux-gradient, 3) accumulation, and 4) mass balance. Each of these approaches are suitable for applications that depend on the scalar of interest and surfaces type, and instrumentation availability. Some of these methods are described in the following sections of this chapter.

2.4.1 HUMIDITY AND TEMPERATURE GRADIENT METHODS

Movement of energy, water and other gases between field surface and atmosphere represent a fundamental process in the soil-plant-atmosphere continuum. The turbulent transport in the surface boundary layer affect the sensible (H) and latent (λE) heat fluxes, which along with the radiation balance, govern the evapotranspiration and canopy temperature [33].

Monteith and Unsworth [70] presented the functional form of the gradient flux equation, and which has applied by Harmsen et al. [36], Ramírez et al. [88] and Harmsen et al. [37]:

$$ET = \left(\frac{\rho_a \cdot c_p}{\gamma \cdot \rho_w}\right) \cdot \frac{(\rho_{vL} - \rho_{vH})}{(r_a + r_s)} \tag{27}$$

where, ρ_w is the density of water, ρ_v is the water vapor density of the air, r_a is the air density, γ is the psychometric constant, c_p is specific heat of air, r_a and r_s are aerodynamic and bulk surfaces resistances (all these variables are discussed in detail below). L and H are vertical positions above the canopy (L: low and H: High positions), for example in small crops like beans or grass, possible values of L and H could be 0.3 m and 2 m above the ground, respectively.

Harmsen et al. [36] developed an automated elevator device (ET Station) for moving a temperature and relative humidity sensor (Temp/RH) between the two vertical positions (Fig. 2). The device consisted of a plastic (PVC) frame with a 12 volt DC motor (1/30 hp) mounted on the base of the frame. One end of a 2-m long chain was attached to a shaft on the motor and the other end to a sprocket at the top of the frame. Waterproof limit switches were located at the top and bottom of the frame to limit the range of vertical movement. For automating the elevator device, a programmable logic controller (PLC) was used which was composed of "n" inputs and "n" relay outputs. To program the device, a ladder logic was used which is a chronological arrangement of tasks to be accomplished in the automation process. The Temp/RH sensor was connected to the elevator device, which measured relative humidity and temperature in the up position for two minutes then changed to the down position where measurements were taken for two minutes. This process started each day at 0600 hours and ended at 1900. When the elevator moves to the up position it activates the limit switch which sends an input signal to the PLC. That input tells the program to stop and remain in that position for two minutes. At the same time it activates an output which sends a 5 volt signal to the control port C2 in the CR10X data logger in which a small subroutine is executed. This subroutine assigns a "1″ in the results matrix which indicates that the temperature and relative humidity corresponding to the up position. At the end of the two minute period, the elevator moves to the down position and repeats the same process, but in this case sending a 5 volts signal to the data logger in the control port C4, which then assigns a "2" in the results matrix. All information was stored in the weather station data-logger CR-10X (Campbell Scientific, Inc.) for later downloading to a personal computer.

FIGURE 2   Automated elevator device developed for moving the Temp/RH sensor between the two vertical positions: (a) Temp/RH sensor in down position and **(b)** Temp/RH sensor in up position. Measuring over common bean (*Phaseolus vulgaris* L.). [Picture obtained by the project sponsored by NOAA-CREST (NA17AE1625), NASA-EPSCoR (NCC5–595), USDA-TSTAR-100, and University of Puerto Rico Experiment Station]

2.4.2   THE BOWEN RATIO ENERGY BALANCE METHOD

The basis for this method is that the local energy balance is closed in such a way that the available net radiative flux (Rn) is strictly composed of the sensible (H), latent (λE), and ground heat (G) fluxes, other stored terms such as those related to canopy heat storage and photosynthesis are negligible [65].

This method combines measurements of certain atmospheric variables (temperature and vapor concentration gradients) and available energy (net radiation and changes in stored thermal energy) to determine estimates of evapotranspiration (ET) [58]. The method incorporates energy-budget principles and turbulent-transfer theory. Bowen showed that the ratio of the sensible- to latent-heat flux (β) could be calculated from the ratio of the vertical gradients of temperature and vapor concentration over a surface under certain conditions.

Often the gradients are approximated from air-temperature and vapor-pressure measurements taken at two heights above de canopy. The Bowen-ratio method assumes that there is no net horizontal advection of energy. With this assumption, the coefficients (eddy diffusivities) for heat and water vapor transport, kh and kw, respectively, are assumed to be equal. Under advective conditions, kh and kw are not equal [112] and the Bowen-ratio method fails to accurately estimate ET.

Based on the assumption that Kh and Kw are equal, and by combining several terms to form the psychometric constants, the Bowen-ratio take the form of the eq. (2). Although the theory for this method was develop in the 1920s by Bowen [11], its practical applications has only been possible in recent decades, due to the availability of accurate instrumentation [77]. The Bowen ratio initial concept is shown below:

$$\beta = \frac{PC_p Kh \dfrac{dT}{dz}}{\lambda \varepsilon Kw \dfrac{de}{dz}} \tag{28}$$

If it is assumed that there is no net horizontal advection of energy, Eq. (28) can be simplified as shown below:

$$\beta = \frac{PC_p \dfrac{dT}{dz}}{\lambda \varepsilon \dfrac{de}{dz}} \tag{29}$$

where, P is the atmospheric pressure (kPa), C_p is the specific heat of air (1.005 J./g°C), ε is the ratio molecular weight of water to air = 0.622 and λ is the latent-heat flux (Jg⁻¹). Once the Bowen ratio is determined, the energy balance can be solved for the sensible-heat flux (H) and latent-heat flux (λE).

$$Rn = \lambda E + H + G \tag{30}$$

where, R_n is the net radiation, λE is the latent-heat flux, H is the sensible-heat flux and G is the soil-heat flux.

$$H = \beta \lambda E \tag{31}$$

$$\lambda E = \frac{(R_n - G)}{(1 + \beta)} \tag{32}$$

The latent heat flux can be separated into two parts: the evaporative flux E (g m^{-1} day^{-1}) and the latent heat of vaporization λ (Jg^{-1}), which can be expressed as a function of air temperature (T) (λ = 2,502.3–2.308 T). The latent-heat of vaporation (λ) is defined as the amount of energy required to convert 1 gram of liquid water to vapor at constant temperature T. Sensible-heat flux (H) is a turbulent, temperaturegradient driven heat flux resulting from differences in temperature between the soil and vegetative surface and the atmosphere.

The soil-heat flux (G) is defined as the amount of energy moving downward through the soil from the land surface, caused by temperature gradient. This flux is considered positive when moving down through the soil from the land surface and negative when moving upward through the soil toward the surface [111]. The soil heat flux is obtained by measuring two soil heat flux plates below the soil surface at 2 and 8 cm, soil moisture at 8 cm, and soil temperature at 6 cm between the two soil heat flux plates [15].

Because the soil-heat flux is measured below the soil surface, some of the energy crossing the soil surface could be stored in, or come from, the layer of soil between the surface and flux plate located closest to the surface, for this reason a change in storage term, S is added to the measured heat flux (eq. (33)). [33]:

$$S = \left[\frac{\Delta Ts}{\Delta t} \right] d\rho_b \left(C_s + (WC_w) \right) \tag{33}$$

where, S is the heat flux going into storage (Wm^{-2}), Δt is the time interval between measurement (sec), ΔTs is the soil temperature interval between measurement, d is the depth to the soil-heat-flux plates (0.08 m), ρ_b is the bulk density of dry soil (1300 kgm^{-3}), C_s is the specific heat of dry soil (840 J./Kg°C), W is the water content of soil (kg the water/Kg the soil) and C_w is the specific heat of water (4.190 J./Kg·C). The soil heat flux (G) at the surface is obtained by including the effect of storage between the surface and depth, d, using equation 11.

$$G = \left(\frac{FX1 + FX2}{2} \right) + S \tag{34}$$

where, FX1 is the soil-heat flux measured 1 (Wm^{-2}), FX2 is the soil-heat flux measured 2 (Wm^{-2}). One of the requirements for using the Bowen-ratio method is that the wind must pass over a sufficient distance of similar vegetation and terrain before it reaches the sensors. This distance is referred to as the fetch, and the fetch requirement is generally considered to be 100 times the height of the sensors above the surface [16]. More detail about determination of the minimum fetch requirement is presented later in this document.

Hanks et al. (1968), described by Frank [27], reported $\lambda E/Rn$ of 0.16 for dry soil conditions and 0.97 for wet soil conditions; On the other hand he found $\lambda E/Rn$ to be lowest in grazed prairie, suggesting that defoliation changes the canopy structure and energy budget components, which may have contributed to increase water loss through evaporation compared with the nongrazed prairie treatment. Hanson and May [34], using the Bowen Ratio Energy Balance Method to measure ET in tomatoes, found that ET rates decreased substantially in respond to drying of the soil surface. Perez et al. [83] proposed a simple model for estimating the Bowen ratio (β) based on the climatic resistance factors:

$$\beta = \frac{\Delta+\gamma}{\Delta} \cdot \frac{1+S}{1+C} - 1 \tag{35}$$

$$C = \frac{\gamma.r_i}{\Delta.r_a} \tag{36}$$

$$S = \frac{\gamma}{\Delta+\gamma} \cdot \frac{r_c}{r_a} \tag{37}$$

where, r_c is the canopy resistance (s m^{-1}) based on the "big leaf" concept, and r_a is the aerodynamic resistance (s m^{-1}). These resistance factors are described in detail in the next section. The factor r_i is the climatological resistance as reported by [66]:

$$r_i = \frac{\rho_a C_p VPD}{\gamma(R_n - G)} \tag{38}$$

where, ρ_a is the air density at constant pressure (Kg.m^{-3}), C_p is the specific heat of moist air at constant pressure (1004 J.Kg^{-1}°C^{-1}), VPD is the vapor pressure deficit of the air (Pa), γ is the pychrometric constant (Pa.°C^{-1}) and R_n and G are in W.m^{-2}. For homogeneous canopies, the effective crop surface and source of water vapor and heat is located at height d + z_{oh}, where d is the zero plane displacement height and z_{oh} is the roughness length governing the transfer of heat and vapor [5].

2.4.3   THE PENMAN-MONTEITH METHOD

The important contribution of Monteith and Penman's original equation was the use of resistances factors, which was based on an electrical analogy for the potential difference needed to drive unit flux systems that involve the transport of momentum, heat, and water vapor [69, 70]. The resistances have dimensions of time per unit length, as will describe later. This methodology calculates the latent heat flux using the vapor pressure deficit, the slope of the saturated vapor-pressure curve and aerodynamic resistance to heat, and canopy resistance in addition to the energy-budget components of the net radiation, soil heat flux, and sensible heat flux. Field measurements of air temperature, relative humidity, and wind speed are needed to determine these variables [11]. Eq. (21) describes the Penman-Monteith (P-M) method to estimate the λE [5, 54]:

$$\lambda E = \frac{\Delta s(R_n - G) + \rho_a C_p \dfrac{VPD}{r_a}}{\Delta s + \gamma\left(1 + \dfrac{r_s}{r_a}\right)} \tag{39}$$

where, λE, R_n, and G in W.m-2, VPD is vapor pressure deficit (kPa), Δs is slope of satu-ration vapor pressure curve (kPa °C-1) at air temperature, ρ is density of air (Kgm-3), Cp in J. Kg-1°C-1, γ in kPa °C-1, r_a is aerodynamic resistant (s m-1) r_s surface resistance to vapor transport (s m-1).

According to Monteith [69], the appearance of a wind-dependent function in the denominator as well as in the numerator implies that the rate of evaporation calculated from the P-M model is always less dependent on wind speed than the rate from corre-sponding the Penman equation when other elements of climate are unchanged. In gen-eral, estimated rates are usually insensitive to the magnitude of r_a and the error gener-ated by neglecting the influence of the buoyancy correction is often small. In contrast, the evapotranspiration rate is usually a strong function of the surfaces resistance (r_s).

Kjelgaard and Stockle [54] say the surface resistance (r_s) parameter in the P-M model is particularly difficult to estimate due to the combined influence of plant, soil and climatic factors that affect its value. The magnitude of the stomatal resistance can be estimated in principle from the number of stomata per unit leaf area and from the diameter and length of pores, which is difficult and therefore rarely measured; there-fore, the stomatal resistance is usually calculated from transpiration rates or estimated gradients of vapor concentration [69].

Knowing the value of the aerodynamic resistance (r_a) permits estimation of the transfer of heat and water vapor from the evaporating surface into the air above the canopy. The aerodynamic resistance for a single leaf to diffusion through the boundary layer surrounding the leaf, within which the transfer of heat, water vapor, *etc.*, occurs, proceeds at a rate governed by molecular diffusion. Provided the wind speed is great enough and the temperature difference between the leaf and air is small enough to en-sure that transfer processes are not affecting by gradients of air density, the boundary layer resistance depends on air velocity and on the size, shape, and altitude of the leaf with respect to the air stream. In very light wind, the rates of transfer are determined mainly by gradients of temperature and therefore by density, so that the r_a depends more on the mean leaf-air temperature difference than on wind speed. According to Thom [109], the r_a for heat transfer can be determined as follows:

$$r_{ah} = \frac{\rho C \rho (T_s - T_a)}{H} \tag{40}$$

At the field level, r_a for homogeneous surfaces, such as bare soil or crop canopies, there is a large-scale analogous boundary layer resistance, which can be estimated or derived from measurements of wind speed and from a knowledge of the aerodynamic properties of the surface as is described later [69]. The r_a can be determined given values of roughness length (Z_o) and zero plane displacement height (d), that depend mainly on crop height, soil cover, leaf area and structure of the canopy [1]:

$$r_a = \frac{Ln\left[\frac{(Z_m - d)}{Z_{om}}\right] Ln\left[\frac{(Z_h - d)}{Z_{oh}}\right]}{K^2 u_z} \tag{41}$$

where, Z_m is height of wind measurements (m), Z_h is height of humidity measurements (m), d is zero displacement height (m), Z_{om} is roughness length governing momentum transfer of heat and vapor (m) is 0.123h, Z_{oh} is roughness length governing transfer of heat and vapor (m) is $0.1Z_{om}$, K is the von Karman's constant (0.41), u_z is wind speed at height z.

The Eq. (41) is restricted for neutral stability conditions, i.e., where temperature, atmospheric pressure, and wind speed velocity distribution follow nearly adiabatic conditions (no heat exchange). The application of the equation for short time periods (hourly or less) may require the inclusion of corrections for stability. However, when predicting ET_o in the well watered reference surface, heat exchange is small, and therefore the stability correction is normally not required [1].

Alves et al. [1] state that though this is the most used expression for r_a, in fact it is not entirely correct, since it assume a logarithmic profile from the source height (d + Z_{oh}) with increasing z in the atmosphere, using the concept to the "big leaf," Eq. (41) can be modified as follows:

$$r_a = \frac{Ln\left(\frac{z-d}{h_c - d}\right) Ln\left(\frac{z-d}{Z_{om}}\right)}{K^2 u_z} \tag{42}$$

where, h_c is the height of the crop canopy. According to Tollk et al. [110], the r_a to momentum transport in the absence of buoyancy effects (neutral stability) follows the Eq. (43):

$$r_{am} = \ln\left[(Zi - d)/Z_{om}\right]^2 / k^2 u_z \tag{43}$$

Under adiabatic conditions, the equations must be corrected using the Richardson number for stability correction, assuming similarity in transport of heat and momentum, yielding:

$$r_{ah} = r_{am}\left(1 + 5R_i\right) \tag{44}$$

The Ri for stability conditions is considered when ($-0.008 \le R_i \le 0.008$) and is calculated by:

$$R_i = \left[g(T_a - T_s)(Z - d)\right]/T_{av}.u^2_z \tag{45}$$

where, g is the acceleration of the gravity (9.8 m.s^{-2}), T_a is the air temperature (K), T_c is the plant canopy temperature (K), T_{av} is the average temperature taken as ((T_a+T_c)/2). The advantage of the R_i over other stability corrections is that it contains only experimentally determined gradients of temperature and wind speed and does not depend directly on sensible heat flux [110].

The bulk surface resistance (r_s) describes the resistance of vapor flow through transpiring crop leaves and evaporation from the soil surface. Where the vegetation does not completely cover the soil, the resistance factor should indeed include the effects of the evaporation from the soil surface. If the crop is not transpiring at a potential rate, the resistance depends also on the water status of the vegetation [5,115], and for this case they proposed the use of the following approximate:

$$r_s = \frac{r_L}{LAI_{active}} \tag{46}$$

where, LAI_{active} is 0.5 times the leaf area index (m² of leaf per m² of soil), and r_L is bulk stomatal resistance, which is the average resistance of an individual leaf, and can be measured using an instrument called a porometry, the first stomatal readings were developed by Francis Darwin who developed horn hygrometer [113].

The r_L readings are highly variable and depend on several factors, such as: crop type and development stage, the weather and soil moisture variability, the atmospheric pollutants and the plant phytohormone balance [113]. Typically to determine minimum r_L using a porometer, fully expanded, sunlit leaves near to the top of the canopy are surveyed during maximum solar irradiance (approximately solar noon under cloudless conditions) and low VPD periods [54]. This "standard" value from literature or porometer measurements are hereafter identified as r_{Lmin}. In addition, r_L has been shown to increase with increasing VPD and/or reduced solar irradiance (R_s). Adjustment factors for VPD (f_{VPD}) and R_s (f_{Rs}) were empirically derived and used as multipliers of r_{Lmin}. The dependence of r_L on VPD can be represented by a linear function [46] as:

$$f_{VPD} = a + b[VPD] \tag{47}$$

where, a and b are linear regression coefficients, and f_{VPD} is equal to 1 (no adjustment) for VPD $\leq$ a threshold value, which can be taken as 1.5 kPa. The same authors presented a calibrated form of equation 47 for corn as, $f_{VPD} = 0.45 + 0.39(VPD)$. Kjelgaard and Stockle [54] presented a modified form of the adjustment factor:

$$f_{Rs} = \frac{R_{smax}}{C_2 + R_s} \tag{48}$$

where, R_s and R_{smax} are the actual and maximum daily solar irradiance (MJ m⁻² day⁻¹) and C_2 is a fitted constant. Taking the maximum of the adjustment factors for VPD and Rs, r_{Lmin} is modified to give the r_L [54]:

$$r_L = r_{Lmin} \max\{f_{VPD}, f_{Rs}\} \tag{49}$$

where, f_{VPD} and f_{Rs}, are equal to or greater than 1. Alves et al. [1] indicated that the surface resistance term (r_s) has been the most discussed in the literature. Several components to be considered here include: a) The resistances to water vapor at the evaporating surfaces: plants and their stomates (r_s^c) and soil (r_s^s); *b*) the resistance to vapor transfer inside the canopy from these evaporating surfaces up to the "big leaf" (r_s^a). The resistance r_s^c, can be approximated using Eq. (50).

$$r_s^c = \frac{\left(\sum_{i=1}^{n} \frac{1}{r_{stj}} \right)^{-1}}{LAI} \tag{50}$$

where, r_{st} is the single leaf stomatal resistance (sm^{-1}), n is a leaf number. The bulk surface resistance can also be calculated using the inversion of the Penman-Monteith equation with incorporation of the Bowen ratio as follow [1, 3]:

$$r_s = r_a \left(\frac{\Delta s}{\gamma} \beta - 1 \right) + \frac{\rho_a C_p VPD}{\gamma \lambda E} \tag{51}$$

Accurate prediction of r_s requires a good estimate of the Bowen ratio (β). Ramirez [87] has used the following inversion form of the Penman-Monteith equation to obtain estimates of r_s:

$$r_s = r_a. \frac{\left[\dfrac{\Delta(R_n - G) + \rho_a C_p \left(\dfrac{VPD}{r_a} \right)}{\lambda E} - \Delta - \gamma \right]}{\gamma} \tag{52}$$

Similarly these authors, analyzing the resistance concepts, concluded that the r_s of dense crops cannot be obtained by simply averaging stomatal resistance because the driving force (vapor pressure deficit) is not constant within the canopy.

Saugier [96] addressed canopy resistance (r_c), stating that it is normally a mixture of soil and plant resistances to evaporation. If the top the soil is very dry, direct soil evaporation may be neglected and r_c is approximately equal to the leaf resistance (r_L) divided by the LAI. Baldocchi et al. [10], indicated that the inverse of the 'big-leaf' model (eg., inverse of the P-M model) will be a good estimate of canopy resistance or surface resistances if certain conditions are met. These conditions include: i) a steady-state environment; ii) a dry, fully developed, horizontally homogeneous canopy situated on level terrain; iii) identical source-sink levels for water vapor, sensible heat and momentum transfer, and negligible cuticular transpiration and soil evaporation. Szeicz and Long [104] described a profile method to estimate r_s as:

$$r_s = \frac{\rho_a. C_p. VPD}{\gamma. \lambda E} \tag{53}$$

These methods can be used in the field when the rate of evapotranspiration is measured by lysimeters or calculated from the Bowen ratio energy balance method, and the temperature, humidity and wind profiles are measured within the boundary layer simultaneously. Ortega-Farias et al. [74], evaluated a methodology for calculating the canopy surface resistance ($r_{cv} \approx r_s$) in soybean and tomatoes, using only meteorological variables and soil moisture readings. The advantage of this method is that it can be used to estimate λE by the general Penman-Monteith model with meteorological reading at one level, and without r_L and LAI measurements.

$$r_s = \frac{\rho_a.c_p.VPD}{\Delta.(R_n - G)} \cdot \frac{\theta_{FC} - \theta_{WP}}{\theta_i - \theta_{WP}} \tag{54}$$

where, θ_{FC} and θ_{WP} are the volumetric moisture content at field capacity (fraction) and wilting point (fraction), respectively, and θ_i is a volumetric soil content in the root zone (fraction) measured each day. Kamal and Hatfield [48] used the Eq. (51) to determine the surface resistance in Potato: and stated that the canopy resistance (r_c in s.m^{-1}; "mean stomatal resistances of crops"), can be determined by dividing the r_s by the effective LAI as defined by other authors such as Hatfield and Allen [38] and for well watered crops, r_c can be can be estimated using Eq. (55).

$$r_c = \frac{0.3LAI + 1.2}{LAI} r_s \tag{55}$$

Kjelgaard and Stockle [54] discussed the estimation of canopy resistance (r_c) from single-leaf resistance (r_L, Eq. (56)), as originally proposed by Szeicz and Long [104]:

$$r_c = \frac{r_L}{LAI_{active}} \tag{56}$$

Kamal and Hatfield [48] divided the surface resistance (r_s) used in the P-M model into two components, and conceptualized an excess resistance (r_o) in series with the canopy stomatal resistance. This excess resistance was linked to the structure of the crop, particularly crop height.

$$r_s = r_c + r_o \tag{57}$$

Pereira et al. [81] stated that the surface resistance (r_s) is the sum of two components: one corresponding mainly to the stomatal resistance (r_{st}), the other to the leaf boundary layer and turbulent transfer inside the canopy (r_{ai}), eq. (58), thus, surface resistance is not a purely physiological parameter:

$$r_s = r_{st} + r_{ai} \tag{58}$$

Stomatal resistance can take values from 80 s.m^{-1} to 90 s.m^{-1} as a common range for agricultural crops suggested a value of 100 s.m^{-1} for most arable crops [67]. Table 1 lists mean average values for various crops under well water conditions.

The r_L is strongly dependent on the time of day (basically due to the temporal nature of climatic conditions), for the soil moisture content and by the genotype. Figure 3a shows how larger differences in r_L occur, with and without drought stress, after 9: 00 am until late in the afternoon, and the most critical point is at 13:00 hours when the highest VPD occurred. For this reason, when this variable (r_L) is not measured, appropriate parameterization is required for good water flux or ET estimation, especially under drought stress conditions. In Figure 3c, it is possible to see in a common bean genotype under drought stress conditions, lower r_L as compared with less drought resistance during several days with drought stress. Perrier (1975), as reported in Kjelgaard and Stockle [54], conceptualized the excess resistance (r_o) as a linear function of crop height and LAI:

$$r_o = ah_c + bLAI \qquad (59a)$$

where, a and b are constants. For corn, Kjelgaard and Stockle [54] parameterized Eq. (59a) as follows:

$$r_o = 16.64h_c + 0.92LAI \qquad (59b)$$

Canopy resistance can also be determined from leaf or canopy temperature since it is affected by plant characteristics, eg. Leaf area index (LAI), height, and maturity. Soil factors (available soil water-ASW, and soil solution salinity) and weather factors (R_n and wind speed) also affect the canopy resistance.

Montheith [66] showed that transpiration rate physically depends on relative changes of surface temperature and r_a, and concluded that r_a depends on the Reynolds number of the air and can be determined from wind speed, the characteristic length of the plant surface, and the kinematic viscosity of the air. An increase in r_c for Wheat was caused by a decrease in total leaf area, by an increase in the resistance of individual leaves due to senescence, or by a combination of both effects; in Sudan grass, r_c increased with plant age and a decrease in soil moisture. Van Bavel [115] studied Alfalfa throughout an irrigation cycle and found that canopy resistance increased linearly with decreasing soil water potential. Kamal and Hatfield [48] found an exponentially inverse relationship between canopy resistance and net radiation, and a linear inverse relationship between canopy resistance and available soil water.

The Drainage and Irrigation Paper-FAO 56 [5] recommended the Szeicz and Long [104] method for calculating r_s (Eq. (56)), where an average of r_L for different positions within the crop canopy, weighted by LAI or $LAI_{effective}$ is used. This method gives good results only in very rough surfaces, like forest and partial cover crops with a dry soil [67]. Alves et al. [1] concluded that r_s of dense crops cannot be obtained by simply averaging stomatal resistance (r_L) because VPD, which is the "driving force," is not constant within the canopy. Alves and Pereira [3] have stated "The PM model can be used to predict ET if accurate methodologies are available for determining the r_s that take into account the energy partitioning."

In addition to the lack of r_s values for crops, questions have been raised relative to the appropriateness of using the PM model for partial or sparse canopies because the source/sink fluxes may be distributed in a nonuniform manner throughout the field [24, 32, 55, 75]. Adequate parameterization of the surface resistance makes the P-M model a good estimator of ET [3, 74–90, 96].

Ramirez [87], reported that the daily ET estimation with the P-M model with r_s based on r_L and $LAI_{effective}$ gave a good estimation in two common bean genotypes with variable LAI, without and with moderate drought stress for both years (2006 and 2007).

Ramirez et al. [88] reported inverse relation between r_a and r_s and r_L in beans (*Phaseolus vulgaris* L), as well as those reported by Alves and Pereira [3] (Fig. 4), which implies that with low r_a (windy conditions), the r_L (and therefore r_s) increases. The Alves and Pereira [3] study did not measure the r_L, rather the r_s was estimated based on micrometeorological parameters.

Disparities in the measured r_s using the P-M inverse model arise from: a) imperfect sampling of leaves and the arbitrary method of averaging leaf resistance over the whole canopy, b) from the dependence of r_s on nonstomatal factors such as evaporation from wet soil or stems, or others and c) the complex aerodynamic behavior of canopies [68].

Lower LAI index (LAI <1.0) and drought stress also affect the precision in the r_s estimation [87]. Use of the $LAI_{effective}$ when LAI < 1.0 is not necessary and tends to overestimate the r_s and under-estimate the ET. Katerji and Perrier [51] found for LAI >1.0 a good agreement between measurement values of evapotranspiration over alfalfa crops using the energy balance method, and values calculated with P-M equation using variable r_s. Katerji and Perrier [50] proposed to simulate r_s using the following relation:

$$\frac{r_s}{r_a} = a\frac{r*}{r_a} + b \tag{60}$$

where, *a* and *b* are linear coefficients that should be determined empirically, $r*$ (s.m^{-1}) is a climatic resistance [52] giving by:

$$r* = \frac{\Delta + \gamma}{\Delta\lambda} \cdot \frac{\rho C_p VPD}{(R_n - G)} \tag{61}$$

TABLE 1   Average values of the stomatal resistance (r_L) for several crops.

Cover crops	r_L s/m	Source	Cover crops	r_L s/m	Source
Corn	200	Kirkham et al. [53]	Cassava	714 Between 476 to 1428	Oguntunde [73]. This data under limited soil water conditions.
Sunflower	400	Kirkham et al. [53]	Eucalyptus	200–400	Pereira and Alves [81]
Soybean and potato	350	Kirkham et al. [53]	Maple	400–700	Pereira and Alves [81]
Sorghum	300	Kirkham et al. [53]	Crops-General	50–320	Pereira and Alves [81]
Millet	300	Kirkham et al. [53]	Grain sorghum	200	Pereira and Alves [81]

TABLE 1   *(Continued)*

Cover crops	r_L s/m	Source	Cover crops	r_L s/m	Source
Aspen	400	Pereira and Alves [81]	Soybean	120	Pereira and Alves [81]
Maize	160	Pereira and Alves [81]	Barley	150–250	Pereira and Alves [81]
Alfalfa	80	Pereira and Alves [81]	Sugar beet	100	Pereira and Alves [81]
Clipped grass (0.15 m)	100– 150	Pereira and Alves [81]	Clipped and Irrigated grass (0.10–10.12 m)	75	Pereira and Alves [81]
Common beans	170– 270	Ramirez [87]	Sorghum	192	Stainer et al. [101]
Corn	264	Ramirez and Harmsen (2007). Unpublished data.	Andes Tropical Forestry	132	Ramirez and Jara-millo [89]. (Calculated)
Coffee	149	Ramirez and Jaramillo [89]. (Calculated)	Coffee	150	Angelocci et al. [7]
Wheat	134	Howell et al. [42]	Corn	252	Howell et al. [42]
			Sorghum	280	Howell et al. [42]

Table 2 presents values of *a* and *b* for several crops. The Penman-Monteith model is considered as a 'single-layer' model, Shuttleworth and Wallace [100] developed a 'double-layer' model, relying on the Penman-Monteith model concept to describe the latent heat flux from the canopy (λT) and from the soil (λE) as follows:

$$\lambda T = \frac{\Delta \left(R_n - R_{ns} \right) + \rho Cp \dfrac{VPDo}{r^c{}_a}}{\Delta + \gamma \left(1 + \dfrac{r^c{}_s}{r^c{}_a} \right)} \qquad (62)$$

$$\lambda E = \frac{\Delta \left(R_{ns} - G \right) + \rho Cp \dfrac{VPDo}{r^s{}_a}}{\Delta + \gamma \left(1 + \dfrac{r^s{}_s}{r^s{}_a} \right)} \qquad (63)$$

where, R_{ns} is the absorbed net radiation at the soil surface, $r^c{}_a$ is the bulk boundary layer resistance of the canopy elements within the canopy, $r^c{}_s$ is the bulk stomatal resistance of the canopy, r^s_a is the aerodynamic resistance between the soil and the mean canopy height, r^s_s is the surfaces resistance of the soil and VPD_o is the vapor pressure deficit at the height of the canopy air stream.

2.4.4 THE DOUBLE-LAYER SHULTTLEWORTH-WALLACE MODEL

The Shulttleworth-Wallace Model (S-W) assumes that there is blending of heat fluxes from the leaves and the soil in the mean canopy airflow at the height of the effective canopy source [100]. The full expression of the Shulttleworth-Wallace Model (S-W) is presented by Zhang et al. [125] as follows:

$$\lambda ET = \lambda E + \lambda T = C^S_{SW} PM^S_{SW} + C^P_{SW} PM^P_{SW} \tag{64}$$

$$PM^S_{SW} = \frac{\Delta A_{SW} + \left[\left(\rho C p D - \Delta r^s_a \right) \left(A_{Sw} - A^s_{SW} \right) / \left(r^a_a + r^s_a \right) \right]}{\Delta + \gamma \left[1 + r^s_s / \left(r^a_a + r^s_a \right) \right]} \tag{65}$$

$$PM^P_{SW} = \frac{\Delta A_{SW} + \left[\left(\rho C p D - \Delta r^P_a A^s_{SW} \right) / \left(r^a_a + r^P_a \right) \right]}{\Delta + \gamma \left[1 + r^P_s / \left(r^a_a + r^P_a \right) \right]} \tag{66}$$

$$C^S_{SW} = \frac{1}{1 + \left[R^S_{SW} R^a_{SW} / R^P_{SW} \left(R^S_{SW} + R^a_{SW} \right) \right]} \tag{67}$$

$$C^P_{SW} = \frac{1}{1 + \left[R^P_{SW} R^a_{SW} / R^S_{SW} \left(R^P_{SW} + R^a_{SW} \right) \right]} \tag{68}$$

$$R^S_{SW} = \left(\Delta + \gamma \right) r^s_a + \gamma r^s_s \tag{69}$$

In Eq. (64)–(70), λE is the latent heat flux of evaporation from the soil surfaces (W/m²), λT the latent heat fluxes of transpiration from canopy (W/m²), r^P_s the canopy resistance (s/m), r^P_a the aerodynamic resistance of the canopy to in-canopy flow (s/m), r^s_s the soil surfaces resistance (s/m), r^a_a and r^s_a the aerodynamic resistance from the reference height to in-canopy heat exchange plane height and from there to the soil surface (s/m), respectively. A_{sw} and A^s_{SW} are the total available energy and the available energy to the soil (W/m²), respectively and defined in Eqs. (70)–(73):

$$R^P_{SW} = \left(\Delta + \gamma \right) r^P_a + \gamma r^P_s \tag{70}$$

$$Ra^P_{SW} = \left(\Delta + \gamma \right) r^a_a \tag{71}$$

$$A_{sw} = R_n - G \tag{72}$$

$$A_{SW}^s = R_{nsw}^s - G \tag{73}$$

In Eq. (73), R_{nsw}^s is the net radiation fluxes into the soil surface (W/m²), and can be calculated using the Beer's law:

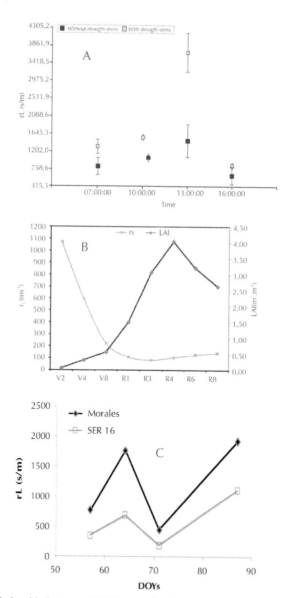

FIGURE 3 Relationship between: (a) Changes in the stomatal resistance during the day with and without drought stress in *Phaseolus vulgaris* L. genotype 'Morales'; **(b)** Surfaces resistance and Leaf area index; and (c) Stomatal behavior represented in stomatal resistance (r_L) under drought stress conditions for two common bean genotypes — 'Morales' lest drought tolerant and 'SER 16' drought stress tolerant.

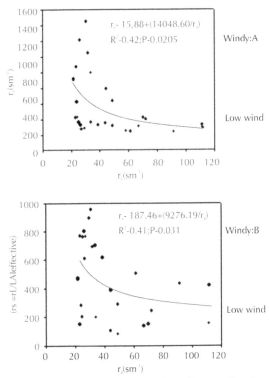

FIGURE 4   (a) Aerodynamic resistance (r_a) as a function of stomatal resistance (r_L); and (b) Aerodynamic resistance (r_a) as a function of measured surface resistance ($r_s = r_L/LAI_{effective}$) [87].

$$R_{nsw}^s = R_n \cdot \exp^{(-c.LAI)} \tag{74}$$

In Eq. (74), c is the extinction coefficient of light attenuation (e.g., Sene, [97] indicate $k=0.68$ for fully grown plant, $k=0$ for bare soil; Zhang et al., [125] use 0.24 for vineyard crops).

The *surfaces resistance* is calculated as follows:

$$r_s^p = \frac{r_{st\,min}}{LAI_{effective} \Pi_i F_i(X_i)} \tag{75}$$

where, $r_{st\,min}$ is the minimal stomatal resistance of individual leaves under optimal conditions. $LAI_{effective}$ is: equal to LAI for LAI ≤ 2.0; LAI/2 for LAI ≥ 4.0 and 2 for intermediate values of LAI, X_i is a specific environmental variable, and $F_i(X_i)$ is the stress function with $0.0 \leq F_i(X_i) \geq 1.0$ [46].

$$F_1(S) = \left(\frac{S}{1100}\right)\left(\frac{1100 + a_1}{S + a_1}\right) \tag{76}$$

$$F_2(T) = \frac{(T - T_L)(T_H - T)^{(TH-a_2)/(a_2-T_L)}}{(a_2 - T_L)(T_H - a_2)^{(TH-a_2)/(a_2-T_L)}}$$
(77)

$$F_2(D) = e - a_3D$$
(78)

$$F_4(\theta) = \begin{bmatrix} 1, if ------ \theta \geq \theta_F \\ \dfrac{\theta - \theta_W}{\theta_F - \theta_W}, if -- \theta_F < \theta < \theta_W \\ 0 -- if ----- \theta \leq \theta_W \end{bmatrix}$$
(79)

where, S is the incoming photosynthetically active radiation flux (W/m²), T is the air temperature (°K), θ_F is the soil moisture at field capacity (cm³/cm³), q_w is the soil moisture at wilting point (cm³/cm³), and θ is the actual soil moisture in the root zone. (cm³/cm³). T_H and T_L are upper and lower temperatures limits outside of which transpiration is assumed to cease (°C) and are set at values of 40 and 0°C [39, 125]. The a_1 = 57.67, a_2 = 25.78, and a_3 = 9.65 that were determined by multivariate optimization [125].

TABLE 2   Coefficients a and b for several crops.

Crop	a	b	Source
Grass	0.16	0.0	Katerji and Rana [52]
Tomato	0.54	2.4	Katerji and Rana [52]
Grain sorghum	0.54	0.61	Katerji and Rana [52]
Soybean	0.95	1.55	Katerji and Rana [52]
Sunflower	0.45	0.2	Katerji and Rana [52]
Sweet sorghum	0.845	1.0	Katerji and Rana [52]
Grass (Tropical climate)	0.18	0.0	Gosse (1976) in Rana et al. [91]
Grass (Mediterranean climate)	0.16	0.0	Rana et al. [91]
Alfalfa	0.24	0.43	Katerji and Perrier (1983) in Rana et al. [91]
Sorghum	0.94	1.1	Rana et al. [92]
Sunflower	0.53	1.2	Rana et a.l [92]

The aerodynamic resistances [r_a^a and r_a^s] are calculated from the vertical wind profile in the field and the eddy diffusion coefficient. Above the canopy height, the eddy diffusion coefficient (K) is given by:

$$K = ku*(z - d)$$
(80)

where, u* is the wind friction velocity (m/s), k is the van-Karman constant (0.41), z is the reference height (m), and d the zero plane displacement (m). The exponential decrease of the eddy diffusion coefficient (K) through the canopy is given as follows:

$$K = k_h \cdot \exp\left[-n\left(1-\frac{z}{n}\right)\right]$$ (81)

where, k_h is the eddy diffusion coefficient at the top of the canopy (m²/s), and n is the extinction coefficient of the eddy diffusion. Brutsaert (1982) cited by [125] indicated that $n = 2.5$ when $h_c < 1$ m; $n = 4.25$ when $h_c > 10$ m, linear interpolation could be used for crops with h between those values. k_h is determined as follows:

$$k_h = ku*(h_c - d)$$ (82)

The aerodynamic resistance r_a^a and r_a^s are obtained by integrating the eddy diffusion coefficients from the soil surface to the level of the "preferred" sink of momentum in the canopy, and from there to the reference height (Shutlleworth and Gurney, 1990, cited [125]) as follows:

$$r_a^a = \frac{1}{Ku*}\ln\left(\frac{z-d}{h_c-d}\right) + \frac{h_c}{nk_h}\left[\exp\left[n\left(1-\frac{zo+d}{h_c}\right)\right]-1\right]$$ (83)

$$r_a^s = \frac{h_c \exp^{(n)}}{nk_h}\left[\exp\left(\frac{-nz_o{}'}{h_c}\right) - \exp\left[-n\left(\frac{z_0+d}{h_c}\right)\right]\right]$$ (84)

The bulk boundary layer resistance of canopy is calculated as follows:

$$r_a^p = \frac{r_b}{2LAI}$$ (85)

where, r_b is the mean boundary layer resistance (s/m) (e.g., Brisson et al., [13], recommend use 50 s/m).

The *soil surface resistance* r_s^s is the resistance to water vapor movement from the interior to the surface of the soil, and is strongly depending of the water content (q_s), and is calculated using the Eq. (86) defined by Anandristakis et al. [6]:

$$r_s^s = r_{s\ min}^s f(\theta_s)$$ (86)

where, q_s is soil volumetric water content (cm³/cm³), and $r_{s\ min}^s$ is the minimum soil surfaces resistance, that correspond with the soil field capacity (θ_{FC}) and is assumed equal to 100 s/m (e.g., [18, 125]). The $f(\theta_s)$ is expressed by Eq. (87) defined by Thompson (1981), cited by [125]

$$f(\theta_s) = 2.5\left(\frac{\theta_{FC}}{\theta_s}\right) - 1.5$$ (87)

2.4.5 CLUMPING MODEL

The Clumping model is based on the Shulttleworth-Wallace model, this model separated the soil surfaces into fractional areas inside and outside the influence of the canopy, and included the fraction of canopy cover (f). Brenner and Incoll [12] and Zhang et al. [125] expressed the model as follows:

$$\lambda E = \lambda E^s + \lambda E^{bs} + \lambda T = f\left(C_c^s PM_c^s + C_c^p PM_c^p\right) + (1-f)C_c^{bs} PM_c^{bs} \qquad (88)$$

where, λE^s is the latent heat of evaporation from soil under the plant (W/m²); λE^{bs} is the latent heat of evaporation from bare soil (W/m²); f is the fractional vegetative cover and the other terms are expressed as follows:

$$PM_c^p = \frac{\Delta A_c + \left[\dfrac{\left(\rho CpD - \Delta r_a^p A_c^s\right)}{r_a^a + r_a^p}\right]}{\Delta + \gamma\left[1 + \dfrac{r_s^p}{r_a^a + r_a^p}\right]} \qquad (89)$$

$$PM_c^s = \frac{\Delta A_c + \left[\dfrac{\left(\rho CpD - \Delta r_a^s A_c^p\right)}{r_a^a + r_a^s}\right]}{\Delta + \gamma\left[1 + \dfrac{r_s^s}{r_a^a + r_a^s}\right]} \qquad (90)$$

$$PM_c^{bs} = \frac{\Delta A_c^{bs} + \left[\dfrac{\left(\rho CpD\right)}{r_a^a + r_a^{bs}}\right]}{\Delta + \gamma\left[1 + \dfrac{r_s^{bs}}{r_a^a + r_a^{bs}}\right]} \qquad (91)$$

$$C_c^s = \frac{R_c^{bs} R_c^p \left(R_c^s + R_c^a\right)}{\left[R_c^s R_c^p R_c^{bs} + (1-f)R_c^s R_c^p R_c^a + fR_c^{bs} R_c^s R_c^a + fR_c^{bs} R_c^p R_c^a\right]} \qquad (92)$$

$$C_c^p = \frac{R_c^{bs} R_c^s \left(R_c^p + R_c^a\right)}{\left[R_c^s R_c^p R_c^{bs} + (1-f)R_c^s R_c^p R_c^a + fR_c^{bs} R_c^s R_c^a + fR_c^{bs} R_c^p R_c^a\right]} \qquad (93)$$

$$C_c^{bs} = \frac{R_c^s R_c^p \left(R_c^{bs} + R_c^a\right)}{\left[R_c^s R_c^p R_c^{bs} + (1-f)R_c^s R_c^p R_c^a + fR_c^{bs} R_c^s R_c^a + fR_c^{bs} R_c^p R_c^a\right]} \qquad (94)$$

$$R_c^s = (\Delta + \gamma)r_a^s + \gamma r_s^s \qquad (95)$$

$$R_c^p = (\Delta + \gamma)r_a^p + \gamma r_s^p \qquad (96)$$

$$R_c^{bs} = (\Delta + \gamma)r_a^{bs} + \gamma r_s^{bs} \qquad (97)$$

$$R_c^a = (\Delta + \gamma) r_a^a \tag{98}$$

where, A_c, A_c^p, A_c^s and A_c^{bs} are energy available to evapotranspiration, to the plant, to soil under shrub and bare soil (W/m²) respectively, r_a^{bs} the eddy diffusion resistance from in-canopy heat exchange plane height to the soil surface (s/m), r_s^{bs} the soil surfaces resistance of bare soil (s/m). The *Available energy* for this model, the net radiation (R_n) is divided into net radiation in the plant (R_n^p) and the net radiation in the soil (R_n^s). If the energy storage in the plant is assumed to be negligible, then:

$$R_{nc}^s = R_n \exp^{(-CLAI/f)} \tag{99}$$

$$R_{nc}^p = R_n - R_{nc}^s \tag{100}$$

$$A_c^s = R_{nc}^s - G^s \tag{101}$$

$$A_c^{bs} = R_n - G^{bs} \tag{102}$$

$$A_c^p = R_{nc}^p \tag{103}$$

where, R_{nc}^p and R_{nc}^s are the radiation absorbed by the plant and the radiation by the soil (W/m²) respectively, G^s and G^{bs} are the soil heat flux under plant and bare soil (W/m²), respectively, C is the extinction coefficient of light attenuation according for Sene [97] is equal to 0.68 for fully grown plant. The resistance for the bare soil surfaces r_s^{bs} can be calculated equally as in the S-W model, mentioned before. The aerodynamic resistance between the bare soil surface and the mean surfaces flow height (r_a^{bs}) can be calculated assuming that the bare soil surface is totally unaffected by adjacent vegetation so that is aerodynamic resistance equal to r_a^b and defined for:

$$r_a^b = \ln \frac{\left(\dfrac{Z_m}{Z'_o}\right)^2}{k^2 U_m} \tag{104}$$

where, Z_m is the mean surface flow height (m), and could be assumed equal to $0.75h_c$, and u_m is the wind speed at the Z_m (m/s). According with Zhang et al. [125], the aerodynamic resistance (r_a^{bs}) varies between r_a^b and r_a^s as f varies from 0 to 1, and the functional relationship of this change is not known.

2.4.6 COMBINATION MODEL

Theoretical approaches to surface evaporation from the energy balance equation combined with sensible heat and latent heat exchange expressions give the following equation for actual evapotranspiration [81]:

$$ET = \frac{\Delta}{\Delta + \gamma} \left[(R_n - G) + \frac{\rho Cp}{\Delta} Hu (VPDa - VPDs) \right] \tag{105}$$

where, $(R_n - G)$ = Available energy (MJ/m²) for the canopy consisting of net radiation, R_n and the soil heat flux, G; $H(u)$ = exchange coefficient (m/s) between the surface level and a reference level above the canopy but taken inside the conservative boundary sublayer; $VPDs$ and $VPDa$ (kPa) = vapor pressure deficits (VPD) for the surface level and the reference level, respectively; ρ = atmospheric density (kg/m³); Cp = specific heat of moist air (J./kg°C); Δ = slope of the vapor pressure curve (Pa/°C); and γ= psychrometric constant (Pa/°C). To obtain evapotranspiration with the Eq. (105), it is not an easy task to estimate $VPDs$, representing the vapor pressure deficit at the evaporative surface. If $VPDs$ can be associated with a surface resistance term (r_s). Therefore, ET can be calculated directly from the flux equation:

$$ET = \frac{\rho Cp}{\gamma} \frac{VPDs}{r_s}$$

(106)

and

$$r_a = \frac{1}{Hu}$$

(107)

where, r_a can be calculated using the equations discussed later in this chapter. Two main solutions can be defined for the Eq. (105) using climatic data:

1. The case of full water availability corresponding to saturation at the evaporative surface. Then $VPDs$ = 0 and r_s becomes null. Eq. (105) then gives the maximum value for ET, the potential evaporation (EP), which depends only on climatic driving forces:

$$EP = \frac{\Delta(R_n - G) + \rho CpF(u)VPDa}{\Delta\lambda}$$

(108)

where, $F(u) = 1/r_a$. The combination the equations can get:

$$ET = \frac{EP}{(1 + \frac{\gamma}{\Delta + \gamma\, r_a} \frac{r_s}{})}$$

(109)

2. The case for equilibrium between the surface and the reference levels corresponds to $VPDs = VPDa$. In this case, the evapotranspiration is referred to as the equilibrium evaporation (Ee):

$$Ee = \frac{\rho Cp}{\gamma} \frac{VPDa}{r_e}$$

(110)

where, r_s was renamed r_e, termed the equilibrium surface resistance, indicating that the term, in this case, represents the surface resistance for equilibrium evaporation. The value for r_e depends predominately on climatic characteristics although these characteristics are influenced by R_n and G of the vegetative surface. For purposes here, the r_e term can be called the climatic resistance for the surface:

$$r_e = \frac{\rho Cp}{\gamma} \frac{\Delta + \gamma}{\Delta} \frac{VPDa}{R_n - G}$$

(111)

EP can be estimated:

$$EP = Ee\left(1+\frac{\gamma}{\Delta+\gamma}+\frac{r_e}{r_a}\right)$$
(112)

and ET can be estimated using:

$$ET = \frac{EP_{(36)}}{\left(1+\frac{\gamma}{\Delta+\gamma}\frac{r_s}{r_a}\right)}$$
(113)

2.4.7 PRISTLEY AND TAYLOR MODEL

Pristley and Taylor [84], proposed to neglect the aerodynamic term and replace the radiation term by a dimensionless coefficient (α):

$$ET = \alpha\frac{\Delta}{\Delta+\gamma}(R_n - G)$$
(114)

where, ET is water flux under references conditions (well watered grass) in mm.day^{-1}; R_n and G are net radiation and soil heat flux respectably in mm.day^{-1}; Δ and γ in kPa.$°C^{-1}$. The term α is given as 1.26 for grass field in humid weather conditions, and was adopted by Pristley and Taylor [84] for wet surfaces. However α ranges from 0.7 to 1.6 for various landscape situations [26]. According with Zhang et al. [124], the term α can be calculated as follows:

$$\alpha = \frac{\lambda E(\Delta+\gamma)}{\Delta(R_n - G)} = \frac{\Delta+\gamma}{\Delta(1+\beta)}$$
(115)

Also the term, α, sensible heat flux at the soil moisture changes [29, 30, 124], and can be estimated using a model given below:

$$\alpha = k\left[1-\exp\left(-c\frac{\theta-d}{\theta_{FC}}\right)\right]$$
(116)

where, k, c and d are parameters of the model, θ is the actual volumetric soil moisture content (cm^3.cm^{-3}) and θ_{FC} is the volumetric moisture content at field capacity (cm^3.cm^{-3}).

2.4.8 EDDY COVARIANCE METHOD

The eddy covariance method is, in general, the most preferred because it provide a direct measure of the vertical turbulent flux across the mean horizontal streamlines, provided by fast sensors (~10 Hz) [65]. Realizing the limitation of the Thornthwaite-Holzman type of approach, Swinbank (1951) cited by Chang [19] was the first to attempt a direct measurement by the so-called eddy correlation technique. The method is based on the assumption that the vertical eddy flux can be determined by simultaneous measurements of the upward eddy velocity and the fluctuation in vapor pressure. Actually is a routinely technique for direct measurement of surfaces layer fluxes of momentum, heat, and traces gases (CO_2, H_2O, O_3) between the surfaces and the turbulent atmosphere [63].

This system recognizes that the transport of heat, moisture, and momentum in the boundary layer is governed almost entirely by turbulence. The eddy correlation method is theoretically simple using an approach to measure the turbulent fluxes of vapor and heat above the canopy surface. The eddy correlation fluxes are calculated and recorded in a 30 min or less temporal resolution. Assuming the net lateral advection of vapor transfer is negligible, the latent heat flux (evapotranspiration) can be calculated from the covariance between the water vapor density (ρ_n) and the vertical wind speed (w):

$$\lambda E = \lambda \overline{w' \rho_v'} \tag{117}$$

where, λE is the latent heat flux (W m^{-2}), l is the latent heat of vaporization (J. kg^{-1}), ρ_n' is the fluctuation in the water vapor density (kg m^{-3}), and w' is the fluctuation in the vertical wind speed (m s^{-1}). The over bar represents the average of the period and primes indicate the deviation from the mean values during the averaging period. According to Weaver [118], the eddy correlation method depends on the relations between the direction of air movement near the land surface and properties of the atmosphere, such as temperature and humidity. The sensible heat flux can be calculated from the covariance of air temperature and the vertical wind speed.

$$H = \rho_a C_p \overline{w'T'} \tag{118}$$

where, H = the sensible heat flux (W m^{-2}), ρ_a the air density (kg m^{-3}), $C_{p=}$ the specific heat of moist air (J. kg^{-1}°C^{-1}) and T' = the fluctuation in the air temperature (°C).

The fine wire thermocouples (0.01 mm diameter) are not included in the eddy correlation system. The air temperature fluctuations, measured by the sonic anemometer, are corrected for air temperature fluctuations in estimation of sensible heat fluxes. The correction is for the effect of wind blowing normal to the sonic acoustic path. The simplified formula by Schotanus et al. [99] is as follows:

$$\overline{w'T'} = \overline{w'T_s'} - 0.51\left(\overline{T + 273.15}\right)\overline{w'q'} \tag{119}$$

where, w'T' is rotated covariance of wind speed and sonic temperature (m°C s^{-1}), T is air temperature (°C) and q is the specific humidity in grams of water vapor per grams of moist air.

Two Eddy covariance systems are used to measure the water vapor fluxes, the open path and close path. According to Anthoni et al. [8] the Open-path eddy covariance systems require corrections for density fluctuations in the sampled air [64, 119] and in general closed-path system require incorporation of a time lag and corrections for the loss of high frequency information, due to the air being drawn through a long sampling tube [64, 71]. The most common correction in the eddy covariance system is described by Wolf et al. [121] as: (1) Coordinate rotation, (2) Air density correction, and (3) Frequency-dependent signal loss.

Estimation of turbulent fluxes is highly dependent on the accuracy of the vertical wind speed measurements. Measurement of wind speed in three orthogonal directions with sonic anemometer requires a refined orientation with respect to the natural co-ordinate system through mathematic coordinate rotations [103]. The vector of wind has three components (u, v, w) in three coordinate directions (x, y, z). The z-direction is oriented with respect to gravity, and the other two are arbitrary. Baldocchi et al. [9] provide procedures to transform the initial coordinate system to the natural coordinate system. Described in details by Sumner [103], the coordinate system is rotated by an angle η about the z-axis to align u into the x-direction on the x-y plane, then rotated by an angle θ about the y-direction to align w along the z-direction. The resultant forces $\bar{v}$ and $\bar{w}$ are equal to zero, and $\bar{u}$ is pointed directly to the air stream. When θ was greater than 10 degrees, the turbulent flux data should be excluded based on the assumption that spurious turbulence was the cause of the excessive amount of the coordinate rotation:

$$\cos\theta = \sqrt{\frac{\left(u^2+v^2\right)}{\left(u^2+v^2+w^2\right)}} \tag{120}$$

$$\sin\theta = \frac{w}{\sqrt{\left(u^2+v^2+w^2\right)}} \tag{121}$$

$$\cos\eta = \frac{u}{\sqrt{\left(u^2+v^2\right)}} \tag{122}$$

$$\sin\eta = \frac{v}{\sqrt{\left(u^2+v^2\right)}} \tag{123}$$

The latent heat and sensible heat fluxes are computed from the coordinate rotation-transformed covariance:

$$\left(\overline{w'\rho_v}\right)_{tr} = \overline{w'\rho_v}\cos\theta - \overline{u'\rho_v}\sin\theta\cos\eta - \overline{v'\rho_v}\sin\theta\sin\eta \tag{124}$$

$$\left(\overline{w'T_s}\right)_{tr} = \overline{w'T_s}\cos\theta - \overline{u'T_s}\sin\theta\cos\eta - \overline{v'T_s}\sin\theta\sin\eta \tag{125}$$

After the coordinate rotation, the final latent heat flux can be estimated from Eq. (117) and the following correction of air density (C_{air}) [119] and correction of oxygen (CO_2) [107]:

$$C_{air} = \frac{\overline{\rho_v H}}{\rho C_p \left(T+273.15\right)}\lambda \tag{126}$$

$$C_{O2} = \frac{F K_o \overline{H}}{K_w \left(T+273.15\right)}\lambda \tag{127}$$

where, F is a factor used in krypton hygrometer correction that accounts for molecular weights of air and oxygen, and atmospheric abundance of oxygen and is equal to 0.229 g°C J.$^{-1}$, K_o is the extinction coefficient of hygrometer for oxygen, estimated as 0.0045 m^3 g^{-1} cm^{-1}, K_w is the extinction coefficient of hygrometer for water and is 0.149 m^3 g^{-1} cm^{-1}, provided by the manufacturer:

With the measured four flux components from the energy balance equation, the energy balance should be closed, however, this is not practically the case. A tendency to underestimate energy and mass fluxes has been a pervasive problem with the eddy covariance technique [33]. Ham and Heilman [33] reported closure of 0.79 for priarine locations and 0.96 for forest. Ramirez and Harmsen (2007-Data without publication) indicated 0.71 for grass and 0.75 for corn.

The errors in eddy covariance method are associated with: 1. Accuracy of the R_n and G measurements (errors are often 5 to 10%); 2. The length scale of the eddies responsible for transport (if is larger, the frequency response and sensor separation error may have been smaller); 3. Sensor separation and inadequate sensor response (can underestimate by 15% of λE by [33] and 10% reported by Laubach and McNauhton [57]; and 4. Ham and Heilman [33] conclude "The inherent tendency to underestimate fluxes when using eddy covariance may be linked to the errors caused by sensor separation and inadequate frequency response of the sensors. The correction proposed by Massman and Lee [64] is difficult to implement for the nonspecialist because they require calculation of cospectra using high-frequency (10 Hz) data, and also is required expertize experience to interpret the cospectra properly."

The "energy balance closure" is corrected using the Bowen ratio [56] as follows:

$$H = \beta * \lambda E \tag{128}$$

$$\lambda E = Rn - G - H \tag{129}$$

where, β and λE are due to eddy covariance system, R_n and G are measured.

The Massman Analytical Formulae for Spectral Corrections to Measured Momentum and Scalar Fluxes for Eddy Covariance Systems: Massman [63] developed an analytical method for frequency response corrections, based on the procedure developed by Horst [43]:

For *Stable atmospheric conditions* (0<z≤ 2):

Fast-response open path system:

$$\frac{Flux_m}{Flux} = \left[\frac{ab}{(a+1)(b+1)}\right]\left[\frac{ab}{(a+p)(b+p)}\right]\left[\frac{1}{(p+1)}\right]\left[1-\frac{p}{(a+1)(a+p)}\right] \tag{130}$$

Scalar instrument with 0.1–0.3s response:

$$\frac{Flux_m}{Flux} = \left[\frac{ab}{(a+1)(b+1)}\right]\left[\frac{ab}{(a+p)(b+p)}\right]\left[\frac{1}{(p+1)}\right]\left[1-\frac{p}{(a+1)(a+p)}\right]\left[\frac{1+0.9p}{1+p}\right] \tag{131}$$

Unstable atmospheric conditions (z≤0):

$$\frac{Flux_m}{Flux} = \left[\frac{a^\alpha b^\alpha}{(a^\alpha +1)(b^\alpha +1)}\right]\left[\frac{a^\alpha b^\alpha}{(a^\alpha + p^\alpha)(b^\alpha + p^\alpha)}\right]\left[\frac{1}{(p^\alpha +1)}\right]\left[1-\frac{p^\alpha}{(a^\alpha +1)(a^\alpha + p^\alpha)}\right] \tag{132}$$

where, the subscript m refers to the measurement flux, $a = 2\pi \int x\tau_h$; $b = 2\pi \int x\tau_b$; $p = 2\pi \int x\tau_c$; and τ_h and τ_b are the equivalent time constants associated with trend removal (τ_h) and block averaging (τ_b). For relatively broad coespectra with relatively shallow peaks, such as the flat terrain neutral/stable, such as flat terrain coespectrum: α=0.925; and for sharper, more peaked coespectra, such as the stable terrain coespectra: α=0.925 [49].

These approximations are clearly easier to employ than numerical approaches and are applicable even when fluxes are so small as to preclude the use of *in situ* methods. Nevertheless, this approach is subject at the next conditions: i) horizontally homogeneous upwind fetch, ii) the validity of the coespectral similarity, iii) sufficiently long averaging periods, and preferably, iv) relatively small corrections [63].

2.4.9   THE INFRARED SURFACE TEMPERATURE METHOD

The infrared surface temperature has also been used for the estimation of the sensible heat flux (H) using the resistance model [2]:

$$H = \rho.Cp \frac{To - Ta}{r_a} \tag{133}$$

where, ρ is air density (Kgm^{-3}), Cp specific heat at constant pressure (Jkg^{-1}°C^{-1}), To is the temperature at surface level (°C), Ta is the temperature at the reference level (°C), and r_a is the aerodynamic resistance to heat flux between the surface and the reference level (s.m^{-1}), the latent heat flux (λE) can be computed as the residual term in the energy balance:

$$\lambda E = Rn - G - H = Rn - G - \rho.Cp \frac{To - Ta}{r_a} \tag{134}$$

Alves et al. [2] say the radioactive surface temperature has a several drawbacks. Thermal radiation received by the instrument can originate from the leaves but also from de soil, and the measurement can be highly dependent on crop cover, inclination of radiometer and sun height and azimuth, especially en partial cover crops, the first one lies in the use of an adequate value of r_a. The variable d is zero plane displacement height (m), Z_{oM} and Z_{OH} are the roughness lengths (m) for momentum and heat respectively, k is the von Karman constant, u_z is the wind speed (ms^{-1}) at the reference height z (m), and ψM and ψH are the integrated stability functions for describing the effects of the buoyancy or stability on momentum transfer and heat between the surface and the reference level.

The necessary instruments are: Wind speed and direction sensor at (0.85 and 1.46 m), psychrometer at the same height that wind sensor, a net radiometer placement a 1.5 m and infrared thermometer perpendicular to the rows the crop, and positioned at an angle of 60° below horizontal to view the top leaves of the plants at 0.40 m distance [2]. The sensible heat flux [H] is calculated with the flux applied to levels Z_1 and Z_2:

$$H = \rho Cp \frac{T1 - T2}{[ra]_1^2} \tag{135}$$

where, $[ra]_1^2$ is the aerodynamic resistance to heat flux between the two levels, and is computed using the Eq. (136):

$$[ra]_1^2 = \frac{\ln\left(\dfrac{Z2-d}{Z1-d}\right)}{ku*} \qquad (136)$$

where, $u*$ = the friction velocity, obtained in the process of determining aerodynamic parameter [d] and Z_{oM} from the win profile measurements.

The air temperature at the surface level (To) is calculated using Eq. (137). The stability conditions can be calculated using the Richardson number.

$$To = Ta + \frac{Hr_a}{\rho Cp} \qquad (137)$$

Fetch requirements: The air that passing over a surface is affected by the field surfaces feature [93]; the minimal fetch requirement was estimated based on the thickness of the internal boundary layer (δ in m) and a roughness parameter (Z_o in m) for each genotype considering the minimal and maximal crop height during the grown season. The δ was calculated using the Eq. (138) proposed by Monteith and Unsworth [70]:

$$\delta = 0.15.L^{4/5}.Z_o^{1/5} \qquad (138)$$

where, L is the distance of traverse (fetch) across a uniform surface with roughness Z_o. The Z_o for crops is approximately one order of magnitude smaller than the crop height h, and can be calculated according with Rosenberg et al. [93] as follows:

$$Log_{10}Z_o = 0.997 \log_{10} h - 0.883 \qquad (139)$$

As a factor of safety, a height to fetch of 1: 50 to 1: 100 is usually considered adequate for studies made over agricultural crop surfaces [5, 93] but may be too conservative and difficult to achieve in practice. Alves et al. [1] obtained full profile development using a 1: 48 fetch relation in Wheat and lettuce. Heilman et al. [40] found that for Bowen-Ratio estimates a fetch 1:20 was sufficient over grass, and Ham and Heilman [32] and Frithschen and Fritschen [28] obtained similar results.

Stability correction: The gradient method need a stability correction, one of the most used is the Monin-Obukhov stability factor (ζ) described by [17, 86, 93]:

$$\xi = \frac{(-k.z.g.H)}{\left(\rho_a.C_p.T_a.u^{*3}\right)} \qquad (140)$$

where, K is von Karman's constant, z is height of wind and air temperature measurements (m), g is the gravitational constant (9.8 m.s^{-2}), H = $\beta.\lambda E$, T_a is air temperature (°K), u× is the friction velocity given by Kjelgaard et al. [55] without the stability correction factor:

$$u* = \frac{k.u_z}{\ln\left(\dfrac{z-d+Z_{om}}{Z_{om}}\right)} \qquad (141)$$

Flux with a negative sign for ζ indicating unstable conditions and needs to be excluded. For flux under unstable conditions the λE is over R_n (Fig. 5a); For the flux with negative ζ are excluded and λE is lower than R_n (Fig. 5b). Payero et al. [77] indicated that fluxes with incorrect sign and $\beta \approx -1$ should not be considered when estimated the energy balance components by the energy balance Bowen ratio method. The negative ζ corresponds to negative β (Fig. 6).

The Richardson number (Ri) is represented by the Eq. (45), also is well known as stability factor [2, 110] and represent the ratio of the buoyancy – "thermal effect" to "mechanical –wind shear" [86]. Negative values indicate instability conditions where surfaces heating enhances buoyancy effects, and positive Ri values indicate a stable conditions where temperature near the surfaces are cooler than away from the surfaces.

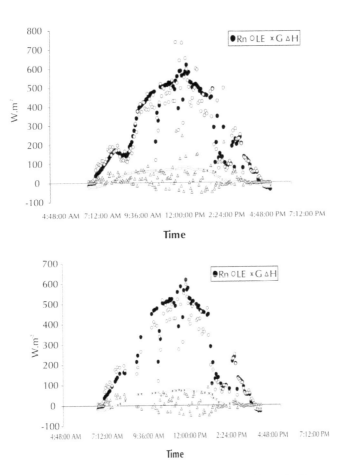

FIGURE 5 Energy balance components measured by Bowen ratio method in grass: (a) without stability correction and (b) with stability correction.

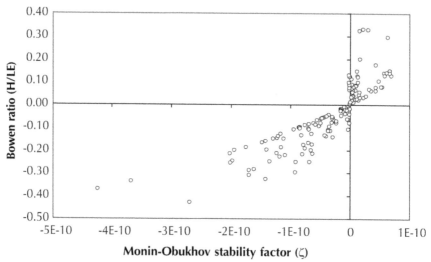

FIGURE 6   Relationship between Bowen ratio (β)) and the Monin-Obukhov stability factor.

2.5   SUMMARY

The water vapor flux in the agroecosystems is the second largest component in the hydrologycal cycle. Water vapor flux from the vegetation to the atmosphere is a widely studied variable throughout the world, due to it applicability in various disciplines such as hydrology, climatology, and agricultural science. The evapotranspiration is important to calculate the water requirement to the crops, to made climatic characterizations and water management. The estimation of evapotranspiration from vegetated area is a basic tool to compute water balances and to estimate water availability and requirements and also to estimate agroclimatic and hydrologic indices. During the last 60 year several methods and models to measure the water flux in agroecosystems has been developed, the aim of this first part of the review is to make a review from de mass balance methods and models in the water flux estimation and the application of the two steps model and the direct transpiration measurements techniques. This chapter provides a revision of these methods and model with special application to crops and covered areas.

KEYWORDS

- aerodynamic resistance
- calibration
- clumping model
- combination model
- crop coefficients
- cumulative growing degree days
- drainage Lysimeter

- eddy covariance method
- energy balances
- evapotranspiration
- fetch requirements
- fluxes
- humidity and temperature method
- latent heat fluxes
- leaf Area Index [LAI]
- micrometeorological method
- Pristley–Taylor model
- resistances
- sensible heat flux
- Shulttleworth-Wallace model
- soil heat flux
- stability correction
- stomatic resistance
- surfaces resistance
- the Bowen ratio energy balance
- the infrared surface temperature method
- the Penman-Monteith reference evapotranspiration method
- the Penman-Monteith general evapotranspiration method
- water stress coefficient
- water vapor deficit
- water vapor flux
- weighing lysimeter

REFERENCES

1. Alves, I.; Perrier, A.; Pereira, L. S.; Aerodynamic and surface resistant of complete cover crops: how good is the "big leaf"? *Trans. Am. Soc. Agric. Eng.;* **1998,** *41(2),* 345–351.
2. Alves, I.; Frontes, J. C.; Pereira, L. S. Evapotranspiration estimation from infrared surface temperature. i: the performance of the flux equation. *Trans. Am. Soc. Agric. Eng.;* **2000,** *43(3),* 591–598.
3. Alves. L.; Pereira, L. S.; Modelling surface resistance from climatic variables? Agric. Water Manag.; **2000,** *42,* 371–385.
4. Allen, G. R.; Cuenca, H. R.; Jensen, E. M.; Blatchley K. R.; Erpenbeck M. J. **1990,** *Evapotranspiration and irrigation water requirements.* American Society of Civil Engineers. New York (USA). 332p.

5. Allen, G. R.; Pereira, S. L.; Raes, D.; Smith, M. **1998,** *Crop Evapotranspiration: Guidelines for Computing Crop Water Requirements.* Food and Agricultural Organization of the United Nations (FAO) Report 56. Rome. 300p.

6. Anandristakis, M.; Liakatas, A.; Kerkides, P.; Rizos, S.; Gavanonis, J.; Poulavassilis. A.; Crop water requirements model tested for crops grown in Greece. Agricultural. Water Management, **2000,** *42,* 371–385.

7. Angelocci, L. R.; BruninI, O.; Magalhaes, A. C. **1983,** Variation of stomatal resistance and water vapor diffusion associated with the water energy status in young coffee farm. 18 pages. In: Ninth Water – Plant Symposium at Federal University of Vicosa, Brazil. July 25–28.

8. Anthoni, A.; Unsworth, M.; Law, B.; Irvine, J.; Baldochi, D.; Kolle, O.; Knohl, A.; Schulze, E. D. **2001,** Comparison of open-path and closed-path eddy covariance system. Max-Planck Institute for Biogeochemie, Jena, D-07745, Germany. Pages 2.

9. Baldocchi, D.; Hicks, B.; Meyers, T. P. Measuring biosphereatmosphere exchanges of biologically related gases with micrometeorological methods. Ecology **1988,** *69(5),* 1331–1340.

10. Baldocchi, D.; Luxmoore, R. J.; Hatfield, J. L. Discerning the forest from the trees: an assay on scaling canopy stomatal conductance. Agric. For. Meteorol. **1991,** *54,* 197–226.

11. Bowen, I. S. **1926,** The ratio of heat losses by conduction and by evaporation from any water surface. Physical Review. *27,* 779–788.

12. Brenner, A. I.; Incoll, L. D. The effect of clumping and stomatal response on evaporation from sparsely vegetation shrunblands. Agric. For. Meteorol.; **1997,** *84,* 187–205.

13. Brisson, N.; Itier, B.; L'Hotel, J. C.; Lorendeau, J. Y. Parameterization of the Shuttleworth-Wallace model to estimate daily maximum transpiration for use in crop models. Ecological Modeling, **1998,** *107,* 159–169.

14. Brown, P. A.; Mancino, C. F.; Joung, M. H.; Thompson, T. L, Wierenga P. J.; Kopec, D. M. Penman Monteith crop coefficients for use with desert turf system. Crop Sci.; **2001,** *41,* 1197–1206.

15. Campbell Scientific, INC (1987–1998). Bowen Ratio Instrumentation Instruction Manuel. 23 p. IN: www.Campbell.com

16. Campbell, G. S. **1977,** *An Introduction to Environmental Biophysics.* Springer-Verlag, New York- USA.

17. Campbell, G. S. **1985,** *Soil physics with basic transport models for soil-plant systems.* Dev. Soil Sci. No. 14. 150 pp. Elsevier.

18. Camilo, P. J.; Gurney, R. J. **1986,** A resistance parameter for bare soil evaporation models. Soil Sci.; *141,* 95–106.

19. Chang, J. H. **1968,** *Climate and Agriculture and Ecological Survey.* University of Hawaii. Ed. Aldine Publishing Company. Chicago. (USA).304p.

20. Coleman, E. A. **1946,** A laboratory of lysimetric drainage under controlled soil moisture tension. Soil Sci.; *62,* 365–382.

21. Conroy, J. W.; Wu.J.; Elliot, W. **2003,** Modification of the evapotranspiration routines in the WEPP model: Part I. ASAE Annual International Meeting. Las Vegas Nevada, USA: 27 to 30 July. 16p. In: http: //www.pubs.asce.org/WWfidisplay.cgi?8801815

22. Doorenbos, J.; Pruitt, W. O. **1977,** *Guidelines for Predicting Crop Water Requirements.* Food and Agricultural Organization of the United Nations (FAO). Publication No. 24. Rome. 300p.

23. Dugas, W. A.; Dan, R.; Ritcchie, J. T. **1985,** A weighing lysimeter for evapotranspiration and root measurements. Agron. J.; *77,* 821–825.

24. Farahami, H. J.; Bausch, W. C. Performance of evapotranspiration models for maize-bare soil to close canopy. Trans. ASAE, **1995,** *38,* 1049–1059.

25. Fank, J. Monolithic field Lysimeter-a precize tool to close the gap between laboratory and field scaled investigations. Geophysical Research Abstracts, **2008,** *10,* 8.

26. Flint, A. L.; Child, S. W. Use the Pristley-Taylor evaporation equation for soil water limited conditions in a small forest clear-cut. Agric. For. Meteorol.; **1991,** *56,* 247–260.

27. Frank, A. B. Evapotranspiration from northern semiarid grasslands. Agron. J.; **2003,** *95,* 1504–1509.

28. Fritschen, J. L.; Fritschen, L. C. **2005,** Bowen ratio energy balance method. In: *Micrometeorology in Agric. Sys.* Agronomy Monograph no, 47. American Society of Agronomy, Crop Sci. Society of America, Soil Sci. Society of America, Madison-W1–53711, USA.; 397–405.

29. Grago, R. D.; Brutsaert, W. **1992,** A comparison of several evaporation equations. Water Resources Research, *32,* 951–954.

30. Grago, R. D. Comparison of the evaporative fraction and the Pristley-Taylor method for parameterization day time evaporation. Water Resources Research, **1996,** *32,* 1403–1409.

31. Grismer, M. E.; Orang, M.; Snyder, R.; Matyac. R. Pan evaporation to reference evapotranspiration conversion methods. Journal of Irrigation and Drainage Engineering, **2002,** *128(3),* 180–184.

32. Ham, M. J.; Heilman, L. J. Aerodynamic and surface resistance affecting energy transport in a sparse crop. Agric. For. Meteorol.; **1991,** *53,* 267–284.

33. Ham, M. J.; Heilman, L. J. Experimental Test of Density and Energy-Balance Corrections on Carbon Dioxide Flux as Measured Using Open-Path Eddy Covariance. Agron. J.; **2003,** *95,* 1393–1403.

34. Hanson, R. B.; May, D. M. **2004,** Crop coefficients for drip-irrigated processing tomatoes. In: ASAE/CSAE Annual International Meeting. Fairmont Chateau Laurier, The Westing, Government Centre, Ottawa, Ontario, Canada.12p.

35. Harmsen, E. W.; Gonzaléz A.; Winter. J. A.; Re-evaluation of Pan Evaporation Coefficients at Seven Locations in Puerto Rico, By Agric. Univ. P. R.; **2004,** *88(3–4): 109–122.*

36. Harmen, W. W.; Ramirez, B. Gonzalez, V. H.:, E. J.; Dukes, D. M.; Jia, X. **2006,** Estimation of short-term actual crop evapotranspiration. Proceedings of the 42nd Annual Meeting of the Caribbean Food Crops Society, July 9–14, **2006,** San Juan, Puerto Rico. Pages 12.

37. Harmsen, E. W.; Ramirez, B. V. H.; Dukes, Jia, M. D.; X.; Perez, A. L.; Vasquez, R. A ground-base method for calibrating remotely sensed temperature for use in estimating evapotranspiration. WSEAS Transactions and Environment and Development, **2009,** *1(5),* 13–23.

38. Hatfield, J. L.; Allen, G. R. Evapotranspiration estimates under deficient water supplies. Journal of Irrigation and Drainage. Engineering, **1996,** *122,* 301–308.

39. Harris, P. P.; Huntingford, C.; Cox, P. M.; Gash, J. H. C.; Malhi, Y. Effect of soil moisture on canopy conductance of Amazonian rainforest. Agric. For. Meteorol.; **2004,** *122,* 215–227.

40. Heilman, J. L.; Brtittin, C. L.; Neale, C. M. U. **1989,** Fetch requirements from Bowen ratio measurements of latent and sensible heat fluxes. Agric. For. Meteorol.; *44,* 261–273.

41. Howell, T. A. **2005,** Lysimetry. In: *Soil and the Environment.* Edited by Elsevier. Pages 379–386.

42. Howell, Steiner, T. A.; Scheider, J. L.;, A. D.; Evett, S. R.; Tolk. A. J. **1994,** Evapotranspiration of irrigated winter wheat, sorghum and corn. ASAE Paper No.94–2081, ASAE, St. Joseph, MI

43. Horst, T. W. **1997,** A simple formula for attenuation of eddy covariance flux measurements. Boundary Layer Meteorology, *82,* 219–233.

44. Hunsaker, D. J. **1999,** Basal crop coefficients and water use for early maturity cotton. Trans. American Society of Agricultural Engineers, *42(4),* 927–936.

45. Hunsaker, D. J.; Pinter, P. J. Jr. Cai, H. Alfalfa basal crop coefficients for FAO-56 procedures in the desert regions of the south western Trans, U. S.; American Society of Agricultural Engineers, **2002,** *45(6),* 1799–1815.

46. Jarvis, P. G. The interpretations of the variations in leaf water potential and stomatal conductance found in canopies in the field. Phil Transaction Royal Society London Bulletin, **1976,** *273,* 593–610.

47. Jhonson, R. S.; Williams, L. E.; Ayars, J. E.; Trout. T. J. **2005,** Weighing lysimeter aid study of water relations in trees and vine crop. Calif. Agric.; *59(2),* 133–136.

48. Kamal, H. A.; Hatfield, L. J. Canopy resistance as affected soil and meteorological factors in potato. Agron. J.; **2004,** *96,* 978–985.

49. Kaimal, J. C.; Wyngaard, Izumi, J. C.; Y.; Cote, O. R. **1972,** Deriving power spectra from a three-component sonic anemometer. Journal Applied Meteorology, *7,* 827–837.

50. Katerji, N.; Perrier, A. **1983,** Modelling the real evapotranspiration in alfalfa crop: role of the crop coefficients. Agronomie, *3(6), 513–521* (in: French).

51. Katerji, N.; Perrier, A. **1985,** Determinations of the canopy resistance to the water vapor diffusion of the six canopy cover. Agricultural Meteorology, *34,* 105–120 (in: French).

52. Katerji, N.; Rana, G. **2006,** Modelling evapotranspiration of six irrigated crops under Mediterranean climate conditions. Agricultural and Forest Meterology, *138,* 142–155.

53. Kirkham, M. B.; Redelfs, M. S.; Stone, L. R.; Kanemasu, E. T. Comparison of water status and evapotranspiration of six row crops. Field Crops Res.; **1985,** *10,* 257–268.

54. Kjelgaard, J. F.; Stokes, C. O. Evaluating surface resistance for estimating corn and potato evapotranspiration with the Penman-Monteith model. Trans. American Society of Agricultural Engineers, **2001,** *44(4),* 797–805.

55. Kjelgaard, J. K.; Stockle, C. O.; Villar Mir, J. M.; Evans, R. G.; Campbell, G. S. Evaluation methods to estimate corn evapotranspiration from short-time interval weather data. Transaction of ASAE, **1994,** *37(6),* 1825–1833,

56. Kosugi, Y.; Katsuyama. M. **2007,** Evapotranspiration over Japanese cypress forest II.; Comparison of the eddy covariance and water budget methods. Journal of Hydrology, *334,* 305–311.

57. Laubach, J.; McNauhton A spectrum-independent procedure for correcting eddy flux measurements with separated sensors. Boundary Layer Meteorology, **1999,***89,* 445–467.

58. Lloyd, W. G. **1992,** Bowen-Ratio measurements. In: *Evapotranspiration measurements of native vegetation, Owens Valley, California.* U. S Geological Survey, Water Resources Investigations Report 91–4159. Pages 5–18

59. López-Urrea, R.; Martin de Santa Olalla, F.; Febeiro, C.; Moratalla. A. **2006,** Testing evapotranspiration equations using lysimeter observations in a semiarid climate. Agric. Water Manag.; 85(1–2): 15–26.

60. Loos, C.; S.; Gainyler.; E.; Priesack. **2007,** Assessment of water balance simulations for large-scale weighing lysimeters. Journal of Hydrology, *335(3–4): 259–270.*

61. Madeiros, G. A.; Arruda, F. B.; Sakai, E.; Fujiwara, M. **2001,** The influence of the crop canopy on evapotranspiration and crop coefficients of beans (Phaseolus vulgaris L.). Agric. Water Manag.; *49,* 211–234.

62. Malone, R. W.; Stewardson, D. J.; Bonta, J. V.; Nelsen, T. **1999,** Calibration and quality control of the coshocton weighing lysimeters. Transaction American Society of Agricultural Engineers, *42(3),* 701–712.

63. Massman, W. J. **2000,** A simple method for estimate frequency response corrections for eddy covariance systems. Agric. For. Meteorol.; *104,* 185–198.

64. Massman, W. J.; Lee, X. **2002,** Eddy covariance flux corrections and uncertainties in long-term studies of carbon and energy exchange. Agric. For. Meteorol.; *113,* 121–144.

65. Meyers, P. T.; Baldocchi, D. D. **2005,** Current micrometeorological flux methodologies with applications in agriculture. Pages 381–396. In: *Micrometeorology in Agric. Sys.* Agronomy Monograph # 47. American Society of Agronomy, Crop Sci. Society of America, Soil Sci. Society of America, 677 S.; Segoe Rd.; Madison, W153711, USA.

66. Monteith, L. J. **1965,** Evaporation and the environment. Symp.Soc.Exper.Biol.; *19,* 205–234.

67. Monteith, L. J. **1981,** Evaporation and surface temperature. Quarterly Journal of the Royal Meteorology Society, *107(451):* 1–27.

68. Monteith, J. L. **1995,** Accommodation between transpiring vegetation and the convective boundary layer. Journal of Hydrology, *166,* 251–263.

69. Monteith, L. J. **1997,** Evaporation Models. Pages 197–234. *In: Agric. Sys. Modelling and Simulation,* Edited by Robert M.; Peart and R.; Bruce Curry. University of Florida, Gainesville, Florida.

70. Monteith, J. L.; Unsworth, M. H. **1990,** *Principles of Environmental Physics.* 2nd ed.; Edward Arnold Publisher. 291 pages.

71. Moore, C. J. **1986,** Frequency response corrections for eddy correlation systems. Boundary Layer Meteorology, *37,* 17–35.

72. Nasab, B. S.; Kashkuli, H. A.; Khaledian, M. R. **2004,** Estimation of crop coefficients for sug-arcane (ratoon) in Haft Tappeh of Iran. ASAE/CSAE Annual International Meeting, Fairmont Chateau Laurier – The Westing, Government Centre Ottawa, Ontario, Canada. 1–4 August. Pages 4.

73. Oguntunde, G. P. **2005,** Whole-plant water use and canopy conductance of cassava under lim-ited available soil water and varying evaporative demand. Plant Soil, *278,* 371–383.

74. Ortega-Farias, S. O.; Antonioletti, R.; Brisson, N. **2004,** Evaluation of the Penman-Monteith model for estimating soybean evapotranspiration. Irrig. Sci.; *23,* 1–9.

75. Ortega-Farias, S. O.; Olioso, A.; Fuentes, S.; Valdes, H. **2006,** Latent heat flux over a furrow-irrigated tomato crop using Penman-Monteith equation with a variable surfaces canopy resis-tance. Agric. Water Manag.; *82,* 421–432

76. Parton, W. J.; Lauenroth, W. K.; Smith, F. M. **1981,** Water loss from a short grass steppe. Agri-cultural Meteorology, *24,* 97–109.

77. Payero, J. O.; Neale, C. M. U.; Wright J. L.; Allen. R. G. Guidelines for validating Bowen ratio data. American Society of Agricultural Engineers, **2003,** *46(4),* 1051–1060.

78. Penman; H. L. **1948,** Natural Evaporation from open water, bare soil and grass. Proceedings of the Royal Society of London. Series, A.; Mathematical and Physical Sciences, 193**1032,**: 120–145.

79. Penman, H. L. **1956,** Evaporation: An introductory survey. Netherlands Journal Agricultural Science, *1,* 9–29, 87–97,151–153.

80. Pereira, S. L.; Perrier, A.; Allen, G.; R.; Alves, I. **1996,** Evapotranspiration: review of concepts and future trends. In: *Evapotranspiration and Irrigation Scheduling.* Proceedings of the Inter-national Conference. November 3–6. San Antonio, TEXAS: 109–115.

81. Pereira, S. L.; Perrier, A.; Allen, G.; R.; Alves, I. Evapotranspiration: Concepts and Future Trends. Journal Irrigation and Drainage Engineering, **1999,** *125(2),* 45–51.

82. Pereira, S. L.; and Alves, I. **2005,** Crop water requirement. Pages 322–334. In: *Soil and Environ-ment,* by Elsevier.

83. Perez, P. J.; Castellvi, F.; Martinez-Cob. A. A simple model for estimation the Bowen ratio from climatic factors for determining latent and sensible heat flux. Agric. For. Meteorol.; **2008,** *148,* 25–37.

84. Priestley, C. H. B.; Taylor, R. J. On the assessment of the surfaces heat and evapotranspiration using large-scale parameters. Monthly Water Review, **1972,** *100(2),* 81–92.

85. Prueger, H. J.; Hatfield, J. L.; Aase, J. K.; Pikul Jr. J. L. Bowen-Ratio comparison with lysimeter evapotranspiration. Agron. J.; **1997,** *89(5),* 730–736.

86. Prueger, H. J.; Kustas, P. W. **2005,** Aerodynamic methods for estimating turbulent fluxes. 407–436. In: *Micrometeorology in Agric. Sys.;* Agronomy Monograph no, 47. American Society of Agronomy, Crop Sci. Society of America, Soil Sci. Society of America, 677 S.; Segoe Rd.; Madison, W1, 53711, USA.

87. Ramírez, B. V. H. **2007,** Plant-Water Relationships for Several Common Bean Genotypes (Phaseolus vulgaris L.) with and Without Drought Stress Conditions. M. Sc. Thesis, Agronomy and Soils Department, University of Puerto Rico – Mayaguez Campus. 190 pages.

88. Ramírez, B. V. H.; Harmsen, W. E.; Porch, G. T. **2008,** Estimation of actual evapotranspiration using measured and calculated values of bulk surfaces resistance. Proceedings of the World Environmental and Water Res. Congress. Honolulu (Hawaii). 10 pages.

89. Ramírez, B. V. H.; Jaramillo, R. A. **2008,** Modifications in the superficial hydrology associated at the cover changes in the colombian's tropical andes. In: XIV Colombian Soil Society Meet-ing, 13 pages. (in Spanish).

90. Rana, G.; Katerji, N.; Mastrorilli, M.; El Moujabber, M El.; Brisson. N.; **1997a.** Validation of a model of actual evapotranspiration for water stressed soybeans. Agric. For. Meteorol.; *86,* 215–224.

91. Rana, G.; Katerji, N.; Mastrorilli, M.; El Moujabber, M. **1997b.** A model for predicting actual evapotranspiration under soil water stress in a Mediterranean region. Theoretical Applied Cli-matology, *56,* 45–55.

92. Ritchie, J. T.; Burnett, E. **1968,** A precision weighing lysimeter for row crop water use studies. Agron. J.; *60(5),* 545–549.

93. Rosenberg, J. N.; Blad, B. L.; Verma, S. B. **1983,** *Microclimate; The Biological Environment.* John Wiley and Sons. 495 pages.

94. Roygard, F. R.; Alley, M. M.; Khosla, R. **2002,** No-Till Corn Yield and Water Balance in the Mid-Atlantic Coastal Plain. Agron. J.; *94,* 612–623.

95. Saugier, B. **1977,** Micrometeorology on crops and grasslands. environmental effects on crop physiology. Proceeding of a Symposium held at Long Ashton Research Station University of Bristol. Academic Press. 39–55.

96. Saugier, B.; Katerji, N. **1991,** Some plant factors controlling evapotranspiration. Agric. For. Meteorol.; *54,* 263–277.

97. Sene, K. J. **1994,** Parameterization for energy transfer from a sparse vine crop. Agric. For. Meteorol.; *71,* 1–18.

98. Seyfried, M. S.; Hanson, C. L.; Murdock, M. D.; Van Vactor. S. **2001,** Long-term lysimeter database, Reynolds Creek Experimental Watershed, Idaho, United States. Water Resources Research, *37(11),* 2853–2856.

99. Schotanus, P.; Nieuwstadt, F. T. M. and de Bruin, H. A. R. **1983,** Temperature measurement with a sonic anemometer and its application to heat and moisture fluxes. Boundary-Layer Meteorology, *50,* 81–93.

100. Shuttleworth, W. J.; and Wallace, J. S. Evaporation from sparse crops-An energy combination theory. Quarterly Journal of the Royal Meteorology Society, **1985,** *111,* 839–855.

101. Steiner, J. L. A.; Howell, T. A.; Schneider, A. D. Lysimeter evaluation of daily potential evapotranspiration models for grain sorghum. Agron. J.; **1991,** *83(1),* 240–247.

102. Suleiman, A.; Crago, R. Hourly and daytime evapotranspiration from grassland using radiometric surface temperature. Agron. J.; **2004,** *96,* 384–390.

103. Sumner, D. M. **2001,** Evapotranspiration from a cypress and pine forest subjected to natural fires in Volusia County, Florida, 1998–99. U. S.; Geological Survey Water-Resources Investigations Report 01–4245. Washington, D. C.

104. Szeicz, C.; and Long, L. F. **1969,** Surface resistance of crop canopies. Water Recourses Research, *5(8),* 622–633.

105. Takhar, H. S.; Rudge, A. J. **1970,** Evaporation studies in standard catchments. Journal of Hydrology, *11(4),* 329–362.

106. Tanner, C. B. **1960,** Energy balance approach to evapotranspiration from crops. Soil Sci. Society of American Proceedings, *24(1),* 1–9.

107. Tanner, B. D.; and Greene, J. P. **1989,** Measurements of sensible heat flux and water vapor fluxes using eddy correlation methods. Final report to U. S.; Army Dugway Proving Grounds. DAAD 09–8, D-0088.

108. Teare, L. D.; H.; Schimmulepfenning.; and Waldren, R. P.; Rainout shelter and drainage lysimeters to quantitatively measure drought stress. Agron. J. **1977,** *65(4),* 544–547.

109. Thom, A. S. **1975,** Momentum, mass and heat exchange of plant communities. 57–109. In: *Vegetation and the atmosphere* edtied by Monteith.; J. L.; Academic Press, New York.

110. Tolk, Howell, A. J.; Steiner, A. T.; L. J.; Krieg, R. D. Aerodynamic characteristics of corn as determined by energy balance techniques. Agron. J.; **1995,** *87,* 464–473.

111. Tomilson, A. E. **1994,** Instrumentation, Methods and preliminary evaluation of evapotranspiration for a grassland in the Arid Lands Ecology Reserve, Beton Country – Washington. May-October **1990,** U. S.; Geological Survey. Water-Resources Investigations Report 93–4081 and Washington State Department of Ecology. Pages 32.

112. Tomilson, A. E. **1997,** Evapotranspiration for three sparce – canopy sites in the Black Rock Valley – Yakima Country – Washington. March **1992,** to October **1995,** U. S.; Geological Survey. Water Resources Investigations Report 96–4207 and Washington State Department of Ecology. Pages 88.

113. Turner, C. N. Measurement and influence of environmental and plant factors on stomatal conductance in the field. Agric. For. Meteorol.; **1991,** *54,* 137–154.

114. Tyagi, N. K.; Sharma, D. K.; Luthra, S. K. Evapotranspiration and crop coefficients of wheat and sorghum. Journal of Irrigation and Drainage Engineering, **2000**, *126(4)*, 215–222.

115. Van Bavel, C. H. M. **1967,** Changes in canopy resistance to water loss from alfalfa induced by soil water depletion. Agric. For. Meteorol.; *4*, 165–176.

116. Vaughan, P. J.; Trout, T. J.; Ayars, J. E. **2007,** A processing method for weighing lysimeter data and comparison to micrometeorological *ETo* predictions. Agric. Water Manag.; 88(1–2): 141–146.

117. Von Unold, G.; Fank, J. Module design of field lysimeter for specific application needs. Water Air Soil Pollution Focus, **2008,** *8,* 233–242

118. Weaver, L. H. **1992,** Eddy-correlation measurements: Evapotranspiration measurements of native vegetation for Ownes Valley – California. U. S.; Geological Survey. Water-Resources Investigations Report 91–4159. Pages 25–33.

119. Webb, E. K.; Pearman, G. I.; Leuning, R. **1980,** Correction of flux measurements for density effects due to heat and water vapor transfer. Quarterly Journal of the Royal Meteorology Society, *106,* 85–100.

120. Wilson, D.; Reginato, R.; Hollet, J. K. **2002,** Evapotranspiration measurements of native vegetation for Ownes Valley – California. U. S.; Geological Survey. Water Resources Investigations Report 91–4159. Pages 1–4.

121. Wolf, A.; Saliendra, N.; Akshalov, K.; Johnson, D. A.; Laca, E. **2008,** Effect of different eddy covariance balance closure and comparisons with the modified Bowen ratio System. Agric. For. Meteorol.; *148,* 942–952.

122. Wright, J. L.; Jensen, M. E. Development and evaluation of evapotranspiration models for irrigation scheduling. Trans ASAE, **1978,** *21(1),* 88–96.

123. Xingfa, H.; Viriyasenakul, V.; Dechao, Z. **1999,** Design, construction and installation of filled-in drainage lysimeter and its applications. Proceeding of 99 International Conference on Agricultural Engineering. Beijing, China, December. II: 162–167.

124. Zhang, Y.; Liu, C.; Yu, Q.; Shen, Y.; Kendy, E.; Kondoh, A.; Tang, C.; Sun, H. Energy fluxes and Pristley-Taylor parameter over winter wheat and maize in the North China Plain. Hydrology Process, **2004,** *18,* 2235–2246.

125. Zhang, B.; Kang, S.; Li, F.; Zhang. L. Comparison of three evapotranspiration models to Bowen ratio-energy balance method for vineyard in an arid desert region of north-west China. Agric. For. Meteorol.; **2008,** *148,* 1629–1640.

CHAPTER 3

DIRECT TRANSPIRATION MEASUREMENT AND ESTIMATION[1,2]

VICTOR H. RAMIREZ

CONTENTS

[1]Victor H Ramirez, Agroclimatology and Crop Science Researcher, National Coffee Research Center (Cenicafe). Chinchina (Caldas, Colombia).
[2]Numbers in parentheses indicate the bibliographical references in the bibliography.

3.1 INTRODUCTION

The transpiration (T) is the water moving from soil layer around the roots to the atmosphere trough the plant by vascular cells and moving forward to the atmosphere as a vapor by the stomas principally. The transpirations is part of the gas interchange of the plant and can be called as the cost that the plant pay by take the carbon dioxide (CO_2) for the photosynthesis. The transpiration is necessary processes in the cycle of live of the plants because by mean of transpiration the plant can take the nutrients from the soil solution, regulate his temperature, and keep growing. In the evapotranspiration (ET) process the transpiration is the water that really uses the plants, but in many cases is not easy separated of the ET complex especially in diverse agroecosystems. In the transpiration process are involve two main factors:

1. **Physical factors:** The physical factors include the available energy (Rn-G), the water vapor pressure deficit (VPD), wind that influence in the VPD, and the stomatical control and available water in the soil layer around the root system; and

2. **Physiological factors:** The physiological factors include the stomas number per unit of area and stomatical control, the leaf area, the height of the plant, and the root depth and density. The stomatical control is defined as the capacity of the stomas to keep open or close as a function of a drought or other physiological stress conditions like wind, salinity, air temperature and others.

The transpiration measurements are useful for direct estimation of water consuming by the plants, as a part of the increasing the water use efficiency in agroecosystems, for estimation of transpiration use efficiency (TUE) especially in plant breeding programs for a biotic stress conditions, for indirect estimation of the gas interchange at physiological level, to evaluate external transpiration regulators, and others. This chapter discusses methods for transpiration measurements directly in field conditions.

3.2 METHODS TO MEASURE TRANSPIRATION

3.2.1 THE SAP FLOW METHOD

The thermal methods based on heat supply to the stem to determine sap flow (SF) represents an advance in the measurements of water consumption by woody plants [5]. The sap flow meter is widely used in the transpiration estimations, in small crops, trees and forest ecosystems [6, 7, 9], according with Bucci et al., [3], the transpiration based on sap flow measurements can be estimated as follow:

$$T = \left(\frac{SF}{BA_i} \right) BA_T \tag{1}$$

where, T is transpiration (mm.day^{-1}), SF is the average daily sap flow per tree (Kg. day^{-1}), BA_i is the mean basal area per tree (cm^2) and BA_T is the total basal area per unit of ground (cm^2.m^{-2}). The SF can be estimated using the method proposed by Sukuratani [11]. A constant power [P_w, watts] related to the insolate section of the steam is divided into the components shown below:

$$P_w = Q_r + Q_v + Q_s + Q_f \tag{2}$$

where, Q_r is the radial heat flow by conduction in the stem; Q_v the heat transported by conduction along the axis of the stem, corresponding to the sum of the heat flows by conduction above (Q_c) and below (Q_b) the heating element; Q_s is the energy stored per time unit in the heated section; Q_f is the heat transported by convection through the moving sap. The Q_r can be determinate as follows:

$$Q_r = K_{sh}\Delta T_r \tag{3}$$

where, K_{sh} is the thermal conductivity of the cork sheath of the radial thermopile (W.K^{-1}), and ΔT_r is the difference of temperature between the inner and the outer surface of sheath, which is determined by the electromotive force generated in the thermopile. The heat flow transported along the axis (Q_v) is calculated as:

$$Q_v = AK_{st}\left[\frac{\Delta T_c + \Delta T_b}{\Delta z}\right] \tag{4}$$

where, A is the cross section area of the heated stem; K_{st} is the thermal conductivity of the stem, assumed as 0.42 W m^{-1} K^{-1} for woody species [10]; ΔT_c and ΔT_p upper and lower temperature differences of the heated stem segment and ΔZ is the difference between pair of junctions fixed just above and below of heating jacket. The sap flow (SF) is estimated below:

$$SF = \frac{Q_f}{C_p \Delta T_{sap}} \tag{5}$$

where, Q_f is determined as the residual of the Eq. (2); In Eq. (2), if the steam diameter is higher than 3 cm, then the Q_s is neglected [5, 13,14]; C_p is the heat capacity of the sap (4.186 kJ kg^{-1} K^{-1}); and ΔT_{sap} is the difference of the sap temperature between the upper and lower limits of the heated segment. Smith and Allen [9] described heat balance method for the trunk sector in the sap flow measurement. The ΔT_{sap} is calculated with Eq. (6).

$$\Delta T_{sap} = \frac{\Delta T_c - \Delta T_b}{2} \tag{6}$$

3.2.2 THE LEAF POROMETER METHOD

The water vapor flux through the stomatas can be measured using the leaf porometer and meteorological measurements based on approach described by Campbell [4] in Eq. (7).

$$T_{ab} = \frac{(X_i - X_a)}{\left(\dfrac{1}{g_s} + \dfrac{1}{g_b}\right)} \tag{7}$$

where, T_{ab} is transpiration from abaxial leaf surfaces (g.m^{-2}.h^{-1}); g_s is the stomatal conductance (m.s^{-1}) measured by the porometer; g_b is the boundary layer conductance (m.s^{-1}); X_i is the leaf absolute humidity, assuming saturation (g.m^{-3}), X_a is the atmospheric absolute humidity. The g_b can be calculated from the sum of the forced and free convection [2].

$$g_{b-forced} = \frac{0.66D^{0.67}u^{0.5}}{d^{0.5}v^{0.17}}$$ (8)

$$g_{b-free} = \frac{0.54D^{0.75}g^{0.25}a^{0.25}\left(T_s - T\right)^{0.25}}{d^{0.25}v^{0.25}}$$ (9)

$$g_b = [g_{b-forced}] + [g_{b-free}]$$ (10)

where, d is the characteristics dimension of the leaf (m); u is the wind speed (m.s^{-1}); $(T_s–T)$ is the difference in temperature (°K) between the leaf and the air. Physical constant at 20°C are: a is a coefficient of thermal expansion of the air (~1/293K^{-1}); D is diffusion coefficient of the air for water vapor (2.4×10–^{5} ^{m2}.s^{-1}); g is the acceleration of the gravity (9.81 m.s^{-1}); v is the kinematic viscosity of the dry air (1. 5×10–^{5} ^{m2}.s^{-1}). In many cases the stomatal density from adaxial leaf surfaces is great, therefore it is necessary to use a correction factor or include adaxial reading for an appropriate transpiration measurement. For a honey mesquite (*Prosopis glandulosa*), Ansley et al. [2] presented following the modified transpiration Eq. (11), where 1.37 is a coefficient derived from adaxial/abaxial g_s ratio.

$$T = \frac{\left(X_i - X_a\right)}{\left(\dfrac{1}{1.37g_s} + \dfrac{1}{g_b}\right)}$$ (11)

3.3   METHODS TO CALCULATE TRANSPIRATION

3.3.1   THE DUAL CROP COEFFICIENTS METHOD

The transpiration, can be estimated using several indirect approaches. In one of the most common method, crop potential transpiration is assumed to be approximately equal to the basal crop transpiration coefficient (ET_{cb}) multiplied by the reference evapotranspiration, as shown in Eq. (12).

$$T \cong ET_{cb} = K_{cb} ET_O$$ (12)

The measurement of the ET_{cb} can be estimated using a method described in the FAO-56 report [1]. FAO-56 method indicates that the the crop coefficient (K_c) can be divided into dual crop coefficient, and the dual crop coefficients are the basal crop coefficient (K_{cb}) and soil evaporation coefficient (K_e). The coefficients K_{bc} and K_e are related to the potential plant transpiration and soil evaporation, respectively. The procedure is summarized in Eqs. (13) and (14).

$$ET_c = (K_{cb} + K_e) \times ET_o$$ (13)

$$K_{cb} = (ET_c/ET_o) - K_e$$ (14)

The K_e is estimated as a function of field surface wetted by irrigation (f_{ew}). K_c in Eq. (15) and the f_{ew} are estimated as a minimum value between the fraction of the soil that is exposed to sunlight and air ventilation and serves as a source of soil evaporation [1–f_c]. The f_c is the soil fraction cover by vegetation and the fraction of soil surface wetted by irrigation or precipitation (f_w) as shown in Eqs. (16) and (17). Both the fractions should be measured throughout the growing season. The Eq. (17) is for a drip irrigation system. If the water source is from a drip irrigation system, the f_w is estimated as a cover crop fraction, Eq. (18). For days with rain, f_w is equal to 1.0.

$$K_e = f_{ew} \times K_c \tag{15}$$

$$f_{ew} = \min [(1-f_c); (f_w)] \tag{16}$$

$$f_{ew} = \min [(1-f_c); (1-0.67 \times f_c) \times (f_w)] \tag{17}$$

$$f_w = 1 - \frac{2}{3} f_c \tag{18}$$

3.3.2 THE EXTINCTION COEFFICIENTS METHOD

The ratio of crop evaporation to crop transpiration without water or other limitations depends on the development stage of the leaf canopy expressed as " δ " the dimensionless fraction of incident beam radiation that penetrates the canopy [15].

$$\delta = \exp(-K.LAI) \tag{19}$$

where, K is the dimensionless canopy extinction coefficient, and LAI is a leaf area index. The evaporation and transpiration can be calculated using Eqs. (20) and (21):

$$E_0 = \delta.ET_0 \tag{20}$$

$$T = (1-\delta).ET_0 \tag{21}$$

3.4 EXAMPLE PROBLEM

One of the utility of the transpiration measurement is the calculation of the transpiration efficiency (TE). The TE is defined as the above ground (aerial) biomass (dry matter of stems, leaves, and fruit) divided by the mass of water transpired during the accumulation of the biomass (T) and was described by the Bierhuizen and Slatyer [12], as shown in Eq. (22).

$$TE = \frac{DM}{T} = \frac{k}{VPD_d} \tag{22}$$

The Eq. (22) indicates that the correlation between DM and T is dependent on k. The k is a species-dependent water-use constant, and VPD is an atmospheric vapor pressure deficit. The VPD is defined as the drying capacity of the air or the driving force for evaporation and transpiration [8]. The T was calculated using the dual crop coefficients approach described in Eq. (12). The Table 1 indicates the differences in the TE for two common bean genotypes and two water conditions (reducted stress and drought stress) during two year trials [8].

The results for TE show relatively large differences between water levels and experiments, with k ranging from 1.0 to 2.2 Pa under drought stress and from 2.2 to 4.2 Pa under nonstress conditions (Table 1). For both years, both genotypes showed the large reductions in k under drought stress conditions. The difference between $k_{nonstress}$ and k_{stress} were: 2.0 Pa in 2006 and 2.3 Pa in 2007 for the common bean genotype Morales, and 1.7 Pa in 2006 and 1.1 Pa in 2007 for the common bean genotype SER 16 [8]. As is indicated in the Table 1, the TE depends of the crop and genotype, the environment and water conditions. For similar water conditions the TE can be uses as indicator of better water uses between genotypes and/or species.

TABLE 1   Estimated transpiration efficiency coefficient (k) for two common bean genotypes [16] under drought stress and reduced stress conditions during two years (2006 and 2007) in Juana Diaz – Puerto Rico [8].

Year	Genotype	Treatment	Mean day time VPD	K
2006	SER 16	Reduced stress	1318.3	4.2
	SER 16	Drought stress	1289.3	2.6
	Morales	Reduced stress	1347.9	3.8
	Morales	Drought stress	1328.8	1.8
2007	SER 16	Reduced stress	1451.9	2.2
	SER 16	Drought stress	1464.2	1.0
	Morales	Reduced stress	1451.9	3.6
	Morales	Drought stress	1498.7	1.3

The transpiration estimation or measurement can be related to crop density and water use by the crop due to transpiration and evaporation from the bare soil. In the case of the coffee crop (*Coffee arabica* L.Var. Colombia), the transpiration and soil evaporation were estimated using the Eqs. (20) and (21) at the same location for different plant densities and leaf area index (LAI). The transpiration fraction (T) with respect to the crop evapotranspiration (ET_c) increased with the plant density (Fig. 1). The ratio, [T/ET_c], was 0.82 in low plant densities (2,500 plants.ha^{-1}) and 0.97 in higher plant densities (near to 10,000 plants.ha^{-1}). When the plant density was increased over 10,000 plants.ha^{-1}, the T/ETc started to reduce. This reduction is related to the biological optimum in yield, which is approximately 10,000 plant.ha^{-1} without water limitations (Fig. 1).

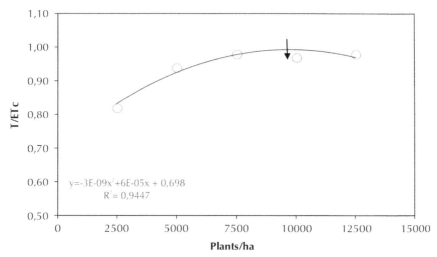

FIGURE 1 Relationship between the transpiration fraction of the crop evapotranspiration with the plant density in a coffee crop (*Coffea arabica* L. Var, Colombia).

3.4 SUMMARY

The transpiration (T) is an important component in the water uses in agricultural system and is related to the crop water use It is an important biophysical component, because the T is a driving force for the gas interchange, the nutrient uptake and the crop growth. The T – process is influenced by physical and physiological factors. The T for crops can be measured or estimated using meteorological methods that are described in this chapter.

KEYWORDS

- basal crop coefficient
- beans
- coffee
- Columbia
- crop coefficient
- drought stress
- extinction coefficient
- genotypes
- leaf area
- root depth
- sap flow
- soil evaporation coefficient
- stomata

- stomatic conductance
- stomatic resistance
- transpiration
- transpiration coefficient
- transpiration use efficiency
- water use efficiency
- water vapor deficit
- wind

REFERENCES

1. Allen, G. R.; Pereira, S. L.; Raes, D.; Smith, M. *Crop evapotranspiration: Guidelines for computing crop water requirements*. Food and Agricultural Organization of the United Nations (FAO) Report 56. Rome. **1998**, 300 pp.

2. Ansley, R. J.; Dugas, W. A.; Hever, M. L.; Trevino, B. A.; Steam flow and porometer measurements of transpiration from honey mesquite (*Prosopis glandulosa*). *J. Experimental Botany,* **1994**, *45(6),* 847–856.

3. Bucci, J. S.; Scholz, F. G.; Goldestein, G.; Hoffman, W. A.; Meinzer, F. C.; Franco, A. C.; Gianbelluca, T.; Miralles-Wilhelm, F. Controls on stand transpiration and soil water utilization along a tree density gradient in a Neotropical savanna. *Agric. For. Meteorol.;* **2008**, *148,* 838–849.

4. Campbell G. S *An Introduction to Environmental Biophysics*. Springer, Berlin, **1977**, 159.

5. Coelho-Filhoi, M. A.; Agelocci, L. R.; de Sauza-Magno, L. F.; Folegatti, M. V.; Bernardes, M. S. Field determination of Young acid lime plants transpiration by the stem heat balance method. Scientia Agricola (Piracicaba, Braz), **2005**, *6(3),* 240–247.

6. Dugas, W. A.; Heuver, M. L.; Mayeux H. S.; Diurnal measurements of honey mesquite transpiration using stem flow gages. J.; Range Manag.; **1992**, *45(1),* 99–102.

7. Dugas, W. A. Comparative measurement of steam flow and transpiration in cotton. *Theor. Appl. Clim.;* **1990**, *42,* 215–221.

8. Ramírez-Builes, V. H.; Porch, T. G.;, Harmsen, E. W.; Genotype differences in water uses efficiency of common bean under drought stress. *Agron. J.;* **2011**, *103,* 1206–1215.

9. Smith, D. M.; Allen, S. J. Measurement of the sap flow in plant stems. *J. Experimental Botany,* **1996**, *47(305),* 1833–1844.

10. Steinberg, S.; van Bavel, C. H. M.; Cornelius, H. M. A gage to measure mass flow rate of sap in stems and trunks of wood plants. *J. Am. Soc. Hortic. Sci.;* **1989**, *114,* 466–472.

11. Sukaratanil, T. A heat balance method for measuring water flux in the stem of intact plants. J.; Agric. Meteorol.; **1981**, *37,* 9–17.

12. Tanner, C. B Sinclair, T. R.; **1983**, Efficient water use in crop production. In: H. M.; Taylor et al. (Eds.) *Limitations to Efficient Water Use in Crop Production*. American Society of Agronomy, Madison, WI.

13. Weibel, F. P.; Vos De, J. A. Transpiration measurements on apple trees with an improved stem heat balance method. Plant Soil, **1994**, *166,* 203–219.

14. Weibel, F.; Boersma, K. An improved stem heat balance method. *Agric. For. Meteorol.;* **1995**, *75,* 191–208.

15. Zhang, Y.; Liu, C.; Yu, Q.; Shen, Y.; Kendy, E.; Kondoh, A.; Tang, C.; Sun, H. Energy fluxes and Pristley-Taylor parameter over winter wheat and maize in the North China Plaint. Hydrology Processes, **2004**, *18,* 2235–2246.

CHAPTER 4

DESIGN OF LYSIMETER FOR TURFGRASS WATER USE[1]

BENJAMIN G. WHERLEY

CONTENTS

[1]Benjamin G. Wherley, Ph.D., Assistant Professor- Turfgrass Physiology & Ecology, Dept. of Soil and Crop Sciences, Texas A&M University, College Station, TX 77843-2474, Email: b-wherley@tamu.edu; T. Sinclair, PhD and M. Dukes, PhD, University of Florida, Gainesville – FL 32611; and G Miller, PhD, North Carolina State University, Raleigh – NC – 27695. Modified and printed with permission from: "*Wherley, B., T. Sinclair, M. Dukes and G. Miller, 2009. Design, construction and field evaluation of a lysimeter system for determining turfgrass water use. Proc. Florida State Hort. Soc., 122:373-377*". © Copyright Florida State Hort. Soc
[2]Numbers in parentheses refer to references in the bibliography.

4.1   INTRODUCTION

Population growth and water concerns are creating an increased need to better understand water use by all plants, including turfgrasses. Lysimeters, which are often used in water-use studies, are containers of soil representing the field environment used to determine the evapotranspiration (ET) of a growing crop or evaporation from bare soil [1]. ET can be estimated using lysimeters, which allow for direct calculation of mass changes due to plant water uptake and soil evaporation [22]. A wide range of lysimeters have been documented in the literature, however there is currently no standard for lysimeter design in turfgrass studies, and consequently, a wide variety of styles and sizes have been used [5]. Lysimeters may be round, square or rectangular, constructed from concrete, steel, fiberglass or plastic, and range in size from 50 mm to 2 m in diameter and from 0.4 to 2 m deep [20].

With regard to turfgrass ET studies, lysimeter volumes ranging from as small as 1.5 L [6] to as large as 20,000 L [21] have been used. Although a smaller lysimeter volume allows for greater ease of handling, replication, and ease of repeated measurements, numerous studies involving a range of species have shown decreased growth and/or water use can occur when plants are grown in too limiting soil volume [13, 15, 16, 19]. Conversely, while very large lysimeter volumes allow for maximal rooting, units cannot be easily weighed and ET is usually estimated indirectly from water balance techniques [4, 7, 12]. Furthermore, the number of replicates that can be easily measured is drastically reduced in studies using large lysimeters [7, 21].

Consequently, a vast number of turfgrass researchers have settled on lysimeters ranging in volume from 6 to 12 L [2, 3, 9–11, 14, 17, 18], because units of this size can be repeatedly lifted from the soil. While not always explicitly detailed, weighing events in these studies typically entail manually lifting units up out of the ground and onto portable scales, or even transporting units to a central location for weighing [8].

This chapter presents the design and construction of an inexpensive lysimeter/weighing system that would be an improvement over past approaches with regard to rooting volume, construction, and weighing technique. The system described here is an improvement over the existing versions reported in the literature because of: (1) the lysimeters provide greater rooting volume than what has generally been provided in past studies, (2) the relatively low cost of materials allows a significant number of replicates to be installed in the field, and (3) the lysimeters can be easily and rapidly weighed by one person in the field (48 measurements in under 3 hours) using a portable tripod and load cell that eliminates the difficulties associated with manually lifting and transporting units to a central location for weighing. It was also of interest to determine whether these lysimeters would produce an environment representative of the soil in ambient plots with respect to soil temperature and volumetric water content.

4.2   MATERIALS AND METHODS

Various materials were considered in designing a lysimeter that would meet our research needs. However, due to the frequent removal from the ground and the need to fabricate a sleeve to maintain a vertical soil wall around the lysimeter, a smooth-sided lysimeter was needed, and polyvinyl chloride (PVC) pipe was chosen.

Final cost for the lysimeter and sleeve design was considered during the prototype stage. Due to the number of lysimeters needed and minimum sizes of available stock materials, many of the stock materials (PVC pipe, foam board, buckets, metal rods, etc.) were purchased as a large unit size or in bulk quantities. Based on the final lysimeter design and materials cost at the time of purchase, the cost per lysimeter was approximately $21. A breakdown of materials and cost for the lysimeter, outside sleeve, and weighing apparatus are presented in Table 1.

TABLE 1 Components and approximate cost of materials used in construction of a single lysimeter, outside sleeve, and cost of items for weighing apparatus (I. D. = inner diameter, PVC = polyvinyl chloride).

Components	Quantity	Cost
Lysimeter materials		$21
10 in I.D. x 13 in length PVC pipe	1	
10 in diameter x 6 mm rigid PVC foam board	1	
½ in PVC threaded plugs	4	
¼ in x 10 ¾ in steel round stock	1	
¼ in x 2 ¼ in steel round stock	6	
4mm x 9 in utility cord	3	
Caulk tube (silicon based)	1	
10 in diameter round landscape fabric (DuPont geotextile)	1	
Outside Sleeve		$5
5 gallon bucket (7 needed to make 6 sleeves)	1	
1/8 in pop rivets	8	
Weighing Apparatus		
Game hoist tripod (Cabela's Inc., Sidney, NE)	1	$135
Load cell (Central Carolina Scale, Sanford, NC)	1	$625
Carabineers	3	$20

I.D. = Internal diameter; PVC = Polyvinly chloride

4.2.1 LYSIMETER CONSTRUCTION

Lysimeters were constructed of PVC pipe and foam board. Schedule 40 PVC of 254 mm (10 in) inside diameter was purchased in standard 6.1 m (20 ft) lengths. The pipe was cut with either a hand-held circular saw with a high-tooth count blade or with an industrial band saw with gates to hold the PVC square (preferred method). The pipe was cut to 330 mm (13 in) lengths to allow adequate room to insert steel rods to support the bottom plate such that the resulting interior depth was 305 mm (12 in). To make the lysimeter bottom, 6-mm thick, rigid PVC foam board was cut using a router or jigsaw after careful measurement using the inside edge of the lysimeter. A router is ideal, as after a template is made with the jigsaw, additional bottoms can be produced rapidly using a router with a flute bit with bearing. Four 17.5 mm (11/16 in) drain holes were drilled in each foam board bottom. These were laid out equally from the center point. The 12.7 mm (1/2 in)-14 NPT (National Pipe Thread) tap was used to thread these holes to accept 12.7 mm (1/2 in) PVC plugs. Four 12.7 mm (1/2 in) plugs were installed.

The foam board bottom was placed into the PVC pipe, resting on steel rod supports (Fig. 1). The steel rods were inserted in the wall of the pipe. Eight 6.4 mm (1/4 in) holes, centered 16 mm (5/8 in) above the bottom edge, were drilled around the bottom edge of PVC tube. Two of these holes were drilled directly across from one another. A 6.4 × 273 mm (1/4 × 10 ¾ in) steel rod was placed into these two opposing holes. The remaining six holes were spaced evenly, three on each side of the steel rod that divided the diameter of the pipe in half. The 6.4 × 57 mm (1/4 × 2 ¼ in) steel rod pieces were placed in these six holes, such that they were flush with the outside surface of the PVC pipe. Once the foam board was placed in the PVC pipe the interior seam was sealed with silicone based caulk. A single layer of landscape fabric was cut to fit the inside diameter of PVC pipe and placed inside the lysimeter to prevent soil from exiting the drainage holes in the foam board bottom.

Three 6.4 mm (1/4 in) holes centered 9.5 mm (3/8 in) below the top edge were drilled around the top edge of PVC tube, evenly spaced around the diameter of the tube (Fig. 1). Braided reinforced nylon cord (4 mm diameter) was threaded through each hole with an overhand knot tied on the inside and a bowline knot tied on the outside of the lysimeter to allow it to be hooked to a carabineer. The carabineer served as the attachment point for lifting the lysimeters out of the ground. After subsequent field testing, some cord attachments in the lysimeters were removed. In place of the rope, steel hooks were made for attaching to the lysimeter via the drilled holes. These hooks were made from the bucket handles that were removed from the 18.9 L (5 gal) buckets used for the outer sleeves. The 'hooked' end works well for attaching to the holes on the lysimeter. These handles were cut at the straight end and curved with pliers to form a closed end.

4.2.2   OUTER SLEEVE CONSTRUCTION

The outer sleeve was constructed using a typical 18.9 L (5 gal) paint bucket with the bottom cut out. The bucket was modified, as the original inside diameter is less than the outside diameter of the lysimeter. Expansion of the bucket was accomplished by pulling the handle out of the bucket and slicing the bucket from top to bottom using a circular saw. Using a router with a flute bit with bearing, the bottoms were cut out of the buckets (a PVC handsaw can also be used).

An additional 140 mm (5½ in) wide vertical bucket slice was required to expand the diameter of the outer sleeve bucket. This piece was riveted onto the outside of the outer sleeve bucket to create an inside diameter that allowed for a 6 mm (1/4 in) gap between the lysimeter and outer sleeve. The inside diameter of

the outer bucket was between 279 and 283 mm (11 and 11–1/8 in). This measurement should be taken on the bottom of the bucket, which will ultimately be the top
when the sleeve is put in the ground. Before riveting the slice onto the outer sleeve
bucket, the protruding rings from the slice added to the bucket were removed using
a PVC handsaw. Using spring clamps, the slice was clamped onto the outer sleeve
bucket, taking care to achieve the proper inside diameter measurement. Four holes
were drilled through both layers on both sides before installing 3.2 mm (1/8 in)
pop rivets. The bottom portion of the paint bucket becomes the top of the outer
sleeve.

4.2.3 FIELD INSTALLATION

Depending on soil type, a two-man auger or tractor-mounted auger may be used to
prepare a hole for the lysimeter plus outer sleeve. For our testing, sod placed in the
lysimeters was removed from the exact spot for the lysimeter before the hole was au-
gured. A piece of 25 cm (10 in) PVC cut to a length of 15 cm (6 in) was used to extract
the sod, so that it fit in the lysimeter. A beveled edge was ground on this piece of PVC
using a hand grinder to allow for easier insertion.

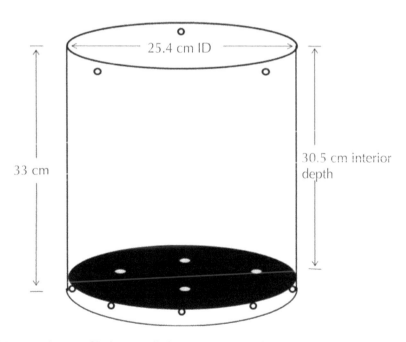

FIGURE 1   Diagram of lysimeter cylinder without the exterior sleeve.

FIGURE 2   Photograph of tripod placement, hoist, load cell, harness, lysimeter, and output display.

Using a piece of wood and a mallet, the PVC was driven into the ground to cut the sod. The soil below the extracted sod was removed to form a hole to insert the lysimeter sleeve. The auger hole was approximately 30 cm (12 in) in diameter and about 38 cm (15 in) deep (depth of the outer sleeve's length). This depth allowed for a base of rock (8–25 mm diameter) under the lysimeter facilitating water percolation below the lyimeter. Once the hole was dug, the outer sleeve was installed with the protruding rings on the bottom. Any gaps in the soil were backfilled around the sleeve with native soil from the site. Rock (8–16 mm diameter) was placed at the bottom inside the lysimeter on top of the geotextile fabric to facilitating drainage from the lysimeter. The lysimeters were filled with desired soil a few centimeters at a time and lightly tamped. After filling, the sod removed from the site was added as the top layer in the lysimeter at the appropriate level. The lysimeter was inserted into the sleeve and its height was adjusted by adding or removing gravel at the bottom of the hole so that it was flush with the surrounding turfgrass. If settling in the hole was experienced later on, it was easily corrected by adding gravel.

4.2.4   LIFTING AND WEIGHING LYSIMETERS

A commercially produced deer hoist (Cabela's Inc, Sidney, NE) was purchased to serve as a tripod to aid in removing the lysimeters during weighing events (Fig. 2). The hoist comes fitted with a crank and cable to raize and lower the lysimeter. A battery-powered load cell (CAS S-Beam, NTEP CoC 96–073A1, Central Carolina Scale, Sanford NC) connecting the hoist cable to the lysimeter was used to weigh the lysimeters. The load cell had a resolution of 5 g, which resulted in a resolution of 0.1 mm water loss for these lysimeters. The S-shaped load cell was attached to the lysimeter via a 3-point rope hitch. The load cell was also connected to a battery- powered digital indicator (Salter 200 SL, Central Carolina Scale, Sanford, NC). When a lysimeter was to be weighed, the hoist with the load cell was centered over the lysimeter. Three carabineers attached to the hitch allowed for quick connect-disconnects to the lysimeter. A shield to deflect wind was fitted around the hoist legs to minimize wind interference but the shield was generally unnecessary.

4.3   RESULTS AND DISCUSSION

4.3.1   UNIFORMITY OF SOIL CONDITIONS: LYSIMETERS VERSUS AMBIENT SOIL

One of the objectives of this study was to design a system that produced a minimal air gap between the lysimeters and plastic sleeves which lined surrounding soil, thus reducing the potential for variation in temperatures between lysimeters and ambient soils. In readings obtained three weeks after installation as well as more recently, soil temperatures (at the 15 cm depth) did not significantly differ between lysimeters and ambient soil (Table 2). Two factors likely contributed to this. First, care was taken to ensure that holes into which lysimeters were installed were only large enough for outer sleeves to fit into, minimizing the amount of bare soil between the lysimeter and surrounding turf immediately following installation. Furthermore, the design, which uses a sleeve that has an inner diameter similar to the outer diameter of the lysimeter, leaves only a 6 mm (1/4 in) air gap between the lysimeter wall and outer sleeve, further minimizing the likelihood of preferential warming or cooling.

Volumetric water content (VWC) of soil was also monitored in lysimeters and ambient plots using time domain reflectometry (Field Scout TDR 300, Spectrum Technologies, Plainfield, IL). Readings, obtained from the 0–20 cm depth 24-hours following rainfall or irrigation, revealed significant differences in VWC between lysimeters and ambient soil. On average, VWC of lysimeter soils was approximately 0.03 m^3 m^{-3} higher than ambient soils (Table 3). This has actually been beneficial in the current project because well-watered treatments are required and drainage holes are sealed.

TABLE 2   Soil temperatures (°C) at 15 cm depth within lysimeters and ambient soil from immediate plots for three dates during the 2008 season. There were no significant differences within species x sampling date based on ANOVA at $\alpha = 0.05$ (n=12).

		3 April	9 April	18 July
Species a	lysimeter	23.0	21.9	30.3
	ambient	23.2	22.1	30.3

TABLE 2 *(Continued)*

		3 April	9 April	18 July
Species b	lysimeter	22.0	21.2	29.7
	ambient	22.0	21.1	29.7
Species c	lysimeter	21.7	21.0	29.2
	ambient	21.6	21.0	29.1
Species d	lysimeter	22.0	21.5	29.6
	ambient	21.9	21.7	29.5
combined	lysimeter	22.2	21.4	29.7
	ambient	22.2	21.5	29.6

TABLE 3   Water content (m^3 m^{-3} soil) of lysimeters and ambient soil from immediate plots for four turfgrass species over three dates during the 2008 season. Measurements were obtained for the 0–25 cm depth using a time domain reflectometry probe. Asterisks denote significant differences within species x sampling date based on ANOVA at $\alpha = 0.05$ (n=12).

		3 April	9 April	18 July
Species a	lysimeter	0.11	0.12 *	0.15 *
	ambient	0.09	0.10	0.11
Species b	lysimeter	0.12 *	0.14	0.14 *
	ambient	0.09	0.10	0.11
Species c	lysimeter	0.13 *	0.14 *	0.16 *
	ambient	0.10	0.10	0.12
Species d	lysimeter	0.10 *	0.11 *	0.14 *
	ambient	0.08	0.08	0.11
combined	lysimeter	0.11 *	0.13 *	0.15 *
	ambient	0.09	0.09	0.11

4.3.2   REPEATABILITY OF LYSIMETER WEIGHT MEASUREMENTS

There was concern that due to the large total mass of the lysimeters combined with the experimental error from using the load cell, that measurement sensitivity would not be sufficient to produce reproducible results. To test the system a study was performed at North Carolina State University Turfgrass Field Research Laboratory, Raleigh, NC, to analyze the reproducibility of the system for measuring lysimeter weights. Eighteen field-installed lysimeters containing dormant Bermuda grass (*Cynodon dactylon* × *C.*

transvaalensis Burtt-Davy) established atop clay loam soil were weighed using the load cell attached to the tripod hoist. Drainage holes in the bottoms of lysimeters had been plugged during the study. Sixty minutes was required to weigh the 18 lysimeters (once). The lysimeters were then immediately weighed again in the same order. The weights (in kg) were recorded and subjected to analysis of group means and paired *t* tests (SAS 9.2, Cary, NC). The mean lysimeter weights were 30,539 ± 224 (S.E.) g for the first weighing and 30,540 ± 223 (S.E.) g for the second (Table 4). Total range in lysimeter weight was 28,990 to 32,170 g. The greatest weight difference noted between the first and second weighing was 25 g, most likely due to some attached mud on the bottom of the lysimeter that was knocked off when reinserting the lysimeter. Otherwise, the mean difference between the two weighings was 5.8 ± 0.8 (S.E.) g, corresponding to the sensitivity of the load cell. Based on analysis of data, no significant difference could be detected between the first and second weighings (P > 0.998), demonstrating the high precision and reproducibility of system and load cell for taking weight measurements.

TABLE 4 Summary of *t* test results analyzing the reproducibility of weight measurements made using the load cell and tripod hoist system (n = 18). Values are means ± standard error.

	Weight 1	Weight 2	Difference (1-2)	P value
	grams			
Minimum	28,990	28,995	--	---
Maximum	32,170	32,165	---	---
Mean	30,539 ± 224	30,540 ± 223	0.83 ± 1.3	0.998

4.4 CONCLUSIONS

These lysimeters have been successfully used for over a year in turfgrass water use studies on both sand and clay soil types. This system is an improvement over the versions previously reported in the literature because of: (1) the increased rooting volume offered relative to past studies, (2) the relatively low cost of materials, allowing a large number of replicates to be installed in the field, and (3) the ease and speed at which portable measurements can be made by a single person in the field. Furthermore, our results demonstrate that the environment produced by the lysimeter is reflective of surrounding soil with regard to temperature. Volumetric water content of lysimeters was slightly higher than surrounding soil; however, this may be advantageous in studies requiring well-watered conditions. We feel this lysimeter system will be of significant benefit to those conducting future studies involving water use.

4.5 SUMMARY

Lysimeters are often used in turfgrass and plant water-use studies; however no detailed description exists for a lysimeter system of moderate volume allowing for rapid, direct measurement of evapotranspiration on a number of replicates. A lysimeter was

developed using 250-mm diameter and 330-mm long polyvinvyl chloride piping resulting in a lysimeter volume of 15.5 L. These lysimeters were installed in the field
by constructing a plastic soil-retention sleeve that was placed in the soil. The sleeve
was matched to the lysimeter diameter so that there was only a 6-mm air gap between
the lysimeter and the sleeve. After turf had filled in around lysimeter edges, there
was no detectable difference in measured soil temperatures between lysimeters and
surrounding plots at the 15 cm depth, however, volumetric soil moisture content of
lysimeters was ~3% higher than that of surrounding soil. The lysimeter was weighed
in the field by positioning a tripod hoist over the lysimeter. A load cell was installed in
the hoist cable assembly so that when the lysimeter was lifted free of the soil-retention
sleeve the weight of the lysimeter could be recorded. The system was shown to provide highly reproducible weight measurement data based on paired t test analysis of
repeated weighing data.

KEYWORDS

- **Bahia grass (Paspalum notatum)**
- **Bermuda grass (Cynodon dactylon × C. transvaalensis)**
- **consumptive water use**
- **evapotranspiration**
- **lysimeters**
- **PVC, polyvinyl chloride**
- **St. Augustine grass (Stenatophrum secundatum)**
- **transpiration**
- **turf coefficient**
- **turfgrass evapotranspiration**
- **warm season turfgrass**
- **Zoysia grass (Zoysia japonica)**

REFERENCES

1. Aboukhaled, A.; Alfaro, A.; Smith, M.; Lysimeters. Food and Agriculture Organization of the United Nations. Rome, Italy. **1982,** 68pp.
2. Aronson, L. J.; Gold, A. J.; Hull, R. J.; Cisar, J. L.; Evapotranspiration of cool-season turf-grasses in the humid north-east. *Agron. J.* **1987,** *79,* 901–905.
3. Beard, J. B.; Green, R. L.; Sifers, S. I.; Evapotranspiration and leaf extension rates of 24 well-watered, turf-type *Cynodon* Genotypes. *HortSci.* **1992,** *27(9),* 986–988.
4. Biran, I.; Bravdo, B.; Bushkin-Harav, I.; Rawitz, E.; Water consumption and growth rate of 11 turfgrasses as affected by mowing height, irrigation frequency, and soil moisture. *Agron. J.* **1981,** *73,* 85–90.
5. Bremer, D. J. Evaluation of lysimeters used in turfgrass evapotranspiration studies using the dual-probe heat-pulse technique. *Agron. J.* **2003,** *95,* 1625–1632.
6. DaCosta, M.; B.; Huang. Deficit irrigation effects on water use characteristics of bentgrass spe-cies. *Crop Sci.* **2006,** *46,* 1779–1786.

7. Devitt, D. A.; Morris, R. L.; Bowman, D. C.; Evapotranspiration, crop coefficients, and leaching fractions of irrigated desert turfgrass systems. *Agron. J.* **1992,** *84,* 717–723.

8. Feldhake, C. M.; Danielson, R. E.; Butler, J. D.; Turfgrass evapotranspiration. I.; Factors influencing rate in urban environments. *Agron. J.* **1983,** *75,* 824–830.

9. Feldhake, C. M.; Danielson, R. E.; Butler, J. D.; Turfgrass evapotranspiration. II.; Responses to deficit irrigation. *Agron. J.* **1984,** *76,* 85–89.

10. Johns, D.; Beard, J. B.; van Bavel, C. H. M. Resistances to evapotranspiration from a St. Augustine grass turf canopy. *Agron. J.* **1983,** *75(3),* 419–422.

11. Kim, K. S.; Beard, J. B.; Comparative turfgrass evapotranspiration rates and associated plant morphological characteristics. *Crop Sci.* **1988,** *28,* 328–331.

12. Kneebone, W. R.; Pepper, I. L.; Luxury water use by bermudagrass turf. *Agron. J.* **1984,** *76,* 999–1002.

13. Peterson, C. M.; B.; Klepper, Pumphrey, F. V.; Rickman, R. W.; Restricted rooting decreases tillering and growth of winter wheat. *Agron. J.* **1984,** *76,* 861–863.

14. Qian, Y. L.; Fry, J. D.; Wiest, S. C.; Upham, W. S.; Estimating turfgrass evapotranspiration using atmometers and the Penman-Monteith model. Crop Sci. **1996,** *36,* 699–704.

15. Ray, J. D.; Sinclair, T. R.; The effect of pot size on growth and transpiration of maize and soybean during water deficit stress. *J. Exp. Bot.* **1998,** *49(325),* 1381–1386.

16. Robbins, N. S.; Pharr, D. M.; Effect of restricted root growth on carbohydrate metabolism and whole plant growth of *Cucumis sativus* L.; *Plant Phys.* **1988,** *87,* 409–413.

17. Rogowski, A. S.; Jacoby, E. L.; Jr. Assessment of water loss patterns with lysimeters. *Agron. J.* **1977,** *69,* 419–424.

18. Salaiz, T. A.; Shearman, R. C.; Riordan, T. P.; Kinbacher, E. J.; Creeping bentgrass cultivar water use and rooting responses.   *Crop Sci.* **1991,** *31,* 1331–1334.

19. Townend, J.; Dickinson, A. L.; A comparison of rooting environments in containers of different sizes. *Plant Soil* **1995,** *175,* 139–146.

20. Winton, K.; Weber, J.; A review of field lysimeter studies to describe the environmental fate of pesticides. *Weed Tech.* **1996,** *10(1),* 202–209.

21. Young, M. H.; Wierenga, P. J.; Mancino, C. F.; Large weighing lysimeters for water use and deep percolation studies. *Soil Sci.* **1996,** *161(8),* 491–501.

22. Young, M. H.; Wierenga, P. J.; Mancino, C. F.; Monitoring near-surface soil water storage in turfgrass using time domain reflectometry and weighing lysimeters. *Soil Sci. Soc. Am. J.* **1997,** *61,* 1138–1146.

CHAPTER 5

EVAPOTRANSPIRATION: METEOROLOGICAL METHODS[1]

MEGH R. GOYAL

CONTENTS

[1]Megh R Goyal, PhD, P.E., Retired Professor in Agricultural and Biomedical Engineering, University of Puerto Rico – Mayaguez Campus, USA; Senior Technical Editor, Apple Academic Press Inc., PO Box 86, Rincon – Puerto Rico – 00677. Email: <goyalmegh@gmail.com>. Printed with permission from: Goyal, Megh R., 2012. Evapotranspiration. Chapter 2, pages 31-70. In: *Management of Drip/Trickle or Micro Irrigation* edited by Megh R. Goyal. New Jersey, USA: © Copyright Apple Academic Press Inc.,

5.1 INTRODUCTION

Due to increased agricultural production, irrigated land has increased in the arid and subhumid zones around the world. Agriculture has started to compete for water use with industries, municipalities and other sectors. This increasing demand along with increments in water and energy costs have absolutely made necessary to develop new technologies for the adequate management of water. The intelligent use of water for crops requires understanding of evapotranspiration processes.

Evapotranspiration (ET) is a combination of two processes: Evaporation and transpiration. Evaporation is a physical process that involves conversion of liquid water into water vapor into the atmosphere (Fig. 1). Evaporation of water into the atmosphere occurs on the surface of rivers, lakes, soils and vegetation. Transpiration is basically a process of evaporation. The transpiration is a physical process that involves flow of liquid water from the soil (root zone) to the surface of leaves/ branches and trunk; and conversion of liquid water from the plant tissue into water vapors into the atmosphere.

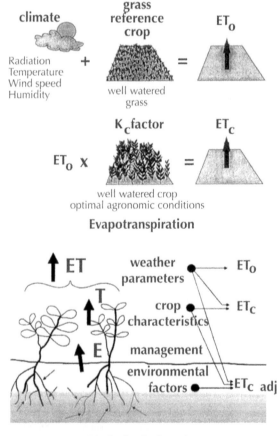

FIGURE 1   Evapotranspiration process.

The water evaporates from the leaves and plant tissue, and the resultant water vapor diffuses into atmosphere through the stomates. An energy gradient is created during the evaporation of water, which causes the water movement into and out of the plant stomates. In the majority of green plants, stomates remain open during the day and stay closed during the night. If the soil is too dry, the stomates will remain closed during the day in order to slow down the transpiration.

Evaporation, transpiration and evapotranspiration processes are important for estimating crop irrigation requirements and for irrigation scheduling [4]. To determine crop irrigation requirements, it is necessary to estimate ET by on site measurements or by using meteorological data. On site measurements are very costly and are mostly employed to calibrate ET methods using climatological data. There are number of proposed equations that require meteorological data and are used to estimate the ET for periods of one day or more. All of these equations are empirical in nature. The simplest methods require only data about average temperature of air, length of the day and the crop. Other equations require daily radiation data, temperature, vapor pressure and wind velocity. Figure 2 shows the instruments for a weather station. None of the equations should be rejected, because the data is not available. Not all methods are equally precize and reliable for different regions of the world. There is no unique meteorological method that can be universally adequate under all climatological conditions.

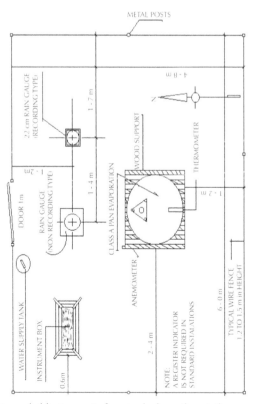

FIGURE 2   Recommended instruments for a typical weather station.

5.2   POTENTIAL EVAPOTRANSPIRATION

Potential evapotranspiration (PET) is a water loss from the soil surface completely covered by vegetation. Meteorological processes determine the evapotranspiration of a crop [7]. Closing of stomates and reduction in transpiration are usually important only under drought or under stress conditions of a plant. The evapotranspiration depends on three factors: (1) Vegetation, (2) Water availability in the soil and (3) Behavior of stomates. Vegetation affects the ET in various forms. It affects ability of soil surface to reflect light. The vegetation changes amount of absorbed energy by the soil surface. Soil properties, including soil moisture, also affect the amount of energy that flows through the soil. The height and density of vegetation influence efficiency of the turbulent heat interchange and the water vapor of the foliage [15].

Changes in the soil moisture affect direct evaporation from the soil surface and available water to the plants. As the plants are water stressed, stomates close resulting in the reduction of a water loss and CO_2 absorption. This is a factor that is not considered in the potential evapotranspiration equation. Under normal conditions (with enough water), there is a variation among stomates of different plant species. Besides, these variations are usually small and the concept of PET results useful for the majority of crop species with complete foliage [10, 13].

5.3   MATHEMATICAL MODELS FOR PET

There are different methods to estimate or measure the ET and the PET. The precision and reliability vary from one method to another, some provide only an approximation. Each technique has been developed with the available climatological data to estimate the ET. The direct measurements of PET are expensive and are only used for local calibration of a given method using climatological data. The most frequently used techniques are: Hydrological method or water balance method, climatic methods and micro meteorological methods. Many investigators have modified the equations that are already established. For example, one may find modification of Blaney-Criddle formula, Hargreaves-Samani, Class A pan evaporation, etc. Allen [1] investigated 13 variations of the Penman equation. He found that the Penman-Monteith formula was most precize. Modified equations are actually recommended by the FAO and the USDA – Soil Conservation Service. Most of investigators agree that Penman, Class A Pan Evaporation, Blaney-Criddle and Hargreaves-Samani equations, can be trusted. High precision can be obtained with local calibration of a given method. Every researcher has its preferred formula that may give good results. Hargreaves and Samani [7] a simplest and practical formulae. I can add that, "There is no evidence of a superior method." Allen and Pruitt [2] presented the FAO modified Blaney-Criddle method, which involves relatively easy calculations and give precize estimates of PET (when it is calibrated for local conditions). Every researcher has preference. However, each formula, depending where it was evaluated, may or may not result in the first or the last place.

5.3.1   HYDROLOGIC METHOD OR WATER BALANCE

This technique employs periodic determination of rainfall, irrigation, drainage, and soil moisture data. The hydrologic method uses water balance equation:

$$PI + SW - RO - D - ET = 0 \tag{1}$$

where, PI =Precipitation and/or irrigation; RO = Runoff; D = Deep percolation; SW = Change in the soil moisture, and ET = Evapotranspiration. In Eq. (1), every variable can be measured with precision with the lysimeters (Figs. 3a, 3b, and 3c; Chapter 3). The ET can be calculated as residual, knowing values of all other parameters (Fig. 4).

5.3.2 CLIMATIC METHODS

Using weather data [3], numerous equations have been proposed. Also, numerous modifications have been made to the available formulae for application to a particular region.

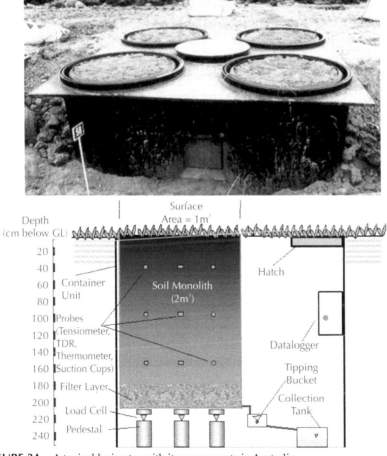

FIGURE 3A   A typical lysimeter with its components in Australia.

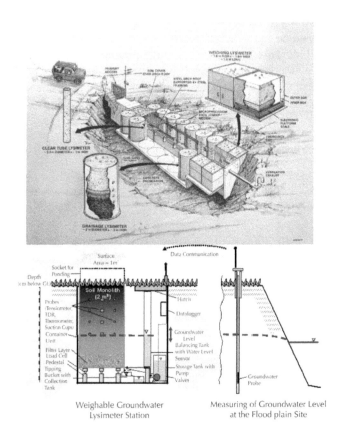

Weighable Groundwater
Lysimeter Station

Measuring of Groundwater Level
at the Flood plain Site

Schematic drawing of lysimeter

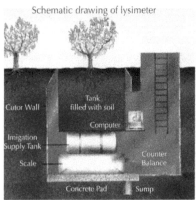

At Universiy of California, Davis

FIGURE 3B   A field lysimetertest facility.

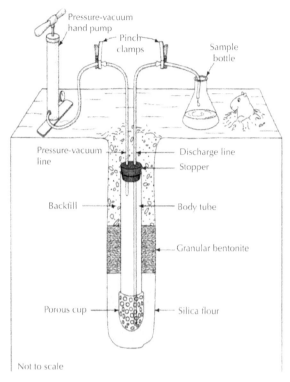

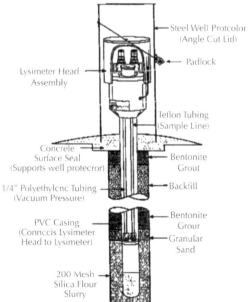

FIGURE 3C   PVC tube lysimeter.

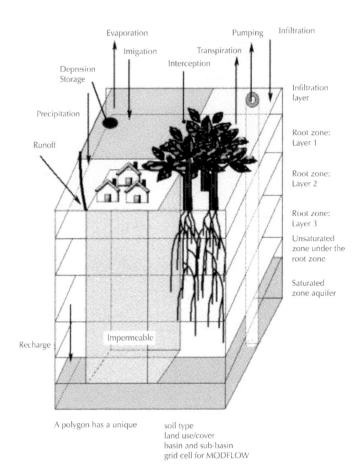

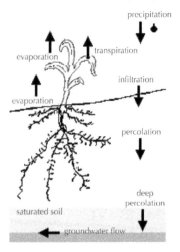

FIGURE 4 Water balance method.

5.3.2.1   PENMAN

The Penman formula was presented in 1948. It employs net radiation, air temperature, wind velocity and deficit in the vapor pressure. He gave the following equation:

$$PET = [R_n/(a + b\,E_a)] \div [c + b] \qquad (2)$$

$$E_a = 0.263\,((e_s - ed) \times (0.5 + 0.0062 \times u_2)) \qquad (2a)$$

where, PET = Daily potential evapotranspiration, mm/day; C = Slope of saturated air vapor pressure curve (See Appendix F), mb/°C; R_n = Net radiation, cal/cm^2 day; a = Latent heat of vaporization of water =(59.59–0.055 (T)), cal/cm^2-mm = 58 cal/cm^2-mm at 29°C; E_a= Average vapor pressure of air, mb = $((e_{max} - e_{min}) / 2)$; e_d = Vapor pressure of air at minimum air temperature, mb; u_2 = Wind velocity a height of 2 meters, km/ day; b = Psychrometric constant = 0.66, mb/°C; $T = ((T_{max} - T_{min}) / 2)$, in degrees°C; $(e_{max} - e_{min})$ = Difference between maximum and minimum vapor pressure of air vapor, mb; and $(T_{max} - T_{min})$ = Difference between maximum and minimum daily temperature,°C.

5.3.2.2   PENMAN MODIFIED BY MONTEITH [14]

After modification, the resultant equation is as follows:

$$LE = - [s\,(R_n - S) + Pa \times Cp\,(e_s - e_a)\,/\,r_a] \div \{((s + b) \times (r_a + r_c))\,/\,r_a\} \qquad (3)$$

where, LE = Latent heat flow; R_n = Net radiation; S = Soil heat flow; Cp = Air specific energy at constant pressure; s = Slope of saturated vapor pressure of air, at air average temperature of wet bulb thermometer; Pa = Density of humid air; e_s= Saturated vapor pressure of water; e_a = Partial water vapor pressure of air; r_a= Air resistance; r_c= Leaf resistance; b = Psychrometric constant (see appendix D).This method has been successfully used to estimate the ET of a crop. The Penman-Monteith equation is limited to research work (experimentation) since the r_a and r_c data are not always available.

5.3.2.3   PENMAN MODIFIED BY DOORENBOS AND PRUITT [4]

$$PET = c \times [\,W \times R_{n + (1 - W)} \times F(u) \times (e_a - e_d)] \qquad (4)$$

where, PET = Potential evapotranspiration, mm/day; W = A factor related to temperature and elevation; R_n= Net radiation, mm/day; F(u) = Wind related function; $(e_a - e_d)$ = Difference between the saturated vapor pressure of air at average temperature and vapor pressure of air, mb; and c = A correction factor. The Penman formula is not popular, because it needs data that is not available at majority of the weather stations. Estimations of PET using Penman formula can be complex. The equation contains too many components, which should be measured or estimated, when data is not available.

5.3.2.4   THORNTHWAITE

This methoduses monthly average temperature and the length of the day.

$$PET = 16\,L_d\,[\,10\,T\,/\,I\,]^{\,a} \qquad (5)$$

where, PET = Estimated evapotranspiration for 30 days, mm; L_d= Hours of the day divided by 12; T = Average monthly temperature, °C; and

$$A = [(6.75 \times 10^{-7}\,I^3) - (7.71 \times 10^5\,I^2) + 0.01792\,I + 0.49239] \qquad (5a)$$

$$I = i_1 + i_2 + \ldots + i_{12}, \qquad (5b)$$

where, $i = [T_m / 5] \times 1.514$

The Thornthwaite method underestimates PET during the summer when maximum radiation of the year occurs. Besides, the application of equation to short periods of time can lead to an error. During short periods, the average temperature is not an adequate measure of the received radiation [10, 14]. During long-terms, the temperature and the ET are similar functions of the net radiation. These are related, when the long periods are considered.

5.3.2.5　BLANEY-CRIDDLE [2, 9, 10, 14]

The original Blaney-Criddle equation was developed to predict the consumptive use of PET in arid climates. This formula uses percentage of monthly sunshine hours and monthly average temperature.

$$PET = K_m F \qquad (6)$$

where, PET = Monthly potential evapotranspiration, mm; K_m= Empirically derived coefficient for the Blaney-Criddle method; T = Monthly average temperature, °C; PD = Monthly percentage daily sunshine hour; and

$$F = \text{Monthly ET factor} = \{[25.4 \times PD \times (1.8\ T + 32)] / 100\} \qquad (6a)$$

This method is easy to use and the necessary data are available. It has been widely used in the Western United States with accurate results, but not the same in Florida, where ET is underestimated during summer months.

5.3.2.6　BLANEY-CRIDDLE MODIFIED BY FAO [4]

$$PET = \{C \times P \times [(0.46 \times T) + 8]\} \qquad (7)$$

where, PET = Potential evapotranspiration, mm/day; T = Monthly average temperature; P = Percentage daily sunshine hours [See Table 1]; and C= Correction factor, which depends on the relative humidity, light hours and wind.Doorenbos and Pruitt [4, 5] recommended individual calculation for each month. They indicated that it may be necessary to increase its value for high elevations.

TABLE 1　Percentage average daily sunshine hours (P) based on annual day light hour for different latitudes.

Latitude, degrees						
North	January	February	March	April	May	June
South*	July	August	September	October	November	December
60	0.15	0.20	0.26	0.32	0.38	0.41
58	0.16	0.21	0.26	0.32	0.37	0.40
56	0.17	0.21	0.26	0.32	0.36	0.39
54	0.18	0.22	0.26	0.31	0.36	0.38
52	0.19	0.22	0.27	0.31	0.35	0.37
50	0.19	0.23	0.27	0.31	0.34	0.36

TABLE 1 *(Continued)*

Latitude, degrees

North	January	February	March	April	May	June
South*	July	August	September	October	November	December
48	0.20	0.23	0.27	0.31	0.34	0.36
46	0.20	0.23	0.27	0.30	0.34	0.35
44	0.21	0.24	0.27	0.30	0.33	0.35
42	0.21	0.24	0.27	0.30	0.33	0.34
40	0.22	0.24	0.27	0.30	0.32	0.34
35	0.23	0.25	0.27	0.29	0.31	0.32
30	0.24	0.25	0.27	0.29	0.31	0.32
25	0.24	0.26	0.27	0.29	0.30	0.31
20	0.25	0.26	0.27	0.28	0.29	0.30
15	0.26	0.27	0.27	0.28	0.29	0.29
10	0.26	0.27	0.27	0.28	0.28	0.29
5	0.27	0.27	0.27	0.28	0.28	0.28
0	0.27	0.27	0.27	0.27	0.27	0.27

Latitude, Degrees

North	July	August	September	October	November	December
South*	Jan	February	March	April	May	June
60	0.40	0.34	0.28	0.22	0.17	0.13
58	0.39	0.34	0.28	0.23	0.18	0.15
56	0.38	0.33	0.28	0.23	0.18	0.16
54	0.37	0.33	0.28	0.23	0.19	0.17
52	0.36	0.33	0.28	0.24	0.20	0.17
50	0.35	0.32	0.28	0.24	0.20	0.18
48	0.35	0.32	0.28	0.24	0.21	0.19
46	0.34	0.32	0.28	0.24	0.21	0.20
44	0.34	0.31	0.28	0.25	0.22	0.20
42	0.33	0.31	0.28	0.25	0.22	0.21
40	0.33	0.31	0.28	0.25	0.22	0.21
35	0.32	0.30	0.28	0.25	0.23	0.22
30	0.31	0.30	0.28	0.26	0.24	0.23
25	0.31	0.29	0.28	0.26	0.25	0.24
20	0.30	0.29	0.28	0.26	0.25	0.25
15	0.29	0.28	0.28	0.27	0.26	0.25
10	0.29	0.28	0.28	0.27	0.26	0.26
5	0.28	0.28	0.28	0.27	0.27	0.27
0	0.27	0.27	0.27	0.27	0.27	0.27

* Southern latitudes have six months of difference as shown in Table 1.

5.3.2.7 BLANEY AND CRIDDLE MODIFIED BY SHIH

$$PET = \{25.4 \times K \times [MR_s \times (1.8\ T + 32)]\} / [TMR_s] \qquad (8)$$

where, PET = Monthly potential evapotranspiration, mm; K = Coefficient for this modified method; MR_s = Monthly solar radiation, cal/cm²; T = Monthly average temperature, °C; and TMR_s = Sum of monthly solar radiation during the year, cal/cm².

5.3.2.8 JENSEN-HAISE
The Jensen-Haise equation [9] resulted from about 3,000 measurements of the ET taken in the Western Regions of the United States for a 35-years period. It is an empirical equation.

$$PET = \{R_s\ (0.025 \times T + 0.08)\} \qquad (9)$$

where, PET = Potential evapotranspiration, mm/day; R_s = Daily total solar radiation, mm of water; T = Average air temperature, °C. This method seriously underestimates ET under conditions of high movements of atmospheric air masses. However, it gives reliable results for calm atmospheres.

5.3.2.9 STEPHENS-STEWART
Stephens-Stewart used solar radiation data. It is similar to the original Jensen-Haise method [9]. The equation is given below:

$$PET = \{0.01476 \times [(T + 4.905) \times MR_s]\}/b \qquad (10)$$

where, PET = Monthly potential evapotranspiration, mm; T = Monthly average temperature, °C; MR_s = Monthly solar radiation, cal/cm²; b = Latent vaporization energy of water = [59.59–0.055 T_m], cal/ cm²ˉmm = 58 cal/cm²ˉmm at 29°C.

5.3.2.10 PAN EVAPORATION
Class A pan instrument is commonly used to measure evaporation. The evaporation pan (Fig. 5) integrates the climate factors and has proven to give accurate estimations of PET. It requires a good service, maintenance and management. Table 2 gives class A pan coefficients under different conditions [4, 5]. The relationship between PET and pan evaporation can be expressed as:

$$PET = K_p \times PE \qquad (11)$$

where, PET = Potential evapotranspiration, mm/day; K_p = Pan coefficient (Table 2); and PE = Class A pan evaporation.

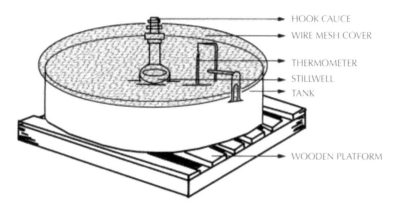

HOOK CAUCE
WIRE MESH COVER
THERMOMETER
STILLWELL
TANK

WOODEN PLATFORM

FIGURE 5   A typical class A pan.

5.3.2.11   HARGREAVES METHOD

Hargreaves method uses a minimum of climatic data. The formula is as below:

$$PET = [MF \times (1.8\ T + 32)] \times CH \tag{12}$$

where, PET = Potential evapotranspiration, mm/month; MF = Monthly factor depending on the latitude; T = Monthly average temperature, °C; and

CH = Correction factor for the relative humidity (HR) = To be used for HR> 64%

$$= 0.166\ [(100 - HR)]^{1/2} \tag{12a}$$

TABLE 2   Pan coefficient (K_p) for the class A pan evaporation under different conditions.

Class A	Condition A			Condition B*		
Pan	**Pan surrounded by grass**			**Pan surrounded by dry uncovered soil**		
Average of HR%	**Low**	**Medium**	**High**		**Low**	**Medium**	**High**
	40	40–70	70		40	40–70	70
Wind**	Distance from			Distance from			
km / day	the green crop,			the dry fallow,			
	m			m			

TABLE 2   *(Continued)*

Light	0	0.55	0.55	0.75	0	0.70	0.80	0.85
175	10	0.65	0.75	0.85	10	0.60	0.70	0.80
	100	9.70	0.80	0.85	100	0.55	0.65	0.75
	1000	0.75	0.85	0.85	1000	0.50	0.60	0.70
Moderate	0	0.50	0.60	0.65	0	0.65	0.75	0.80
175–425	10	0.60	0.70	0.75	10	0.55	0.65*	0.70
	100	0.65	0.75	0.80	100	0.50	0.60	0.65
	1000	0.70	0.80	0.80	1000	0.45	0.55	0.60
Strong	0	0.45	0.50	0.60	0	0.60	0.65	0.70
425–700	10	0.55	0.60	0.65	10	0.50	0.55	0.65
	100	0.60	0.65	0.70	100	0.45	0.45	0.60
	1000	0.65	0.70	0.75	1000	0.40	0.45	0.55
Very	0	0.40	0.45	0.50	0	0.50	0.60	0.65
Strong	10	0.45	0.55	0.60	10	0.45	0.50	0.55
	100	0.50	0.60	0.65	100	0.40	0.45	0.50
	1000	0.55	0.60	0.65	1000	0.35	0.40	0.45

* For areas of extensive uncovered and not developed agricultural soils. Reduce values of K_p by 20% under hot wind conditions and by 5 to 10 % for moderatewind conditions, temperature and humidity.

** Total wind movement in km/day.

The Hargreaves original formula for the PET was based on radiation and temperature as given below:

$$PET = [(0.0135 \times RS)] \times [T + 17.8] \tag{13}$$

where, RS= Solar radiation, mm/day; and T = Average temperature, °C.To estimate solar radiation (RS) using extraterrestrial radiation (RA), Hargreaves and Samani [7, 8] formulated the following equation:

$$RS = K_{rs} \times RA \times TD^{0.50} \tag{13a}$$

where, T = Average temperature, °C; RS = Solar radiation; RA = Extraterrestrial radiation; K_{rs} = Calibration coefficient; and TD = Difference between maximum and minimum temperatures, °C.

5.3.2.12   HARGREAVES AND SAMANI MODIFIED METHOD

Finally after several years of calibration, Eq. (13) was modified as follows:

$$PET = 0.0023 \, R_a \times [T + 17.8] \times (TD)^{0.50} \tag{14}$$

where, PET = Potential evapotranspiration, mm/day; R_a = Extraterrestrial radiation, mm/day; T = Average temperature, °C; and TD = Difference between maximum and minimum temperatures, °C. The Eq. (14) requires only maximum and minimum temperature data. This data is normally available. This formula is precize and reliable.

5.3.2.13   LINACRE METHOD

The Linacre equation is given below:

$$PET = \{700\ T_m\ /\ [100-L_a] + 15\ [T-T_d]\} \div \{80 - T\} \tag{15}$$

where, PET = Potential evapotranspiration, mm; Z = Elevation, m; T = Average temperature, °C; L_a = Latitude, degrees; T_d = Daily average temperature, °C; and

$$T_m = (T_a + 0.0062 \times Z) \tag{15a}$$

The variations in PET values by this formula are 0.3 mm/day based annual data and 1.7 mm/day based on daily data.

5.3.2.14   MAKKINK METHOD

This formula provides good results in humid and cold climates, and in arid regions. Makkink used radiation measurements to develop a following regression equation:

$$PET = \{R_s \times [s / (a + b)] + 0.12\} \tag{16}$$

where, PET = Potential evapotranspiration, mm/day; R_s = Total daily solar radiation, mm/day; b = Psychrometric constant [See Appendix D]; and s = Slope of saturated vapor pressure curve at average air temperature [See Appendix F].

5.3.2.15   RADIATION METHOD

Doorenbos and Pruitt [4] presented following radiation equation, which is a modified Makkink formula [16]:

$$PET = c \times (W \times R_s) \tag{17}$$

where, PET = Potential evapotranspiration for the considered period, mm/day; R_s = Solar radiation, mm/day; W = Correction factor related to temperature and elevation; and C = Correction factor, which depends on the average humidity and average wind speed. This method was employed in the Equator zone, in small islands and in high latitudes. Solar radiation maps provide the necessary data for the formula.

5.3.2.16   REGRESSION METHOD

The simple lineal regression equation is given as below:

$$PET = \{[a \times R_s] + b\} \tag{18}$$

where, PET = Potential evapotranspiration, mm/day; a and b = Empirical constants (regression coefficients), which depend on the location and season; and R_s = Solar radiation, mm/day. This regression method is simple and easy to use. However, it is not frequently used because of highly empirical nature.

5.3.2.17   PRIESTLY TAYLOR METHOD

In the absence of atmospheric air mass movement, Priestly and Taylor showed that the PET is directly related to evaporation equilibrium:

$$PET = \{A \times [s/(S + B)] \times [(R_n + S)]\} \tag{19}$$

where, PET = Potential evapotranspiration, mm/day; A = Empirically derived constant; s = Slope of saturated vapor pressure curve, at average air temperature [*see*

Appendix F]; B = Psychrometric constant [*see* Appendix D]; and R_n = Net radiation, mm/day. This method is of semiempirical in nature. It is reliable in humid zones, and is not adequate in arid regions. Table 3 shows advantages and disadvantages of different methods of estimation of potential evapotranspiration.

5.4   CALIBRATION OF PET METHOD FOR LOCAL CONDITIONS

The methods that use weather data are not adequate for all the locations, especially in tropical areas and at high elevations. Local calibration is always necessary to obtain reliable and good estimates of the crop water requirements. Table 4 shows the data used in different equations in this chapter. For the Blaney- Criddle method, ET can be estimated using measurements of the soil moisture in lysimeters, and that can measure water entering and going out. Only ambient temperatures and rainfall data are necessary for complete calibration when determining appropriate monthly crop coefficient. The Jensen-Haise method [9] is recommended for periods of 5 to 30 days. To make a calibration, local field data or lysimeters can be used for periods of 5 days. For a monthly calibration, the ET can be estimated by soil moisture measurements, inlet and outlet flows, in lysimeters, *etc.* The Penman equation can provide precize estimations from a month to an hour depending on the calibration method. For short periods, lysimeters can provide the necessary data for the ET. Usually local calibration is completed through a calibration correction factor.

5.5   CROP EVAPOTRANSPIRATION (ET$_C$)

To obtain the ET_c (consumptive use), it is necessary to know crop and ambient conditions. This includes climate, soil moisture, crop type, growth stage and the amount soil coverage by the crop. The ET_c indicates amount of water consumed for a given crop stage and the irrigation requirements can be determined. The procedure involves use of PET estimations and the experimental crop coefficients for the ET. This method is extensively used to schedule irrigation. The Blaney-Criddle method does not need a crop coefficient. The estimate of the ET_c is only made in one step. Doorenbos and Pruitt [4] provided an appropriate crop coefficient to estimate the ET for specific crops. These procedures resulted in precize estimates for periods of ten days to a month.

TABLE 3   Advantages and disadvantages of different methods to estimate potential evapotranspiration [PET].

Method	Advantages	Disadvantages
1. Penman	Easy to apply.	Underestimates ET under high movement conditions of atmospheric air masses.
2. Penman (FAO)	Provide satisfactory results.	The formula contains many components, which may result complex calculations.

TABLE 3 *(Continued)*

Method		Advantages	Disadvantages
3.	Water balance	Easy to process the data and integrate with the observations.	Low precision on the daily measures and difficult to obtain the ET when it is raining.
4.	Thornthwaite	Is reliable for long-terms.	Underestimate the ET during the summer. Is not precize for short-terms.
5.	Blaney- Criddle	Easy to use, The data is usually available.	The crop coefficient depends a lot on the climate.
6.	Blaney- Criddle (FAO)	The given crop coefficient depend less on the climate.	In high elevations, coasts and small islands there is no relation between temperature and solar radiation.
7.	Stephens-Stewart	Is reliable on the western side of the United States (where it was developed).	Need to be evaluated in other locations.
8.	Jensen- Haise	Is reliable under calm atmospheric conditions.	Underestimates ET under conditions of high movement of atmospheric air masses.
9.	Evaporation pan	Integrate all climatological factors.	Evaporation continues during the night in the pan, which affects the PET estimates.
10.	Hargreaves	Requires a minimum of climatological data.	Underestimates PET on the coasts and under high movements of air masses.
11.	Hargreaves and Samani	Requires only maximum and minimum temperature data.	Needs to be evaluated in many locations for its acceptance.
12.	Radiation	Is reliable in Equator, small islands and high altitudes.	Monthly estimates are often necessary outside Equator.
13.	Makkink	Good for humid and cold climates.	It is not reliable in arid regions.

TABLE 3   *(Continued)*

Method	Advantages	Disadvantages
14. Linacre	Is precize on annual basis.	Precision decreases on daily base.
15. Priestly Taylor	Reliable on humid areas.	Not adequate for arid zones.

TABLE 4   Parameters used in different formulas to estimate the PET.

Method	Temp.	HR	Wind	Sun	Radiation	Evaporation Class A	Ambient
1 Blaney – Criddle	*	0	0	0		0	
2 Radiation	*	0	0	*		0	(*)
3 Penman	*	*	*	*		0	(*)
4 Class A Pan		0	0		*	*	
5 Thornthwaite	*			*			
6 Hargreaves	*	0				0	
7 Linacre	*					*	
8 Jensen-Haise	*						*
9 Makkink	*						*
10 Priestly-Taylor	*						*
11 Regression							*

* = Measured data essential, 0 = Estimation required, (*) = Available data, but not essential.

5.6 CROP COEFICIENTS

The crop coefficients (K_c) are related to crop species, crop physiology, crop growth stage, days after planting, degree of ground coverage and the PET. When using the coefficients, it is important to know, how these were obtained. The empirical relation between ET_c and PET is given in Eq. (20):

$$K_c = [ET_c / PET] \qquad\qquad (20)$$

The combined K_c includes evaporation from the soil surface and the plant surface. The evaporation from the soil surface depends on the soil moisture and soil characteristics. The transpiration depends on the amount and nature of leaf area index of a plant and the available soil moisture to the root zone. The K_c can be adjusted to the available soil moisture and evaporation on the surface. Crop coefficient curve shows variation of K_c with days after planting (Fig. 6).

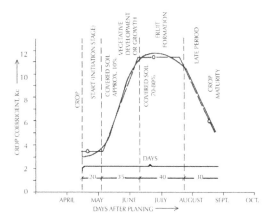

FIGURE 6   A typical crop coefficient curve.

5.6.1   REFERENCE CROP (ALFALFA)

The alfalfa is frequently selected as a reference crop, because it has high ET rates in arid regions [9]. Under these conditions, the PET is equal to the daily ET. When the crop is actively growing, alfalfa has a height of about 20 cm. and there is sufficient available soil moisture. The PET obtained with alfalfa is usually higher for the Bermuda grass, particularly in windy arid regions. The daily ET rates can be measured with the sensitive lysimeters.

5.6.2   CROP COEFFICIENT

Crop coefficients are given in Table 5. It is possible to estimate the consumptive water use (ET_c) using the crop coefficient and the calculated PET relation:

$$ET_c = Kc \cdot PET \tag{21}$$

where, ET_c=Crop evapotranspiration (consumptive water use), mm/day; K_c= Crop coefficient (Table 5); and PET = Potential evapotranspiration, mm/day.

5.7   SUMMARY

Evapotranspiration (ET) is a combination of two processes: Evaporation and transpiration. Evaporation is a conversion of liquid water into water vapor. The transpiration is the flow of liquid water from the soil to the surface of leaves/branches and trunk; and conversion of liquid water from the plant tissue into water vapor. Evapotranspiration processes are important for estimating crop irrigation requirements and for irrigation scheduling. Potential evapotranspiration (PET) is a water loss from the soil surface completely covered by vegetation. This chapter includes various Methods to estimate the ET and PET: Hydrologic method or water balance, climatic method, Penman equation, the Penman-Monteith equation, the Penman equation modified by Doorebos and Pruitt, Thornthwaite, Blaney-Criddle, the Blaney-Criddle equation modified by FAO, the Blaney-Criddle equation modified by Shih, Jensen-Haise, Stephens-Stewart, the Pan evaporation, Hargreaves method, Hargreaves and Samani modified method, Linacre method, Makkink method, Radiation method, Regression method and Priestly Taylor method. The methods that use weather data like Blaney-Criddle, Jensen-Haise and Penman are not adecuate for all the locations, especially in tropical areas and at high elevation. To obtain the crop water use (ET$_c$), it is necessary to know crop and ambient conditions. The crop coefficients (K$_c$) are related to the crop type, the crop physiology, crop stage, days after planting, the degree of coverage and the PET.

TABLE 5   Crop coefficients (K$_c$) at different growth stages of a given crop.

	Growth stage					
Crop	Initial[1,2]	Vegetative Development	Fruit Formation	Late Period	Crop Maturity	Total Growth Period
Banana	—	0.70–0.85	1.00–1.10	1.90–1.00	0.75–0.85	0.70–0.80
Beans: Green	0.30–0.40	0.65–0.75	0.95–1.05	0.90–0.95	0.85–0.95	0.85–0.90
Dry	0.30–0.40	0.70–0.80	1.05–1.12	0.65–0.75	0.25–0.30	0.70–0.80
Cabbage	0.40–0.50	0.70–0.80	0.95–1.11	0.90–1.00	0.80–0.95	0.70–0.80
Grape	0.35–0.55	0.60–0.80	0.70–0.90	0.60–0.80	0.55–0.70	0.55–0.75
Corn: Sweet	0.30–0.50	0.70–0.90	1.05–1.20	1.00–1.15	0.95–1.10	0.80–0.90
Corn: Field	0.30–0.50	0.70–0.85	1.05–1.20	0.8–0.95	0.55–0.60	0.75–0.90
Onion: Dry	0.40–0.60	0.70–0.80	0.95–1.10	0.85–0.90	0.75–0.85	0.80–0.90
Green	0.40–0.60	0.60–0.75	0.95–1.05	0.95–1.05	0.95–1.05	0.65–0.80
Pepper	0.30–0.40	0.60–0.75	0.95–1.10	0.85–1.00	0.80–0.90	0.70–0.80
Potato	0.40–0.50	0.70–0.80	1.05–1.20	0.85–0.95	0.70–0.75	0.75–0.90
Rice	1.10–1.15	1.10–1.50	1.10–1.30	0.95–1.05	0.95–1.05	1.05–1.02
Sorghum	0.30–0.40	0.70–0.75	1.00–1.15	0.75–0.80	0.50–0.55	0.75–0.85
Soy	0.30–0.40	0.70–0.80	1.00–1.05	0.70–0.80	0.40–0.30	0.75–0.90
Sugarcane, (stalk)	0.40–0.50	0.70–1.00	1.00–1.30	0.75–0.80	0.50–0.60	0.85–1.05
Tobacco	0.30–0.40	0.70–0.80	1.00–1.20	0.90–1.00	0.75–0.85	0.85–0.95
Tomato	0.40–0.50	0.70–0.80	1.05–1.25	0.80–0.95	0.60–0.65	0.75–0.90
Watermelon	0.40–0.50	0.70–0.80	0.95–1.05	0.80–0.90	0.65–0.75	0.75–0.85

[1] Value for high humidity of HR 70% and low wind velocity, U = 5 meters per second.
[2] Value for low humidity of HR 20% and high wind velocity, U = 5 meters per second.

KEYWORDS

- atmosphere
- available water
- blaney-criddle method
- class a pan
- class A pan coefficient
- consumptive use
- crop coefficient
- crop evapotranspiration
- crop water requirement
- consumptive water use
- drainage
- evaporation
- evapotranspiration
- ground cover
- irrigation requirement
- percolation
- potential evapotranspiration
- precipitation
- relative humidity
- root zone
- soil moisture
- solar radiation
- thornthwaite method
- transpiration

REFERENCES

1. Allen, R. G.; A Penman for all seasons. J. Irrig. Drain. Div. ASCE, **1986**, 112(4), 348–368.
2. Allen, R. G.; Pruitt, W. O.; Rational use of the FAO Blaney- Criddle formula. J. Irrig. Drain. Div. ASCE, 1986, 112 (2), 139–155.
3. *Climatic Summary of the United States*. Dept, US of Commerce, US Printing Office Washington, D.C.; Report 86-45, page. 40.
4. Doorenbos, J.; Pruitt, W. O. *Crop Water Requirements*. FAO Irrigation and Drainage Division of ASCE.; Paper 24, Food and Agriculture Organization of the United Nations, Rome. **1977**, 1–156.
5. Doorenbos, J.; Pruitt, W. O. Yield response to water. FAO Irrigation and Drainage. **1979**, Paper 33.
6. Goyal, M. R.; Potential evapotranspiration for the South Coast of Puerto Rico with the Hargreaves–Samani technique. J. Agric. Univ. P.R.; **1988**, 72 (1), 41–50

7. Hargreaves, G. H.; Estimation of potential and crop evapotranspiration. Trans. ASAE **1974**, 17, 701–704.

8. Hargreaves, G. H.; Samani, S. A.; Reference crop evapotranspiration from temperature. Applied Eng. in Agric. ASAE, 1985, 1(2), 96–99.

9. Hargreaves, G. H.; Samani, Z. A.; *World Water for Agriculture Precipitation Management*. International Irrigation Center, Utah State University, Logan - Utah, USA, **1986**, 1–617.

10. Jensen, M. E.; Design and Operation of Farm Irrigation Systems. ASAE Monograph #3, American Society of Agricultural Engineers, Chapter 6, 1980, 189–225.

11. Jones, J. W.; Allen, L. H.; Shih, S. F.; Rogers, J. S.; Hammond, L. C.; Smajstrala A. G.; Martsolf, J. D.; *Estimated and Measured Evapotranspiration for Florida Climatic, Crops and Soils.* Agric. Exp. Sta. Inst. of Food and Agric. Sci. Univ. of Florida, Gainesville. Bulletin No. 840, **1984**, 1–65.

12. Linacre, E. T. **1977**, A simple formula for estimating evapotranspiration rates in various climates, using temperature data along. Agricultural Meteorology, 18, 409–424.

13. Michael, A. M.; *Irrigation Theory and Practice*. New Delhi: Vikas Publishing House Pvt. Ltd.; Chapter 7, **1978**, 448–584.

14. Penman, H. L.; The dependence of transpiring on climate and soil conditions. J. Soil Sci.; **1949**, 1, 74–89.

15. Rosenberg, N. J.; Blad B. L.; Verma, S. B.; *Microclimate: the Biological Environment*. A Wiley-Interscience Publ. Chapter **1983**, 7, 209–287.

16. Salih, A. M.; Sendil, U.; Evapotranspiration under extremely arid climates. J. Irrig. Drain. Div. ASCE, **1984**, 110(IR3), 289–303.

17. Shih, S. F.; Data rquirement for evapotranspiration estimation. J. Irrig. Drain. Div. ASCE, **1984**, 110(IR3), 263–274.

18. Wright, J. L.; New evapotranspiration crop coefficients. J. Irrig. Drain. Div. ASCE, **1982**, 108 (IR1), 57–74.

CHAPTER 6

EVAPORATION ESTIMATIONS WITH NEURAL NETWORKS[1]

P. S. SHIRGURE and G. S. RAJPUT

CONTENTS

[1]Published with permission. Modified from P. S. Shirgure and G. S. Rajput, *Evaporation modeling with neural networks – A research review.* International Journal of Research and Reviews in Soft and Intelligent Computing, United Kingdom, 2011, *1(2)*, 37–47. <www.sciacademypublisher.com> Copyright © Science Academy Publisher, United Kingdom.

6.1   INTRODUCTION

Evaporation is a complex and nonlinear phenomenon because it depends on several interacting climatological factors, such as temperature, humidity, winds speed, bright sunshine hours, etc. An Artificial Neural Networks (ANN) is a flexible mathematical structure, which is capable of identifying complex nonlinear relationships between input and output datasets. The ANN models have been found useful and efficient, particularly in problems for which the characteristics of the processes are difficult to describe using physical equations. An ANN model can compute complex nonlinear problems, which may be too difficult to represent by conventional mathematical equations. These models are well suited to situations where the relationship between the input variable and the output is not explicit. Instead, ANN, map the implicit relationship between inputs and outputs through training by field observations. The model may require significantly less input data than a similar conventional mathematical model, since variables that remain fixed from one simulation to another do not need to be considered as inputs. The ANN is useful, requiring fewer input and computational effort and less real time control. An ANN can quickly present sensitive responses to tiny input changes in a dynamic environment. ANN are effective tools to model nonlinear systems [52–81]. A neural network model is a mathematical construct whose architecture is essentially analogous to the human brain. Basically, the highly interconnected processing elements, arranged in layers are similar to the arrangement of neurons in the brain. The ANN has found successful applications in the areas of science, engineering, industry, business, economics and agriculture. Recently, ANN have been applied in meteorological and agro ecological modeling and applications [37]. Most of the applications reported in literature concern estimation, prediction and classification problems. Neural network applications have diffused rapidly due to their functional characteristics, which provide many advantages over traditional analytical approaches [38–57].

6.1.1   HISTORIC REVIEW OF EVAPORATION MODELS WITH NEURAL NETWORKS

The ANN provide better modeling flexibility than the other statistical approaches with its successive adaptive features of error propagation, where each meteorological variable takes its share proportionally. Numerous researchers have shown applicability of multiple linear regression (MLR) for estimating the evaporation by Baier and Robertson [7], Hanson [35], Sharma [78] and Jhajharia [45], but very few have been seen on ANN in agricultural and hydrological processes in India. For instance multilayered feed forward ANN with error back propagation techniques has been used for estimating air temperature by Cook and Wolfe [19], Dimri [25], Smith [80], wind speed by Mohandes [62], rainfall by Lee [53] and Chattopadhya, [15], rainfall-runoff by Minns and Hall [61], Braddock [9], Dawon and Wilbye [21], Sajikumar [75], Rajurkar [72], Mamdouh [58], and Sarkar [76], runoff by Jagdesh [42], solar radiation by Elizondo [28], Dorvlo [27], Irmak [41], Reddy and Ranjan [71] and Bocco [8], evapo-transpiration by Kumar [52], soil water content by Schaap and Bouten [77], Pachepsky [68], soil temperature by Mehuys [60] and Tasadduq [84], soil water evaporation by Han and Felkar [34], and various neuro-computing techniques for predicting the various atmospheric processes and parameters [5, 16, 32, 49, 59]. The ANN is also widely used in number of diversified fields of soil and water engineering. The neural networks is used in river flow prediction by Imrie [40], Kisi, [51], river flow forecasting by Dibike and Solomatine [24], stream flow by Zealand [94], remote sensing and land use classification by Ashis D [6], land drainage engineering by Yang [92], prediction and estimation of sediment concentration by Kisi, [50] prediction of water quality parameters by Maier and Dandy [56], and for control of frost damage by Mort and Robinson [64] and Jain [43].

A number of researchers have attempted to estimate the evaporation values from the climatic variables and most of these methods require data that are not easily available. The simpler methods that are reported to fit a linear relationship between variables are multiple linear regression. However, the process of evaporation (pan) is highly nonlinear in nature, as it is evidenced by many of the estimation procedures. Many researchers have emphasized the need for accurate estimates of evaporation modeling using better models that will consider the inherent nonlinearity in the evaporation process.

The comparison of automatically and manually collected pan evaporation data was done by Bruton [12]. Recent researchers have reported that ANN may offer a promising alternative to the conventional methods for estimating the evaporation by Clayton [18]; Arca [4]; Gavin and Agnew [31]; Ozlem and Evolkesk [67]; Terzi and Keskin [86]; Keskin and Terzi [48] and the lake evaporation by Bruin [11]; Anderson and Jobson [3]; Reis and Dias [74]; Coulomb [20] and Murthy and Gawande [60].

6.2   RESEARCH REVIEW

Evaporation reflects the influence of several meteorological parameters like air temperature, sunshine hours, wind velocity, relative humidity, solar radiation, evaporating power of the air and vapor pressure deficit of a locality. But measurement of evaporation with accuracy is difficult task. In such cases, it becomes assertive to use formulae or neural network model that can estimate pan evaporation from available climatic data, may give more accurate results than the measured pan evaporation. In this regard, a number of models have therefore been proposed and developed by several investigators for different locations in India and abroad are reviewed in following sections.

1.   General Artificial Neural Network (ANN) Models
2.   Artificial Neural Network Evaporation Models

In this chapter, the evaporation from open pan as well as surface evaporation modeling is discussed in brief. The process of evaporation is very much complex and nonlinear in nature with respect to the meteorological parameter, which influences the evaporation. The review related to general evaporation models is discussed. The neural network is a new tool that can solve the more complex modeling problems like estimating evaporation from pan, which may be difficult to solve by conventional mathematical equations and multiple linear regression. It is observed from this review that the prediction model for of evaporation is superior with neural networks.

6.2.1 GENERAL ARTIFICIAL NEURAL NETWORK MODELS

An ANN is a capable of identifying complex nonlinear relationships between input and output datasets. ANN models have been found useful and efficient, particularly in problems for which the characteristics of the processes are difficult to describe using physical equations. The conventional model requires many input parameters and variables, some of which cannot be obtained easily from the field and may vary from site to site even within the same geographical region. An artificial neural network has found successful applications in the areas of science, engineering, industry, business, economics and agriculture. For example, ANN have improved the technique of satellite images and patterns recognition, wastewater treatment, remote sensing and seawater pollution classification. Neural Network (NN) have been used to model a variety of biological and environmental processes by Altendorf [2] and [30].

ANN models compute complex nonlinear problems, which may be difficult to represent by conventional mathematical equations. These models are well suited to situations, where the relationships between the input variable and the output are not explicit. The model may require significantly less input data than a similar conventional mathematical model, since variables that remain constant from one simulation to another, do not need to be considered as inputs. The advantage of the ANN approach in estimating evaporation is that it requires only limited climatic data. Pan evaporation data are limited and methods are required to estimate evaporation with a minimum of climatic data. Most of the available models for estimation of evaporation may be reliable and most appropriate for use in climates similar to where they were developed. It is likely that errors may occur when these models are used under climatic conditions that are different under which they were developed. Since no evaluation of the different ANN evaporation models has been undertaken there is a need to determine the applicability of these models under the different agro-climatic regions of India [36].

The different methods of estimating pan evaporation approaches reviewed generally performed better when solar radiation and relative humidity were included as input variables. However, these data are often not available. Models of pan evaporation are required to be established that corresponds to the relatively minimum weather data variable for most of the locations. In earlier research work of empirical modeling and equation were fit to the data and the correlation was determined with same dataset. No attempt was made to apply those models to the other independent dataset of different location. The evaporation prediction also highly depends upon the quality of the pan evaporation measurements used in the ANN modeling. Large number of researchers have been established the applicability of ANNs to the problems in agricultural, hydrological, meteorological and environmental fields. Hu [39] initiated the implementation of ANN, an important soft computing methodology in weather forecasting.

Allen and Marshall [1] compared the performance of ANN approach and discrimination analysis method for operational forecasting of rainfall and established the superiority of neural network approach over conventional statistical approach in forecasting rainfall over Australia. Elizondo [28] developed an ANN model to estimate daily solar radiation for locations in south-eastern US based on daily maximum and minimum air temperature, precipitation, clear sky radiation, and daylength. Of these inputs, only temperature and precipitation were measured under field conditions. The researchers found r^2 values as high as 0.74 and a root error as low as 2.92 MJ/m². Schaap and Bouten [77] developed an ANN to model water retention curves of sandy soils using particle size distribution, bulk density, and organic matter content as inputs. From this model they estimated soil water contents at different tensions with a root mean square error as low as 0.020 cm³/cm³.

Yang [92] developed and trained an artificial neural network by using the simulated mid span water table depths from DRAINMOD, a conventional water table management model. Compared to DRAINMOD, the model was very simple to run and required only a small amount of data, such as precipitation, ET, and initial mid span water table depth. The results indicated that the artificial neural network model could make predictions similar to DRAINMOD, with the least root mean square error of 0.1193 and doing this significantly faster and with fewer input data. Generally, the artificial neural network structure with six processing elements and one hidden layer was sufficient for the study. It was found that the network should be trained with at least 145,000 cycles but more than 200,000 cycles are unnecessary. A lag procedure was suggested which improved the performance of ANN under irregular situations, such as sudden and large rainstorms. A 3-day lag of all input parameters was the best choice when the weather conditions were irregular.

Daily soil water evaporation has been estimated using a radial basis function ANN by Han and Felkar [34]. The ANN models were implemented to establish daily soil water evaporation from average relative humidity, air temperature, wind speed and soil water content in cactus field study. This ANN had an average absolute percent error of 21.0 % and a root mean square error (RMSE) of 0.17 mm/day. This was better than a multiple linear regression models with values of 30.1 % and 0.28 mm/day for the same parameters. They used daily values of average temperatures, relative humidity and wind speed as inputs to the model. This study was based on only 40 daily evaporation observations. It was also limited in that the weather data were obtained from a weather station 40 km from the site of evaporation measurements. They concluded that the ANN technique appeared to be an improvement over multilinear regression technique for estimating soil temperature.

Mohandes [62] also applied ANN approach for prediction of wind speed. Gardner and Dorling (1998) discussed the proficiency of the Multi-layer Perceptron as a suitable model for atmospheric prediction. Lee [53] applied ANN in rainfall prediction by splitting the available data into homogeneous subpopulations.

Tokar and Johnson [87] used ANN methodology was employed to forecast daily runoff as a function of daily precipitation, temperature, and snowmelt for the Little Patuxent River watershed in Maryland. The sensitivity of the prediction accuracy to the content and length of training data was investigated. The ANN rainfall-runoff model compared favorably with results obtained using existing techniques including statistical regression and a simple conceptual model. The ANN model provides a more systematic approach, reduces the length of calibration data, and shortens the time spent in calibration of the models. At the same time, it represents an improvement upon the prediction accuracy and flexibility of current methods. Trajkovic [89] presented the application of radial basis function (RBF) network to estimate the FAO Blaney-Criddle *b* factor. Tabular *b* values are given in the United Nations FAO Irrigation and

Drainage Paper Number 24. The *b* values obtained by the RBF network compared to the appropriate *b* values produced using regression equations. An example was given to illustrate the simplicity and accuracy of the RBF network for *b* factor.

Yang [93] developed a back propagation ANN model that could distinguish young corn plants from weeds. Although only the color indices associated with image pixels were used as inputs, it was assumed that the ANN model could develop the ability to use other information, such as shapes, implicit in these data. The 756 × 504 pixel images were taken in the field and were then cropped to 100×100-pixel images depicting only one plant, either a corn plant or weeds. There were 40 images of corn and 40 of weeds. The ability of the ANNs to discriminate weeds from corn was then tested on 20 other images. A total of 80 images of corn plants and weeds were used for training purposes. For some ANNs, the success rate for classifying corn plants was as high as 100%, whereas the highest success rate for weed recognition was 80%. This is considered satisfactory, given the limited amount of training data and the computer hardware limitations. Therefore, it is concluded that an ANN-based weed recognition system can potentially be used in the precision spraying of herbicides in agricultural fields.

Perez and Reyes [71] applied Multi-layer Perceptron model in predicting the particulate air pollution. Maqsood [59] established the usefulness of ANN in atmospheric modeling explained its potential over conventional weather prediction model. Patel [69] studied the applicability of ANNs for predicting salt build-up in the crop root zone. ANN models were developed with salinity data from field lysimeters subirrigated with brackish water. Different ANN architectures were explored by varying the number of processing elements (PEs) (from 1 to 30) for replicate data from a 0.4 in water table, 0.8 in water table, and both 0.4 and 0.8 in water table lysimeters. Different ANN models were developed by using individual replicate treatment values as well as the mean value for each treatment. For replicate data, the models with 20, seven, and six PEs were found to be the best for the water tables at 0.4 in, 0.8 in and both water tables combined, respectively The correlation coefficients between observed salinity and ANN predicted salinity of the test data with these models were 0.89, 0.91, and 0.89, respectively. The performance of the ANNs developed using mean salinity values of the replicates was found to be similar to those with replicate data. Not only was there agreement between observed and ANN predicted salinity values, the results clearly indicated the potential use of ANN models for predicting salt build-up in soil profile at a specific site.

Chaloulakou [14] made a comparative study between multiple regression models and feed forward ANN with respect their predictive potential for PM_{10} over Athens and established that neural network approach has and edge over regression method expressed both in terms of prediction error and of episodic prediction ability.

Jain [43] developed ANN models to forecast air temperature in hourly increments from 1 to 12 hours for Alma, Fort Valley and Blairsville in Georgia, USA. However, this study was limited by the fact that the model was not specifically developed to predict frosts. So, even though the model could give a good overall performance, a dedicated model might be able to perform better on the near freezing and freeing temperatures. Jian-Ping Suen and Wayland Eheart [46] studied the ANN (ANNs) are applied to estimating nitrate concentrations in a typical Midwestern river, i.e., the

Upper Sangamon River in Illinois. Throughout the Midwestern United States, nitrate in raw water has recently become an increasingly important problem. This is due to recent changes in the U.S. EPA nitrate standard and to the increasingly widespread use of chemical fertilizers in agriculture. Back-propagation neural networks (BPNNs) and radial basis function neural networks (RBFNNs) are compared as to their effectiveness in water quality modeling. Training of the RBFNNs is much faster than that of the BPNNs, and yields more robust results. These two types of ANNs are compared to traditional regression and mechanistic water quality modeling, based on overall accuracy and on the frequency of false-negative prediction. The RBFNN achieves the best results of all models in terms of overall accuracy, and both BPNN and RBFNN yield the same false-negative frequency, which is better than that of the traditional models.

Li [54] established the suitability of ANN model in establishing maximum surface temperature, minimum surface temperature and solar radiation over regression method at Tifton, Georgia and Griffin. Jef [44] developed an ANN model to predict daily PM_{10} concentration over Belgium.

Trajkovic [90] studied the application of RBF (Radial Basis Function) networks to estimate the FAO Penman c factor. The values of the c factors obtained by RBF networks were compared to the appropriate c values produced using regression expressions. It was shown that the RBF networks ensure a better agreement with table c values, thus improving the accuracy of the estimation of reference crop evapo-transpiration. An example that demonstrated the simplicity of the use of RBF networks and the accuracy of the c factor estimation was presented.

Lopez [55] studied a series of algorithms developed to estimate hydraulic roughness coefficients for overland flow. The algorithms are combinations of neural networks that use surface configuration parameters and the local flow Reynolds number as inputs and provide an estimate of the roughness coefficient. The results showed that as new neural networks were combined into the stacked algorithm, the estimate errors become gradually smaller. The scarcity of data points in some regions of the output space for the Darcy-Weisbach and Manning models caused a reduction in the predictability of the algorithms for these regions and prevented the use of more complex neural networks.

Kumar [52] investigated the utility of ANNs for estimation of daily grass reference crop Evapo-transpiration (ET_0) and compared the performance of ANNs with the conventional method (Penman-Monteith) used to estimate ET_0. Several issues associated with the use of ANNs were examined, including different learning methods, number of processing elements in the hidden layer(s), and the number of hidden layers. Three learning methods, namely, the standard back-propagation with learning rates of 0.2 and 0.8, and back propagation with momentum were considered. The networks were trained with climatic data (solar radiation, maximum and minimum temperature, maximum and minimum relative humidity and wind speed) as input and the Penman-Monteith estimated ET_0 as output. The best ANN architecture was selected on the basis of weighed standard error of estimate (WSEE) and minimal ANN architecture. The ANN architecture of 6–7–1 (six, seven and one neuron(s) in the input, hidden and output layers, respectively) gave the minimum WSEE (less than 0.3 mm/day) for all learning methods. That value was lower than the WSEE (0.74 mm/day) between the

Penman-Monteith and lysimeter measured ET_0 as reported by Jensen et al., in 1990. Similarly, ANNs were trained, validated and tested using the lysimeter measured ET_0 and corresponding climatic data. Again, all learning methods gave less WSEE (less than 0.6 mm/day) as compared to the Penman-Monteith method (0.97 mm/day). Based on these results, it can be concluded the ANN can predict ET_0 better than the conventional method. Sudheer [82] examined the potential of ANN in estimating the actual crop evapotranspiration (ET_c) from limited climatic data. The study employed RBF type ANN for computing the daily values of evapotranspiration for rice crop. Six RBF networks, each using varied input combinations of climatic variables, had been trained and tested. The model estimates were compared with measured lysimeter evapotranspiration. The results of the study clearly demonstrated the proficiency of the ANN method in estimating the evapotranspiration. The analyzes suggest that the crop ET could be computed from air temperature using the ANN approach. However, the study used a single crop data for a limited period, therefore further studies using more crops as well as weather conditions may be required to strengthen these conclusions.

Trajkovic [91] applied a sequentially adaptive radial basis function network to the forecasting of reference evapotranspiration (ET_o). The sequential adaptation of parameters and structure was achieved using an extended Kalman filter. The criterion for network growing was obtained from the Kalman filter's consistency test, while the criteria for neuron/connection pruning were based on the statistical parameter significance test. The weather parameter data (air temperature, relative humidity, wind speed, and sunshine) were available at Nis, Serbia and Montenegro, from January 1977 to December 1996. The monthly reference evapotranspiration data were obtained by the Penman-Monteith method, which was proposed as the sole standard method for the computation of reference evapotranspiration. The network learned to forecast $ET_{o, t+1}$ based on $ET_{o, t-11}$ and $ET_{o, t-23}$. The results showed that ANNs can be used for forecasting reference evapotranspiration with high reliability.

Rajurkar [72] presented artificial neural network for modeling daily flows during flood events. The rainfall-runoff process was modeled by coupling a simple linear (black box) model with the ANN. The study uses data from two large size catchments in India and five other catchments used earlier by the World Meteorological Organization (WMO) for intercomparison of the operational hydrological models. The study demonstrated that the approach adopted herein for modeling produced reasonably satisfactory results for data of catchments from different geographical locations, which thus proves its versatility. Most importantly, the substitution of the previous days runoff (being used as one of the input to the ANN by most of the previous researchers), by a term that represents the runoff estimated from a linear model and coupling the simple linear model with the ANN may prove to be very much useful in modeling the rainfall-runoff relationship in the nonupdating mode.

Sarkar [76] studied the rainfall-runoff modeling through the use of ANN approach with reasonable accuracy. In the study back propagation ANN runoff models have been developed to simulate and forecast daily runoff for a part of the Satluj basin of India. It is observed that only rainfall and temperature are considered as input are not adequate to develop a model for the simulation as well as forecasting of the catchments runoff resulting from rainfall and snowmelt contribution. In order to improve upon the performance of the models, the runoff of the upstream site are also included as an additional input to the model.

Patil [70] developed simulation models for predicting the soil hydraulic properties in which indirect estimation can be obtained using pedotransfer functions (PTF). Empirical relationship between water retention at hydrolimits i.e. field capacity (FC) and permanent wilting point (PWP) and physical properties of soil was explored to develop pedotransfer functions. Textural data and bulk density input was used for developing PTF's. where as regression analysis and artificialANN was employed for the study. The study pertains to clay and associated soils of Madhya Pradesh. PTF's developed by other researchers were also used for comparative evaluation. Results showed that the use of ANN improved the accuracy of prediction of field capacity by 25 % in relative terms.

Ferentinos and Albright [29] presented an ANN model that predicts pH and EC changes in root zone of lettuce grown in hydrophonic system. A feed forward NN is the basis of modeling. The NN model has nine inputs (pH, EC, nutrient solution temperature, air temperature, relative humidity, light intensity, plant age, amount of added acid and amount of added base.) and two outputs (pH and EC at the next time step). The most suitable and accurate combination of network architecture and training method was one hidden layer with nine hidden nodes trained with given quasi-Newton back propagation algorithm. The model proved capable of predicting pH at the net 20-minute time step with in 0.01 pH units and EC with in 5 m S cm^{-1}. Simpler prediction methods like linear extrapolation performed poorly in situations where control actions of the system activated and produced relatively rapid changes in predicted parameters. In those cases, the neural network model did not encounter in difficulties predicting the rapid changes.

Bocco [8] developed neural network models of back propagation type to estimate solar radiation based on extraterrestrial radiation data, daily temperature range, precipitation, cloudiness and relative sunshine duration. Data from Cordoba, Argentina were used for the development and validation. The behavior and adjustment between values observed and obtained by neural networks for different combinations of inputs were assessed. These estimates showed root mean square error between 3.15 and 3.88 MJ m^{-2} d^{-1}. The latter corresponds to the model that calculates radiation using only precipitation and daily temperature range. In all models, results show good adjustment to seasonal solar radiation. These results allow inferring the adequate performance and pertinence of this methodology to estimate complex phenomenon, such as solar radiation.

Brian [10] created the Models to predict air temperature at hourly intervals from one to 12 hours ahead. Each ANN model, consisting of a network architecture and set of associated parameters, was evaluated by instantiating and training 30 networks and calculating the mean absolute error (MAE) of the resulting networks for some set of input patterns. The inclusion of seasonal input terms, up to 24 hours of prior weather information, and a larger number of processing nodes were some of the improvements that reduced average prediction error compared to previous research across all horizons. For example, the four-hour MAE of 1.40°C was 0.20°C, or 12.5%, less than the previous model. Prediction MAEs eight and 12 hours ahead improved by 0.17°C and 0.16°C, respectively, improvements of 7.4% and 5.9% over the existing model at these horizons. Networks instantiating the same model but with different initial random weights often led to different prediction errors. These results strongly suggest that ANN model developers should consider instantiating and training multiple networks with different initial weights to establish preferred model parameters.

Muleta and Nicklow [65] stated that ANNs have become common data driven tools for modeling complex, nonlinear problems in science and engineering. Many previous applications have relied on gradient-based search techniques, such as the back propagation (BP) algorithm, for ANN training. Such techniques, however, are highly susceptible to premature convergence to local optima and require a trial-and-error process for effective design of ANN architecture and connection weights. This paper investigates the use of evolutionary programming (EP), a robust search technique, and a hybrid EP–BP training algorithm for improved ANN design. Application results indicate that the EP–BP algorithm may limit the drawbacks of using local search algorithms alone and that the hybrid performs better than EP from the perspective of both training accuracy and efficiency. In addition, the resulting ANN is used to replace the hydrologic simulation component of a previously developed multiobjective decision support model for watershed management. Due to the efficiency of the trained ANN with respect to the traditional simulation model, the replacement reduced the overall computational time required to generate preferred watershed management policies by 75%. The reduction is likely to improve the practical utility of the management model from a typical user perspective. Moreover, the results reveal the potential role of properly trained ANNs in addressing computational demands of various problems without sacrificing the accuracy of solutions.

Mamdouh [89] is presented a rainfall-runoff model based on an ANN for the Blue Nile catchment. The best geometry of the ANN rainfall-runoff model in terms of number of hidden layers and nodes is identified through a sensitivity analysis. The Blue Nile catchment (about 300,000 km^2) in the Nile basin is selected here as a case study. The catchment is classified into seven subcatchments, and the mean aerial precipitation over those subcatchments is computed as a main input to the ANN model. The available daily data (1992–1999) are divided into two sets for model calibration (1992–1996) and for validation (1997–1999). The results of the ANN model are compared with one of physical distributed rainfall-runoff models that apply hydraulic and hydrologic fundamental equations in a grid base. The results over the case study area and the comparative analysis with the physically based distributed model show that the ANN technique has great potential in simulating the rainfall-runoff process adequately. Because the available record used in the calibration of the ANN model is too short, the ANN model is biased compared with the distributed model, especially for high flows.

Tokar and Johnson [87] was employed An ANN methodology to forecast daily runoff as a function of daily precipitation, temperature, and snowmelt for the Little Patuxent River watershed in Maryland. The sensitivity of the prediction accuracy to the content and length of training data was investigated. The ANN rainfall-runoff model compared favorably with results obtained using existing techniques including statistical regression and a simple conceptual model. The ANN model provides a more systematic approach, reduces the length of calibration data, and shortens the time spent in calibration of the models. At the same time, it represents an improvement upon the prediction accuracy and flexibility of current methods.

Tokar and Markus [88] inspired by the functioning of the brain and biological nervous systems, ANNs have been applied to various hydrologic problems in the last 10 years. In this study, ANN models are compared with traditional conceptual models in predicting watershed runoff as a function of rainfall, snow water equivalent, and temperature. The ANN technique was applied to model watershed runoff in three basins with different climatic and physiographic characteristics—the Fraser River in Colorado, Raccoon Creek in Iowa, and Little Patuxent River in Maryland. In the Fraser River watershed, the ANN technique was applied to model monthly stream flow and was compared to a conceptual water balance (WATBAL) model. The ANN technique was used to model the daily rainfall-runoff process and was compared to the Sacramento soil moisture accounting (SAC-SMA) model in the Raccoon River watershed. The daily rainfall-runoff process was also modeled using the ANN technique in the Little Patuxent River basin, and the training and testing results were compared to those of a simple conceptual rainfall-runoff (SCRR) model. In all cases, the ANN models provided higher accuracy, a more systematic approach, and shortened the time spent in training of the models. For the Fraser River, the accuracy of monthly stream flow forecasts by the ANN model was significantly higher compared to the accuracy of the Watbal model. The best-fit ANN model performed as well as the SAC-SMA model in the Raccoon River. The testing and training accuracy of the ANN model in Little Patuxent River was comparatively higher than that the SCRR model. The initial results indicate that ANNs can be powerful tools in modeling the precipitation-runoff process for various time scales, topography, and climate patterns.

Grivas and Chaloulakou [33] evaluated the potential of various developed neural network models to provide reliable predictions of PM_{10} hourly concentrations, a task that is known to present certain difficulties. The modeling study involves 4 measurement locations within the Greater Athens Area, which experiences a significant PM-related air pollution problem. The PM_{10} data used cover the period of 2001–2002. Artificial neural network models were developed using a combination of meteorological and time-scale input variables. A genetic algorithm optimization procedure for the selection of the input variables was also evaluated. The results of the neural network models were rather satisfactory, with values of the coefficient of determination (r^2) for independent test sets ranging between 0.50 and 0.67 for the four sites and values of the index of agreement between 0.80 and 0.89. The performance of examined neural network models was superior in comparison with multiple linear regression models that were developed in parallel (r^2 for regression models ranging between 0.29 and 0.35). Their performance was also found adequate in the case of high-concentration events, with acceptable probabilities of detection and low false alarm rates. The suitability of the developed neural network models for use at real-time conditions was further evaluated for PM_{10} hourly concentrations recorded during the days of the 2004 Athens Olympic Games.

Chokmani [17] studied to assess the ability of the ANN models in estimating river ice thickness using easy available climate data. A site specific ANN models were developed for two hydrometric stations at two rivers in Alberta (Canada). The ANN models were found to adequately estimate ice thickness. Ways to improve the performances of the ANN models are proposed.

6.2.2 ARTIFICIAL NEURAL NETWORK EVAPORATION MODELS

Evaporation involves the transformation of water from its liquid state into a gas and the subsequent diffusion of water vapor into the atmosphere. There is growing demand for evaporation data for studies of surface water and energy fluxes, especially for the studies, which address the impacts of global warming. However, the measurement of evaporation in the open environment is difficult and is usually done by proxy. Potential evaporation is the variable most often used. Potential evaporation is a measure of the ability of the atmosphere to remove water from a surface assuming no limit to water availability, whereas actual evaporation is the quantity of water that is removed from that surface by evaporation [13]. The review of literature on artificial neural network and evaporation models is presented here.

Bruton [12] developed ANN models to estimate daily pan evaporation using measured weather variables as inputs. Weather data from Rome, Plains and Watkinville, Georgia, consisting of 2044 daily records from 1992 to 1996 were used to develop the models of daily pan evaporation. Additional weather from these locations, which included 720 daily records from 1997 and 1998, served as an independent evaluation dataset for the models. The measured variables included daily observations of rainfall, temperature, relative humidity, solar radiation, and wind speed. Daily pan evaporation was also estimated using multiple linear regression and compared to the results of the ANN models. The ANN models of daily pan evaporation with all available variables as an input was the most accurate model delivering an r^2 of 0.717 and a root mean square error 1.11 mm for the independent evaluation dataset. ANN models were developed with some of the observed variables eliminated to correspond to different levels of data collection as well as for minimal datasets. The accuracy of the models was reduced considerably when variables were eliminated to correspond to weather stations. Pan evaporation estimated with ANN models was slightly more accurate than the pan evaporation estimated with a multiple linear regression models.

Sudheer [81] investigated the prediction of Class A pan evaporation using the ANN technique. The ANN back propagation algorithm has been evaluated for its applicability for predicting evaporation from minimum climatic data. Four combinations of input data were considered and the resulting values of evaporation were analyzed and compared with those of existing models. The results from this study suggest that the neural computing technique could be employed successfully in modeling the evaporation process from the available climatic dataset. However, an analysis of the residuals from the ANN models developed revealed that the models showed significant error in predictions during the validation, implying loss of generalization properties of ANN models unless trained carefully. The study indicated that evaporation values could be reasonably estimated using temperature data only through the ANN technique. This would be of much use in instances where data availability is limited.

Taher [85] estimated potential evaporation, especially in arid regions such as Saudi Arabia, has been of a great concern to many researchers. Its importance is obvious in many water resources applications such as management of hydrologic, hydraulic and agricultural systems. For this purpose, four three-layer back propagation neural networks were developed to forecast monthly potential evaporation in Riyadh, Saudi Arabia, based on four explanatory climatic factors. Observations of relative humidity, solar radiation, temperature, wind speed and evaporation for the past 22 years have been used to train and test the developed networks. Results revealed that the networks were able to well learn the events they were trained to recognize. Moreover, they were capable of effectively generalizing their training by predicting evaporation for sets of unseen cases. These encouraging results were supported by high values of coefficient of correlation and low mean square errors reaching 0.98 and 0.00015, respectively. The study has also evolved a comparison with traditional methods and has proven that the developed neural networks were superior.

Keskin [47] concluded that evaporation is one of the fundamental elements in the hydrological cycle, which affects the yield of river basins, the capacity of reservoirs, the consumptive use of water by crops and the yield of underground supplies. In general, there are two approaches in the evaporation estimation, namely, direct and indirect. The indirect methods such as the Penman and Priestley-Taylor methods are based on meteorological variables, whereas the direct methods include the class A pan evaporation measurement as well as others such as class GGI-3000 pan and class U pan. The major difficulty in using a class A pan for the direct measurements arizes because of the subsequent application of coefficients based on the measurements from a small tank to large bodies of open water. Such difficulties can be accommodated by fuzzy logic reasoning and models as alternative approaches to classical evaporation estimation formulations were applied to Lake Egirdir in the western part of Turkey. This study has three objectives: to develop fuzzy models for daily pan evaporation estimation from measured meteorological data, to compare the fuzzy models with the widely used Penman method, and finally to evaluate the potential of fuzzy models in such applications. Among the measured meteorological variables used to implement the models of daily pan evaporation prediction are the daily observations of air and water temperatures, sunshine hours, solar radiation, air pressure, relative humidity and wind speed. Comparison of the classical and fuzzy logic models shows a better agreement between the fuzzy model estimations and measurements of daily pan evaporation than the Penman method.

Ozlem [67] estimated daily pan evaporation is achieved by a suitable ANN model for the meteorological data recorded from the Automated *GroWheather* meteorological station near Lake Egirdir, Turkey. In this station six meteorological variables are measured simultaneously, namely, air temperature, water temperature, solar radiation, air pressure, wind speed and relative humidity. Since the purpose in the estimation of evaporation the ANN architecture has only one output neuron with up to 4 input neurons representing air and water temperature, air pressure and solar radiation. Prior to ANN model construction the classical correlation study indicated that the insignificance of the wind speed and the relative humidity in the Lake Egirdir area. Hence the final ANN model has 4 input neurons in the input layer with one at the output layer. The hidden layer neuron number is found 3 after various trial and error models running. The ANN model provides good estimate with the least Mean Square Error (MSE).

Molina Martinez [63] developed and validated a simulation model of the evaporation rate of a Class A evaporimeter pan (E_{pan}). A multilayer model was first developed, based on the discretization of the pan water volume into several layers. The energy balance equations established at the water surface and within the successive in-depth layers were solved using an iterative numerical scheme. The wind function at the pan surface was identified from previous experiments, and the convective processes within the tank were accounted for by introducing an internal 'mixing' function, which depends on the wind velocity. The model was calibrated and validated using hourly averaged measurements of the evaporation rate and water temperature, collected in a Class A pan located near Cartagena (South-east Spain). The simulated outputs of both water temperature and E_{pan} proved to be realistic when compared to the observed values. Experimental data evidenced that the convective mixing process within the water volume induced a rapid homogenization of the temperature field within the whole water body. This result led us to propose a simplified version of the multilayer model, assuming an isothermal behavior of the pan. The outputs of the single layer model are similar to those supplied by the multilayer model although slightly less accurate. Due to its good predictive performances, facility of use and implementation, the simplified model may be proposed for applied purposes, such as routine prediction of Class A pan evaporation, while the multilayer model appears to be more appropriate for research purposes.

Keskin and Terzi [48] studied the ANN models and proposed as an alternative approach of evaporation estimation for Lake Eirdir. This study has three objectives: (1) to develop ANN models to estimate daily pan evaporation from measured meteorological data; (2) to compare the ANN models to the Penman model; and (3) to evaluate the potential of ANN models. Meteorological data from Lake Eirdir consisting of 490 daily records from 2001 to 2002 are used to develop the model for daily pan evaporation estimation. The measured meteorological variables include daily observations of air and water temperature, sunshine hours, solar radiation, air pressure, relative humidity, and wind speed. The results of the Penman method and ANN models are compared to pan evaporation values. The comparison shows that there is better agreement between the ANN estimations and measurements of daily pan evaporation than for other model.

Dogan and Demir [26] investigated that the evaporation amount from the lake surface is important in terms of drinking water, irrigation, demand of industrial water, cultivated plant. Generally, daily evaporation amount is calculated two ways. First way is directly evaporation pan estimation. Secondly way is indirectly depending on meteorological data like Penman-Monteith model (PM model). There are some difficulties in this method, such as; long measurement times, difficulties in measurement, evaporation calculation equations are not universal, *etc.* In this study, Genetic Algorithm (GA) and Back propagation Feed Forward Neural Network (FFNN) have been adapted to estimate daily evaporation amount for Lake Sapanca. FFNN and GA models have been applied to daily evaporation estimation depending on daily min and max temperature, wind speed, relative humidity, real solar period and maximum solar period. When performances of the ANN and GA models compared, it has been seen that ANN model yields best result.

Tan Stephen Boon Kean [83] studied evaporation rate estimation is important for water resource studies. Previous studies have shown that the radiation-based models, mass transfer models, temperaturebased models and ANN models generally perform well for areas with a temperate climate. This study evaluates the applicability of these models in estimating hourly and daily evaporation rates for an area with an equatorial climate. Unlike in temperate regions, solar radiation was found to correlate best with pan evaporation on both the hourly and daily time-scales. Relative humidity becomes a significant factor on a daily time-scale. Among the simplified models, only the radiation-based models were found to be applicable for modeling the hourly and daily evaporations. ANN models are generally more accurate than the simplified models if appropriate network architecture is selected and a sufficient number of data points are used for training the network. ANN modeling becomes more relevant when both the energy- and aerodynamics-driven mechanisms dominate, as the radiation and the mass transfer models are incapable of producing reliable evaporation estimates under this circumstance.

Deswal and Mahesh Pal [23] studied an Artificial Neural Network based modeling technique has been used to study the influence of different combinations of meteorological parameters on evaporation from a reservoir. The dataset used is taken from an earlier reported study. Several input combination were tried so as to find out the importance of different input parameters in predicting the evaporation. The prediction accuracy of Artificial Neural Network has also been compared with the accuracy of linear regression for predicting evaporation. The comparison demonstrated superior performance of Artificial Neural Network over linear regression approach. The findings of the study also revealed the requirement of all input parameters considered together, instead of individual parameters taken one at a time as reported in earlier studies, in predicting the evaporation. The highest correlation coefficient (0.960) along with lowest root mean square error (0.865) was obtained with the input combination of air temperature, wind speed, sunshine hours and mean relative humidity. A graph between the actual and predicted values of evaporation suggests that most of the values lie within a scatter of $\pm 15\%$ with all input parameters. The findings of this study suggest the usefulness of ANN technique in predicting the evaporation losses from reservoirs.

Shirgure and Rajput [79] developed the model, which can generalize for the diversified Indian conditions. The investigation was carried out to develop and test the daily pan evaporation prediction models using various weather parameters as input variables with ANN and validated with the independent subset of data for five different locations in India. The measured variables included daily observations of maximum and minimum temperature, maximum and minimum relative humidity, wind speed, sunshine hours and rainfall. In this GM model development and evaluation has been done for the five locations viz. Nagpur; Jabalpur; Akola; Hyderabad and Udaipur. The daily data of pan evaporation and other inputs for two years was considered for model development and subsequent 1–2 years data for validation. Weather data consisting of 3305 daily records from 2002–2006 were used to develop the GM models of daily pan evaporation. Additional weather from these five locations, which included 2066 daily records from 2004–2007, served as an independent evaluation dataset for the performance of the models. From the studies it is concluded that three layered feed forward neural networks with *Levenberg Marquardt* minimization training function gave the best network training when used in the back propagation algorithm with hidden nodes as 2^n+1 for GM modeling. The General ANN models of daily pan evaporation with all available variables as an input was the most accurate model delivering an R^2 of 0.84 and a root mean square error 1.44 mm/day for the model development dataset. The GM evaluation with NM model development data shown lowest RMSE (1.961 mm/d), MAE (0.038 mm) and MARE (2.30 %) and highest r (0.848), R^2 (0.719) and d (0.919) with ANN GM with all input variables. The GM evaluation data has shown the lowest RMSE (1.615 mm/d) and highest R^2 (0.781) with ANN GM model consisting of all inputs except sunshine hours (Model M-3). The General model evaluation with NRCC, Nagpur data has shown the lowest RMSE as 1.86 mm/day; with JNKVV, Jabalpur has shown the lowest RMSE as 1.547 mm/day; with PDKV, Akola as 1.572 mm/day RMSE; with ICRISAT, Hyderabad as 1.481 mm/day RMSE and with MPUAT, Udaipur as 2.069 mm/day.

6.3 SUMMARY

Evaporation is an essential component of the hydrological cycle and influenced by several meteorological parameters. But measurement of evaporation with accuracy is and continuous is a difficult operation. In such situations, it becomes an imperative to use neural network models that can estimate evaporation from available climatic data and may give more accurate results than the measured pan evaporation. In this regard, a number of models for predicting the pan evaporation have been developed by several investigators for different locations of India and abroad. Most of the current models for predicting evaporation use the principles of the deterministically based combined energy balance – vapor transfer approach or empirical relationships based on climatological variables. This resulted in relationships that were often subjected to rigorous local calibrations and therefore proved to have limited global validity. Due to these limitations the conventionally applied regression modeling techniques need to be further refined to achieve improved performance by adopting new and advanced technique like neural networks. Evaporation process is complex and needs nonlinear modeling and hence, can be modeled through ANN. Large number of researchers have been established the applicability of ANN to the problems in agricultural, hydrological, meteorological and environmental fields. The research review related to evaporation modeling using neural networks is discussed here in brief.

KEY WORDS

- **Artificial Neural Network**
- **bias weight**
- **coefficient of determination**
- **correlation coefficient**
- **epochs**
- **error back propagation algorithm**

- evaporation
- evaporation modeling
- evapotranspiration
- feed forward neural networks
- hidden layer
- hydrological process
- hydrology
- index of agreement
- input variable
- input weight
- irrigation management
- land resources management
- layer weight
- learning rate
- *Levenberg Marquardt* minimization training function
- linear regression methods
- linear regression model
- logistic sigmoid
- MATLAB
- maximum temperature
- mean absolute error
- mean absolute relative error
- mean square error
- minimum temperature
- model evaluation
- model architecture
- model development
- model strategies
- model validation
- modeling
- momentum
- multi-layer preceptron
- multi-layered feed forward ANN
- multiple linear regression
- network function
- non-linear regression models

- observed
- output variable
- pan evaporation
- predicted
- prediction of evaporation
- rainfall
- relative humidity
- root mean square error
- soil water evaporation
- sunshine hours
- the class 'A' pan evaporimeter
- water balance assessment
- water resources planning
- weather forecasting
- wind speed

REFERENCES

1. Allen, G.; Le Marshall, J. F. "An evaluation of neural networks and discrimination analysis methods for application in operational rain forecasting," Australian Meteorological Magazine, **1994**, *43,* 17–28,

2. Altendorf, C. T.; Elliott, R. C.; Stevens E. W. M. L, "Stone. Development and validation of a neural network model for soil water content prediction with comparison to regression techniques," Trans. ASAE, **1999**, *42(3)*, 691–699,

3. Anderson, M. E.; Jobson, H. E.; "Comparison of techniques for estimating annual lake evaporation using climatological data," Water Resources Res.; **1982**, *18,* 630–636,

4. Arca, B.; Benincasa F.; De Vincenzi M.; Ventura, A.; "Neural network to simulate evaporation from Class A pan," In: Proc. 23rd Conference on Agric. For. Meteorol.; 2–6 November **1998**, Albuquerque, New Mexico, Am. Meteorol. Soc.; Boston, MA. **1998.**

5. Ashrafzadeh, A., "Application of ANN for prediction of evaporation from evaporative ponds," M. S. thesis, University of Tehran, Tehran (in Persian), **1999**, 131.

6. Ashish, D.; Hoogenboom G.; McClendon, R. W. "Land-Use Classification of Gray Scale Aerial Images using ANN," Transactions of the ASAE **2004**, *47(5)*, 1813–1819.

7. Baier, W.; Robertson, G. W.; "Evaluation of meteorological factors influencing evaporation," J.; Hydrology, **1965**, *45,* 276–284.

8. Bocco, M.; Ovando, G.; Sayago, S, "Development and evaluation of neural network models to estimate daily solar radiation at Cordoba, Argentina," Pesq. Agropec. Bras.; Brasilia, **2006**, *41(2),* 179–184.

9. Braddock, R. D.; Kremmer, M. L.; Sanzogni, L.; "Feed-forward artificial neural network model for forecasting rainfall run-off," Environmetrics, **1998**, *9,* 419–432.

10. Brian A.; Smith, Ronald W. McClendon, Hoogenboom, G.; "Improving air temperature prediction with ANN," International Jr. of Computational Intelligence, **2006**, *3(3)*, 179–186.

11. Bruin, H. A. R. D, "A simple model for shallow lake evaporation," Applied Meterol.; **1978**, *17,* 1132–1134.

12. Bruton, J. M.; McClendon, R. W.; Hoogenboom, G. "Estimating daily pan evaporation with ANN," Trans. ASAE, **2000,** *43(2),* 491–496.
13. Brutsaert, W, "Evaporation into the atmosphere: Theory, history, and application," D.; Reidel, **1982,** 299.
14. Chaloulakou, A.; Grivas G.; Spyrellis, N.; "Neural network and multiple regression models for PM_{10} prediction in Athens: A comparative assessment," J. of Air Waste Manag. Association, **2003,** *53,* 183–190.
15. Chattopadhyay, S, "Multilayered feed forward artificial neural network model to predict the average summer monsoon rainfall in India," URL: http://www.arxiv.org/ftp/nlin/papers/0609/0609014.pdf. **2006.**
16. Chaudhuri, S.; Chattopadhyay, S., "Neuro-Computing based short range prediction of some meteorological parameters during premonsoon season," Soft-computing – A fusion of foundations, Methodologies and Applications, **2005,** *9(5),* 349–354.
17. Chokmani, K.; Khalil, B. M.; Taha Ouarda B. M. J.; Bourdages, R., "Estimation of river ice thickness using ANN," CGU HS Committee on River Ice Processes and the Environment, 14th Workshop on the Hydraulics of Ice Covered Rivers Quebec City, June 19–22. **2007.**
18. Clayton, L. H., "Prediction of class A pan evaporation in south Idaho," ASCE J. of Irrig. Drain. Eng.; **1989,** *115(2),* 166–171.
19. Cook, D. F.; Wolfe, M. L., "A back propagation neural network to predict average air temperatures," AI applications, **1991,** *5,* 40–46.
20. Coulomb, C. V.; Legesse, D.; Gasse, F.; Travi, Y.; Cherner, T, "Lake evaporation estimates in tropical Africa (Lake Ziway, Ethopia)," J. of Hydrology, **2001,** *245,* 1–18.
21. Dawon, W. C.; Wilby, R., "An artificial neural network approach to rainfall-runoff modeling," Hydrol. Sci. J.; **1998,** *43(1),* 47–66.
22. de Villiers, J.; Barnard, E., "Backpropagation neural nets with one and two hidden layers," IEEE Transactions on Neural Networks, **1993,** *4(1),* 136–141.
23. Deswal S.; Mahesh Pal, "Artificial Neural Network based Modeling of Evaporation Losses in Reservoirs," Proc. of World Academy of Science, Engineering and Technology, **2008,** *29,* 279–283.
24. Dibike Y. B.; Solomatine D. P., "River flow forecasting using ANN," Phys. Chem. Earth (B), **2001,** *26,* 1–7.
25. Dimri, A. P.; Mohanty, U. C.; Madan, O. P.; Ravi N.; "Statistical model-based forecast of minimum and maximum temperatures at *Manali,"* Current Science, **2002,** *82(8),* 25–27.
26. Dogan, E.; Demir, A. S, "Evaporation amount estimation using Genetic algorithm and Neural networks," Proceedings of 5th International Symposium on Intelligent Manufacturing Systems, May 29–31, **2006,** 1239–1250.
27. Dorvlo, A. S. S.; Jervase J. A.; Al-Lawati, A. "Solar radiation estimation using ANN," Applied Energy, **2002,** *71,* 307–319.
28. Elizondo, D.; Hoogenboom, G.; McClendon, R. W, "Development of a neural network model to predict daily solar radiation," Agric. For. Met.; **1994,** *71,* 1–2, 115–132.
29. Ferentinos, K. P.; Albright, L. D.; "Predictive neural network modeling of pH and electrical conductivity in deep through hydrophonics," Trans. ASAE, 2007, *45(6),* 2007–2015.
30. Gallant, S. T, "Neural Network learning and Expert systems," Cambridge, Mass: MIT press. 1993.
31. Gavin, H.; Agnew, C. A, "Modeling actual, reference and equilibrium evaporation from a temperate wet grassland," Hydrol. Processes, **2004,** *18,* 229–246.
32. Gardner, M. W.; Dorling, S. R, "Artificial neural network (Multilayer Perceptron) a review of applications in atmospheric sciences," Atmos. Environ.; **1998,** *32,* 2627–2636.
33. Grivas, G.; Chaloulakou, A, "Artificial neural network models for prediction of PM_{10} hourly concentrations, in the Greater Area of Athens, Greece," Atmospheric environment, **2006,** *40(7),* 1216–1229.
34. Han, H.; Felkar, P.; "Estimation of daily soil water evaporation using an artificial neural network," J.; Arid Environ.; **1997,** *37(2),* 251–260.

35. Hanson, C. L., "Prediction of class A pan evaporation in South-west Idaho," J.; Irrig. Drain. Engg.; ASCE, **1989**, *115(2)*, 166–171.

36. Haykin, S., "Neural Networks: A comprehensive Foundation," 2nd Edition, Pearson Education Inc.; New Delhi. 2001.

37. Hoogenboom, G., "Contribution of agro-meteorology to the simulation of crop production and its applications," Agril. Forest Met.; **2000**, *103*, 1–2, 137–157.

38. Hornik, K.; Stinchcombe, M.; White, H., "Multilayer feed forward networks are universal approximators," Neural Networks, **1989**, *2(5)*, 359–366.

39. Hu, M. J. C., "Application of ADLINE system to weather forecasting," Technical Report, Stanford Electron. **1964,**

40. Imrie, C. E.; Durucan S.; Korre, A.; "River flow prediction using ANN: generalization beyond the calibration range," Hydrol. **2000**, *233*, 138–153.

41. Irmak, S.; A.; Irmak, Jones, J. W.; Howell, T. A.; Jacobs, J. M.; R. G.; Allen and G.; Hoogenboom, "Predicting daily net radiation using minimum climatological data," J.; Irri. Drain. Engg.; ASCE, **2003**, *129(4)*, 256–269.

42. Jagadesh, A., "Comparison of ANN and other empirical approaches for predicting watershed runoff," J. of Water Resources Planning and Management, **2000**, *126*, 156–166.

43. Jain, A.; McClendon R. W.; Hoogenboom, G.; "Frost prediction using ANN: A temperature prediction approach," ASAE Paper 033075, ASAE, St. Joseph, MI. **2003.**

44. Jef, H.; Clemens, M.; Gerwin, D.; Frans, F.; Oliver, B., "A neural network forecast for daily average PM_{10} concentration in Belgium," Atmos. Environ.; **2005**, *39*, 3279–3289.

45. Jhajharia, D.; Fancon, A. K.; Kithan, S. B., "Relationship between USWB class A pan evaporation and meteorological parameters under humid climatic conditions of *Umiam, Meghalaya*," Proc. of International Conference on Recent advances in Water Resources Development and Management, Nov. 23–25, **2005**, IIT Roorkee, Allied Pubs. PVT. **2005**, 71–83.

46. Jian-Ping Suen J.; Wayland Heart, "Evaluation of Neural Networks for Modeling Nitrate Concentrations in Rivers," J.; Water Resour. Plng. and Mgmt.; 2003, *129(6)*, 505–510.

47. **Keskin, M. E.; Özlem Terzi; Dilek Taylan,** "Fuzzy logic model approaches to daily pan evaporation estimation in western Turkey," Hydrological Sciences Journal, **2004**, *49(6)*, 1001–1010.

48. Keskin, M. E.; Ozlem Terzi, "Artificial Neural Network Models of Daily Pan Evaporation," J.; Hydrologic Engrg.; **2006**, *11(1)*, 65–70.

49. Khan, M. A., "Evaporation of water from free water surface," Ind. J.; Soil Cons.; **1992**, *20*, 1–2, 22–27.

50. Kisi, O., "Multilayer perceptrons with Levenberg-Marquardt training algorithm for suspended sediment concentration prediction and estimation," Hydrol. Sci. J. **2004**, *49(6)*, 1025–40.

51. Kisi, O., "Daily river flow forecasting using ANN and autoregressive models," Turk. J.; Eng. Environ. Sci.; **2005**, *29*, 9–20.

52. Kumar, M.; Raghuwanshi, N. S.; Singh, R.; Wallender W. W.; Pruitt, W. O. "Estimating evapotranspiration using artificial neural network," J.; Irri. Drain. Engg. ASCE, **2002**, *128(4)*, 224–233.

53. Lee, S.; Cho, S.; Wong, P. M., "Rainfall prediction using Artificial neural network," J. of Geographic Information and Decision Analysis, **1998**, *2*, 254–264.

54. Li, B. R.; McClendon, W.; Hoogenboom, G.; "Spatial interpolation of weather data for single locations using ANN," Transactions of the ASAE, **2004**, *47(2)*, 629–637.

55. Lopez-Sataber, C. J.; Renard, V, K. G.; Lopes, L.; "Neural network based algorithms of hydraulic roughness for overland flow," Trans. ASAE.; **2002**, *45(3)*, 661–667.

56. Maier. H. R.; Dandy, G. C., "The use of ANN for prediction of water quality parameters," Wat. Resour. Res.; **1996**, *32(4)*, 1013–1022.

57. Maier, H. R.; Dandy, G. C.; "Neural network based modeling of environmental variables: a systematic approach," Mathematical and Computer Modeling, **2001**, *33*, 669–682.

58. Mamdouh, A. A.; Ibrahim, E.; Mohamed N. A., "Rainfall-runoff modeling using ANN technique: a Blue Nile catchment case study," Hydrological Processes, **2006**, *20(5)*, 1201–1216.

59. Maqsood, I.; Muhammad, R. K.; Abraham, A, "Neuro-computing based Canadian weather analysis," Computational Intelligence and applications, Dynamic Publishers, Inc. USA, **2002**, 39–44.

60. Mehuys, G. R.; Patni, N. K.; Prasher S. O.; Yang, C. C.; "Application of ANN for simulation of soil temperature," Trans. of the ASAE, **1997**, *40*, 3, 649–656.

61. Minns, A. W.; Hall, M. J.; "ANN as rainfall runoff models," Hydrol. Sci. J.; **1996**, *41(3)*, 99–117.

62. Mohandes, M. A.; Rehman, S.; Halawani, T. O., "A neural networks approach for wind speed prediction," Renewable Energy, **1998**, *13*, 345–354.

63. Molina J. M.; Martinez, M.; Alvarezv, M.; Gonzalez Real M.; Baille, A., "A simulation model for predicting hourly pan evaporation from meteorological data," Journal of hydrology, **2006**, *318*, 1–4, 250–261.

64. Mort, N.; Robinson, C.; "A neural network system for the protection of Citrus crops from frost damage," Computer and Electronics in Agriculture, **1996**, *16(3)*, 177–187.

65. Muleta, M. K.; John W.; Nicklow, "Joint Application of ANN Evolutionary Algorithms to Watershed Management," Water Resources Management, **2004**, *18(5)*, 459–482.

66. Murthy, S.; S.; Gawande, "Effect of metrological parameters on evaporation in small reservoirs 'Anand Sagar' Shegaon – a case study," J.; *Prudushan Nirmulan*, **2006**, *3(2)*, 52–56.

67. Ozlem, T.; Evolkesk, M., "Modeling of daily pan evaporation," J. of Applied Sciences, *5(2)*, 368–372. **2005**,

68. Pachepsky, Y. A.; Timlin, D. J.; Varallvay, G, "ANN to estimate soil water retension from easily measurable data," Soil sciences Soc. of American Journal, 1996, *60*, 727–733.

69. Patel, R M, Prasher, S. O.; Goel, *P* K, Bassi, R, "Soil salinity prediction using ANN," Jr. of the American Water Reso. Asso., **2002**, *38(1)*, 91–100.

70. Patil, N. G.; Rajput, G. S.; Nema, R. K.; Singh, R. B, "Indirect estimation of soil hydrolimits from physical properties," Proc. of National seminar on technological options for improving the water productivity in agriculture, November 15–17, 2006, J. N.; K. V. V.; Jabalpur, 2006, 85–90.

71. Perez, P.; Reyes, J.; "Prediction of particulate air pollution using neural techniques," Neural Computing and Applications, **2001**, *10*, 165–171.

72. Rajurkar, M. P.; Kothyari U. C.; Chaube, U. C.; "Modeling of the daily rainfall-runoff relationship with artificial neural network," J.; Hydrology, **2004**, *285(1)*, 96–113.

73. Reddy, K. S.; Ranjan, M, "Solar resource estimation using ANN and comparison with other correlation models," Energy Conversion and Management, **2003**, *44*, 2519–2530.

74. Reis, R. J.; Dias, N. L, "Multi-season lake evaporation: energy budget estimates and CRLE model assessment with limited meteorological observations," J. of Hydrology, **1998**, *208*, 135–147.

75. Sajikumar, N.; Thandaveswara, B. S., "A nonlinear rainfall-runoff model using an artificial neural network," J.; Hydrology, **1999**, *216*, 32–55.

76. Sarkar, A.; Agrawal, A.; Singh, R. D, "Artificial neural network models for rainfall-runoff forecasting in hilly catchment," J. of Ind. Water Resources Soci.; **2006**, *26*, 3–4, 5–12.

77. Schaap, M. G.; W.; Bouten, "Modeling water retention curves of sandy soils using neural networks," Water Resour. Res.; **1996**, *32(10)*, 3033–3040.

78. Sharma, Mukesh Kumar, "Estimation of pan evaporation using meteorological parameters," Hydrology. J.; **1995**, *18*, 3–4, 1–9.

79. Shirgure, P. S.; Rajput G. S.; Seth, N. K.; "Artificial neural network models for estimating daily pan evaporation," Abstract, In 45th Annual Convention of ISAE International Symposium on Water for Agriculture, held at College of Agriculture, Dr. PDKV, Nagpur during 17–19th Janurary, **2011**, 81.

80. Smith, B. A.; McClendon, R. W.; Hoogenboom, G.; "Improving Air Temperature Prediction with ANN," International Journal of Computational Intelligence, **2006**, *3(3)*, 179–186.

81. Sudheer, K. P.; Gosain, A. K.; Mohana, R. D.; Saheb, S. M., "Modeling evaporation using an artificial neural network algorithm," Hyd. Processes, **2002**, *16*, 3189–3202.

82. Sudheer, K. P.; Gosain A. K.; Ramasastri, K. S.; "Estimating actual evapotranspiration from limited climatic data using neural computing technique," J.; Irri. Drain. Engg. ASCE, **2003,** *129(3),* 214–218.

83. Tan Stephen Boon Kean· Eng Ban Shuy and Chua Lloyd Hock Chye, "Modeling hourly and daily open-water evaporation rates in areas with an equatorial climate," Hydrological processes, **2007,** *21(4),* 486–499.

84. Tasadduq, I.; Rehman S.; Bubshait, K.; "Application of neural networks for the prediction of hourly mean surface temperatures in Saudi Arabia," Renewable Energy, **2002,** *25,* 545–554.

85. Taher, S. A., "Estimation of potential evaporation: ANN versus Conventional methods," J.; King Saud Univ.; **2003,** *17(1),* 1–14.

86. Terzi, O.; Keskn, M. E.; "Modeling of daily pan evaporation," *Appl. Sci.;* **2005,** *5,* 368–372.

87. Tokar, A. S.; Johnson, P. A, "Rainfall-Runoff Modeling Using ANN," J.; Hydrologic Engineering, 1999, *4(3),* 232–239.

88. Tokar, A. S.; and Momcilo Markus, "Precipitation-runoff modeling using ANN and conceptual models," Journal of Hydrologic Engineering, 2000, *5(2),* 56–161.

89. Trajkovic, Slavisa, Miomir Stankovic and Branimir Todorovic, "Estimation of FAO Blaney-Criddle *b* factor by RBF network," *J. Irri. Drain. Engg. ASCE.* **2000,** *126(4),* 268–274.

90. Trajkovic, Slaves, Branimir Todorovic and Miomir Stankovic, "Estimation of FAO Penman C factor by RBF networks," Arch.itecture and Civil Engineering, **2001,** *2(3),* 185–191.

91. Trajkovic, Slavisa, Branimir Todorovic and Miomir Stankovic, "Forecasting of reference evapotranspiration using artificial neural network," J.; Irri. Drain. Engg. ASCE, **2003,** *129(6),* 454–457.

92. Yang, C. C.; Prasher S. O.; Lacroix, R.; "Applications of artificial neural network to land drainage engineering," Trans. ASAE, **1996,** *39(2),* 525–533.

93. Yang, C. -C.; Prasher, S. O.; Landry, J.-A.; Ramaswamy, H. S.; DiTommaso, "Application of ANN in image recognition and classification of crop and weeds," Can. Agric. Eng.; **2000,** *42,* 147–152.

94. Zealand, C. M.; Burn, D. H.; Simonovic, S. P., "Short term streamflow forecasting using ANN," *Hydrol.;* **1999,** *214,* 32–48.

CHAPTER 7

PAN EVAPORATION MODELING: INDIAN AGRICULTURE[1]

P. S. SHIRGURE and G. S. RAJPUT

CONTENTS

[1]P. S. Shirgure, Senior Scientist (Soil and Water Conservation Engineering) and Agricultural Engineer, National Research Center for Citrus – Indian Council of Agricultural Research, Amravati Road, Nagpur – India – 440010. EMAIL: <shirgure@gmail.com> and G. S. Rajput, Principal Scientist, J.N.K.V.V., Jabalpur-India.

7.1 INTRODUCTION

Artificial neural networks are much easier to set-up than many of the statistical models, such as regression (with only a small amount of time an ANN can be set-up; with learning and practice they can provide better results, since it is more of an art to setting one up, than a science). Most of the research on the application of ANN in agriculture science has been done in the last decade, mainly in the area of yield prediction, spatial modeling, and spatial-temporal forecasting. This chapter will introduce the basic concepts about neural network. Also the work done on the application of ANN in agriculture research will be reviewed. The application of ANN in agriculture research using real data will be presented.

7.2 ARTIFICIAL NEURAL NETWORK (ANN) MODELING

Artificial neural networks (ANN) are among the newest signal-processing technologies in the engineer's toolbox. The field is highly interdisciplinary, but our approach will restrict to the engineering perspective. In engineering, neural networks serve two important functions: as pattern classifiers and as nonlinear adaptive filters. This chapter will provide a brief overview of the theory, learning rules, and applications of the most important neural network models. Definitions and Style of Computation an Artificial Neural Network is an adaptive, most often nonlinear system that learns to perform a function (an input/output map) from data. Adaptive means that the system parameters are changed during operation, normally called the training phase. After the training phase the Artificial Neural Network parameters are fixed, and the system is deployed to solve the problem at hand (the testing phase). The Artificial Neural Network is built with a systematic step-by-step procedure to optimize a performance criterion or to follow some implicit internal constraint, which is commonly referred to as the learning rule.

The input/output training data are fundamental in neural network technology, because they convey the necessary information to "discover" the optimal operating point. The nonlinear nature of the neural network processing elements (PEs) provides the system with lots of flexibility to achieve practically any desired input/output map, i.e., some Artificial Neural Netw. are universal mappers. There is a style in neural computation that is worth describing.

An input is presented to the neural network and a corresponding desired or target response set at the output (when this is the case the training is called supervized). An error is composed from the difference between the desired response and the system output. This error information is fed back to the system and adjusts the system parameters in a systematic fashion (the learning rule). The process is repeated until the performance is acceptable. It is clear from this description that the performance hinges heavily on the data. If one does not have data that cover a significant portion of the operating conditions or if they are noisy, then neural network technology is probably not the right solution. On the other hand, if there is plenty of data and the problem is poorly understood to derive an approximate model, then neural network technology is a good choice. This operating procedure should be contrasted with the traditional engineering design, made of exhaustive subsystem specifications and intercommunication protocols. In artificial neural networks, the designer chooses the network topology, the performance function, the learning rule, and the criterion to stop the training phase, but the system automatically adjusts the parameters. So, it is difficult to bring a priori information into the design, and when the system does not work properly it is also hard to incrementally refine the solution. But ANN-based solutions are extremely efficient in terms of development time and resources, and in many difficult problems artificial neural networks provide performance that is difficult to match with other technologies. Denker indicated that "artificial neural networks are the second best way to implement a solution" motivated by the simplicity of their design and because of their universality, only shadowed by the traditional design obtained by studying the physics of the problem. At present, artificial neural networks are emerging as the technology of choice for many applications, such as pattern recognition, prediction, system identification, and control.

7.3   THE BIOLOGICAL MODEL

Artificial neural networks emerged after the introduction of simplified neurons. These neurons were presented as models of biological neurons and as conceptual components for circuits that could perform computational tasks. The basic model of the neuron is founded upon the functionality of a biological neuron. "Neurons are the basic signaling units of the nervous system" and "each neuron is a discrete cell whose several processes arize from its cell body."

The neuron has four main regions to its structure: The cell body, or soma, has two offshoots from it, the dendrites, and the axon, which end in presynaptic terminals. The cell body is the heart of the cell, containing the nucleus and maintaining protein synthesis. A neuron may have many dendrites, which branch out in a treelike structure, and receive signals from other neurons. A neuron usually only has one axon which grows out from a part of the cell body called the axon hillock. The axon conducts electric signals generated at the axon hillock down its length.

These electric signals are called action potentials. The other end of the axon may split into several branches, which end in a presynaptic terminal. Action potentials are the electric signals that neurons use to convey information to the brain. All these signals are identical. Therefore, the brain determines what type of information is being received based on the path that the signal took. The brain analyzes the patterns of signals being sent and from that information it can interpret the type of information being received. Myelin is the fatty tissue that surrounds and insulates the axon. Often short axons do not need this insulation. There are un-insulated parts of the axon. These areas are called Nodes of Ranvier. At these nodes, the signal traveling down the axon is regenerated. This ensures that the signal traveling down the axon travels fast and remains constant (i.e., very short propagation delay and no weakening of the signal). The synapse is the area of contact between two neurons. The neurons do not actually physically touch. They are separated by the synaptic cleft, and electric signals are sent through chemical 13 interaction. The neuron sending the signal is called the presynaptic cell and the neuron receiving the signal is called the postsynaptic cell. The signals are generated by the membrane potential, which is based on the differences in concentration of sodium and potassium ions inside and outside the cell membrane. Neurons can be classified by their number of processes (or appendages), or by their function. Their processes are classified into three categories. Uni-polar neurons have a single process (dendrites and axon are located on the same stem), and are most common in invertebrates. In bipolar neurons, the dendrite and axon are the neuron's two separate processes. Bipolar neurons have a subclass called pseudobipolar neurons, which are used to send sensory information to the spinal cord. Finally, multipolar neurons are most common in mammals. Examples of these neurons are spinal motor neurons, pyramidal cells and Purkinje cells (in the cerebellum). If classified by function, neurons again fall into three separate categories. The first group is sensory, or afferent, neurons, which provide information for perception and motor coordination. The second group provides information (or instructions) to muscles and glands and is therefore called motor neurons. The last group, interneuronal, contains all other neurons and has two subclasses. One group called relay or projection interneurons have long axons and connect different parts of the brain. The other group called local interneurons are only used in local circuits.

7.4 THE NEURAL NETWORK MODEL

Neural networks are composed of simple elements operating in parallel. These elements are inspired by biological nervous systems. As in nature, the network function is determined largely by the connections between elements. A neural network can be trained to perform a particular function by adjusting the values of the connections (weights) between the elements. Commonly neural networks are adjusted, or trained, so that a particular input leads to a specific target output.

Neural networks use a complex combination of weights and functions to convert input variables into an output (the prediction). The Multi-Layer Perceptron (MLP) consists of a system a simple processing inter connected elements called neurons, cells or nodes. Each of the various inputs to the network is multiplied by a connection weight. These products are simply summed, fed through a transfer function to generate a result, and then output. This is a gradient-decent algorithm that is normally used to train a MLP network. Errors in the output of this procedure are assumed to be due to all processing element and connections, and these errors are reduced by propagating the output error backward to the connections in the previous layers.

Feed forward ANN models comprize a system of neurons, which are arranged in successive layers, namely input and output layers in addition to one or more hidden layers. The neurons in each layer are connected to the neurons in the subsequent layer by a weight w, which may be adjusted during training. A data pattern comprizing the values x_i presented at the input layer i, is propagated forward through the network towards the hidden layer j. Each hidden neuron receives the weighted outputs $w_{ji} x_i$, from the neurons in the previous layer. These are summed to produce a net value (NET_j), which is then transformed to an output value upon the application of an activation function.

A typical three-layer feed-forward ANN consists of three layers namely input, hidden and output layers. Input layer neurons are $x_1, x_2, x_3, ...x_n$; hidden layer neurons are $h_1, h_2, h_3, ...h_n$; and finally output layer neurons are $o_1, o_2, o_3, ...o_n$.

A neuron consists of multiple inputs and single output. The sum of inputs and their weights lead to a summation operation as,

$$NET_j = \Sigma W_{ij} X_i \tag{1}$$

where, W_{ij} is established weight, X_i is input value and NET_j is input to a node in a layer j.

The output of a neuron is decided by an activation function. There are number of activation functions that can be used in ANNs such as step, sigmoid, threshold, linear etc. The logistic sigmoid function, $f(x)$, commonly used, can be formulated mathematically as:

$$f(x) = 1/[1+\exp(-x)]$$

$$OUTPUT_j = f(NET_j) = 1/[1+\exp(-NET_j)] \tag{2}$$

When creating a functional model of the biological neuron, there are three basic components. First, the synapses of the neuron are modeled as weights. The strength of the connection between an input and a neuron is noted by the value of the weight. Negative weight values reflect inhibitory connections, while positive values designate excitatory connections. The next two components model the actual activity within the neuron cell. An adder sums up all the inputs modified by their respective weights. This activity is referred to as linear combination. Finally, an activation function controls the amplitude of the output of the neuron. An acceptable range of output is usually between 0 and 1, or −1 and 1. Mathematically, this process is described in the Fig. 1.

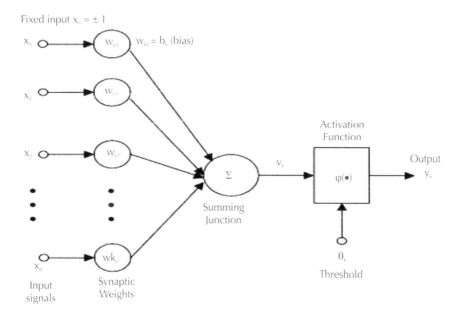

FIGURE 1   The mathematical process.

From this model the interval activity of the neuron is shown below:

$$v_k = \sum_{j=1}^{p} w_{kj} x_j \qquad (3)$$

The output of the neuron, y_k, is therefore be the outcome of some activation function on the value of v_k.

7.4.1 ACTIVATION FUNCTIONS

As mentioned previously, the activation function acts as a squashing function, such that the output of a neuron in a neural network is between certain values (usually 0 and 1, or −1 and 1). In general, there are three types of activation functions, denoted by . First, there is the Threshold Function, which takes on a value of 0 if the summed input is less than a certain threshold value (), and the value 1 if the summed input is greater than or equal to the threshold value.

$$\varphi(v) = \begin{cases} 1 \text{ if } v \geq 0 \\ 0 \text{ if } v < 0 \end{cases} \qquad (4)$$

Secondly, there is the Piecewise-Linear function. This function again can take on the values of 0 or 1, but can also take on values between that depending on the amplification factor in a certain region of linear operation.

$$\varphi(v) = \begin{cases} 1 & v \geq \frac{1}{2} \\ v - \frac{1}{2} > v > -\frac{1}{2} \\ 0 & v \leq -\frac{1}{2} \end{cases} \qquad (5)$$

Thirdly, there is the sigmoid function (Fig. 2). This function can range between 0 and 1, but it is also sometimes useful to use the −1 to 1 range. An example of the sigmoid function is the hyperbolic tangent function.

$$\varphi(v) = \tanh\left(\frac{v}{2}\right) = \frac{1 - \exp(-v)}{1 + \exp(-v)} \qquad (6)$$

The artificial neural networks that we describe are all variations on the parallel-distributed processing (PDP) idea. The architecture of each neural network is based on very similar building blocks, which perform the processing. In this chapter we first discuss these processing units and discuss different neural network topologies. Learning strategies as a basis for an adaptive system

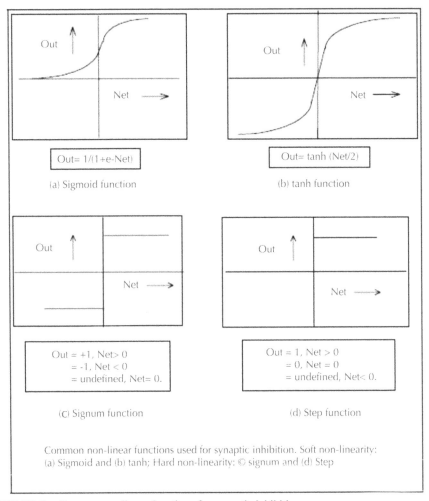

Out= 1/(1+e-Net)

(a) Sigmoid function

Out= tanh (Net/2)

(b) tanh function

Out = +1, Net> 0
= -1, Net < 0
= undefined, Net= 0.

(C) Signum function

Out = 1, Net > 0
= 0, Net = 0
= undefined, Net< 0.

(d) Step function

Common non-linear functions used for synaptic inhibition. Soft non-linearity:
(a) Sigmoid and (b) tanh; Hard non-linearity: © signum and (d) Step

FIGURE 2   Common nonlinear functions for synaptic inhibition.

7.5   NEURAL NETWORK TOPOLOGIES

In the previous section, we have discussed the properties of the basic processing unit
in an artificial neural network. This section focuses on the pattern of connections between the units and the propagation of data. As for this pattern of connections, the
main distinction we can make is between:

- **Feed-forward neural networks**, where the data flow from input to output units
  is strictly feed forward. The data processing can extend over multiple (layers
  of) units, but no feedback connections are present, that is, connections extending from outputs of units to inputs of units in the same layer or previous layers.
- **Recurrent neural networks** that do contain feedback connections. Contrary to
  feed-forward networks, the dynamical properties of the network are important.
  In some cases, the activation values of the units undergo a relaxation process

such that the neural network will evolve to a stable state in which these activations do not change anymore. In other applications, the change of the activation values of the output neurons is significant, so that the dynamical behavior constitutes the output of the neural network.

7.5.1   THE FEED-FORWARD NEURAL NETWORK MODELS

If we consider the human brain to be the 'ultimate' neural network, then ideally we would like to build a device, which imitates the brain's functions. However, because of limits in our technology, we must settle for a much simpler design. The obvious approach is to design a small electronic device, which has a transfer function similar to a biological neuron, and then connect each neuron to many other neurons, using RLC networks to imitate the dendrites, axons, and synapses. This type of electronic model is still rather complex to implement, and we may have difficulty 'teaching' the network to do anything useful. Further constraints are needed to make the design more manageable. First, we change the connectivity between the neurons so that they are in distinct layers, such that each neuron in one layer is connected to every neuron in the next layer. Further, we define that signals flow only in one direction across the network, and we simplify the neuron and synapse design to behave as analog comparators being driven by the other neurons through simple resistors. We now have a feed-forward neural network model that may actually be practical to build and use.

As shown in Figs. 3 and 4, the network functions as follows: Each neuron receives a signal from the neurons in the previous layer, and each of those signals is multiplied by a separate weight value. The weighted inputs are summed, and passed through a limiting function which scales the output to a fixed range of values. The output of the limiter is then broadcast to all of the neurons in the next layer. So, to use the network to solve a problem, we apply the input values to the inputs of the first layer, allow the signals to propagate through the network, and read the output values.

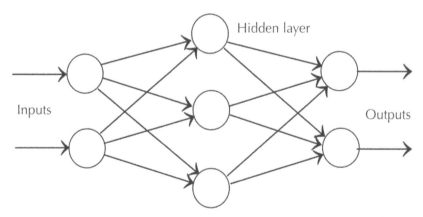

FIGURE 3   A Generalized neural network.

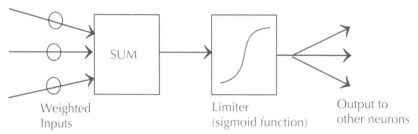

FIGURE 4   The structure of a neuron.

Stimulation is applied to the inputs of the first layer, and signals propagate through the middle (hidden) layer(s) to the output layer. Each link between neurons has a unique weighting value. Inputs from one or more previous neurons are individually weighted, then summed. The result is nonlinearly scaled between 0 and +1, and the output value is passed on to the neurons in the next layer.

Since the real uniqueness or 'intelligence' of the network exists in the values of the weights between neurons, we need a method of adjusting the weights to solve a particular problem. For this type of network, the most common learning algorithm is called Back Propagation (BP). A BP network learns by example, that is, we must provide a learning set that consists of some input examples and the known-correct output for each case. So, we use these input-output examples to show the network what type of behavior is expected, and the BP algorithm allows the network to adapt. The BP learning process works in small iterative steps: one of the example cases is applied to the network, and the network produces some output based on the current state of it's synaptic weights (initially, the output will be random). This output is compared to the known-good output, and a mean-squared error signal is calculated. The error value is then propagated backwards through the network, and small changes are made to the weights in each layer. The weight changes are calculated to reduce the error signal for the case in question. The whole process is repeated for each of the example cases, then back to the first case again, and so on. The cycle is repeated until the overall error value drops below some predetermined threshold. At this point we say that the network has learned the problem "well enough" – the network will never exactly learn the ideal function, but rather it will asymptotically approach the ideal function.

7.5.2   MULTI-LAYER FEED FORWARD NETWORKS

A single-layer network has severe restrictions: the class of tasks that can be accomplished is very limited. In this chapter we will focus on feed-forward networks with layers of processing units. A two layer feed-forward network can overcome many restrictions, but did not present a solution to the problem of how to adjust the weights from input to hidden units. The errors for the units of the hidden layer are determined by back-propagating the errors of the units of the output layer. For this reason the method is often called the back-propagation learning rule. Back-propagation can also be considered as a generalization of the delta rule for nonlinear activation functions and multilayer networks.

7.5.2.1 MULTI-LAYER FEED-FORWARD NETWORK ALGORITHM

A feed-forward network has a layered structure. Each layer consists of units, which receive their input from units from a layer directly below and send their output to units in a layer directly above the unit. There are no connections within a layer. The Ni inputs are fed into the first layer of Nh;1 hidden units. The input units are merely 'fan-out' units; no processing takes place in these units. The activation of a hidden unit is a function Fi of the weighted inputs plus a bias, as given in in equation below.

$$y_k(t+1) = \mathcal{F}_k(s_k(t)) = \mathcal{F}_k\left(\sum_j w_{jk}(t)y_j(t) + \theta_k(t)\right) \qquad (7)$$

The output of the hidden units is distributed over the next layer of Nh; two hidden units, until the last layer of hidden units, of which the outputs are fed into a layer of No output units [Fig. 5].

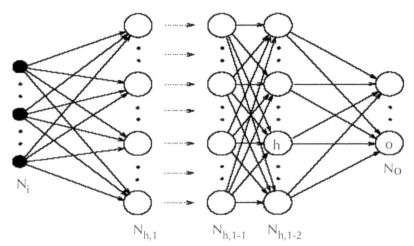

FIGURE 5   The output of hidden units in multilayer feed-forward network algorithm.

Although back-propagation can be applied to networks with any number of layers, just as for networks with binary units it has been shown that only one layer of hidden units sufficient to approximate any function with finitely many discontinuities to arbitrary precision, provided the activation functions of the hidden units are nonlinear (the universal approximation theorem). In most applications a feed-forward network with a single layer of hidden units is used with a sigmoid activation function for the units.

7.6 DELTA LEARNING RULE

Since we are now using units with nonlinear activation functions, we have to generalize the delta rule: The activation is a differentiable function of the total input, given by

$$\text{where: } s_k^p = \sum_j w_{jk} y_j^p + \theta_k, \text{ where: } y_k^p = \mathcal{F}_k(s_k^p), \qquad (8)$$

To get the correct generalization of the delta rule, we must set:

$$\Delta_p w_{jk} = -\gamma \frac{\partial E^p}{\partial w_{jk}} \tag{9}$$

The error measure (E^p) is defined as the total quadratic error for pattern p at the output units:

$$E^p = \frac{1}{2} \sum_{o=1}^{N_o} \left(d_o^p - y_o^p \right)^2 \tag{10}$$

where: is the desired output for unit o when pattern p is clamped. We further set the equation (11) for summed squared error.

$$E = \sum_p E^p \tag{11}$$

We can write:

$$\frac{\partial E^p}{\partial w_{jk}} = \frac{\partial E^p}{\partial s_k^p} \frac{\partial s_k^p}{\partial w_{jk}} \tag{12}$$

In Eq. (12), the second factor is defined as:

$$\frac{\partial s_k^p}{\partial w_{jk}} = y_j^p \tag{13a}$$

and

$$\delta_k^p = -\frac{\partial E^p}{\partial s_k^p} \tag{13b}$$

From Eqs. (11)–(13), we will get an update rule which is equivalent to the delta rule as described above, resulting in a gradient descent on the error surface if we make the weight changes according to Eq. (14):

$$\Delta_p w_{jk} = \gamma \delta_k^p y_j^p \tag{14}$$

The procedure involves finding out what should be for each unit k in the network. The Eq. (14) is a simple recursive computation of these δ's, which can be implemented by propagating error signals backward through the network.

To compute we apply the chain rule to write this partial derivative as the product of two factors, one factor reflecting the change in error as a function of the output of the unit and one reacting the change in the output as a function of changes in the input. Thus, we have:

$$\delta_k^p = -\frac{\partial E^p}{\partial s_k^p} = -\frac{\partial E^p}{\partial y_k^p} \frac{\partial y_k^p}{\partial s_k^p} \tag{15}$$

Now, we shall compute the second factor as follows:

$$y_k^p = \mathcal{F}_k \left(s_k^p \right) \tag{16}$$

$$\frac{\partial y_k^p}{\partial s_k^p} = \mathcal{F}' \left(s_k^p \right) \tag{17}$$

The Eq. (17) is the same result as we obtained with the standard delta rule. Substituting this and Eq. (18) in Eq. (17), we get Eq. (19) for any output unit o:

$$\delta_k^p = -\frac{\partial E^p}{\partial s_k^p} = -\frac{\partial E^p}{\partial y_k^p}\frac{\partial y_k^p}{\partial s_k^p} \tag{18}$$

$$\delta_k^p = (d_o^p - y_o^p)\mathcal{F}_o'(s_o^p) \tag{19}$$

Secondly, if k is not an output unit but a hidden unit $k = h$, we do not readily know the contribution of the unit to the output error of the network. However, the error measure can be written as a function of the net inputs from hidden to output layer; $=$ (sp1, sp2,........, spj.....) and we use the chain rule to write equation (20):

$$\frac{\partial E^p}{\partial y_h^p} = \sum_{o=1}^{N_o}\frac{\partial E^p}{\partial s_o^p}\frac{\partial s_o^p}{\partial s_h^p} = \sum_{o=1}^{N_o}\frac{\partial E^p}{\partial s_o^p}\frac{\partial}{\partial y_h^p}\sum_{j=1}^{N_h} w_{ko}y_j^p = \sum_{o=1}^{N_o}\frac{\partial E^p}{\partial s_o^p}w_{ho} = -\sum_{o=1}^{N_o}\delta_o^p w_{ho} \tag{20}$$

From eqs (15) and (20), we get:

$$\delta_h^p = \mathcal{F}'(s_h^p)\sum_{o=1}^{N_o}\delta_o^p w_{ho} \tag{21}$$

Eqs. (15) and (21) give a recursive procedure for computing the δ›s for all units in the network, which are then used to compute the weight changes according to equation. This procedure constitutes the generalized delta rule for a feed-forward network of non-linear units.

7.6.1   UNDERSTANDING BACK-PROPAGATION NETWORKS

The Eqs. (21) may be mathematically correct, but what do they actually mean? Is there a way of understanding back-propagation other than reciting the necessary equations? The answer is, of course, yes. In fact, the whole back-propagation process is intuitively very clear. What happens in the above equations is the following:

When a learning pattern is clamped, the activation values are propagated to the output units, and the actual network output is compared with the desired output values, we usually end up with an error in each of the output units. Let's call this error e_o for a particular output unit o. We have to bring e_o to zero The simplest method to do this is the greedy method: we strive to change the connections in the neural network in such a way that, next time around, the error $[e_o]$ will be zero for this particular pattern. We know from the delta rule that, in order to reduce an error, we have to adapt its incoming weights according to:

$$\Delta w_{ho} = (d_o - y_o)y_h \tag{22}$$

This is step one. However, this is not enough: when we only apply this rule, the weights from input to hidden units are never changed, and we do not have the full representational power of the feed-forward network as promized by the universal approximation theorem. In order to adapt the weights from input to hidden units, we again want to apply the delta rule. In this case, however, we do not have a value for δ for the hidden units. This is solved by the chain rule which does the following: distribute the error of an output unit o to all the hidden units that is it connected to, weighted by this connection. Differently put, a hidden unit h receives a delta from each output unit o equal to the delta of that output unit weighted with (= multiplied by) the weight of the connection between those units.

The application of the generalized delta rule thus involves two phases: During the first phase the input x is presented and propagated forward through the network to compute the output values $[y^p_o]$ for each output unit. This output is compared with its desired value do, resulting in an error signal δp o for each output unit. The second phase involves a backward pass through the network during which the error signal is passed to each unit in the network and appropriate weight changes are calculated.

7.6.2 WEIGHT ADJUSTMENTS WITH SIGMOID ACTIVATION FUNCTION

- The weight of a connection is adjusted by an amount proportional to the product of an error signal δ, on the unit k receiving the input and the output of the unit j sending this signal along the connection:

$$\Delta_p w_{jk} = \gamma \delta^p_k y^p_j \tag{23}$$

- If the unit is an output unit, the error signal is given by:

$$\delta^p_o = (d^p_o - y^p_o)\mathcal{F}'(s^p_o) \tag{24}$$

- Take as the activation function the 'sigmoid' function as defined:

$$y^p = \mathcal{F}'(s^p) = \frac{1}{1+e^{-s^p}} \tag{25}$$

In this case the derivative is equal to:

$$\mathcal{F}'(s^p) = \frac{\partial}{\partial s^p}\frac{1}{1+e^{-s^p}}$$

$$= \frac{1}{\left(1+e^{-s^p}\right)^2}\left(-e^{s^p}\right)$$

$$= \frac{1}{\left(1+e^{-s^p}\right)}\frac{e^{-s^p}}{\left(1+e^{-s^p}\right)} \tag{26}$$

$$= y^p(1 - y^p)$$

Such that the error signal for an output unit can be written as:

$$\delta^p_o = (d^p_o - y^p_o)y^p_o(1 - y^p_o) \tag{27}$$

- The error signal for a hidden unit is determined recursively in terms of error signals of the units to which it directly connects and the weights of those connections. For the sigmoid activation function:

$$\delta_h^p = \mathcal{F}'(s_h^p) \sum_{o=1}^{N_o} \delta_o^p \, w_{ho} = y_h^p (1 - y_h^p) \sum_{o=1}^{N_o} \delta_o^p \, w_{ho} \qquad (28)$$

7.6.2.1   LEARNING RATE AND MOMENTUM

The learning procedure requires that the change in weight is proportional to $\frac{\partial E^p}{\partial w}$.

True gradient descent requires that infinitesimal steps are taken. The constant of proportionality is the learning rate. For practical purposes we choose a learning rate that is as large as possible without leading to oscillation. One way to avoid oscillation at large, is to make the change in weight dependent of the past weight change by adding a momentum term:

$$\Delta w_{jk}(t + 1) = \gamma \delta_k^p y_j^p + \alpha \Delta w_{jk}(t), \qquad (29)$$

where: t indexes the presentation number and F is a constant which determines the effect of the previous weight change.

Although, theoretically, the back-propagation algorithm performs gradient descent on the total error only if the weights are adjusted after the full set of learning patterns has been presented, more often than not the learning rule is applied to each pattern separately, i.e., a pattern p is applied, E^p is calculated, and the weights are adapted (p = 1, 2, ... P). There exists empirical indication that this results in faster convergence. Care has to be taken, however, with the order in which the patterns are taught. For example, when using the same sequence over and over again the network may become focused on the first few patterns. This problem can be overcome by using a permuted training method.

The back propagation-learning algorithm is applied to multilayered feed-forward networks consisting of processing element with continuous and differentiable activation functions. Given a training set of input-output pairs, the algorithm provides a procedure for changing the weights in a back-propagation learning algorithm to classify the given input patterns correctly. The basis for this weight update algorithm is simply the gradient descent method as used for simple perceptron with differentiable neurons.

For a given input-output pair, the back propagation algorithm performs two phases of data flow. First, the input pattern is propagated from the input layer to the output layer, and as a result of this forward flow of data, it produces an actual output. Then the error signals resulting from the difference between output pattern and an actual output are back-propagated from the output layer to the previous layers for them to update their weights. The type of neural network used in this study is three-layer back-propagation network, consisting of an input layer, a hidden layer and an output layer. The learning algorithm used in this present research work *Levenberg-Marquardt* back-propagation of MATLAB Neural Network Tool Box (version 7.4, 2007). The transfer function selected for the layers will be sigmoid, for the hidden layer and linear for the output layer.

7.6.3   TRAINING AND TESTING OF NEURAL NETWORKS

A neural network has to be configured such that the application of a set of inputs produces (either 'direct' or via a relaxation process) the desired set of outputs. Various methods to set the strengths of the connections exist. One way is to set the weights explicitly, using a priori knowledge. Another way is to 'train' the neural network by feeding it teaching patterns and letting it change its weights according to some learning rule. We can categorize the learning situations in two distinct sorts. These are:

- **Supervized learning** or Associative learning (Fig. 6) in which the network is trained by providing it with input and matching output patterns. These input-output pairs can be provided by an external teacher, or by the system which contains the neural network (self-supervized).

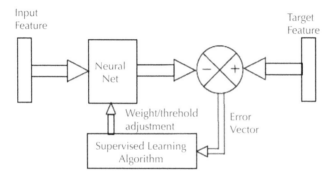

FIGURE 6   Training the networks with supervized learning.

- **Unsupervized learning** or Self-organization in which an (output) unit is trained to respond to clusters of pattern within the input. In this paradigm the system is supposed to discover statistically salient features of the input population. Unlike the supervized-learning paradigm, there is no a priori set of categories into which the patterns are to be classified; rather the system must develop its own representation of the input stimuli.
- **Reinforcement Learning** This type of learning may be considered as an intermediate form of the above two types of learning. Here the learning machine does some action on the environment and gets a feedback response from the environment. The learning system grades its action good (rewarding) or bad (punishable) based on the environmental response and accordingly adjusts its parameters. Generally, parameter adjustment is continued until an equilibrium state occurs, following which there will be no more changes in its parameters. The self-organizing neural learning may be categorized under this type of learning.

7.6.3.1   NEURAL NETWORK TRAINING AND ANN MODEL DESCRIPTION

A multilayer neural network consists of a number of interconnected nodes arranged into layers where each node (neuron) operates as a simple processing element. Each node in the network is interconnected with all nodes in the preceding and following layers. There are no interconnections within nodes of the same layer. The number of layers and the number of nodes by layer represent the network architecture. The input layer serves as an entry for the vector of input data presented to the network where each node corresponds to one explanatory variable (input). The output layer represents the output data (target). All layers between the input and output layers are referred to hidden layers.

The input (Ij) to a node (j) is the weighted sum (using w_{ij}) of the outputs (O_i) from the nodes of the preceding layer (i). This sum is then passed through an activation function (f) to produce the node's output (O_j) within the range of the activation function. The activation function is usually a sigmoid or hyperbolic tangent, which is a nonlinear function with an asymptotic behavior. In our case, we used the hyperbolic tangent sigmoid function. The interconnecting weight values are adjusted and updated during the training phase to achieve minimal overall training error between the desired and calculated output vectors. All data should be rescaled to ensure they receive equal attention during the training process. In this study, data were rescaled to the interval (–1, 1).

All parameters in neural network (weight and biases) can be denoted by the vector:

$$w\varepsilon\, W \subset R^P \tag{30}$$

where, P is the total number of parameters of the network. Training a neural network involves selecting the 'best' set of network parameters (w) that minimize a training error estimator. The error estimator used here was the sum-squared error (SSE). Preliminary, training explored how possible architectures performed when training with different training algorithms. Next, the best combination of network architecture and training algorithm was selected. Finally the chosen combination was further tested and trained to obtain the best generalization capacity. The training methodology used was back propagation training algorithm.

While back propagation with gradient decent technique is a steepest descent algorithm, the Levenberg – Marquardt algorithm is an approximation to Newton's method. The minimization algorithm used was the Levenberg – Marquardt algorithm. In back propagation algorithm, the vector of the network parameters (w) is updated in each epoch (training step) using,

$$W_{k+1} = W_k - \alpha_k M_k g_k \tag{31}$$

where, a = learning rate; g = gradient of the error functions; M = approximation of the inverse of the Hessian matrix. This matrix is positive definite in order to ensure the descent. All quantities are for the kth iteration. In the *Levenberg-Marquardt* algorithm, the inverse of the Hessian matrix approximated by:

$$(\epsilon.\, I + J. - J^T)^{-1} \tag{32}$$

where, ϵ = small quantity; I = identity matrix; $J.$ = Jacbian matrix (J_{ij}) with:

$$J_{ij} = \frac{\delta e_j}{\delta w_i} \tag{33}$$

where, e_j is the error $(y_j - t_j)$ from the target value. A crucial point in the back propagation algorithm is the chose of the learning rate (a), which minimizes the error of the nest step (epoch). In the large majority of the works in the literature a fixed value of the learning rate is used during the training of the neural network.

7.6.4   MODEL ARCHITECTURE

The best neural network architecture can only be determined experimentally for each particular application. According to de Villiers and Barnard (3), ANN models with one hidden layer can approximate any continuous function. Thus, in this study, model architecture of one hidden layer was used. The number of nodes in the hidden layer(s) should be large enough to ensure a sufficient number of degrees of freedom for the network function and small enough to minimize the risk of loss of the network's generalization ability. Furthermore, it's important to note that a useless increase of the neural network size will lead to a significant increase in training and running time. As recommended by Maier and Dandy [12], the node number upper limit in the hidden layer(s) was fixed as follows:

$$N^H = \min (2N^I+1; N^{TR} /N^I+1) \qquad (34)$$

where, N^H is the number of nodes in the hidden layer(s), N^I number of input nodes and N^{TR} the number of training sample.

7.7   PREDICTION OF DAILY PAN EVAPORATION USING NEURAL NETWORKS MODELS: INDIAN AGRICULTURE

The evaporation is a complex and nonlinear phenomenon involving several weather factors in the hydrological process. It is the necessary components of any water balance assessments for different water resources planning, design, operation and management studies including hydrology, agronomy, horticulture, forestry, land resources, irrigation management, flood forecasting, investigation of agro-ecosystem and modeling *etc.* and perhaps the most difficult to estimate owing to complex interactions between the components of the soil-water-plant-atmosphere system. Evaporation is an important variable in making the crop management decision and modeling crop response to weather conditions and climate change. It has been extensively used for estimating the potential and reference evapotranspiration of various crops [5]. Based on the limitations in measuring the pan evaporation research has been performed to model pan evaporation using various meteorological variables. It is therefore necessary to develop approaches to estimate the evaporation rates from other available meteorology variables, which are comparatively easier for measurements. The class A pan evaporation is direct method. Indirect method includes those that use meteorological data to estimate evaporation from other weather variables through empirically developed methodologies or artificial neural network and statistical approaches.

The study was conducted to develop and test the daily pan evaporation prediction models using various weather parameters as input variables with artificial neural network (ANN) and validated with the independent subset of data for five different locations in India. The measured variables included daily observations of maximum and minimum temperature, maximum and minimum relative humidity, wind speed, sunshine hours, rainfall and open pan evaporation.

In this general model (GM), model development and evaluation has been done for the five locations in India:

- NRCC, Nagpur (Maharashtra)
- JNKVV, Jabalpur (Madhya Pradesh)
- PDKV, Akola (Maharashtra)
- ICRISAT, Hyderabad (Andhra Pradesh) and
- MPUAT, Udaipur (Rajasthan).

The daily data of pan evaporation and other inputs for two years was considered for model development and subsequent 1–2 years data for validation. Weather data consisting of 3305 daily records from 2002 to 2006 were used to develop the GM models of daily pan evaporation. Additional weather data of Nagpur station, which included 2139 daily records from 1996–2004, served as an independent evaluation dataset for the performance of the models. The model plan strategy with all inputs has shown better performance than the reduced number of inputs. The General ANN models of daily pan evaporation with all available variables as inputs was the most accurate model delivering an R^2 of 0.84 and a root mean square error 1.44 mm for the model development dataset. The GM evaluation with Nagpur model development (NMD) data shown lowest RMSE (1.961 mm), MAE (0.038 mm) and MARE (2.30%) and highest r (0.848), R^2 (0.719) and d (0.919) with ANN GM M-1with all input variables.

The basic aim of this research was to improve the effectiveness and operational ability of classic and sophisticated methods and discover perspectives of modern and sophisticated approaches such as artificial neural networks (ANN) for the prediction and estimation of pan evaporation. The regression models developed from meteorological data involve empirical relationships to some extent accounts for many local conditions. Therefore, most regression models gave reliable results when applied to climatic conditions similar to those for which they were developed. The correlations and regression studies between different meteorological parameters and evaporation measured from US Class A open pan evaporimeter under Indian conditions. Linear regression methods for prediction of evaporation using weather variables has been applied for decades and are well and understood [8, 16]. The highest correlation was found with maximum temperature followed by wind speed. The coefficient of determination for maximum temperature, minimum temperature, wind speed and soil temperature at 5 cm depth was positively correlated. The relative humidity was negatively correlated [10]. The pan evaporation and the meteorological parameters recorded at Jabalpur revealed that the morning relative humidity and the maximum temperature have a significant influence on the rate of evaporation [17]. There was a need for developing the model, which can generalize for the diversified Indian conditions. Artificial neural network (ANN) is effective tool to model nonlinear process like evaporation from open pan. Pan evaporation has been widely used as an estimate for irrigation scheduling of vegetables and horticultural crops [15].

The artificial neural networks provide better modeling flexibility than the other statistical approaches. Numerous researchers have shown applicability of multiple linear regression technique for estimating the evaporation, but very few have been seen on artificial neural networks in agricultural and hydrological processes in India. For instance multilayered feed forward ANN with error back propagation techniques has been used for estimating evaporation [1, 9, 14, 18], evapo-transpiration [11, 18], soil water evaporation [7] and various neuro-computing techniques for predicting the various atmospheric processes and parameters [6]. Bruton et al. [2] developed ANN models to estimate daily pan evaporation using measured weather variables as inputs. The measured variables included daily observations of rainfall, temperature, relative humidity, solar radiation, and wind speed. Daily pan evaporation was also estimated using multiple linear regression and compared to the results of the ANN models. The ANN models of daily pan evaporation with all available variables as a inputs was the most accurate model delivering an r^2 of 0.717 and a root mean square error 1.11 mm for the independent evaluation dataset. The accuracy of the models was reduced considerably when variables were eliminated to correspond to weather stations. Pan evaporation estimated with ANN models was slightly more accurate than the pan evaporation estimated with a multiple linear regression models. Sudheer et al. [19] investigated the prediction of Class A pan evaporation using the ANN technique. The ANN back propagation algorithm has been evaluated for its applicability for predicting evaporation from minimum climatic data. The study indicated that evaporation values could be reasonably estimated using temperature data only through the ANN technique. This would be of much use in instances where data availability is limited. These features provide neural networks the potential to model complex nonlinear phenomenon like prediction of daily pan evaporation using meteorological measured variables. Observations of relative humidity, solar radiation, temperature, wind speed and evaporation for the past 22 years have been used to train and test the developed networks. Results revealed that the networks were able to well learn the events they were trained to recognize. These encouraging results were supported by high values of coefficient of correlation and low mean square errors reaching 0.98 and 0.00015 respectively [20]. Daily pan evaporation was estimated by a suitable ANN model for the meteorological data recorded from the Automated *GroWheather* meteorological station near Lake Egirdir, Turkey [14]. The ANN models were developed and validated a simulation model of the evaporation rate of a Class A evaporimeter pan located near Cartagena (South-east Spain) [13]. Keskin and Terzi [9] studied the ANN models and proposed as an alternative approach of evaporation estimation for prediction of daily pan evaporation estimation. The comparison shows that there is better agreement between the ANN estimations and measurements of daily pan evaporation than for other model. Deswal and Mahesh Pal [4] studied an ANN based modeling and the influence of meteorological parameters on evaporation from a reservoir. The findings of the study also revealed the requirement of all input parameters considered together, instead of individual parameters taken one at a time. The highest correlation coefficient (0.960) along with lowest root mean square error (0.865) was obtained with the input combination of air temperature, wind speed, sunshine hours and mean relative humidity.

The basic objectives of this research work are to develop three-layered feed forward with error back propagation neural network models to estimate daily pan evaporation values based on different meteorological data. The other objectives are to test the suitability of the artificial neural networks for modeling daily pan evaporation and to test and validate the developed ANN evaporation models for the input variables using the independent subset of data for other locations.

7.7.1   STUDY AREA AND DATASET

Data were taken from agro-meteorological stations at: Nagpur (National Research Centre for Citrus, Experiment Station, lat. 21.09°N., long. 79.22°W. 311.3 m amsl), Jabalpur (Jawaharlal Nehru Agricultural University, lat. 23.10°N., long. 79.58°W, 410 m amsl), Akola (Dr. Panjabrao Deshmukh Agricultural University, lat. 20.42°N. long. 77.04°W. 309 m amsl), Hyderabad (International Crop Research Institute for Semi Arid Tropics, lat. 17.53°N. long. 78.27°W. 545 m amsl) and Udaipur (Maharana Pratap University of Agriculture and Technology, lat. 24.35 °N. long.73.42 °W. 582 m amsl).   These locations represent the dry subhumid, subhumid, semiarid and arid climatic regions within the state of Maharashtra, Madhya Pradesh, Andhra Pradesh and Rajasthan. The intent was to develop a single model that could be used for any of the agro-climatic zones of India. The data were partitioned into a model development dataset and an independent evaluation dataset. The model development dataset consisted of 607, 731, 507, 730 and 730 days of data (3305 total observations) for Nagpur, Jabalpur, Akola, Hyderabad and Udaipur respectively, from 2002 to 2006. The model development dataset was further divided by randomly placing 80% of the observations in a training dataset and the remainder 20% in a testing dataset. The training dataset was used to develop the neural network models. The testing set was used to evaluate the accuracy of the ANN models during training in order to determine when to stop the training, Training was continued as long as the minimum goal of error of the ANN estimate of pan evaporation on was kept as 0.001. The model development dataset was used to choose preferred ANN training parameters and to develop ANN models of pan evaporation based on various numbers of inputs. With this procedure, all model development and parameter selection were done with the model development dataset. The Nagpur model development (NMD) data size was 2139. The model GM evaluation dataset consisted of 2066 observations for Nagpur, Jabalpur, Akola, Hyderabad and Udaipur respectively, from 2004 to 2007 for the five locations. Weather variables included in the evaluation dataset were the same as those in the model development dataset. The evaluation dataset was not used in model development in any way and was only used to assess the accuracy of the ANN models.

A three-layer back propagation ANN architecture was employed in all models. The MATLAB (Ver.7.4, 2007) software package was used to develop the ANN pan evaporation models, The dataset included daily records of eight measured variables viz. maximum temperature (TMAX), minimum temperature (TMIN), maximum relative humidity (RMAX), minimum relative humidity (RMIN), wind speed (WISP), sunshine hours (SNHR), rainfall (RAIN) and pan evaporation (EPAN). The number of measurable weather variables used as inputs in the development of the ANN pan evaporation models (M-1 to M-7) was varied with emphasis placed on investigating models with reduced numbers of inputs, based on positive and negative effects of the parameters on evaporation process. Seven model strategies (combination of variable inputs) to be evaluated in this study are shown in Table 1.

Modeling strategy in M-1 included all seven available daily weather input variables. Model strategy M-2 considered the effect of removing rainfall from the variables in modeling strategy M-1creating the situation of nonrainy period evaporation process. Model strategy M-3 considered removing sunshine hours and inclusion of rainfall crating the weather with rainy season and without sunshine hours. The effect of absence of maximum and minimum relative humidity in the evaporation process is tested in model strategy M-4. This is like without humidity days. In modeling plan M-5 the maximum and minimum temperature as well as relative humidity was as variables, modeling the situation with temperature and humidity. Modeling strategy M-6 included maximum temperature, wind speed and sunshine hours as input variables. The model M-7 included maximum temperature and wind speed as the only observed weather values.

TABLE 1   The modeling strategies in development of ANN pan evaporation models.

Variables	Modeling strategies						
	M-1	M-2	M-3	M-4	M-5	M-6	M-7
TMAX	√	√	√	√	√	√	√
TMIN	√	√	√	√	√	—	—
RMAX	√	√	√	—	√	—	—
RMIN	√	√	√	—	√	—	—
WISP	√	√	√	√	—	√	√
SNHR	√	√	—	√	—	√	—
RAIN	√	—	√	√	—	—	—

√ — indicates the inclusion of the variable in neural network modeling.

7.7.2   MODEL ARCHITECTURE

The best neural network architecture according to de Villiers and Barnard [3], ANN models with one hidden layer can approximate any continuous function. Thus, in this study, model architecture of one hidden layer was used. The number of nodes in the hidden layer(s) should be large enough to ensure a sufficient number of degrees of freedom for the network function and small enough to minimize the risk of loss of the network's generalization ability. Furthermore, it's important to note that a useless increase of the neural network size will lead to a significant increase in training and running time. As recommended by Maier and Dandy [12], the node number upper limit in the hidden layer(s) was fixed as follows:

$$N^H = \min (2N^I+1; N^{TR} /N^I+1) \tag{35}$$

where, N^H is the number of nodes in the hidden layer(s), N^I number of input nodes and N^{TR} the number of training sample. The number of hidden nodes is therefore taken from 5 to 20 as the inputs vary from seven in model M-1 to two in model M-7. The number of training samples is large. So, the ratio of the training samples to training parameters is large and $2N^I+1$ is minimum number of hidden nodes of the networks. The accuracy increase with increasing the nodes, but the generalization ability of the model goes down. So, the optimization of nodes is done with minimum squared and absolute error. The final GM models were developed with 15 hidden nodes, which gave minimum errors. The developed model evaluation was done with the help of the Nagpur model development dataset and GM evaluation dataset. The number of records used for evaluation is 484 for NRCC, Nagpur; 487 for JNKVV, Jabalpur; 365 each for PDKV, Akola, ICRISAT, Hyderabad and MPUAT, Udaipur.

7.7.3   ANN MODELING SOFTWARE SOURCE CODE PROGRAM USING MATLAB

The neural network utility file was edited in MATLAB (version 7.4, 2007) source code program. The main GUI consisting of all the input variables selection, input data source file, network options, training and testing functions, setting the data for training, evaluation, plotting the predicted values and saving the network is created and run in MATLAB software. The mainframe is executed by running the MainGUI.mat file in MATLAB program. The Graphical User Interface (GUI) appears by changing the directory option, which gives the details of the ANN modeling selections. The program displays the title, the input variables to be selected by making the tick marks on/off. The source data input file (.csv) is browsed for the training, testing and evaluation. The variables were selected according to the model plan M-1 to M-7 for developing and evaluating the ANN models. The ANN model architecture is three-layered feed forward with error back propagation. Which is most commonly used neural network for the prediction of the nonlinear process like prediction of pan evaporation. The number of hidden nodes selection is done from 5 to 20. The transfer function from input to hidden layer is logistic sigmoid and from hidden layer to output layer is linear (as the output is one i.e., EPAN). The training function is *Levenberg-Marquardt*, which is most common and accurate is selected.

The performance functions for training and testing the networks used are MAE (mean absolute error) and MSE (mean squared error). The learning rate, which decides the accuracy of the training is used for optimization are 0.1, 0.3, 0.5 and 0.7. The momentum of the training the network, which gives the speed of the training varied from 0.1, 0.3, 0.5, 0.7 and 0.9. The various combinations of hidden nodes, learning rate and momentum is done to arrive at optimum combinations to give less error. The network iterations (epochs) were kept at 1000. The ANN model development and evaluation with saved model plans is done as follows. The mainframe of the ANN modeling with MATLAB is shown in Fig. 7. The GUI.mat file is opened in MATLAB software. The model development input data file (saved in.csv format) is browsed to the main GUI frame. The hidden nodes, training and transfer functions are selected. The model is generated on clicking the button Generate model. This generated model is displayed in command window of the MATLAB program. This gives the network architecture, input, hidden and output layer weights and bias weights. The generated ANN model is trained by clicking on the Train model option. Training is undertaken by setting the percentage training, testing and validation values as 80%, 20% and 100% respectively.

The network training process is displayed on the screen with minimum error in comparison with the set 0.001 goal. The program gives the graphical output with set goal value. The model when gives the minimum value of MSE/MAE is saved as a model with particular identification name. The GM models (GM) were saved in the name as GM 15 M-1 to GM 15 M-7. This also indicates the General model with 15 hidden nodes, 0.1 learning rate and momentum each and for the model plans.

FIGURE 7 The Graphical User Interface (GUI) of ANN modeling for EPAN prediction.

These saved models were used for validation of the same data on screen or the other source data of the independent locations. This was done by selecting the source file and clicking on the button Evaluate model on mainframe of the program. The saved neural network model was used for evaluation of new input file, which is browsed in evaluation module of the mainframe. The output of the pan evaporation is predicted in command window of the MATLAB program. The same is copied and used for estimating the model performance like RMSE, MAE, MARE, r, R² and d. The graphical output of the training, testing and validation of the network as well as the plotting of the observed and ANN predicted EPAN is displayed on the desktop, which was saved for the comparison of the different ANN model plans. In this way model plans M-1 to

M-7 of the General model was developed and evaluated with five different locations in India, as shown in this section.

7.7.4   EVALUATION OF PERFORMANCE PARAMETERS FOR THE MODEL

The ANN model development and evaluation using the independent datasets of various locations was done with help of the MATLAB source code program and the datasets concerned. The developed and evaluated models were verified with the squared and absolute statistics of different performance functions. Mean squared error (MSE), Root mean squared error (RMSE), Mean absolute relative error (MARE), Correlation coefficient (r) and coefficient of determination (R^2) are used as model development as well as evaluation criteria's. Based on these criteria's the superiority of the model is judged. The R^2 measures the degree to which two variables are linearly related. MSE and MARE provide different types of information about the predictive capabilities of the model. The MSE and RMSE measures the goodness of fit relevant to high evaporation values, whereas the MARE yields a more balanced perspective of the goodness of fit at moderate evaporation. In addition an index of agreement (d) between the observed and predicted values was calculated.

7.7.5   GENERAL MODEL DEVELOPMENT

The General model (GM) is developed using daily records of the weather inputs from Nagpur, Jabalpur; Akola; Hyderabad and Udaipur. The program in MATLAB codes is opened and run for the main frame of the ANN modeling. The main frame which helped in selecting the input data file, selecting the input variables, number of the hidden nodes, training, learning and transfer functions, learning rate and momentum. The model input file which is saved in spreadsheet (with csv format) is browsed and selected. The number of hidden nodes, learning rate and momentum is kept as 15, 0.1 and 0.1 respectively, as it has shown optimum parameters. The training function is back propagation with *Levenberg-Marquardt* algorithm and transfer function is logistic sigmoid. The number of iterations (epochs) were kept 1000. The network performance functions are kept as MAE and MSE. The goal set for testing the neural networks was 0.001 (0.1%). The data was set to train, test and validate in the ratios of 80%, 20% and 100% respectively. The training network figure for lowest MSE / MAE is also saved. The process of training the network was repeated many times and when the MSE/MAE was shown minimum the ultimate network is saved in .mat file with some name. Like this GM models with 15 hidden nodes and 0.1 learning rate and 0.1 momentum were saved optimally in the names of GM 15 M-1 to GM 15 M-7. The ANN General models M-1 to M-7 has input layer and bias weights. The input weights, layer weights and bias weights of these ANN models are explored in window of the MATLAB program and saved.

7.8   RESULTS AND DISCUSSION

The model development dataset was used to determine preferred values for number of hidden nodes, learning rate, and momentum. Data for all five sites were included to obtain the best model parameters, as well as all seven input parameters. The ANN parameters for learning rate and momentum have values, which are typically less than one. Learning rate and momentum combinations using 0.1, 0.3, 0.5, and 0.9 were tested with hidden node numbers ranging from 5 to 20. The training data were used for training and the testing data were used to determine when to stop training. The accuracy of the model on the model development dataset (training data and testing data) was then determined. In the initial experimentation, it was found that the accuracy of the models was reduced for number of hidden nodes above 15, and Learning rate and momentum below 0.1. Otherwise, there was little variation in accuracy for the various combinations of parameters. The accuracy was higher for 15 hidden nodes, a learning rate of 0.1 and a momentum of 0.1. Experimentation was performed due to the limited effect of the ANN parameters on model accuracy. The accuracies of the models in feed forward mode for the development and evaluation datasets are shown in Table 2. The results for the evaluation dataset provide an independent evaluation of the prediction accuracy of the models.

TABLE 2 RMSE, coefficient of determination (R^2) and index of agreement (d) of ANN General model development and evaluation dataset.

ANN GM models	GM model development			GM model evaluation with NMD data		
Model	RMSE mm/d	R^2	d	RMSE mm/d	R^2	d
GM M-1	1.44	0.84	0.95	1.96	0.72	0.92
GM M-2	1.48	0.83	0.95	1.97	0.71	0.91
GM M-3	1.56	0.81	0.94	1.98	0.71	0.91
GM M-4	1.62	0.79	0.94	2.08	0.68	0.90
GM M-5	1.68	0.78	0.93	2.06	0.69	0.91
GM M-6	1.70	0.77	0.93	2.12	0.67	0.90
GM M-7	1.76	0.76	0.92	2.11	0.67	0.90

7.8.1 ERROR STATISTICS OF GENERAL MODEL (GM) DEVELOPMENT

The ANN General model development data shows that the RMSE, MAE and MARE is increased and r, R^2 and d is decreased when the input variables were sequentially removed from seven variables in M-1 to two variables in M-7. The lowest RMSE (1.439 mm), MAE (0.0207 mm) and MARE (0.915%) and highest r (0.916), R^2 (0.839) and d (0.956) was observed with GM M-1 model, in which all the input variable are taken into consideration. In model M-2, where all the six input variables excluding RAIN was considered in ANN model has also shown lower RMSE, MAE and MARE and higher r, R^2 and d. The RMSE, MAE and MARE was 1.478 mm, 0.0218 mm and 1.076%, which is slightly more than the model M-1, but not a significant difference exists. The r, R^2 and d in model M-2 is 0.911, 0.831 and 0.953, respectively. These values are less than the values of model M-1. The RMSE, MAE and MARE of GM M-3 model are higher and r, R^2 and d are lower as compared to ANN model M-1 and M-2. This is mainly due to the replacement of sunshine hours input of model M-2 with RAIN variable. This indicates that the performance of all variables without SNHR (but with RAIN variable) shows higher RMSE, MAE and MARE and lower r, R^2 and d as compared to model M-2. The RMSE, MAE and MARE of model M-3 are 1.559 mm, 0.024 mm and 1.362%, respectively. The r, R^2 and d of model M-3 are 0.901, 0.812 and 0.947, respectively. The error statistics of model GM M-4 in which the relative humidity (RMAX and RMIN) are not the input variables. The RMSE, MAE and MARE of model M-4 is 1.625 mm, 0.026 mm and 1.506%, which is higher than the models M-1, M-2 and M-3. The r, R^2 and d of M-4 is 0.892, 0.796 and 0.942, respectively and these values are higher than model M-1 M-2 and M-3. It is mainly due to the absence of RMAX and RMIN input variables and have higher negative correlation with the pan evaporation. Although five variables are there in model M-4, but the important variables i.e., RMAX and RMIN are not considered for modeling has shown more error.

The model M-5 performance is better than the model M-4. The RMSE, MAE and MARE in model M-5 is increased to 1.685 mm, 0.0284 mm and 1.549% from 1.625 mm, 0.026 mm and 1.50% in model M-4. The r, R^2 and d is also decreased to 0.883, 0.78 and 0.935 in model M-5 as compared to model M-4. The input variables are TMAX, WISP and SNHR in model M-6. The RMSE, MAE and MARE of model M-6 is 1.702 mm, 0.0289 mm and 1.996%, respectively. The r, R^2 and d of model M-6 was 0.88, 0.776 and 0.935 which are lower as compared to M-5 and higher than model M-7. The RMSE, MAE and MARE is increased to 1.759 mm, 0.0309 mm and 2.52% in GM model M-7, in which only TMAX and WISP are the input variables of the model. The r, R^2 and d were also lowest in model M-7 as compared to M-1 to M-6. The r, R^2 and d in model M-7 were 0.872, 0.76 and 0.929, respectively. The evaluation of general model development is graphically presented (Fig. 8).

7.8.2   GENERAL MODEL EVALUATION WITH NAGPUR MODEL DEVELOPMENT (NMD) DATASET

The Nagpur model development (NMD) data from April, 1996 to March, 2004 (2139 observations) was used for evaluating the ANN General model with M-1 to M-7 model strategies. The optimum number of hidden nodes, learning rate and momentum for General model (GM) is 15, 0.1 and 0.1, respectively. The trained ANN models M-1 to M-7 with minimum error reaching towards the goal. Each saved model architecture indicates the input variables, hidden layers and number of nodes in it, biases, input, hidden and output layer weights, bias weights. The corresponding model output values of the predicted EPAN as used for evaluation of these developed models. The output resulted from the ANN models at minimum MSE / MAE was compared with observed EPAN and the various model evaluation parameters were analyzed to decide the best predictive model for pan evaporation. The absolute and squared statistics of the ANN models is tabulated and presented in Table 2. ANN General model evaluation using Nagpur model development data shows that the RMSE, MAE and MARE is increased and r, R^2 and d is decreased when the input variables were sequentially removed from seven inputs in model M-1 to six inputs in model M-2 and M-3. The lowest RMSE

(1.961 mm), MAE (0.038 mm) and MARE (2.30%) and highest r (0.848), R^2 (0.719) and d (0.919) was observed with GM M-1 model, in which all the input variables are taken into consideration. However, in model M-2 and M-3, where six input variables are considered in GM modeling has also shown lower RMSE, MAE and MARE and higher r, R^2 and d. This shows that the GM model generalization capacity is better when tested with Nagpur model development data. The RMSE, MAE and MARE in model plan M-2 was 1.973 mm, 0.0389 mm and 2.49%, which is slightly more than the model M-1. The r, R^2 and d in model M-2 is 0.845, 0.715 and 0.917, respectively. These values are also less to the values of model M-1. The RMSE, MAE and MARE of GM M-3 model are higher and r, R^2 and d are lower as compared to model M-1 and M-2. This is mainly due to the replacement of sunshine hours input of model M-2 with RAIN variable. This indicates that the performance of all variables without SNHR (but with RAIN variable) shows higher RMSE, MAE and MARE and lower r, R^2 and d as compared to model M-2. The RMSE, MAE and MARE of model M-3 are 1.984 mm, 0.0394 mm and 2.86%, respectively. The r, R^2 and d of model M-3 are 0.844, 0.712 and 0.915%, respectively. The error statistics of model GM M-4 in which the relative humidity (RMAX and RMIN) are not the input variables. The RMSE, MAE and MARE of model M-4 is 2.081 mm, 0.0433 mm and 3.27%, which is higher than the models M-1, M-2 and M-3. The r, R^2 and d of M-4 is 0.826, 0.683 and 0.906, respectively and these values are higher than model M-3. It is mainly due to the absence of RMAX and RMIN input variables and have higher negative correlation with the pan evaporation.

The model M-5 performance is better than the model M-4. The RMSE, MAE and MARE in model M-5 is reduced to 2.061 mm, 0.0425 mm and 3.24% from 2.081 mm, 0.0433 mm and 3.27% in model M-4. The r, R^2 and d is also increased to 0.830, 0.689 and 0.910 in model M-5 as compared to model M-4. The model M-5, in which four input variables is assumed to be one of the better models with minimum number of input variables. The input variables are TMAX, WISP and SNHR in model M-6. The RMSE, MAE and MARE of model M-6 is 2.127 mm, 0.0453 mm and 4.24%, respectively. The r, R^2 and d of model M-6 was 0.818, 0.669 and 0.904, which are lower as compared to M-5. The RMSE, MAE and MARE is decreased to 2.111 mm, 0.0445 mm and 5.01% in GM model M-7, in which only TMAX and WISP are the input variables of the model. The r, R^2 and d was also slightly higher in model M-7 as compared to M-6. The r, R^2 and d in model M-7 is 0.821, 0.674 and 0.902, respectively. The evaluation of General models using Nagpur model development data with ANN model M-1 to M-7 are graphically presented (Fig. 9). These results indicated that maximum temperature and relative humidity observations were very beneficial in accurate estimation of pan evaporation, when compared to any other single variable.

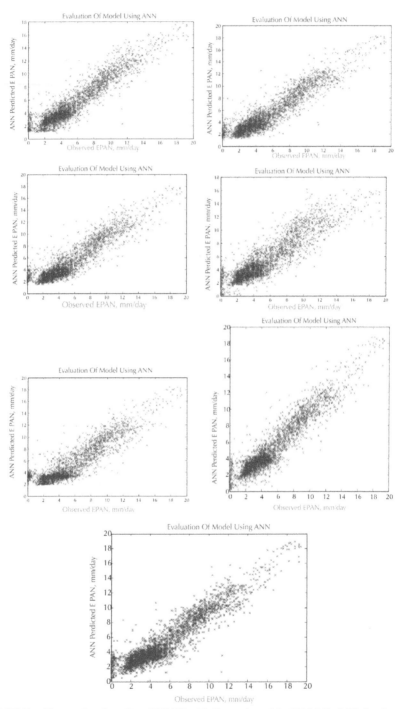

FIGURE 8   Observed and predicted EPAN with General models (GM M1- M7) development
data.

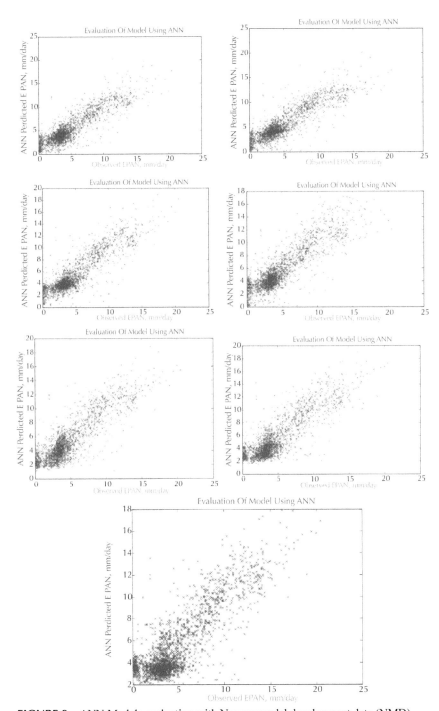

FIGURE 9 ANN Models evaluation with Nagpur model development data (NMD).

7.9   SUMMARY

The three-layered feed forward neural network with back propagation prediction method that uses Artificial Intelligence (AI) to model the daily pan evaporation was developed and evaluated using independent datasets. From these research results, it is concluded that the constructed ANN models successfully identifies the evaporation process and accurately predict the pan evaporation in a next time step. Three layered feed forward neural networks with error back propagation was found suitable for modeling the evaporation. Once the network is trained, with a satisfactory amount of input data, it can accurately predict the daily pan evaporation. The network architectures performed better with one hidden layer and 15 hidden nodes (2^{n+1}) for GM modeling problems. Single hidden layer performed well for both the models. The *Levenberg Marquardt* minimization training function gave the best network training results when used in the back propagation algorithms. While evaluating the data the model plan strategy with all inputs has shown better performance than the reduced number of inputs. The daily pan evaporation is estimated for the five locations using three layers feed forward neural network with error back propagation. Seven model strategies with combinations of input variables to develop the ANN models of pan evaporation were developed. Strategy M-1, which includes all 7 input variables, had the highest accuracy. This ANN model had an R^2 of 0.78 and an RMSE of 1.61 mm on the independent GM evaluation dataset. The models evaluation with Nagpur model development (NMD) dataset had an R^2 of 0.72 and an RMSE of 1.96 mm. The ANN model has a strong correlation with observed pan evaporation. Differences between the ANN model prediction and observed pan evaporation do not appear to be biased as the observation of pan evaporation changes.

KEYWORDS

- **Artificial Neural Network**
- **coefficient of determination**
- **correlation coefficient**
- **Dr. Panjabrao Deshmukh Agricultural University, Akola**
- **Epochs**
- **error back propagation algorithm**
- **evaporation**
- **evaporation modeling**
- **evapotranspiration**
- **feed forward neural networks**
- **hidden layer**
- **hydrological processes**
- **hydrology**
- **index of agreement**

- **network function**
- **neural network**
- **non-linear regression models**
- **observed values**
- **output variable**
- **pan evaporation**
- **predicted values**
- **prediction of pan evaporation**
- **rainfall**
- **root mean square error**
- **soil water evaporation**
- **sunshine hours**
- **the class 'A' pan evaporimeter**
- **Training function**
- **water management**
- **water resources planning and management**
- **weather forecasting**
- **wind speed**

REFERENCES

1. Arca, B.; Benincasa F.; De Vincenzi M.; Ventura A. **1998,** Neural network to simulate evaporation from Class A pan. *In: Proc. 23rd Conf. on Agric. For. Meteorol.;* 2–6 November **1998,** Albuquerque, New Mexico, Am. Meteorol. Soc.; Boston, MA.
2. Bruton, J. M.; McClendon R. W.; Hoogenboom, G. Estimating daily pan evaporation with artificial neural networks. *Trans. ASAE,* **2000,** *43(2),* 491–496.
3. de Villiers, J.; Barnard, E. Back propagation neural nets with one and two hidden layers. IEEE *Transactions on Neural Netw.;* **1993,** *4(1),* 136–141.
4. Deswal S.; Mahesh Pal. Artificial Neural Network based Modeling of Evaporation Losses in Reservoirs. *Proc. of World Academy of Science, Engineering and Technology,* **2008,** *29,* 279–283.
5. Gavin, H.; Agnew, C. A. Modeling actual, reference and equilibrium evaporation from a temperate wet grassland. *Hydrol. Processes,* **2004,** *18,* 229–246.
6. Gardner, M. W.; Dorling, S. R. **1998,** Artificial neural network (Multilayer Perceptron) a review of applications in atmospheric sciences. *Atmos. Environ.;* 32, 2627–2636.
7. Han, H.; Felkar. P.; Estimation of daily soil water evaporation using an artificial neural network. *J.; Arid Environ.* **1997,** *37 (2),* 251–260.
8. Jhajharia, D.; Kithan, S. B.; Fancon, A. K. Correlation between pan evaporation and metrological parameters under the climatic conditions of Jorhat (Assam). *J. Indian Water Res.,* **2006,** *26(1–2),* 39–42.
9. Keskin, M. E.; Terzi, O. Artificial neural network models of daily pan evaporation. *ASCE J. Hydrologic Engineering,* **2006,** *11(1),* 65–70.
10. Khanikar, P. G.; Nath K. K.; Relationship of open pan evaporation rate with some important meteorological parameters. *J.; Agril. Sci. Soc. N. – E India* **1998,** *11 (1),* 46–50.

11. Kumar, M.; Raghuwanshi, N. S.; Singh, R.; Wallender W. W.; Pruitt. W. O Estimating evapotranspiration using artificial neural network. *J.; Irri. Drain. Engg. ASCE,* **2002,** *128(4),* 224–233.

12. Maier, H. R.; Dandy, G. C.; Neural network based modeling of environmental variables: a systematic approach. *Mathematical and Computer Modelling,* **2001,** *33,* 669–682.

13. Molina Martinez, J. M.; Alvarezv, M.; Gonzalez Real, M. M.; Baille, A. A simulation model for predicting hourly pan evaporation from meteorological data. *J. hydrology,* **2006,** *318 (1–4),* 250–261.

14. Ozlem, T.; Evolkesk, M. Modeling of daily pan evaporation. *J. Applied Sciences,* **2005,** *5 (2),* 368–372.

15. Shirgure, P. S.; Srivastava A. K.; Shyam Singh. Effect of pan evaporation based irrigation scheduling on yield and quality of drip irrigated Nagpur mandarin. *Indian J. Agric. Sci.;* **2001,** *71 (4),* 264–266.

16. Shirgure, P. S.; Rajput, G. S. Evaporation modeling with neural networks – A Research review. *Int. J. Res. Reviews in Soft Intelligent Computing.* **2011,** *1* (2): 37–47.

17. Shrivastava, S. K.; Sahu, A. K.; Dewangan, K. N.; Mishra, S. K.; Upadhyay A. P.; Dubey, A. K.; Estimating pan evaporation from meteorological data for Jabalpur. *Indian J.; Soil Cons.;* **2001,** *29* (3): 224–228.

18. Sudheer, K. P.; Gosain, A. K.; Mohana, R. D.; Saheb, S. M. Modeling evaporation using an artificial neural network algorithm. *Hyd. Processes,* **2002,** *16,* 3189–3202.

19. Sudheer, K. P.; A. K.; Gosain and Ramasastri, K. S.; Estimating actual evapotranspiration from limited climatic data using neural computing technique. *J. Irri. Drain. Engg.* ASCE. **2003,** *129 (3): 214–218.*

20. Taher, S. A. Estimation of Potential Evaporation: Artificial Neural Networks versus Conventional methods. *J. King Saud Univ. Eng. Sci.;* **2003,** *1,* 17.

CHAPTER 8

EVAPOTRANSPIRATION FOR CYPRESS AND PINE FORESTS: FLORIDA, USA[1]

DAVID M. SUMNER

CONTENTS

[1]The author gratefully extends his appreciation to Catherine Lowenstein and her staff at the Tiger Bay State Forest for providing assistance during this study. The contributions of Timothy Curran and Jerome Kelly of the USGS in construction and maintenance of the evapotranspiration station are gratefully acknowledged. This chapter is an edited version of: "Sumner, D.M., 2001. Evapotranspiration from a Cypress and Pine Forest Subjected to Natural Fires, Volusia County, Florida, 1998–99: U.S. Geological Survey Water-Resources Investigations Report 01–4245, 56 p. available at: <http://fl.water.usgs.gov/Abstracts/wri01_4245_sumner.html>."
© Copyright fl.water.usgs.gov

8.1 INTRODUCTION

The importance of evapotranspiration (ET) in the hydrologic cycle has long been recognized; in central Florida, evapotranspiration is second only to precipitation in magnitude. Of the approximately 1,320 millimeters (mm) of mean annual rainfall in central Florida, 680 to 1,220 mm have been estimated to return to the atmosphere as evapotranspiration [44, 49]. Despite the importance of evapotranspiration in the hydrologic cycle, the magnitude, seasonal and diurnal distributions, and relation to environmental variables of evapotranspiration remain relatively unknown. Uncertainty in evapotranspiration from nonagricultural vegetation is particularly apparent. The mixed cypress wetland and pine flat wood forest cover examined in the present investigation is common in central Florida, as are the fires that burned much of the forest during the study. Accurate estimates of evapotranspiration from commonly occurring land covers are fundamental to the quantitative understanding necessary for prudent management of Florida's water resources.

The eddy correlation method (or eddy covariance method) has been used successfully to directly measure evapotranspiration in Florida by Bidlake and others [3]; Knowles [22]; and Sumner [44]. This micrometeorological method offers several advantages to alternative water-budget approaches (lysimeter or regional water budget) by providing more areal integration and less site disruption than lysimeters, by eliminating the need to estimate other terms of a water budget (precipitation, deep percolation, runoff, and storage), and by allowing relatively fine temporal resolution (less than 1 hour).

Evapotranspiration can be estimated by using evapotranspiration models. These models also provide insight into the relative importance of individual environmental variables in the evapotranspiration process. The Priestley-Taylor model [34] for evaporation from a wet surface (potential evapotranspiration) was modified to allow for nonpotential conditions [13], and has successfully simulated evapotranspiration in the Florida environment [15, 22, 44].

The U.S. Geological Survey (USGS), in cooperation with the St. Johns River Water Management District and the County of Volusia, began a 4-year study in 1996 to estimate the temporal pattern of evapotranspiration in the Tiger Bay watershed, Volusia County, Fla., a forested watershed, and to develop a quantitative description of the effect of environmental variability on evapotranspiration from forested areas in Florida. This analysis can provide guidance in the estimation of evapotranspiration and the description of the relation between the environment and evapotranspiration in other areas with similar environmental characteristics. During the study period, the watershed experienced a severe drought and natural fires, which provided the opportunity to study the effects of such extreme events on the evapotranspiration process.

This chapter presents daily estimates of evapotranspiration during a 2-year period from a forested watershed (Tiger Bay, Volusia County, Fla.), which was subjected to natural fires, and provides evaluations of the causal relations between the environment and evapotranspiration. Measurements were made on a nearly continuous basis from January 1998 through December 1999 at an evapotranspiration station just outside the watershed, using eddy correlation and meteorological instrumentation. An evapotranspiration model based on the Priestley-Taylor equation was used to estimate evapotranspiration for burned and unburned areas and to quantify the relation between evapotranspiration and the environment. A water budget of the watershed was constructed to assess the validity of the eddy correlation-measured evapotranspiration totals for the 2-year period.

8.1.1   DESCRIPTION OF THE STUDY AREA

The study area is the approximately 7,500-hectare Tiger Bay watershed within Volusia County, Fla. (Fig. 1). The watershed was almost completely forested in January 1998, but was subjected to extensive burning and logging during the study period. The watershed characteristics are typical of many areas within the lower coastal plain of the south-eastern United States—nearly flat, slowly draining land with a vegetative cover consisting primarily of pine flat wood uplands interspersed within cypress wetlands. The northern part of the watershed mostly is within the 9,500-hectare Tiger Bay State Forest; the southern part of the watershed primarily is privately owned land used for timber production. The watershed is within the relatively flat Talbott Terrace physiographic area [37]. More than 90% of the watershed is at an altitude of 11 to 13 meters (m). Small variations in local topography result in areal variations in hydroperiod. A low-lying wetland can be inundated much of the year, whereas an adjacent upland, less than a few tens of centimeters (cm) elevated above the wetland, may only occasionally or never exhibit standing water. Most of the surface runoff from the watershed is through interconnected wetlands [36].

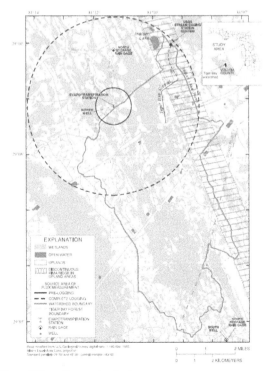

FIGURE 1   Location of Tiger Bay watershed.

More than 95% of the watershed is forested. Two tree species dominate the forest cover in the watershed: slash pine (evergreen) and pond cypress (deciduous; leaves drop in November-December with regrowth in March–April). The distribution of vegetation in the vicinity of the evapotranspiration station is shown in Fig. 2.

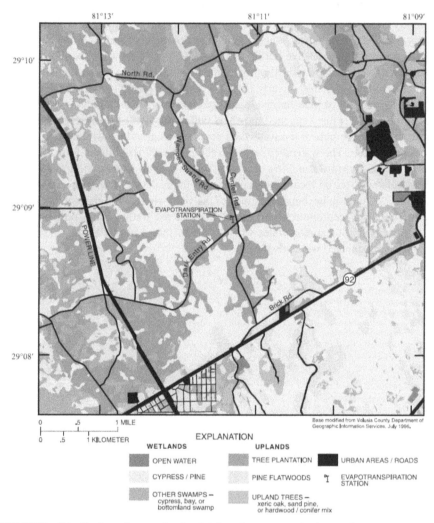

FIGURE 2   Distribution of vegetation in vicinity of evapotranspiration station.

Vegetation in the watershed reflects the variation in hydroperiod [40]. Wetlands are dominated by pond cypress (*Taxodium ascendens*), with lesser amounts of other wetland tree species including blackgum (*Nyssa biflora*), loblolly bay (*Gordonia lasianthus*), and red maple (*Acer rubrum*). The under story of wetlands consists of a wide variety of plants including leather fern (*Acrostichum danaeifolium*), marsh fern

(*Thelypteris palustris*), cinnamon fern (*Osmunda cinnamomea*), swamp lily (*Crinum americanum*), maidencane (*Panicum hemitomon*), red root (*Lachnanthes caroliniana*), hooded pitcher plant (*Sarracenia minor*), St. John's Wort (*Hypericum fasciculatum*), yellow colic root (*Aletris lutea*), pipewort (*Eriocaulon decangulare*), and white-topped sedge (*Rhynchospora colorata*). Water level varies from about 0.3 m above land surface to as much as 1 m below land surface in low-lying areas, although these areas are inundated more than 50% of the time [40].

Uplands generally are either slash pine tree (*Pinus elliottii*) plantations or naturally seeded pine flatwoods (primarily slash pine with some longleaf pine (*Pinus palustris*)). These areas have an under story including saw palmetto (*Serenoa repens*), gallberry (*Ilex glabra*), wax myrtle (*Myrica cerifera*), red root (*Lachnanthes caroliniana*), and broomsedge (*Andropogon virginicus*). Under story vegetation in the pine plantations is control-burned about every 3 years. Water level varies from about 0.1 m above land surface to as much as 2 m below land surface in uplands; however, water levels are always greater than 2 m below land surface in the small part of the uplands within the Rima Ridge (Fig. 1). The Rima Ridge consists of discontinuous remnants of terrace deposits parallel to the present-day coastline [37]. Vegetation on the ridge areas includes sand live oak (*Quercus geminata*) and sand pine (*Pinus clausa*). Most of the limited urbanization within the Tiger Bay watershed is on the Rima Ridge.

Brush fires burned extensively throughout peninsular Florida during spring 1998 as a result of a severe drought. A high-pressure system remained stationary over the State, blocking the normal pattern of convective thunderstorms [48]. During the 3-month period, April-June, National Oceanic and Atmospheric Administration (NOAA) stations at Daytona Beach and DeLand recorded about 10 and 30% of long-term, average precipitation, respectively. Brush fires, ignited by lightning strikes, began in Volusia County on June 19, 1998, and continued until rainfall resumed in late June and early July, burning about 55,000 hectares (one-fifth of the County) and about 40% of the watershed (Fig. 3). Although areas of both wetlands and uplands were burned during the June-July fires, a comparison of Figures 1 and 3 reveals that upland areas were burned more extensively than wetland areas. Re-growth of under story vegetation occurred rapidly after the fires ceased and the rains began. Emergent growth of red root (*Lachnanthes caroliniana*) in burned areas was particularly evident. Some trees were killed by the fire, whereas other burned trees were merely damaged and exhibited leaf regrowth soon after the fire (Fig. 4). Large-scale harvesting of insect-infested, fire-damaged trees (both living and dead trees) occurred during the months following the fires. Of the approximately 4,800 hectares that burned within the 9,500-hectare Tiger Bay State Forest, about 3,200 hectares were logged (Catherine Lowenstein, Tiger Bay State Forest, oral communication, 2000).

Fires moved from west-to-east through the area of the evapotranspiration station on June 25, 1998. Damaged trees in the vicinity of the evapotranspiration station were logged during November–December 1998.

The two dominant soil groups of the watershed also reflect the areal variation in hydroperiod and vegetation [2]. Wetlands tend to be underlain by organic soils (hyperthermic family of Terric Medisaprists) of the Samsula-Terra Ceia-Tomoka group, that are very poorly drained. The uplands tend to be underlain by poorly drained soils (sandy, siliceous, hyperthermic family of Ultic Haplaquods) of the Pomona-Wauchula group that have a dark, organic-stained subsoil underlain by loamy material.

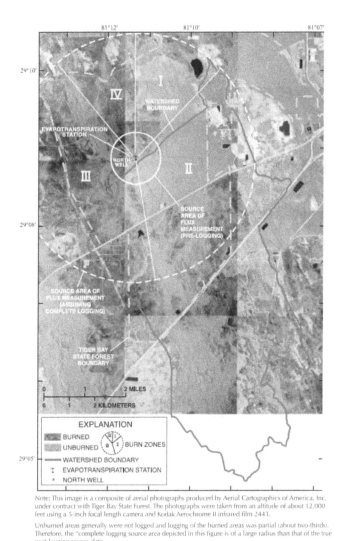

Note: This image is a composite of aerial photographs produced by Aerial Cartographics of America, Inc. under contract with Tiger Bay State Forest. The photographs were taken from an altitude of about 12,000 feet using a 5-inch focal length camera and Kodak Aerochrome II infrared film 2443.

Unburned areas generally were not logged and logging of the burned areas was partial (about two-thirds). Therefore, the "complete logging source area depicted in this figure is of a large radius than that of the true post-logging source data.

FIGURE 3 Infrared photograph (July 7, 1998) of vicinity of evapotranspiration station showing area: burned during fires of June 1998.

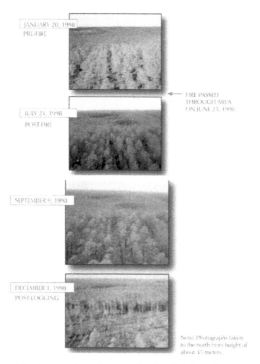

JANUARY 20, 1998
PRE-FIRE

FIRE PASSED
THROUGH AREA
ON JUNE 25, 1998

JULY 21, 1998
POST-FIRE

SEPTEMBER 9, 1998

DECEMBER 1, 1998
POST-LOGGING

Note: Photographs taken
to the north from height of
about 15 meters.

FIGURE 4   Photographic times series of vegetation in vicinity of evapotranspiration station.

The climate of central Florida is humid subtropical and is characterized by a warm, wet season (June–September) and a mild, relatively dry season (October–May). During the dry season, precipitation commonly is associated with frontal systems. Rainfall averages about 1,350 mm/yr in Volusia County [37]. More than 50% of the annual rainfall generally occurs during the wet season when diurnal thunderstorm activity is common. Mean air temperature in the study area is about 21 °C, ranging from occasional winter temperatures below 0°C to summer temperatures approaching 35°C. Diurnal temperature variations average about 12°C.

FIGURE 5   Krypton hygrometer (foreground) and sonic anemometer (background) mounted at top of tower at evapotranspiration station.

FIGURE 6   Evapotranspiration station being serviced by hydrologic technician.

TABLE 1   Study instrumentation [CSI, Campbell Scientific, Inc.; REBS, Radiation and Energy Balance Systems, Inc.; RMY, R. M. Young, Inc.; TE, Texas Electronics, Inc.]. *Note:* Negative height is depth below land surface.

Parameter	Instrument	Height(s) above land surface, meters
Evapotranspiration	CSI eddy correlation system including model CSAT3 3-D sonic anemometer and model KH20 krypton hygrometer.	36.5
Air temperature/ Relative humidity	CSI model HMP35C temperature and relative humidity probe.	1.5, 9.1, 18.3 and 35
Net radiation	REBS model Q-7.1 net radiometer.	35
Wind speed & direction	RMY model 05305-5 wind monitor - AQ	35
Photosynthetically active radiation (PAR)	LI-COR, Inc., Model LI-190SB quantum sensor.	35
Soil moisture	CSI model CS615 water content reflectometer.	0.0 to -0.3
Precipitation	TE model 525 tipping bucket rain gage and No-vaLynx model 260-2520 forester's (storage) rain gages (two)	18.3 (tipping bucket) and 1.0 (storage)
Water level in well	Druck, Inc., Model PDCR950 pressure trans-ducer.	- 2.0
Datalogging	CSI model 21X and model 10X data loggers; 12 volt deep-cycle batteries (two); 20 watt solar panels (two)	0 to 1.0

Rainfall to the watershed leaves the basin as runoff, evapotranspiration, or deep leakage from the surficial aquifer system to the underlying Upper Floridan aquifer [21, 33]. Intermittent runoff gaged at Tiger Bay canal along the northern edge of the watershed (Fig. 1) averaged 0.47 cubic meters per second (m³/s) or about 200 millimeters per year (mm/yr) from 1978 to 1999 [53]. Evapotranspiration has been estimated to average about 990 mm/yr over Volusia County [37] and about 890 mm/yr in the Tiger Bay watershed [7]. Previous researchers have documented relatively small differences in the annual evapotranspiration rates from the two primary land covers. Bidlake and others [3] estimated annual cypress evapotranspiration (970 mm) to be only 8.5% less than that from pine flatwoods (1,060 mm), based on studies conducted in Sarasota and Pasco Counties, Fla. Liu [24] estimated average annual evapotranspiration from both covers to be 1,080 mm, based on a study conducted in Alachua County, Fla.

The hydraulic head in the surficial aquifer system within the watershed generally is above that of the underlying Upper Floridan aquifer. Consequently, water leaks downward from the surficial aquifer system, through the intermediate confining unit, to the Upper Floridan aquifer. Deep leakage was estimated (based on ground-water flow simulations) to have been about 56 mm/yr prior to ground-water development, but in 1995, the rate was estimated to have doubled to 112 mm/yr, as a result of lowering the hydraulic head in the Upper Floridan aquifer by pumping (Stan Williams, St. Johns River Water Management District, oral communication, 2000).

8.2 METHODS AND MATERIALS: MEASUREMENT AND SIMULATION OF EVAPOTRANSPIRATION

Evapotranspiration was measured at a site just outside the study area (Fig. 1) using the eddy correlation method in a manner similar to that described by Sumner [44]. The site chosen for the evapotranspiration station was within an 18.3-m-tall, 30-year-old pine plantation (Fig. 2). Eddy correlation instrumentation was mounted on a 36.5-m-tall Rohn 45G communications- type tower at the site (Figs. 5 and 6), and data were collected for a 2-year period from January 1, 1998, to December 31, 1999. Other meteorological instrumentation also was deployed on or around the tower to collect data for evapotranspiration modeling and to provide ancillary data for the eddy correlation analysis. Instrumentation used in the study is described in Table 1. Measured daily values of evapotranspiration were used to calibrate evapotranspiration models (modified Priestley-Taylor). Evapotranspiration was estimated for burned and unburned areas using the calibrated evapotranspiration models. A water budget for the watershed over the study period was constructed based on measured or estimated values of precipitation, evapotranspiration, runoff, leakage, and storage.

8.2.1 MEASUREMENT OF EVAPOTRANSPIRATION

8.2.1.1 EDDY-CORRELATION METHOD

The eddy correlation method [9, 45] was used to measure two components of the energy budget of the plant canopy: latent and sensible heat fluxes. Latent heat flux (λE) is the energy removed from the canopy in the liquid-to- vapor phase change of water, and is the product of the heat of vaporization of water (λ) and the evapotranspiration rate (E). Sensible heat (H) is the heat energy removed from the canopy as a result of a temperature gradient between the canopy and the air.

Both latent and sensible heat fluxes are transported by turbulent eddies in the air. Turbulence is generated by a combination of frictional and convective forces. The energy available to generate turbulent fluxes of vapor and heat is equal to the net radiation (R_n) minus the sum of the heat flux into the soil surface (G) and the change in storage (S) of energy in the biomass and air. The energy involved in fixation of carbon dioxide usually is negligible [5]. Net radiation is the difference between incoming radiation (shortwave solar radiation and long wave atmospheric radiation) and outgoing radiation (reflected shortwave and long- wave radiation; and emitted long wave canopy radiation). Energy is transported to and from the base of the canopy by conduction through the soil.

Assuming that net horizontal advection of energy is negligible, the energy-budget equation, for a control volume extending from land surface to a height z_s at which the turbulent fluxes are measured, is given in Eq. (1):

$$R_n - G - S = H + \lambda E \tag{1}$$

$$E = \overline{w\rho_v} = \overline{(\overline{w} + w')(\overline{\rho_v} + \rho_v')} \tag{2}$$

$$E = (\overline{\overline{w}\overline{\rho_v}} + \overline{\overline{w}\rho_v'} + \overline{w'\overline{\rho_v}} + \overline{w'\rho_v'}) \tag{3}$$

$$\overline{w'\rho'}_v = covariance(w, \rho_v) \tag{4}$$

In the energy-budget eq. (1): the left side of represents the available energy and the right side represents the turbulent flux of energy; R_n is net radiation to or from plant canopy, in watts per square meter; G is soil heat flux at land surface, in watts per square meter; S is change in storage of energy in the biomass and air, in watts per square meter; H is sensible heat flux at height z_s above land surface, in watts per square meter; and ΔE is latent heat flux at height z_s above land surface, in watts per square meter. The sign convention is such that R_n and G are positive downwards; and H and ΔE are positive upwards.

The eddy correlation method is a conceptually simple, one-dimensional approach for measuring the turbulent fluxes of vapor and heat above a surface. For the case of vapor transport above a flat, level land- scape, the time-averaged product of measured values of vertical wind speed (w) and vapor density (ρ_v) is the estimated vapor flux (evapotranspiration rate) during the averaging period, assuming that the net lateral advection of vapor is negligible. Because of the insufficient accuracy of instrumentation available for measurement of actual values of wind speed and vapor density, this procedure generally is performed by monitoring the fluctuations of wind speed and vapor density about their means, rather than monitoring their actual values.

This formulation is represented in eqs. (2) to (4), where: E is evapotranspiration rate, in grams per square meter per second; w is vertical wind speed, in meters per second; ρ_v is vapor density, in grams per cubic meter; and overbars and primes indicate means over the averaging period and deviations from means, respectively.

The first term of the right side of eq. (3) is approximately zero because mass-balance considerations dictate that mean vertical wind speed perpendicular to the surface is zero; this conclusion is based on an assumption of constant air density (correction for temperature-induced air-density fluctuations is discussed later in this chapter). The second and third terms are zero based on the definition that the mean fluctuation of a variable is zero. Therefore, it is apparent from eq. (4) that vertical wind speed and vapor density must be correlated in order for the value of vapor flux to be non-zero. The turbulent eddies that transport water vapor (and sensible heat) produce fluctuations in both the direction and magnitude of vertical wind speed. The ascending eddies must on average be more moist than the descending eddies for evapotranspiration to occur, that is, upward air movement must be positively correlated with vapor density and downward air movement must be negatively correlated with vapor density.

8.2.2   SOURCE AREA OF MEASUREMENTS

The source area for a turbulent flux measurement defines the area (upwind of measurement location) contributing to the measurement. The source area can consist of a single vegetative cover if that cover is adequately extensive. This condition is met if the given cover extends sufficiently upwind such that the atmospheric boundary layer has equilibrated with the cover from ground surface to at least the height of the instrumentation. If this condition is not met, the flux measurement is a composite of fluxes from two or more covers within the source area. The source area is defined in this report as the area contributing to 90% of the sensor measurement. Schuepp and others [39] provide an estimate of the source area, and the relative contributions within the source area, based on an analytical solution of a one-dimensional (upwind) diffusion equation for a uniform surface cover. In this approach, source area varies with instrument height (z_s), zero displacement height (d), roughness length for momentum (z_m), and atmospheric stability. The instrument height in this study was 36.5 m. Campbell and Norman [8] proposed empirical relations based on canopy height (h) for zero displacement height ($d \sim 0.65h$) and roughness length for momentum ($z_m \sim 0.10h$.). Uniform canopy heights of 18.3 m (prelogging) and 0.3 m (assuming complete logging) were assumed in this analysis. The source area estimates were made assuming mildly unstable conditions; the Obukhov stability length [6] was set equal to −10 m. The source area increases as the height of the instrument above the vegetative canopy increases and as the roughness length for momentum decreases; therefore, the extensive logging that occurred following the fires enlarged the source area. The source area for the turbulent flux measurements (Fig. 7) was estimated to be within an upwind distance of about 1,000 m (prelogging) or 4,800 m (assuming complete logging). As stated earlier, unburned areas generally were not logged and logging of the burned areas was partial (about two-thirds). Therefore, the "complete logging" source area depicted in Fig. 7 is of a larger radius than that of the true postlogging source area.

   The site of the evapotranspiration station was chosen such that the source area of the turbulent flux measurements would be representative of the relative mix of wetlands and uplands in the prefire watershed (Fig. 1). Before the fire and associated logging, the source area of the turbulent flux measurement (Fig. 1) consisted of: 43.7% upland, 56.1% wetland, and 0.2% lake. These relative fractions of wetland and upland were very close to those of the entire Tiger Bay watershed (43.8% upland, 55.5% wetland, and 0.7% lake) before the fires. Also, areas of wetland and upland within the pre fire source area were interspersed, indicating that turbulent flux measurements approximated a representative value of the composite mix of wetlands and uplands, regardless of the wind direction.

Fires within the watershed during spring 1998 changed the primary components of source area heterogeneity from wetland/upland to burned/unburned (Fig. 3) and complicated interpretation of the turbulent flux measurements. Burned and unburned areas were not well-interspersed, resulting in measurements that reflected varying fractions of burned and unburned areas, depending on the wind direction. Following the fires, turbulent fluxes representative of burned areas were measured, both pre and postlogging, when the wind was from the north-west (zone IV in Fig. 3).

Turbulent fluxes representative of unburned areas were measured when the wind was from the east (zone II) throughout the study period. The absence of near-station burning in zone II, and therefore a lack of subsequent near-station logging in this zone, resulted in a consistently small (radius of 1,000 m), and unburned, source area throughout the study period when the wind was from zone II. Turbulent fluxes representative of burned areas were measured following the fires and prior to logging when the wind was from the north-east (zone I). With the expansion of the source area associated with logging, however, the postlogging turbulent flux measurements were representative of a composite of burned and unburned areas when the wind was from zone I. Examination of the estimated [39] cumulative fractional contribution to the turbulent flux measurement as a function of upwind distance from the measurement (Fig. 7) provided information to approximate the relative degree of burned/unburned area compositing. Based on this approach, an estimate was made that postlogging turbulent flux measurements made when the wind was from zone I reflected a surface cover that was 75% burned and 25% unburned. Burned and unburned areas within zone III were relatively well interspersed and in approximately equal relative amounts following the fires. Therefore, postfire turbulent fluxes measured when the wind was from zone III were assumed to reflect a surface cover that was 50% burned and 50% unburned. Estimates of the relative contribution (as a function of wind direction and status of the surface cover) of burned vegetation to the measured turbulent flux signal are summarized in Table 2. These estimates were used to develop weighting coefficients indicative of the fraction of the turbulent flux measurement for a given day that reflected burned vegetation, which is further discussed later in this chapter.

8.2.3   INSTRUMENTATION

Instrumentation capable of high-frequency resolution must be used in an application of the eddy correlation method because of the relatively high frequency of the turbulent eddies that transport water vapor. Instrumentation included a three-axis sonic anemometer and a krypton hygrometer to measure variations in wind speed and vapor density, respectively (Fig. 5). The sonic anemometer relies on three pairs of sonic transducers to detect wind-induced changes in the transit time of emitted sound waves and to infer fluctuations in wind speed in three orthogonal directions. The measurement path length between transducer pairs is 10.0 cm (vertical) and 5.8 cm (horizontal); the transducer path angle from the horizontal is 60 degrees. In contrast to some sonic anemometers used previously [44], the transducers of this improved anemometer are not permanently destroyed by expo- sure to moisture, and thus are suitable for long-term deployment. Operation of the anemometer used in this study ceases when moisture on the transducers disrupts the sonic signal, but recommences upon drying of the transducers.

The hygrometer relies on the attenuation of ultra-violet radiation, emitted from a source tube, by water vapor in the air along the 1-cm path to the detector tube. The instrument path line was laterally displaced 10 cm from the midpoint of the sonic transducer path lines. Hygrometer voltage output is proportional to the attenuated radiation signal, and fluctuations in this signal can be related to fluctuations in vapor density by Beer's Law [59]. Similar to the anemometer, the hygrometer ceases data collection when moisture obscures the windows on the source or detector tubes. Also, the tube windows become "scaled" with exposure to the atmosphere, resulting in a loss of signal strength. The hygrometer is designed such that vapor density fluctuations are accurately measured in spite of variable signal strength; however, if signal strength declines to near-zero values, the fluctuations cannot be discerned. Periodic cleaning of the windows (performed monthly in this study) with a cotton swab and distiled water restored the signal strength. Eddy correlation instrument-sampling frequency was 8 Hertz with 30-minute averaging periods. The eddy correlation instrumentation was placed about 18.2 m above the tree canopy (Fig. 6). Data were processed and stored in a data logger near ground-level.

To be representative of the surface cover, flux measurements must be made in the inertial sublayer, where vertical flux is constant with height and lateral variations in vertical flux are negligible [28]. Measurements made in the underlying roughness sublayer can reflect individual roughness elements (for example, individual trees or gaps between trees), rather than the composite surface cover. Garrat [14] defines the lower boundary of the inertial sublayer to be at a height such that the difference of this height and the zero displacement height (d) is much greater than the roughness length for momentum (z_m). Employing Campbell and Norman's [8] empirical relations and assuming that "much greater than" implies greater by a factor of ten [10], leads to an instrument height (z_s) requirement of $z_s >[1.65*h]$. A factor of about two was used in this chapter as a conservative measure. As a conservative measure, the instrument height (36.5 m) used in this chapter was about twice canopy height.

8.2.4 CALCULATION OF TURBULENT FLUXES

Latent heat flux [see Eq. (5)] was estimated based on a modified form of Eq. (4). In Eq. (5): λE is latent heat flux, in watts per m²; λ is latent heat of vaporization of water, estimated as a function of temperature [43], in joules per gram; ρ is air density that is estimated as a function of air temperature, total air pressure, and vapor pressure [28], in grams per cubic meter; H is sensible heat flux, in watts per m²; C_p is specific heat capacity of air, estimated as a function of temperature and relative humidity [43], in joules per gram per degree Celsius; T_a is air temperature, in °C; F is a factor that accounts for molecular weights of air and atmospheric abundance of oxygen, and is equal to 0.229 gram-degree Celsius per joule; K_o is an extinction coefficient of hygrometer for oxygen, estimated as 0.0045 cubic meters per gram per centimeter [46]; K_w is an extinction coefficient of hygrometer for water, equal to the manufacturer-calibrated value, in cubic meters per gram per centimeter; and overbars and primes indicate means over the averaging period and deviations from the means, respectively. The second and third terms of the right side of Eq. (5) account for temperatureinduced fluctuations in air density [58] and for the sensitivity of the hygrometer to oxygen [45], respectively.

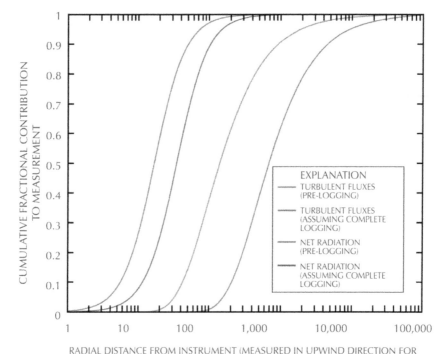

RADIAL DISTANCE FROM INSTRUMENT (MEASURED IN UPWIND DIRECTION FOR TURBULENT FLUXES AND RADIALLY OMNI-DIRECTIONAL FOR NET RADIATION). IN METERS

FIGURE 7   Radial extent of source areas of turbulent flux and net radiation measurements.

TABLE 2   Relative fraction of burned vegetation sensed by eddy correlation instrumentation [The sector is in degrees measured clockwise from north (Fig. 3); g_i is the fractional contribution of burned area within burn zone i to the measured latent heat flux when wind direction is from burn zone i].

Burn zone i	Sector	g_i = Fractional contribution		
		Pre-fire	Post – fire/pre-logging	Post-logging
I	0 to 45	0.0	1.0	0.75
II	45 to170	0.0	0.0	0.00
III	170 to 320	0.0	0.5	0.50
IV	320 to 360	0.0	1.0	1.00

Similarly to vapor transport, sensible heat can be estimated by using Eq. (6). The sonic anemometer is capable of measuring "sonic" temperature based on the dependence of the speed of sound on this variable [19, 20]. Schotanus and others [38] related the sonic sensible heat based on measurement of sonic temperature fluctuations to the true sensible heat given in Eq. (6). Those researchers included a correction, for the effect of wind blowing normal to the sonic acoustic path that has been incorporated directly into the anemometer measurement by the manufacturer (Swiatek, E., 1998. Campbell Scientific, Inc., written communication), leading to a simplified form of the Schotanus and others [38] formulation given in Eq. (7).

In Eq. (7): T_s is the sonic temperature, in °C; and q is specific humidity, in grams of water vapor per grams of moist air. Fleagle and Businger [12] defined the specific humidity [q] in Eq. (8), based on the relation between specific humidity and vapor density.

In Eq. (8): ρ_v is vapor density, in grams per cubic meter; R_d is the gas constant for dry air ($= 0.28704$ joules per degree Celsius per gram); and P_a is atmospheric pressure, in pascals (assumed to remain constant at 100.7 kilopascals at top of tower at about 48 meters above sea level). Eq. (7) can be expressed in terms of fluctuations in the hygrometer-measured water vapor density rather than fluctuations in specific humidity as shown in Eq. (9).

$$\lambda E = \lambda \left(\overline{w' \rho'}_v + \left[\frac{\rho_v H}{\rho C_p} \right] * \left[\frac{1}{T_a + 273.15} \right] + \frac{F K_o H}{K_w (T_a + 273.15)} \right) \tag{5}$$

$$H = \rho C_p \overline{w' T_a'} \tag{6}$$

$$\overline{w' T_a'} = \overline{w' T_s'} - 0.51 \overline{(T_a + 273.15) w' q'} \tag{7}$$

$$q \approx \frac{\rho_v R_d (T_a + 273.15)}{P_a} \tag{8}$$

$$\overline{w' T_a'} = \frac{(T_a + 273.15)}{(T_s + 273.15)} \left(\overline{w' T_s'} - 0.51 R_d \overline{(T_a + 273.15)^2 w' \rho'}_v / P_a \right) \tag{9}$$

$$(\overline{w'c'})_r = \overline{w'c'} \cos\theta - \overline{u'c'} \sin\theta \cos\eta - \overline{v'c'} \sin\theta \sin\eta \tag{10}$$

$$\cos\theta = \sqrt{\frac{(u^2 + v^2)}{(u^2 + v^2 + w^2)}} \tag{11}$$

$$\sin\theta = \frac{w}{\sqrt{(u^2 + v^2 + w^2)}} \tag{12}$$

$$\cos \eta = \frac{u}{\sqrt{(u^2 + v^2)}} \tag{13}$$

$$\sin \eta = \frac{v}{\sqrt{(u^2 + v^2)}} \tag{14}$$

Estimation of turbulent fluxes (eqs. (5) and (6)) relies on an accurate measurement of velocity fluctuations perpendicular to the lateral airstream. The study area is relatively flat and level, indicating that the air stream is approximately perpendicular to gravity and the sonic anemometer was oriented with respect to gravity with a bubble level. Measurement of wind speed in three orthogonal directions with the sonic anemometer allows for a more refined orientation of the collected data with the natural coordinate system through mathematical coordinate rotations. The magnitudes of the coordinate rotations are determined by the components of the wind vector in each 30-minute averaging period.

The wind vector is composed of three time-averaged components (u, v, w) in three initial coordinate directions (x, y, z). Using a bubble level, direction z initially was approximately oriented with respect to gravity, and the other two directions were arbitrary. Tanner and Thurtell [47] and Baldocchi and others [1] outline a procedure in which measurements made in the initial coordinate system are transformed into values consistent with the natural coordinate system. First, the coordinate system is rotated by an angle η about the z-axis to align u along a transformed x-direction on the x-y plane. Next, rotation by an angle θ is performed about the y-direction to align w along a transformed z-direction. These rotations result in a natural coordinate system with mean values of wind speed along the transformed y and z axes equal to zero and the mean airstream pointed directly along the transformed x axis.

The coordinate rotation-transformed covariances needed to compute turbulent fluxes are described in eq. (10), where:

$(\overline{w'c'})_r$	= is the rotated covariance;
$(\overline{w'c'})_r$	= is the rotated covariance;
c'	= is the fluctuation in either vapor density (r v) or virtual temperature (T s); and
$\overline{w'c'}, \overline{u'c'},$ and $\overline{v'c'}$	= are covariances measured in the original coordinate system.

The [cos θ], [sin θ], [sin η], and [cos η] are defined in eqs. (11) to (14), respectively. The presence of the tower and the anemometer produced spurious turbulence which possibly impacted measured velocity fluctuations, particularly when the wind was from the tower-side of the sensor. Turbulent flux data for which "the inferred mis-leveling angle θ greater than 10°" were excluded based on the assumption that spurious turbulence was the cause of the excessive amount of coordinate rotation.

8.2.5 CONSISTENCY OF MEASUREMENTS WITH ENERGY BUDGET

Previous investigators [3, 15, 17, 23, 29, 44, 50] have described a recurring problem with the eddy correlation method: A common discrepancy of the measured latent and sensible heat fluxes with the energy-budget equations (Eq. (1)). The usual case is that measured turbulent fluxes $(H + \lambda E)$ are less than the measured available energy $(R_n - S)$. Bidlake et al. [3] accounted for only 49 and 80% of the measured available energy with measured turbulent fluxes $(H + \lambda E)$ at cypress swamp and pine flat- wood sites, respectively. Turbulent fluxes measured above a coniferous forest by Lee and Black [23] accounted for only 83% of available energy. Several researchers [15, 17, 29] have shown that the eddy correlation method performs best in windy conditions (relatively high friction velocity, u^*). Friction velocity is directly proportional to wind speed, but also incorporated rates the frictional effects of the plant canopy and land surface on the wind and the effects of atmospheric stability [8]. Friction velocity can be computed with three dimensional sonic anemometer measurements of velocity fluctuations as [43] shown in Eq. (15).

Goulden and others [17] concluded that eddy correlation-measured values of carbon flux from a forest were underestimated when u^* was less than 0.17 m/s. German [15] noted that at u^* greater than 0.3 m/s, little discrepancy existed between measured available energy and measured turbulent fluxes. Possible explanations for the observed discrepancy between the measured turbulent fluxes and the measured available energy include: a sensor frequency response that is insufficient to capture high-frequency eddies; an averaging period insufficient to capture low-frequency eddies, resulting in a nonzero mean wind speed perpendicular to the airstream; drift in the absolute values of anemometer and hygrometer measurements resulting in statistical nonstationarity within the averaging period; lateral advection of energy; and overestimation of available energy. Lateral advection of energy is not a likely explanation because most of the studies reporting underestimation of turbulent fluxes were conducted at sites with adequately extensive surface covers. Measurement of the soil heat flux and storage terms of the available energy can be problematic, given the difficulty in making representative measurements of these terms; however, the turbulent flux underestimation occurs even with a daily composite of fluxes (in which case these terms generally are negligible).

Likewise, overestimation of net radiation seems unlikely, given the relative simplicity and laboratory calibration of net radiometers. For these reasons, it was assumed in this study that the available energy was accurately measured and that any error in energy-budget closure was associated with errors in measurement of turbulent fluxes. Moore [29] also noticed an under-estimation of turbulent fluxes and suggested that this under-estimation would likely apply equally to each of the turbulent fluxes (sensible and latent heat flux), leading to the conclusion that the ratio of the fluxes can be measured adequately. This assumption seems reasonable, given that the same turbulent eddies transport both sensible and latent heat, and therefore, any eddies that are missed by the instrumentation because of anemometer response or averaging period would have a proportionally equal effect on both turbulent fluxes. German [15] provided empirical support for this assumption at a saw-grass site in south Florida where simultaneous measurement of the ratio of fluxes was based on two approaches: the eddy correlation method (using instrumentation identical to that used in the present study) and the measurement of temperature and vapor pressure differentials between vertically separated sensors [4]. These independent approaches for estimating the ratio of turbulent fluxes

were in reasonable agreement during the daylight hours when evapotranspiration predominated. Assuming that the ratio of turbulent fluxes is adequately measured by the
eddy correlation method, the energy budget equation (Eq. (1)), along with turbulent
fluxes (H and λE) measured using the standard eddy correlation technique, can be used
to produce corrected (H_{cor} and λE_{cor}) turbulent fluxes in an energy-budget variant of the
eddy correlation method, as shown in Eq. (16). As shown below, we get Eq. (19) for
H_{cor}, after introducing the Bowen ratio [Eq. (17)].

$$u^* = \sqrt{\sqrt{\overline{u'c'}^2 + \overline{v'w'}^2}} \tag{15}$$

$$R_n - G - S = H_{cor} + \lambda E_{cor} = \lambda E_{cor}(1 + B) \tag{16}$$

Bowen ratio is defined in eq. (17):

$$B = \frac{H}{\lambda E} \tag{17}$$

Rearranging eq. (16), we get:

$$\lambda E_{cor} = \frac{R_n - G - S}{1 - B} \tag{18}$$

Combining eq. (17) and (18):

$$H_{cor} = R_n - G - S - \lambda E_{cor} \tag{19}$$

Instrumentation was installed at the evapotranspiration station to provide estimates of
soil heat flux (G) and changes in stored energy (S) in the biomass and air. Soil heat flux
at a depth of 8 cm was measured at two representative locations using soil heat-flux
plates. An estimate of the soil heat flux at land surface was computed based on the
estimated change in stored energy in the soil above the heat flux plates. The changes
in stored energy in the soil above the heat flux plates were estimated based on thermo-
couple-measured changes in soil temperature and estimates of soil heat capacity. The
estimates of soil heat capacity were based on mineralogy, soil bulk density, and soil
moisture content. Soil moisture content was measured using time-domain reflectometry (TDR) probes placed within the upper 8 cm of soil. Thermocouples were installed
at multiple locations within the trunks of representative trees to allow for estimation of
changes in storage of energy within the biomass. Estimates of biomass density (based
on tree surveys) and biomass heat capacity (available from previous studies) also are
required for calculation of changes in biomass stored energy. Changes in storage of
energy in the air generally are small in comparison with soil heat flux and biomass heat
storage, but were estimated based on measurement of the temperature and relative humidity profile below the turbulent flux sensors. With the exception of the temperature
and relative humidity sensors, all of the instrumentation intended to provide data to
estimate soil heat flux and changes in stored energy was destroyed by earth-moving

equipment used to construct a fire break around the evapotranspiration station a few hours before a fire passed through the area of the station.

Energy generally enters the soil surface and is stored in the biomass and air during the day and released at night. It was assumed that soil heat flux and changes in energy storage in the biomass and air were negligible over a diurnal cycle. This facilitated the evaluation of Eqs. (18) and (19), using daily composites of terms in these equations. This approach allowed for neglect of those terms of the energy budget that were not measured as a result of firedamaged instrumentation.

During periods of rapid temperature changes (for example, cold front passage), however, the net soil heat flux and the net change in energy stored in the biomass and air over a diurnal cycle may not be negligible. As mentioned previously, problems such as scaling of hygrometer windows, moisture on anemometer or hygrometer, or excessive coordinate rotation can result in missing 30-minute turbulent flux data. These data must be estimated prior to construction of daily composites of turbulent fluxes. In the present study, regression analysis of measured turbulent flux data and photosynthetically active radiation (PAR) was used to estimate unmeasured values of turbulent fluxes. These regression-estimated values of turbulent fluxes are not as reliable as measured values. Therefore, the fraction of daily composited turbulent flux data derived from regression estimates was limited to 25% (up to 6 hours per day). The procedure outlined above for culling, estimating, and compositing 30-minute turbulent flux data still resulted in missing values for some days.

8.2.6   SIMULATION OF EVAPOTRANSPIRATION

An evapotranspiration model was developed for estimating daily values of evapotranspiration representative of both burned and unburned areas. Post-fire measurements of evapotranspiration generally reflected a composite of evapotranspiration from burned and unburned vegetation. A model was developed that reflected the mixture of source area characteristics and allowed calculation of the evapotranspiration from each source area.

8.2.7   EVAPOTRANSPIRATION MODELS

The eddy correlation instrumentation can have extended periods of inoperation, as discussed previously. However, more robust meteorological and hydrologic instrumentation (sensors for measurement of net radiation, air temperature, relative humidity, PAR, wind speed, soil moisture, and water-table depth) can provide nearly uninterrupted data collection. Evapotranspiration models, calibrated to measured turbulent flux data and based on continuous meteorological and hydrologic data, can provide continuous estimates of evapotranspiration. Evapotranspiration models also can provide insight into the cause-and-effect relation between the environment and evapotranspiration.

Physics-based evapotranspiration models generally rely on the work of Penman [32], who developed an equation for evaporation from wet surfaces based on energy budget and aerodynamic principles. Penman equation has been used to estimate evapotranspiration from well-watered, dense agricultural crops (reference or potential evapotranspiration). In Penman's equation, the transport of latent and sensible heat fluxes from a "big leaf" to the sensor height is subject to an aerodynamic resistance.

The big leaf assumption implies that the plant canopy can be conceptualized as a single source of both latent and sensible heat at a given height and temperature. Inherent in the Penman approach is the assumption of a net one-dimensional, vertical transport of vapor and heat from the canopy. The Penman equation is shown in Eq. (20).

$$\lambda E = \frac{\Delta(R_n - G - S) + \frac{\rho C_p (e_s - e)}{r_a}}{\Delta + \gamma} \tag{20}$$

$$\lambda E = \frac{\Delta(R_n - S)}{\Delta + \gamma} \quad \text{for } e_s = e \tag{21}$$

$$\lambda E = \alpha \frac{\Delta(R_n - S)}{\Delta + \gamma} \quad \text{with Priestley-Taylor coefficient, } \alpha \tag{22}$$

$$\lambda E = (1 - w_b)\lambda E_u + w_b \lambda E_b \tag{23}$$

In eq. (20): λE is latent heat flux, in watts per square meter; Δ is slope of the saturation vapor-pressure curve, in kilopascals per degree Celsius; G is soil heat flux at land surface, in watts per square meter; S is change in storage of energy in the biomass and air, in watts per square meter; C_p is specific heat capacity of the air, in joules per gram per degree Celsius; e_s is saturation vapor pressure, in kilopascals; e is vapor pressure, in kilopascals; r_h is aerodynamic resistance, in seconds per meter; and γ is the psychrometric constant = approximately 0.067 kilopascals per degree Celsius, but varying slightly with atmospheric pressure and temperature. The first term is known as the energy term; the second term is known as the aerodynamic term. The eq. (21) is simplification of the Penman equation for the case of saturated atmosphere ($e = e_s$), for which the aerodynamic term is zero.

However, Priestley and Taylor [34] noted that empirical evidence suggests that evaporation from extensive wet surfaces is greater than this amount, presumably because the atmosphere generally does not attain saturation. Therefore, the Priestley-Taylor coefficient [, eq. (22)] was introduced as an empirical correction to the theoretical expression [eq. (21)]. This formulation assumes that the energy and aerodynamic terms of the Penman equation are proportional to each other. The value of has been estimated to be 1.26, which indicates that under potential evapotranspiration conditions, the aerodynamic term of the Penman equation is about 21 percent of the total latent heat flux. Eichinger and others [10] have shown that the empirical value of has a theoretical basis: A nearly constant value of is expected under the existing range of Earth-atmospheric conditions.

Previous studies [13, 41, 44] have applied a modified form of the Priestley-Taylor equation. The approach in these studies relaxs the Penman assumption of a free-water surface or a dense, well-watered canopy by allowing α to be less than 1.26 and to vary as a function of environmental factors. The Penman-Monteith equation [27] is a more theoretically rigorous generalization of the Penman equation that also accounts for a relaxation of the these Penman assumptions. However, Stannard [] noted that the modified Priestley-Taylor approach to simulation of observed evapotranspiration rates was superior to the Penman-Monteith approach for a sparsely vegetated site in the semiarid rangeland of Colorado. Similarly, Sumner [44] noted that the modified Priestley-Taylor approach performed better than did that of Penman-Monteith for a site of herbaceous, successional vegetation in central Florida. Therefore, the modified Priestley-Taylor approach was chosen for the present investigation.

8.2.8 PARTITIONING OF MEASURED EVAPOTRANSPIRATION

An evapotranspiration model (daily resolution) was developed to partition the measured evapotranspiration into two components characteristic of the primary types of surface cover (burned and unburned) of the watershed during the study period. As mentioned previously, upland areas were more likely to have been burned during the June-July 1998 fires than wetland areas. Therefore, to some extent, the model results also reflect the variation between upland and wetland evapotranspiration. The model was of the form, as shown in Eq. (23), where: λE is measured latent heat flux at the station, in watts per m^2; w_b is the fraction of the measured latent heat flux originating from burned areas, dimensionless; λE_u is latent heat flux from unburned areas, in watts per m^2; and λE_b is latent heat flux from burned areas, in watts per m^2.

$$w_b = \Sigma_{i=I}^{IV} g_i f_i \tag{24}$$

$$f_i = \frac{\Sigma_{i=1}^{48} PAR_k \delta_i \psi_k}{\Sigma_{i=1}^{48} PAR_k} \tag{25}$$

In Eq. (24): the weighting coefficient (w_b) for a given day must incorporate the spatial distribution of surface cover types near the point of flux measurement (Fig. 3 and Table 2), the changing (upwind) source area for the measurement associated with changes in wind direction, and the diurnal changes in evapotranspiration. If the relative fraction of burned surface cover in the upwind source area remained constant for a given day (that is, the wind direction remained from a given zone of a relatively uniform mixture of surface cover types), w_b would be simply the fraction of burned surface cover within the zone. Also, if evapotranspiration from each surface cover type remained constant during a given day, w_b would be simply the time-weighted average of the fraction of burned surface cover within the upwind source areas. However, intra-day changes in source area composition, associated with changes in wind direction, and the strong diurnal cycle in evapotranspiration had to be considered during computation of day-by-day values of w_b. For example, suppose that the wind were from the west during the night and from the east during the day. In this situation, the measured daily evapotranspiration would be much more representative of the surface cover to the east because day- time evapotranspiration generally is much higher than nighttime evapotranspiration. Strong diurnal biases in wind direction (Fig. 8) exist in the study area, which can lead to situations such as that described. Therefore, weighting coefficients must reflect these diurnal patterns in evapotranspiration.

The diurnal pattern of evapotranspiration during a given day generally is strongly correlated with the diurnal pattern of incoming radiation, as can be inferred from the

Priestley-Taylor equation (Eq. (22)) or seen empirically [44]. *PAR* was used as a surrogate for the factors that produce intraday variations in evapotranspiration for both surface cover types. Nighttime *PAR* is equal to zero, implying that only daytime winds from a given zone are assumed to contribute to the measured latent heat flux for a given day. Other factors (such as variations in air temperature) that contribute to the diurnal pattern of evapotranspiration were considered minor, compared to the effect of *PAR*, and were not considered in the determination of weights for use in the Eq. (23). The computation for the day-by-day values of w_b is derived in Appendix I at the end of this chapter and is shown in Eq. (24), where: g_i is the fractional contribution of burned area within burn zone i to the measured latent heat flux when wind direction is from burn zone i (Table 2); i is an index for the burn zones (Fig. 3); and f_i is the *PAR*-weighted fraction of the day that wind direction is from burn zone i and is computed using Eq. (25).

In Eq. (25): k is an index for the 48 measurements of 30-minute averages within a given day; $\delta_i(\Psi_k)$ is a binary function equal to 1, if Ψ_k is within burn zone i and otherwize equals 0; PAR_k is the measured *PAR* for time period k within a given day; and Ψ_k is the wind direction for time period k within a given day.

In the evapotranspiration model (Eq. (23)): Both λE_u and λE_b are simulated by the modified Priestley- Taylor equation (Eq. (22)) with individual Priestley-Taylor α functions. The α function for λE_u was assumed to remain unchanged throughout the 2-year study period; however, the α function for λE_b was divided into multiple time periods to reflect the radical change in surface cover of the burned areas following the fire, logging, and regrowth of vegetation. The measurements of average, daily evapotranspiration provided a standard with which to calibrate the Priestley-Taylor evapotranspiration model. Calibration of the Priestley- Taylor model involved quantification of the functional relations between the Priestley-Taylor's α and environmental variables. This quantification was achieved through identification of the form of the functional relation (trial-and-error approach) and estimation of the parameters of that relation (regression analysis) that produced optimal correspondence between measured and simulated values of latent heat flux.

The form of the calibrated model (Eq. (23)) allowed for evapotranspiration to be estimated for any mix of burned and unburned areas through appropriate specification of w_b. Daily values of evapotranspiration for burned and unburned areas were estimated with w_b equal to 1 and 0, respectively. Evapotranspiration from the watershed was estimated with w_b equal to 0 and 0.4 (burned fraction of watershed) prior to and following the fires, respectively. The potential evapotranspiration from the watershed was estimated with similar weighting, but with a Priestley-Taylor α equal to a constant value of 1.26.

8.2.9 MEASUREMENT OF ENVIRONMENTAL VARIABLES

Meteorological, hydrologic, and vegetative data were collected in the study area for several reasons: (1) as ancillary data required by the energy- budget variant of the eddy correlation method, (2) as independent variables within the evapotranspiration model, and (3) to construct a water budget for the Tiger Bay watershed. Meteorological variables monitored included net radiation, air temperature, relative humidity, wind speed, and *PAR*. These data were recorded by data loggers at 15-second intervals, using instrumentation summarized in Table 1, and the resulting 30-minute means were stored.

Two net radiometers, each deployed at a height of 35 m, provided redundant measurements of net radiation at the evapotranspiration station. Measured values of net radiation were corrected for wind-speed effects as suggested by the instrument manual for the Radiation and Energy Balance Systems, Inc., Model Q-7.1 net radiometer. In late 1999, missing net radiation data necessitated an estimate of net radiation based on a regression of *PAR* and net radiation. *PAR* consists of that part of incoming solar radiation that is used in plant photosynthesis and is highly correlated with incoming solar radiation. Based on data collected during 1993–1994 in Orange County – Florida, solar radiation (in watts per m^2) can be approximated (standard error of estimate = 11 watts per m^2) as 0.49 times *PAR* (in micromoles per second per m^2).

The source area of the net radiation measurement was estimated by using the approach of Reifsnyder [35] and Stannard [42]. The measurement of net radiation had a much smaller source area than the turbulent flux measurement (Fig. 7). About 90% of the source area for the net radiometers was within a radial distance of 55 m (prelogging) or 110 m (post logging). Therefore, the source area for the net radiometer in the near-vicinity of the evapotranspiration station was one of the following: (1) pine plantation (prelogging), (2) burned pine plantation (postfire, but prelogging), or (3) clear-cut, with under story regrowth (postlogging). Other covers also existed within the watershed, primarily wetlands and unburned pine lands. Lacking net radiation measurements over more than one cover, the assumption was made that net radiation measured at the unburned pine plantation was representative of all unburned surface covers. The period of record prior to the fire (the initial 175 days of 1998) was used to develop a regression-based predictor of net radiation as a function of *PAR*. This relation was used to estimate net radiation in unburned areas following the burning of the area around the evapotranspiration station. The net radiation measured at the evapotranspiration station following burning was assumed to be representative of all burned areas. Logging of the burned area near the evapotranspiration station occurred during a period of extensive logging through- out the watershed. Some error is introduced to the estimation of net radiation over burned areas because the logging was not simultaneous for all burned areas and because the logging over burned areas was not complete (as mentioned previously, two-thirds of the burned forest within Tiger Bay State Forest was logged). Estimates of daily net radiation for burned and unburned areas were composited as shown in Eq. (26) into a value consistent with the turbulent flux measurements (Eqs. (18) and (19)) using the weighting coefficient (w_b) previously defined (Eq. (24)).

$$R_N = (1 - w_b)R_{nu} + w_b R_{nb} \qquad (26)$$

$$NDVI = \frac{NIR - Vis}{NIR + Vis} \qquad (27)$$

In Eq. (26): R_n is composited net radiation, in watts per square meter; R_{nu} is net radiation for unburned areas, in watts per square meter; and R_{nb} is net radiation for burned areas, in watts per square meter.

A regression between postlogging, daily values of net radiation and *PAR* was used to estimate net radiation from burned and logged surfaces during the latter part of 1999 after net radiometer domes were damaged, perhaps by birds. Vegetation within the study area was mapped previously by Volusia County Department of Geographic Information Systems [56, 57] and Simonds and others [40]. Post-fire, infrared, aerial photographs were used to identify the areal distribution of burned vegetation in the watershed.

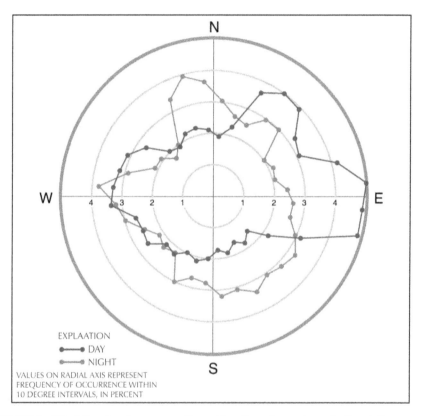

FIGURE 8   Wind direction frequency pattern at location of evapotranspiration station.

Temporal variations in vegetation were documented with monthly photographs taken from the tower at the evapotranspiration station and with normalized difference vegetation index (NDVI) data. NDVI data were provided by the USGS Earth Resources Observation Systems (EROS) Data Center through analysis of the Advanced Very High Resolution Radiometer (AVHRR) data [11, 52, 54] from operational National Oceanic and Atmospheric Administration (NOAA) polar-orbiting satellites. NDVI is defined in Eq. (27), where: NIR is near-infrared reflectance measured in AVHRR band 2 (725–1100 nanometers); and V is is visible reflectance measured in AVHRR band 1 (580–680 nanometers).

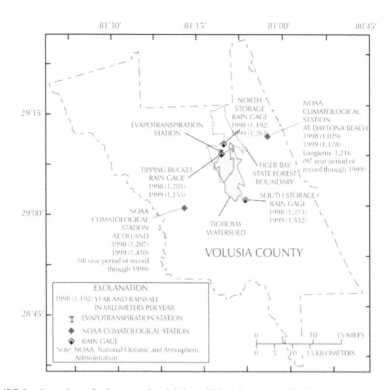

FIGURE 9   Location of rain gages in vicinity of Tiger Bay watershed.

NDVI is highly correlated with the density of living, leafy vegetation. The physical basis for this correlation is the sharp contrast in the absorptivity of visible and
near-infrared radiation by leaves, which absorb approximately 85% of incident visible
radiation, but only 15% of near-infrared radiation [8]. Other ground covers (dead plant
material, soil, and water) do not exhibit this extreme spectral differential in absorption. The AVHRR-computed NDVI data are provided at 2-week and 1-kilometer (km)
by 1-km resolution. For the present study, NDVI data, within a 3-km by 3-km square
and approximately centered on the location of the evapotranspiration station, were
composited to quantify temporal trends in the density of living, leafy vegetation in the
vicinity of the turbulent flux measurements during the study period.

Air temperature and relative humidity were monitored at the evapotranspiration
station at heights of 1.5, 9.1, 18.3, and 35 m. The slope of the saturation vapor pressure
curve (a function of air temperature) and vapor pressure deficit were computed in the
manner of Lowe [25] using the average of air temperature and relative humidity values
measured at these four heights. A propeller-type anemometer to monitor wind speed
and direction and an upward-facing quantum sensor to measure incoming *PAR* were
deployed at a height of 35 m at the evapotranspiration station.

Hydrologic variables that were monitored included precipitation, water-table
depth, stream discharge, and soil moisture. Precipitation records were obtained from

a tipping bucket rain gage mounted at a height of about 18.3 m at the evapotranspiration station and from two storage rain gages installed in forest clearings and monitored weekly (Fig. 9). Spatial variability in annual rainfall can be substantial within Volusia County, based on the long-term NOAA stations at DeLand and Daytona Beach (Fig. 9). The Daytona Beach area, on average, receives about 15% less annual rainfall than does the DeLand area [30, 31]. The uncertainty associated with the rainfall distribution between these two stations precluded the use of both stations for estimation of rainfall to the Tiger Bay watershed during the study period. Rather, the rainfall totals from the two storage rain gages located near the watershed were averaged to provide estimates of rain- fall to the watershed. Tipping bucket rain gages can underestimate rainfall, particularly during high-intensity events; therefore, the tipping bucket gage monitored at the evapotranspiration station was used primarily to provide a high-resolution description of the temporal rainfall pattern, and the storage rain gages were used primarily to estimate cumulative rainfall.

Water-table depth was monitored at two surficial- aquifer system wells at opposite ends of the watershed. Water-level measurements were obtained at 30-minute intervals using a pressure transducer in the north well (USGS site identification number 290813081111801), located at the evapotranspiration station. The south well (USGS site identification number 290119081074001), at the location of the south storage rain gage (Fig. 1), was measured weekly using an electric tape. Although the two wells monitored were located at opposite ends of the watershed (Fig. 1), both wells were within similar upland settings. Although the water-table depth in wet- land areas would be expected to be less than that measured in upland wells, water levels are expected to change at the same rate in the low relief environment of this watershed. Therefore, changes in the measured upland water-table depths can be regarded as indicators of changes in the representative water-table depth of the watershed.

Daily values of stream discharge for the only surface-water outflow from the Tiger Bay watershed, Tiger Bay canal near Daytona Beach (Fig. 1; USGS station number 02247480), were obtained from the USGS database [51, 53, 55]. Soil moisture at two representative locations at the evapotranspiration station was monitored using time-domain reflectometry (TDR) probes installed to provide an averaged volumetric soil moisture content within the upper 30 cm of the soil. Soil moisture measurements were made and recorded on the data logger every 30 minutes. The TDR probes were damaged by a fire in late June 1998, but were replaced in early August 1998. The soil moisture measurements made at the evapotranspiration station probably are indicative of only the uplands; wetlands commonly are inundated at times when shallow upland soils are not.

8.3   RESULTS AND DISCUSSION: EVAPOTRANSPIRATION MEASUREMENT AND SIMULATION

Most (73%) of the 30-minute resolution eddy correlation measurements made during the 2-year study period were acceptable and could be used to develop an evapotranspiration model to estimate missing data and to discern the effects of environmental variables on evapotranspiration. Unacceptable measurements resulted from failure of the krypton hygrometer or sonic anemometer, or because of excessive (more than

10 degrees) coordinate rotation in the postprocessing "leveling" of the anemometer data. Unacceptable data were most extensive in the evening and early morning hours (Fig. 10) because dew formation on the sensors during these times of day was common. This diurnal pattern of missing data was fortunate because turbulent fluxes are expected to be relatively small during the evening and early morning, when solar radiation is low. Missing data were estimated based on linear regression between the turbulent fluxes and PAR (Figs. 11 and 12). Because PAR is zero at night, this approach assigned constant values of latent and sensible heat flux to missing night time data. The assumed constant value of night time latent heat flux assigned to missing data was 9.04 watts per m^2 (Fig. 11). This value generally was small relative to daytime values of latent heat flux, and therefore, not significantly inconsistent with the assumption of negligible nighttime latent heat flux inherent in the development of weighting coefficients (Eqs. (23)–(25)). Examples of measured and PAR-estimated turbulent fluxes are shown for a period in late February 1998 in Fig. 13.

Turbulent flux data exhibited pronounced diurnal patterns. The average diurnal pattern of turbulent fluxes and PAR (Fig. 14) indicates that the vast majority of evapotranspiration occurs in daytime, driven by incoming solar radiation. During average daytime conditions, both latent and sensible heat flux are upward, with most of the available energy partitioned to latent heat flux. At night, the land or canopy surface cools below air temperature, producing a reversal in the direction of sensible heat flux (Fig. 14). Although the average, nighttime latent heat flux is upward (Fig. 14), dew formation (downward latent heat flux) commonly occurs.

The relation between net radiation and PAR varied as a result of the fire, logging, and regrowth. Regressions between daily values of net radiation and PAR are shown in Fig. 15 for three periods: prefire, postfire/prelogging, and postlogging. The measured and estimated values of daily net radiation for burned and unburned areas are shown in Fig. 16. Measured values of PAR, a quantity highly correlated with incoming solar radiation, are shown in Fig. 17. The strong seasonality of net radiation evident in Fig. 16 was a consequence of the yearly solar cycle, which produces a sinusoidal input of solar radiation to the upper atmosphere. Deviations from the sinusoidal pattern (such as during September–October 1999) were largely the result of cloudy conditions that produced periods of low PAR. The cloudy and rainy period immediately after the fire resulted in relatively low values of PAR and low estimated values of net radiation in unburned areas. The measured (burned) net radiation, however, was relatively high, indicating that the surface reflectance of burned areas decreased markedly after the fire blackened much of the landscape. The measured net radiation for burned areas was about 20% higher than the estimated net radiation for unburned areas in the 6 months following the June 1998 fire. With the regrowth of vegetation, reflectance gradually increased to near prefire values in the postlogging period, and the differences between values of net radiation for burned and unburned areas were less distinct.

As described previously, daily composites of measured turbulent fluxes were constructed with the restriction that no more than 6 hours of data for a given day could be missing and subject to estimation using the gross PAR-based relations (Figs. 11 and 12). This restriction limited the number of acceptable daily values of measured turbulent fluxes to 449 during the 2-year (730 days) study period. Only a small amount of the total turbulent flux (5.6 and 5.1% for latent and sensible heat flux, respectively) comprizing the acceptable daily values was estimated by the PAR-based relation. As expected from previous studies, the available energy tended to be greater (measured turbulent fluxes accounted for only about 84.7% of estimated available energy) than the turbulent fluxes derived from the standard eddy correlation method (Fig. 18), and the energy-budget closure tended to improve with increasing friction velocity (Fig. 19). The measured turbulent fluxes generally accounted for estimated available energy at friction velocity values greater than about 0.6 m/s. The acceptable daily values of turbulent fluxes, computed by both the standard eddy correlation method (Eqs. (5) and (6)) and the energy-budget variant of the eddy correlation method (Eqs. (18) and (19)), are presented in Figs. 20 and 21.

These values represent the fluxes measured at the evapotranspiration station, and therefore, represent varying proportions of burned and unburned source areas. The relative proportions varied widely following the fire (Fig. 22), with values ranging from those that were almost completely representative of unburned areas ($w_b = 0$) to those with 80% representative of burned areas ($w_b = 0.8$).

As a consequence of the previously mentioned discrepancy between available energy and measured turbulent fluxes, the standard eddy correlation method produced turbulent flux values that were, on average, only 84.7% of those produced by the energy-budget variant.

8.3.1 CALIBRATION OF EVAPOTRANSPIRATION MODEL

Calibration of the evapotranspiration model was essentially a process of determining the best functional form of the modified Priestley-Taylor coefficient, α. The environmental variables considered as possible predictors of Priestley-Taylor's α (Eq. (22)) included: water-table depth, soil moisture, *PAR*, air temperature, vapor-pressure deficit, daily rainfall, NDVI, and wind speed. Of these variables, only water-table depth, soil moisture, and *PAR* were identified as significant determinants of Priestley-Taylor's α. Soil moisture was highly correlated with water-table depth (Fig. 23), and there fore, one of these variables can be excluded from the α function to avoid redundancy. To enhance the transfer value of this study, water-table depth was retained as a variable in the α function, and soil moisture was eliminated, because water-level data are more commonly available than soil moisture data. In other environmental settings, such as areas with a relatively deep water table or coarse-textured soils, the water table may be hydraulically de-coupled from the shallow soil moisture much of the time, and a different functional representation of α than was used in this study would be appropriate.

Priestley-Taylor's α was initially simulated with a three-part model incorporating the three different surface covers: (1) unburned areas; (2) postfire/pre logging, burned areas (June 25 to December 16, 1998); and (3) postlogging, burned areas (December 17, 1998 to December 31, 1999). The time divisions for the burned areas grossly approximated the observed variation in NDVI over the study period (Fig. 24). The effects of the fire and transient regrowth of vegetation (Fig. 4) on NDVI were evident (Fig. 24). In the almost 6 months prior to the fire (January 1–June 24, 1998), NDVI maintained a relatively constant value of about 0.5. NDVI sharply declined at the time of the fire, but recovered within 4 months to a value of about 0.4, which was maintained throughout the remainder of the study. As a simplification, the effect of the transient aspect of vegetative regrowth within the 4-month recovery period was not incorporated into the model for α. Instead, the function of α for this recovery period, as for all time periods, was a function solely of water-table depth and *PAR*.

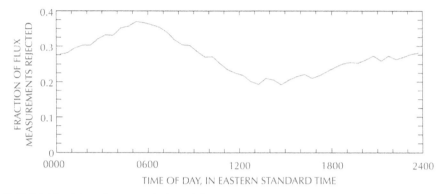

FIGURE 10 Diurnal pattern of rejected flux measurements.

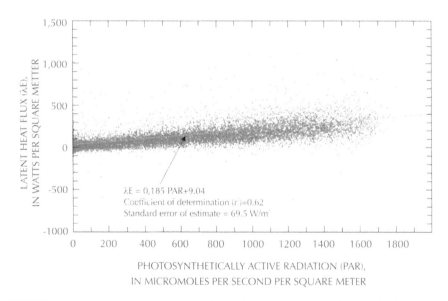

FIGURE 11 Relation between measured 30-minute averages of photosynthetically active radiation (PAR) and latent heat flux (λE).

Surprizingly, the annual pattern of leaf growth and drop for the deciduous cypress trees within the watershed was not apparent in values of NDVI, perhaps because of the exposure of under story vegetation following leaf drop. Simulations that attempted to use NDVI directly as an explanatory variable for variations in evapotranspiration were unsuccessful. This failure is perhaps related to erratic variations in NDVI (Fig. 24), which are a product of sensor and data registration limitations (Kevin Gallo, NOAA, written communication, 2001).

An analysis of error in the preliminary model showed a seasonal pattern in the residuals (difference of measured and simulated latent heat fluxes) within the postlogging period (Fig. 25). Measured evapotranspiration generally was overestimated in the early part of this period and underestimated in the late part of the period. The bias was apparently unrelated to changes in green leaf density, based on the relatively constant value of NDVI following logging (Fig. 24). Possible explanations for the model bias include factors not clearly identified by NDVI: phenological changes associated with maturation or seasonality of plants that emerged after the fire or successional changes in composition of the plant community within burned areas. To reflect the apparent change in system function during the postlogging period, this period was further subdivided into an early period (December 17, 1998 through April 22, 1999) and a late period (April 23 through December 31, 1999). This subdivision of the postlogging period resulted in an improved model (standard error of estimate = 9.67 watts per m²), compared to the model with a single postlogging period (standard error of estimate = 10.82 watts per m²) and reduced the seasonal bias in residuals (Fig. 25).

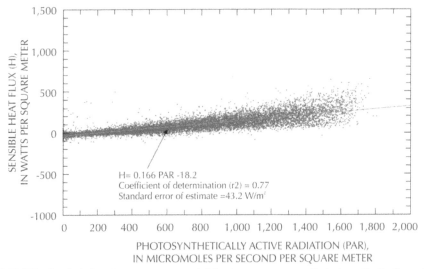

$$H = 0.166\ PAR - 18.2$$
Coefficient of determination (r2) = 0.77
Standard error of estimate = 43.2 W/m²

FIGURE 12   Relation between measured 30-minute averages of photosynthetically active radiation (PAR) and sensible heat flux (H).

The general form of α was identical for all surface covers (Eq. (28)), although model parameter values varied with surface cover (Table 3):

$$\alpha_j = C_{1j}h_{wt} + C_{2j}PAR + C_{3j} \tag{28}$$

In Eq. (28): α_j is the Priestley-Taylor coefficient for the jth surface cover; j is an index denoting the surface cover; $j = 1$ (unburned areas); $j = 2$ (burned areas during postfire/prelogging period; $j = 3$ (burned areas during initial postlogging period); and $j = 4$ (burned areas during final post logging period); C_{1j}, C_{2j}, and C_{3j} are empirical parameters that are estimated through regression, within the context of Eqs. (22) to (25); and h_{wt} is water table depth below a reference level placed at the highest water level measured (0.11 meters above land surface) at the evapotranspiration station (uplands environment) during the study period, in meters. h_{wt} is constrained to be greater than zero.

Regressions to estimate the model parameters within Eq. (28) were designed to minimize the sum of squares of error residuals between measured and simulated latent heat fluxes. Measured latent heat flux was used as the dependent variable of the regression; the right side of Eq. (22) contained the independent variables, as well as the unknown parameter (C_{1j}, C_{2j}, and C_{3j}; and $j = 1$ to 4). The values of λEu and λEb were estimated with Eq. (22), using the appropriate values of net radiation (R_{nu} and R_{nb} of Eq. (26) for λEu and λEb, respectively), and Eq. (28). The variable w_b was estimated with the Eqs. (24) and (25).

The form of α used in this study is similar to that used by German [15] for south Florida wetlands, where water level and incoming solar radiation were the sole determinants of α. In the study by German, however, the form of α involved both first and second order terms of incoming solar radiation. In this study, addition of the second-order PAR term added negligible improvement to simulation of evapotranspiration.

A comparison between simulated and measured values of latent heat flux is shown in Fig. 26 and regression statistics are shown in Table 3. The model exhibited little temporal bias (Fig. 25), even in the postfire/prelogging period when substantial transient changes (regrowth) in vegetative cover occurred in the burned areas. The lack of significant temporal bias supports the utilization of the particular discretization of time used in the model. More than 95% of the values of latent heat flux were within 25% of the measured values.

8.3.2   APPLICATION OF EVAPOTRANSPIRATION MODEL

The calibrated evapotranspiration model (Eqs. (22) and (23), with α values given by Eq. (28) and regression-derived parameters given in Table 3) described in the previous section was used to estimate average, daily values of evapotranspiration for both burned and unburned areas of the watershed during the 2-year study period. The model also provided a quantitative framework to examine the relation between evapotranspiration and the environment. The input variables for the model included daily values of net radiation (Fig. 16), PAR (Fig. 17), water-table depth at the evapotranspiration station (Fig. 27), and air temperature (Fig. 28).

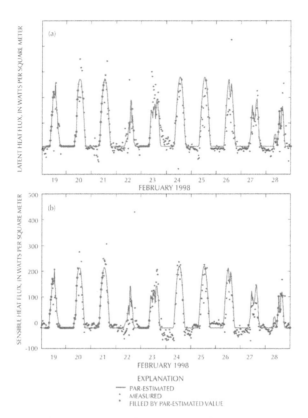

FIGURE 13 Measured and photosynthetically active radiation (PAR)-estimated values of (a) latent heat flux and (b) sensible heat flux during 10-day period in late February 1998.

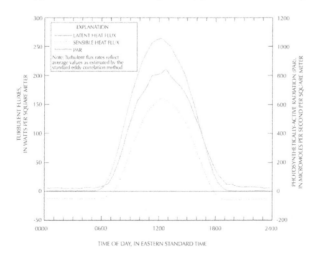

FIGURE 14 Average diurnal pattern of energy fluxes and photosynthetically active radiation (PAR).

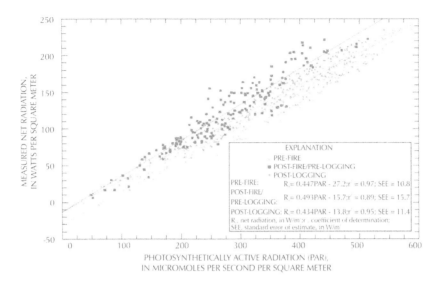

FIGURE 15   Relation between daily values of measured net radiation and photosynthetically active radiation (PAR).

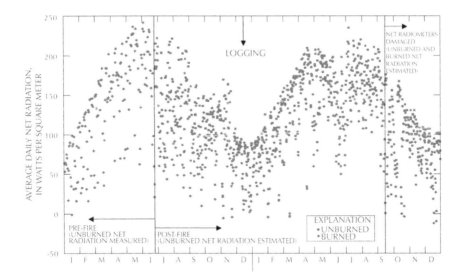

FIGURE 16   Average daily net radiation for burned and unburned areas.

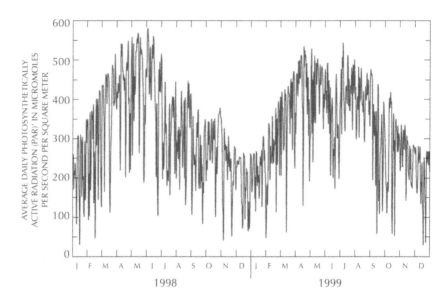

FIGURE 17   Average daily photosynthetically active radiation (PAR).

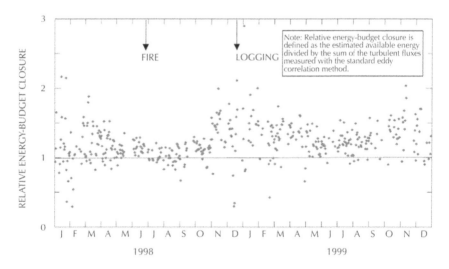

FIGURE 18   Temporal distribution of daily relative energy-budget closure.

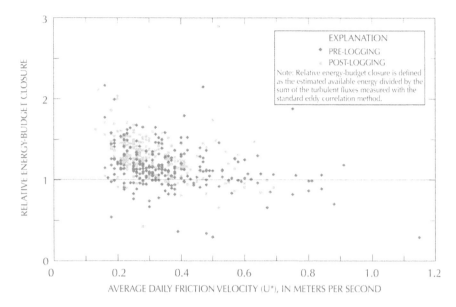

FIGURE 19 Relation between daily energy-budget closure and average daily friction velocity.

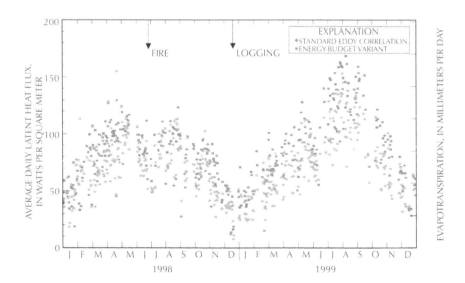

FIGURE 20 Average daily latent heat flux measured by the eddy correlation method and the energy-budget variant.

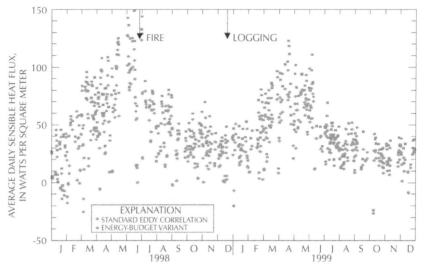

FIGURE 21   Average daily sensible heat flux measured by the eddy correlation method and the energy-budget variant.

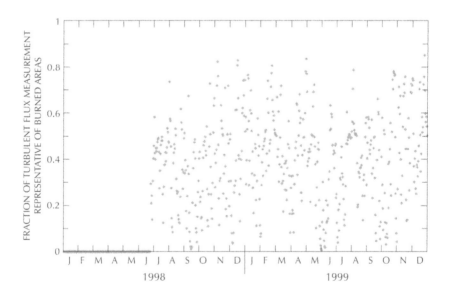

FIGURE 22   Daily values of fraction of burned fraction of turbulent flux measurement.

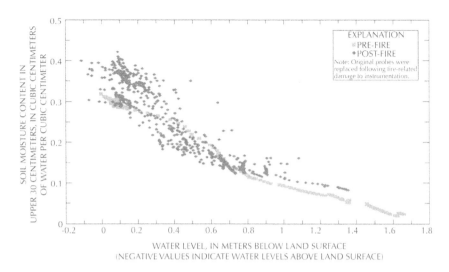

FIGURE 23   Relation between soil moisture content and water level.

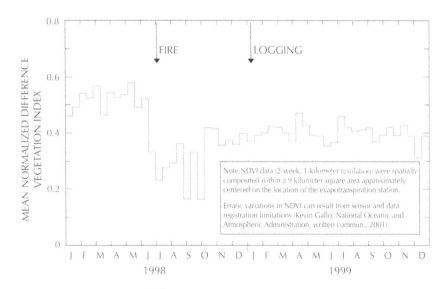

FIGURE 24   Temporal variability of normalized difference vegetation index (NDVI).

TABLE 3 Summary of parameters and error statistics for daily evapotranspiration models [Parameters $C1j$, $C2j$, and $C3j$ are defined by the equation: $\alpha j = C1\,j\,hwt + C2\,j\,PA\,R + C3\,j$ where: j is an index denoting the surface cover; hwt is water-table depth below a reference level placed at the highest water level measured (0.11 m above land surface) at the evapotranspiration station (uplands environment), in meters; and PAR is photosynthetically active radiation, in micromoles per m^2 per second. Error statistics: r^2, coefficient of determination of measured and simulated values of latent heat flux, dimensionless; SEE, standard error of estimate (in watts per m^2); CV, coefficient of variation, dimensionless, equal to SEE divided by the mean of the measured values of latent heat flux]

Parameters	Unburned	Three-part model for burned area		
	area (j = 1)	Post-fire/pre-logging (j=2)	Post-logging I (j=3)	Post-logging II (j=4)
Time period				
	01-1-1998 to	25-06-1998 to	17-12-1998 to	23-04-1999 to
	31-12-1999	16-12-1998	22-04-1999	31-12-1999
C_{1j}	-0.175	-0.167	-0.312	-0.508
C_{2j}	-0.00102	-0.00147	-0.00031	0.00013
C_{3j}	+1.42	1.26	1.03	1.36

Error statistics: r^2 = 0.90; SEE = 9.67 and CV= 0.11

Values of latent heat flux and evapotranspiration for January 1998 through December 1999 were estimated using the calibrated model (Fig. 29). Despite the relatively high net radiation in burned areas (Fig. 16), evapotranspiration from burned areas generally remained lower than that from unburned areas until spring 1999. This effect presumably was a result of destruction of transpiring vegetation by fire and then logging. Beginning in spring 1999 (postlogging II period for burned areas), evapotranspiration from burned areas increased sharply relative to unburned areas, sometimes exceeding evapotranspiration from unburned areas by almost 100 percent. From a simulation perspective, this change in evapotranspiration in spring 1999 was clearly the result of the change in Priestley-Taylor α model parameters between the two postlogging periods. From a physics perspective, the possible explanation(s) for the change in evapotranspiration is identical to those described in the earlier discussion of the differentiation of the early and late post logging periods within the evapotranspiration model. Evapotranspiration from burned areas for the 10-month period after the fire (July 1998–April 1999) averaged about 17% less than that from unburned areas and, for the following 8-month period (May 1999–December 1999), averaged about 31% higher than from unburned areas. During the 554-day period after the fire, the average evapotranspiration for burned areas (1,043 mm/yr) averaged 8.6% higher than that for unburned areas (960 mm/yr).

Annual evapotranspiration from the watershed was 916 mm for 1998 and 1,070 mm for 1999, and averaged 993 mm. The extensive burning and logging that occurred during the study produced a landscape that was not typical of forested areas of Florida. The estimated evapotranspiration from unburned areas can be considered representative of more typical forest cover. Annual evapotranspiration from unburned areas was

937 and 999 mm for 1998 and 1999, respectively, and averaged 968 mm. Both actual and potential evapotranspiration showed strong seasonal patterns and day-to-day variability (Figs. 29 and 30). Actual evapotranspiration from the watershed averaged only 72% of potential evapotranspiration.

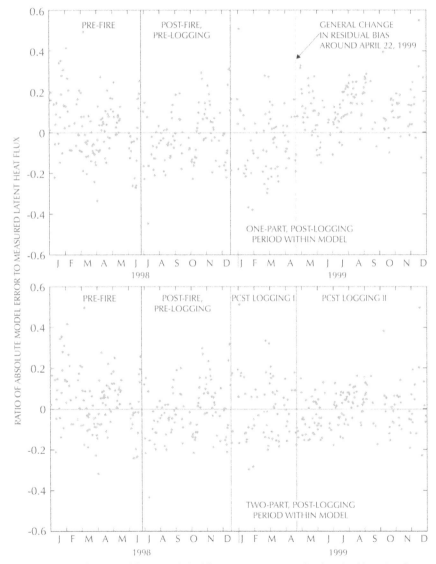

Note: Absolute Model Error equals the difference between measured and simulated latent heat flux.

FIGURE 25   Temporal variability in relative error of evapotranspiration model.

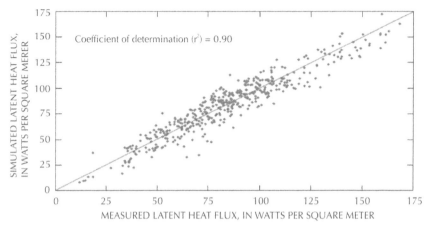

FIGURE 26   Comparison of simulated and measured values of daily latent heat flux.

The effect of the extreme drought period in spring 1998 (Fig. 27) on turbulent fluxes was substantial (Figs. 29, 31, and 32). Turbulent fluxes usually emulate the general sinusoidal, seasonal pattern of solar radiation and air temperature [15, 22, 44]. The usual sinusoidal pattern of latent heat flux was truncated in spring 1998 (Fig. 29) because of a lack of available moisture (Figs. 27 and 33). The drought-induced reduction in latent heat flux was compensated by an increase in sensible heat flux (Fig. 31) with an associated increase in the Bowen ratio. Comparison of the Bowen ratio (Fig. 32) with the water-table and soil moisture records (Figs. 27 and 33) indicates that the moisture status of the watershed has a major role in the partitioning of the available energy. Relative evapotranspiration is a ratio of actual to potential evapotranspiration and was computed as α /1.26; and it decreased from about 1 in the early, wet part of 1998 to less than 0.50 during the drought (Fig. 34). After the drought ended in late June and early July 1998 and water levels quickly returned to near land surface, evapotranspiration increased sharply. The evapotranspiration rate, however, averaged only about 60% of the potential rate in the burned areas, as compared to about 90% in the unburned areas. This discrepancy can be explained as a result of fire damage to vegetation.

Potential evapotranspiration rates for burned and unburned areas were similar (Fig. 30), although actual evapotranspiration rates for the two areas were quite distinct from each other (Fig. 29). The relation between actual and potential evapotranspiration was not a simple constant multiplier (for example, a crop factor), but rather was time-varying as a function of water-table depth, PAR, and surface cover (Fig. 34). Relative evapotranspiration exceeded a value of 1 at times, probably as a result of experimental error, as well as the approximate and empirically derived nature of the assumed potential value of 1.26 for α. The potential evapotranspiration rates (Fig. 30) did not strongly reflect either the drought or surface burning and logging, as does the actual evapotranspiration.

Within the framework of the calibrated model, variations in the environmental variables contained in α (water-table depth and *PAR*) reduce actual evapotranspiration below potential evapotranspiration for a given surface cover. The evapotranspiration model indicated that relative evapotranspiration decreased as the depth to the water table increased (Fig. 35). The range of water-table depths prevalent during the study period was slightly above land surface to about 1.75 m below land surface. Presumably, at some water-table depth greater than 1.75 m, relative evapotranspiration would reach an asymptotic constant value as vegetation becomes unable to access moisture below the water table. The rate of decline of relative evapotranspiration with water-table depth was greater for the postlogging period than for the prelogging period. This result is perhaps a manifestation of the replacement of many deep- rooted trees by shallow-rooted under story vegetation following the fires. Shallow-rooted plants would be less able to tap into deep soil moisture or the water table than would deep-rooted vegetation.

Water-table depth has been considered an important predictor of evapotranspiration in hydrologic analysis [49], but little empirical evidence has been available to define the relation between these two environmental variables. The USGS modular finite – difference ground-water flow model (**MODFLOW**) simulates relative evapotranspiration as a unique, piece-wise, linear function of water-table depth, where evapotranspiration declines from a potential rate when the water table is at or above land surface to zero at the "extinction depth" [26]. Contrary to the MODFLOW conceptualization of evapotranspiration, this study indicates that the variation in relative evapotranspiration is explained not only by water-table depth, but also by PAR. Relative evapotranspiration decreased with increasing *PAR* (Fig. 36), with the exception of the late postlogging period, which showed a slight increase in relative evapotranspiration with increasing *PAR*. This observation perhaps can be explained by assumptions within the Priestley- Taylor formulation that the energy and aerodynamic terms of the Penman equation are proportional to each other. Under nonpotential conditions, these two terms might deviate from the assumption of proportionality, but in such a manner that can be "corrected" through a functional relation between the multiplier α and a term (*PAR*) strongly correlated with the energy term.

Within the model developed in this study, net radiation and air temperature do not directly affect the Priestley-Taylor α and relative evapotranspiration, although net radiation has an indirect effect through the correlation of this variable with *PAR*. These variables, however, are important in the determination of evapotranspiration, as can be seen in Eq. (22). Evapotranspiration is directly proportional to $[\Delta / (\Delta + \gamma)]$, a term that is a function of temperature (Fig. 37).

For example, a change in air temperature from 20 to 30°C will produce about a 14-percent increase in evapotranspiration, assuming the environment is otherwize unchanged. The relation of net radiation and evapotranspiration is one of direct proportionality. Net radiation displayed dramatic temporal variations, both day-to-day (as a result of variations in cloud cover) and seasonally (Fig. 16), making this variable the most important determinant of evapotranspiration. This conclusion is supported by a sensitivity analysis (Table 4) based on perturbing each environmental variable of the evapotranspiration model by an amount equal to the observed standard deviation of the daily values of that variable. All unperturbed variables were assumed equal to mean values. This analysis indicated that variations in net radiation explained the greatest amount of the variation in evapotranspiration. Variations in *PAR*, closely correlated with net radiation, explained a large amount of the variation in evapotranspiration prior to logging, but explained little of the variation after logging. Evapotranspiration was moderately sensitive to variations in air temperature. Variations in water-table depth explained a moderate amount of the variation in evapotranspiration prior to the fire; however, evapotranspiration became more sensitive to variations in water-table depth after logging.

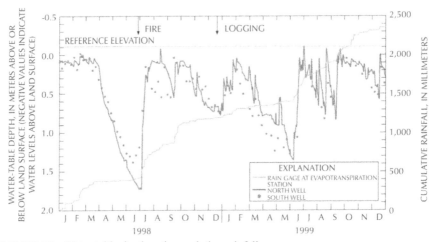

FIGURE 27   Water-table depth and cumulative rainfall.

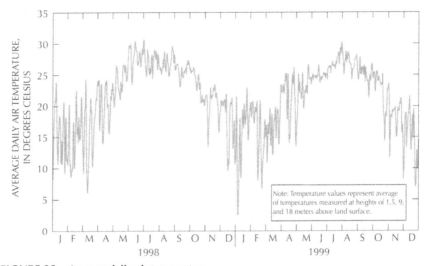

FIGURE 28   Average daily air temperature.

The model developed in this study is subject to several qualifications. The form of the equation developed for α was empirical, rather than physics-based, and was simply designed to reproduce measured values of evapotranspiration as accurately as possible. The correlation between environmental variables complicates a unique determination of parameters.

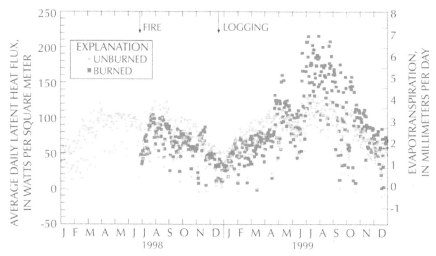

FIGURE 29 Average daily latent heat flux and evapotranspiration.

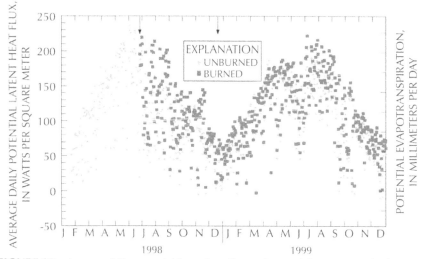

FIGURE 30 Average daily potential latent heat flux and potential evapotranspiration.

The model was developed for a limited range of environmental conditions, and therefore, extrapolation of the model to conditions not encountered in this study should be done with caution. The measured (upland) water-table depth at the evapotranspiration station, used as an independent variable in the model, explained some of the variation in evapotranspiration from the mixed upland/wetland watershed. However, water-table depth is not uniform within the watershed and, in particular, water-table depth in wetland areas usually is less than in upland areas. Therefore, caution should be used in applying the model to estimate evapotranspiration based on water-table depth measurements made at other locations in the watershed. For these reasons, the evapotranspiration model described in this report should be viewed as a general guide, rather than as a definitive description of the relation of evapotranspiration to environmental variables. The fact that the model successfully ($r^2 = 0.90$) reproduced 449 daily measurements of site evapotranspiration over a wide range of seasonal and surface-cover values lends credence to the ability of the model to estimate evapotranspiration at the site.

8.3.3 WATER BUDGET

Construction of a water budget for the Tiger Bay watershed serves to provide a tool for watershed management and for assessing the integrity of the eddy correlation evapotranspiration measurements. The water budget for the watershed is given in Eq. (29), where: P is precipitation, in millimeters per year; ET is evapotranspiration, in millimeters per year; R is runoff, in millimeters per year; L is leakage to the Upper Floridan aquifer, in millimeters per year; and ΔS is rate of change in storage, in millimeters per year.

$$P - (ET + R + L - \Delta S) = 0 \qquad (29)$$

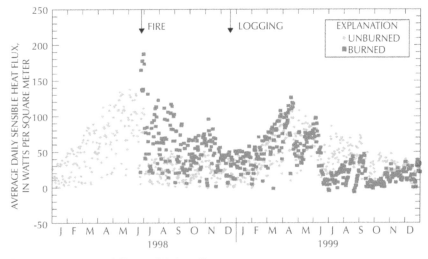

FIGURE 31 Average daily sensible heat flux.

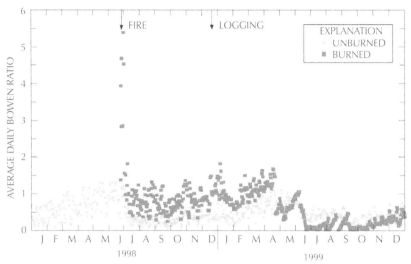

FIGURE 32   Average daily Bowen ratio.

A water budget for the Tiger Bay watershed during the 1998–1999 study period is shown in Table 5 and Fig. 38. Precipitation (Figs. 9 and 27), evapotranspiration (Fig. 29), and runoff (Fig. 39) were measured or obtained as described previously in this chapter. The estimated value of deep leakage to the Upper Floridan aquifer (112 mm/yr) during 1995 (Stan Williams, St. Johns River Water Management District, oral communication, 2000) also was assumed to be appropriate for the study period (1998–1999). The rate of change in water- shed storage over the study period was not directly measured, but was estimated as the water-budget residual. The water budget (Tables 5 and 6; Fig. 38) indicated that about 76% of watershed rainfall was lost as evapotranspiration during the 2-year study. The ratio of evapotranspiration to rainfall was remarkably stable from year-to-year (74% in 1998 and 77% in 1999). This stability occurred despite the very different environmental conditions prevailing during the study. Rainfall was a more consistent predictor of evapotranspiration than was potential evapotranspiration. The relative evapotranspiration varied rather greatly (67% in 1998 to 77% in 1999).

Runoff removed about 18% of the rainfall during the study period, but this percentage varied widely from year-to-year (29% in 1998 and 8% in 1999) as shown in Fig. 39. The runoff for 1998 was over three times that of 1999, despite the greater rainfall in 1999. This disparity can be explained largely by the antecedent water-table conditions for individual rain periods (Fig. 27).

A relatively large fraction of precipitation in 1998 occurred when the water- table depth was shallow, leading to relatively high rejection of infiltration and subsequent runoff. Additionally, the temporal distribution of precipitation affects the amount of watershed runoff. Runoff is maximized following short, but intense, rainfall during which the infiltration capacity of the soil is exceeded. This phenomenon may explain the disparate runoff responses in July 1998 (very intense rainfall and significant runoff) and June–July 1999 (less intense rainfall and no runoff). This disparity was noted despite similar total amounts of precipitation with similar antecedent water-table conditions for each of the two periods. An alternative explanation may be that the soils became hydrophobic as a result of the fire, contributing to relatively more runoff in July 1998. Also, seasonal or firerelated variations in evapotranspiration can result in variations in the amount of precipitation available as runoff. Deep leakage was a relatively small fraction of the rainfall (about 9 percent), although this water-budget term could increase (at the expense of runoff and evapotranspiration) if continued development of the Upper Floridan aquifer in the area increases the hydraulic gradient between the surficial aquifer system and the underlying Upper Floridan aquifer.

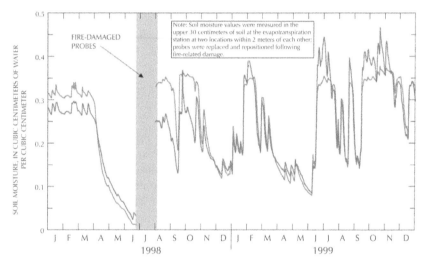

FIGURE 33   Shallow, volumetric soil moisture at evapotranspiration station.

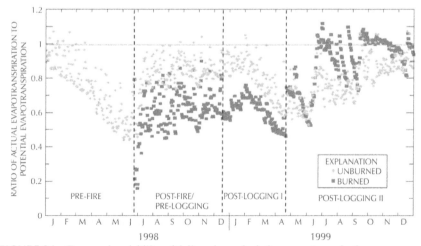

FIGURE 34   Temporal variability of daily values of relative evapotranspiration.

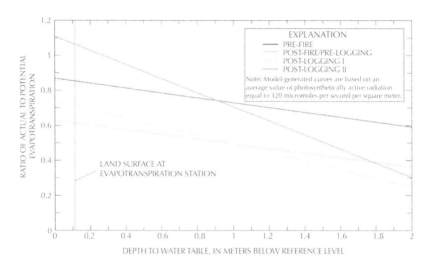

FIGURE 35 Relation between relative evapotranspiration and depth to water table.

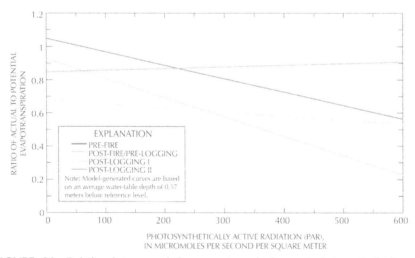

FIGURE 36 Relation between relative evapotranspiration and photosynthetically active radiation (PAR).

The consistency of the water-budget terms can be expressed by the absolute and relative water-budget closures as shown in Eqs. (31) and (32), where: C_a is absolute water-budget closure, in millimeters per year; C_r is relative water budget closure, in percent; and P, ET, R, L, and ΔS are the same as defined in Eq. (29).

$$C_a = P - (ET + R + L + \Delta S) \qquad (30)$$

$$C_r = \frac{100C_a}{P} \qquad (31)$$

Watershed storage (ΔS) was an unmeasured quantity within the water budget. There-fore, evaluation of water-budget closure was facilitated by the judicious choice of a time period when negligible change in storage occurred within the watershed. Based on the measured water levels in the watershed (Figure 27), the time period from March 3, 1998, through September 23, 1999, was selected as an interval when change in wa-tershed storage could be assumed to be zero.

TABLE 4   Sensitivity analysis of evapotranspiration models to environmental variables.

Environmental variables, X	Mean value	Standard deviation, σ
R_n for unburned	118.3	50.0
R_n for burned	127.6	49.6
PAR	320.0	118.3
T_a	21.7	5.4
h_{wt}	0.57	0.42

	ET1 (mean + σ)	ET2 (mean - σ)	% change, (+)	% change (-)
Unburned model				
R_n for unburned	4.15	1.69	42	-42
PAR	2.58	3.27	-12	12
T_a	3.16	2.64	8	-10
h_{wt}	2.71	3.14	-7	7
Post-fire/pre-logging model				
R_n for burned	3.05	1.33	39	-39
PAR	1.64	2.74	-25	25
T_a	2.37	1.98	8	-10
h_{wt}	1.97	2.41	-10	10
Post-logging I model				
R_n for burned	3.32	1.45	39	-39
PAR	2.26	2.50	-5	5
T_a	2.58	2.15	8	-10
h_{wt}	1.97	2.80	-17	17
Post-logging II model				
R_n for burned	4.86	2.12	39	-39
PAR	3.53	3.44	1	-1
T_a	3.78	3.15	8	-10
h_{wt}	2.80	4.16	-19	19

Mean ET, mm/day = 2.92 for unburned; 2.19 for post-fire/pre-logging; 2.38 for post – logging I; and 3.49 for post-logging II models. R_n = Net radiation in watts per m²; PAR = Photosynthetically active radiation in μmoles per m² per sec.; T_a = Air temperature in °C; h_{wt} = Water table depth below the reference level in meters.
Note: Values in table were computed using each of the four ET models defined in Table 3. Mean and standard deviation values are representative of daily values during the 2-year period of record and the R_n values are only representative of 1999.

TABLE 5   Water budget for Tiger Bay watershed.

Year	P	ET	R	L	ΔS
1998	1233	916	357	112	-152
1999	1396	1070	114	112	100
1998-9	1315	993	236	112	-26

P, mm/year = Precipitation, (average of north & south rain gages: See Fig. 1); ET, mm/year = Evapotranspiration; R, mm/year = Runoff from watershed at Tiger Bay canal; L, mm/year = Estimated leakage to the upper Floridian aquifer; and ΔS, mm/year = Rate of change in watershed storage estimated as a water-budget residual.

TABLE 6   Potential evapotranspiration [PET, mm/year] and relative rates of annual water-budget terms for Tiger Bay watershed.

Year	PET	Relative rates			
		ET/PET	ET/P	R/P	L/P
1998	1356	0.67	0.74	0.29	0.09
1999	1391	0.77	0.77	0.08	0.08
1998-99	1374	0.72	0.76	0.18	0.09

ET = Evapotranspiration, mm/year; P = Precipitation, mm/year [average of North and South storage rain gages]; R, mm/year = Runoff from the watershed; L, mm/year = Estimated leakage to the upper Floridian aquifer for 1995.

TABLE 7   Average rate of change in water table depth at monitor wells.

Year	Δh_{North}	Δh_{South}	Δh_{avg}
1998	-660	-616	-638
1999	+432	+308	+370
1998-99	-114	-154	-134

Δh_{North} = Rate of change in water-table depth at the ET North station in mm/year; Δh_{South} = Rate of change in water-table depth at the south rain gage in mm/year; and Δh_{avg} = average rate of change in water-table depth in mm/year = $[\Delta h_{North} + \Delta h_{South}]/2$.

The beginning and ending of this interval occurred at times when temporal changes in water level were relatively slight, implying that the water levels measured at the two monitor wells at the beginning and ending dates of the interval were probably representative of the watershed. The absolute value of the measured rate of change in water level was less than 6 mm/yr at both monitor wells over this 570-day interval.

Based on measured or estimated values of P (1,245 mm/yr), ET (1,048 mm/yr), R (132 mm/yr), and L (112 mm/yr), the absolute and relative water-budget closures were –47 mm/yr and 3.8 percent, respectively. The consistency of these independently measured water-budget terms provides support for, but not confirmation of, the reliability of the measured evapotranspiration. Compensating errors among water-budget terms or compensating errors within the temporal pattern of estimated evapotranspiration also could produce a consistent water budget.

Evapotranspiration was estimated during the present study using an energy-budget variant (Eq. (18)) of the eddy correlation method, rather than the standard eddy correlation method (Eq. (5)). The water-budget analysis provided an independent means to evaluate the relative accuracies of the two eddy correlation methods. The standard method produced turbulent flux estimates that were, on average, about 84.7% of those produced by the energy-budget variant. Applying this fraction to the evapotranspiration total for the water budget period from March 3, 1998, to September 23, 1999, the absolute and the relative budget closures corresponding to the standard eddy correlation method are 113 mm/yr and 9.1 percent, respectively. These closure values are greater than the values reported for the energy-budget variant, consistent with the assumption that the energy-budget variant was more accurate than the standard eddy correlation method.

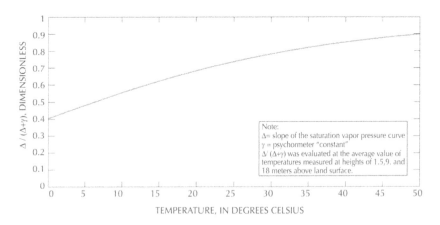

FIGURE 37   The Priestley-Taylor variable $\Delta/(\Delta + \gamma)$ as function of temperature.

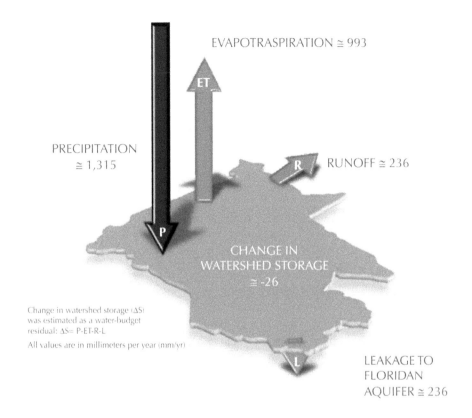

FIGURE 38   Water budget for Tiger Bay watershed during calendar years 1998-99.

TABLE 8   Comparison of estimates of specific yield based on ET estimated with energy budget variant and with the standard eddy correlation method.

Year	Energy-budget variant		Standard eddy correlation method	
	ΔS	S_y	ΔS	S_y
1998	−152	0.24	−12	0.02
1999	100	0.27	268	0.71
1998-99	−26	0.19	126	−0.94

Note: ΔS, mm/year = Rate of change in the watershed storage computed as a residual of the water balance method [$\Delta S = P - (ET + R + L)$]; P, mm/year = Average watershed precipitation; ET, mm/year = Evapotranspiration estimated by the energy-budget variant of the eddy correlation method is about 84.7% of the ET [Table 5] estimated by the energy-budget of the eddy correlation method; R, mm/year = Average watershed runoff; L, mm/year = Estimated leakage in 1995 to the Upper Floridian aquifer; S_y = Specific yield = [ΔS/Δh_{avg}],no units; Δh_{avg} = Estimated average rate of change in water-table depth [Table 7, mm/ year].

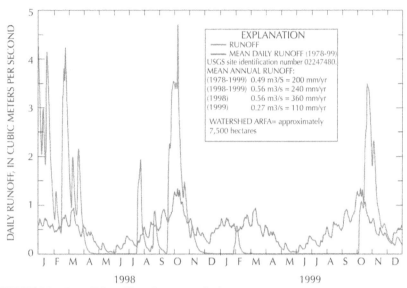

FIGURE 39   Runoff from Tiger Bay watershed.

Additional support for the assumption that the energy-budget variant was preferable to the standard eddy correlation method could be discerned from a residual analysis that assumed that precipitation, leakage, runoff, and evapotranspiration were accurately measured and that a lack of water-budget closure can be explained solely by the residual-calculated storage term. The specific yield representative of the watershed was then computed as the rate of change of watershed storage divided by a representative rate of change in water level within the watershed. The specific yield, estimated in this manner, was evaluated for credibility as a means of identifying the preferred variant of the eddy correlation method. Specific yield is defined as the volume of water yielded per unit area per unit change in water level. Specific yield can range from near zero if the capillary fringe intersects land surface [16] to near unity for standing water. The specific yield of sandy soils (such as those in the uplands) ranges from 0.10 to 0.35 [18]. In this analysis, the representative rate of change in water-table depth for the watershed was assumed equal to the average rate of change in water-table depth at the two upland monitor wells (Table 7). As mentioned previously, upland and wetland water levels are expected to change at the same rate in the low relief environment of this watershed.

Results of the residual analysis, using evapotranspiration estimated by both approaches, are shown in table 8. The energy-budget variant of the eddy correlation method produced specific yield estimates (0.24 in 1998, 0.27 in 1999, and 0.19 in 1998–1999) that were somewhat consistent between each of the three time periods and were within the range of possible values. The standard eddy correlation method produced estimates of specific yield that were inconsistent between each of the three time periods and were unreasonable (0.02 in 1998, 0.71 in 1999, and −0.94 in 1998–

1999). The residual analysis of water budgets further supports the assumption that the energy-budget variant of the eddy correlation method is more accurate than the standard method.

Based on this research, followings can be concluded: A 2-year (1998–1999) study was conducted to estimate evapotranspiration (ET) from a forested watershed (Tiger Bay, Volusia County, Florida), which was subjected to natural fires, and to evaluate the causal relations between the environment and ET. The watershed characteristics are typical of many areas within the lower coastal plain of the south-eastern United States – nearly flat, slowly draining land with a vegetative cover consisting primarily of pine flat wood uplands interspersed within cypress wetlands. Drought-induced fires in spring 1998 burned about 40% of the watershed and most of the burned area was logged in late-fall 1998. ET was measured using eddy correlation sensors placed on a tower 36.5-meter (m) high within an 18.3-m-high forest. About 27% of the 30-minute eddy correlation data were missing as a result of either inoperation of the sensors related to scaling of the hygrometer windows, collection of rainfall or dew on the sensors, or spurious turbulence created by the sensor mounting arm and the attached tower. These missing data generally occurred during periods (evening to early morning) when ET was relatively low. Linear relations between photosynthetically active radiation (PAR) and the fluxes of ET and sensible heat were used to estimate missing 30-minute values. Data were composited into daily values if the turbulent fluxes for more than 18 hours of a given day were directly measured, rather than being estimated with the PAR-based relation. Daily values for which more than 6 hours of data were missing were considered nonmeasured. This procedure resulted in 449 measurements of daily ET over the 2-year (730-day) period. An energy-budget variant of the standard eddy correlation method that accounts for the common underestimation of ET by the standard method was computed.

Following the fires, the daily measurements of ET were a composite of rates representative of burned and unburned areas of the watershed. The fraction of a given daily measurement derived from burned areas was estimated based on the diurnal pattern of wind direction and PAR for that day and on the transpiration station. The daily values of ET were used to calibrate a Priestley-Taylor model. The model was used to estimate ET for burned and unburned areas and to identify and quantify the environmental controls on ET.

The ET model successfully ($r^2 = 0.90$) reproduced daily measurements of site ET over a wide range of environmental conditions, giving credence to the ability of the model to estimate ET at the site. Estimation of ET from the watershed was based on an area-weighted composite of estimated values for burned and unburned areas. Annual ET from the watershed was 916 and 1,070 millimeters (mm) for 1998 and 1999, respectively, and averaged 993 mm. These values are comparable to those reported by previous researchers. ET has been estimated to average about 990 millimeters per year (mm/yr) over Volusia County [37] and to average about 890 mm/yr in the Tiger Bay watershed [7]. Bidlake and others [3] estimated annual cypress ET (970 mm) to be only 8.5% less than that of pine flat woods (1,060 mm) based on studies conducted in Sarasota and Pasco Counties, Florida. Liu [24] estimated average, annual ET of both cypress and pine flatwoods to be 1,080 mm based on a study in Alachua County, Florida.

The extensive burning and logging that occurred during the study produced a landscape that was not typical of forested areas of Florida. The estimated ET from unburned areas can be considered more representative of typical forest cover. Annual ET from unburned areas was 937 and 999 mm for 1998 and 1999, respectively, and averaged 968 mm. ET from burned areas for the 10-month period after the fire (July 1998–April 1999) averaged about 17% less than that from unburned areas and, for the following 8-month period (May–December 1999), averaged about 31% higher than from unburned areas. During the 554-day period after the fire, the average ET for burned areas (1,043 mm/yr) averaged 8.6% higher than that for unburned areas (960 mm/yr). Both actual and potential ET showed strong seasonal patterns and day-to-day variability. Actual ET from the watershed averaged only 72% of potential ET. ET declined from near potential rates in the wet conditions of January 1998 to less than 50% of potential ET after the fire and at the peak of the drought in June 1998. After the drought ended in early July 1998 and water levels returned to near land surface, ET increased sharply. The ET rate, however, was only about 60% of the potential rate in the burned areas, as compared to about 90% of the potential rate in the unburned areas. This discrepancy can be explained as a result of fire damage to vegetation. Beginning in spring 1999, ET from burned areas increased sharply relative to unburned areas, sometimes exceeding unburned ET by almost 100 percent. Possible explanations for the dramatic increase in ET from burned areas are not clear at this time, but may include phenological changes associated with maturation or seasonality of plants that emerged after the fire or successional changes in composition of plant community within burned areas.

Within the framework of the Priestley-Taylor model developed during this study, variations in daily ET were the result of variations in: surface cover, net radiation, PAR, air temperature, and water-table depth. Potential ET depended solely on net radiation and air temperature and increased as each of these variables increased. The extent to which potential ET was approached was determined by the Priestley-Taylor coefficient α. In this study, Priestley-Taylor α was a linear function of water-table depth and PAR. Unique parameters within the α function were estimated for each of four surface covers or time periods: unburned; burned, but unlogged; and both burned and logged (early postlogging and late postlogging). The ET model indicated that relative ET (the ratio of actual to potential ET) decreased as the depth to the water table increased. The rate of decline of relative ET with water-table depth was greater for the postlogging period than for the prelogging period, perhaps indicative of the replacement of many deeply rooted trees by shallow-rooted under story vegetation following the fires. Shallow-rooted plants would be less able to tap into deep soil moisture or the water table than deep-rooted trees. Relative ET decreased with increasing PAR, with the exception of the late postlogging period, which showed a slight increase in relative ET with increasing PAR.

A water budget for the watershed supported the validity of the estimates of ET produced with the energy-budget variant of the eddy correlation method. Independent estimates of average rates of rainfall (1,245 mm/yr), runoff (132 mm/yr), deep leakage (112 mm/yr), as well as ET (1,048 mm/yr) were compiled for a 570-day period over which the change in watershed storage was negligible. Water-budget closure was 47 mm/yr or 3.8% of rainfall, indicating good consistency between the estimated ET and estimates of the other terms of the water budget. Estimates of ET produced by the standard eddy correlation method were relatively inconsistent with the water budget (water-budget closure was 113 mm/yr or 9.1% of rainfall), indicating that the ener-gy- budget variant is superior to the standard eddy correlation method. Specific yield was estimated based on estimated changes in watershed storage and water level. The change in watershed storage was estimated as a residual of the water budget. Specific yield values produced using ET estimated by the energy-bud- get variant of the eddy correlation method were reasonable and relatively consistent from year-to-year (0.19 to 0.27). However, specific yield values based on ET estimated by the standard eddy correlation method were unreasonable and inconsistent from year-to-year (–0.94 to 0.72). These results further support the premize that the energy – budget variant is more accurate than the standard eddy correlation method.

ET rates were about 74 and 77% of rainfall for 1998 and 1999, respectively, rela-tively constant considering the variability in surface cover and rainfall patterns be-tween the 2 years. Potential ET was less consistent as an indicator of actual ET; ET was 67 and 77% of potential ET for years 1998 and 1999, respectively.

8.4 SUMMARY

Daily values of evapotranspiration [ET] from a watershed in Volusia County, Florida, were estimated for a 2-year period (January 1998 through December 1999) by us-ing an energy-budget variant of the eddy correlation method and a Priestley-Taylor model. The watershed consisted primarily of pine flat wood uplands interspersed within cypress wetlands. A drought-induced fire in spring 1998 burned about 40% of the watershed, most of which was subsequently logged. The model reproduced the 449 measured values of ET reasonably well (r^2= 0.90) over a wide range of seasonal and surface-cover conditions. Annual ET from the water- shed was estimated to be 916 millimeters (36 inches) for 1998 and 1,070 millimeters (42 inches) for 1999. ET declined from near potential rates in the wet conditions of January 1998 to less than 50% of potential ET after the fire and at the peak of the drought in June 1998. After the drought ended in early July 1998 and water levels returned to near land-surface, ET increased sharply; however, the ET rate was only about 60% of the potential rate in the burned areas, compared to about 90% of the potential rate in the unburned areas. This discrepancy can be explained as a result of fire damage to vegetation. Beginning in spring 1999, ET from burned areas increased sharply relative to unburned areas, sometimes exceeding unburned ET by almost 100 percent. Possible explanations for the dramatic increase in ET from burned areas could include phenological changes associated with maturation or seasonality of plants that emerged after the fire or suc-cessional changes in composition of plant community within burned areas.

Variations in daily ET are primarily the result of variations in surface cover, net radiation, photosynthetically active radiation (PAR), air temperature, and water-table depth. A water budget for the watershed supports the validity of the daily measure-ments and estimates of ET. A water budget constructed using independent estimates of average rates of rainfall, runoff, and deep leakage, as well as ET, was consistent within 3.8 percent. An alternative water budget constructed using ET estimated by the standard eddy correlation method was consistent only within 9.1 percent. This result indicates that the standard eddy correlation method is not as accurate as the energy-budget variant.

KEYWORDS

- aerodynamic resistance
- anemometer
- aquifer
- atmospheric pressure
- big leaf assumption
- bowen ratio
- bowen ratio energy-budget variant (of eddy covariance method)
- cypress
- eddy covariance
- eddy covariance method
- energy-budget closure
- energy-budget variant
- evaporation
- evaporative fraction
- evapotranspiration
- evapotranspiration, potential
- evapotranspiration, reference
- fire
- Florida
- Florida pinelands
- Floridan aquifer
- Hargreaves equation
- hygrometer
- inertial sublayer
- insect defoliationlatent heat flux
- National Centers for Environmental Prediction (NCEP)
- net radiation
- North American Regional Reanalysis (NARR)
- Penman equation

- **Penman-Monteith equation**
- **Photosynthetically Active Radiation (PAR)**
- **pinelands**
- **precipitation**
- **Priestley-Taylor equation**
- **residual energy-budget variant**
- **residual energy-budget variant (of eddy covariance method)**
- **roughness sublayer**
- **runoff**
- **sensible heat flux**
- **soil moisture**
- **source area**
- **source area (eddy covariance)**
- **Time-Domain Reflectometry (TDR)**
- **turbulent fluxes**
- **United States Geological Survey**
- **uplands**
- **USGS**
- **vapor flux**
- **water balance**
- **wetlands**

REFERENCES

1. Baldocchi, D. D.; Hicks, B. B.; Meyers, T. P.; Measuring biosphere-atmosphere exchanges of biologically related gases with micrometeorological methods: Ecology, **1988,** *69(5),* 1331–1340.
2. Baldwin, R.; Bush, C. L.; Hinton, R. B.; Huckle, H. F.; Nichols, P.; Watts, F. C.; Wolfe, J. A.; Soil Survey of Volusia County, Florida: US Soil Conservation Service, **1980,** 207 p. and 106 pls.
3. Bidlake, W. R.; Woodham, W. M.; Lopez, M. A.; Evapotranspiration from areas of native vegetation in west-central Florida: US Geological Survey Open-File Report 93-415, **1993,** 35 p.
4. Bowen, I. S.; The ratio of heat losses by conduction and by evaporation from any water surface: Physical Review, 2nd series, **1926,** *27(6),* 779–787.
5. Brutsaert, W.; Evaporation into the atmosphere-Theory, history, and applications: Boston, Kluwer Academic Publishers, **1982,** 299 p.
6. Businger, J. A.; Yaglom, A. M.; 'Introduction to Obukhov's paper on "Turbulence in an atmosphere with a non-uniform temperature"', Boundary-Layer Meteorology, **1971,** *2,* 3–6.
7. Camp, Dresser and McKee, Inc.; **1996,** Volusia County, Florida-Tiger Bay water conservation and aquifer recharge evaluation-Phase I: Volusia County, Florida, Technical Report.
8. Campbell, G. S.; Norman, J. M.; An introduction to environmental biophysics: New York, Springer, **1998,** 286 p.

9. Dyer, A. J. Measurements of evaporation and heat transfer in the lower atmosphere by an automatic eddy-correlation technique: Quarterly Journal of the Royal Meteorological Society, **1961,** *87, 401*–412.

10. Eichinger, W. E.; Parlange, M. B.; Stricker, H.; On the concept of equilibrium evaporation and the value of the Priestley-Taylor coefficient: Water Resources Research, **1996,** *32(1), 161*–164.

11. Eidenshink, J. C.; The 1990 conterminous U. S. AVHRR data set: J. Photogrammetry and Remote Sensing, **1992,** *58,* 809–813.

12. Fleagle, R. G.; Businger, J. A.; *An introduction to atmospheric physics*. New York, Academic Press, **1980,** 432 p.

13. Flint, A. L.; Childs, S. W.; Use of the Priestley-Taylor evaporation equation for soil water limited conditions in a small forest clearcut: Agricultural and Forest Meteorology, **1991,** *56,* 247–260.

14. Garratt, J. R.; Surface influence upon vertical profiles in the atmospheric near-surface layer: Quarterly Journal of the Royal Meteorological Society, **1980,** *106,* 803–819.

15. German, E. R.; Regional evaluation of evapotranspiration in the Everglades: US Geological Survey Water-Resources Investigations Report 00-4217, **2000,** 48 p.

16. Gillham, R. W.; The capillary fringe and its effect on water-table response: **1984,** *67,* 307–324.

17. Goulden, M. L.; Munger, J. W.; Fan, S-M, Daube, B. C.; Wofsy, S. C.; Measurements of carbon sequestration by long-term eddy covariance: methods and a critical evaluation of accuracy: Global Change Biology, **1996,** *2,* 169–182.

18. Johnson, A. I.; **1967,** Specific yield-Compilation of specific yields for various materials: US Geological Survey Water-Supply Paper 1662-D, 74 p.

19. Kaimal, J. C.; Businger, J. A.; A continuous wave sonic anemometer-thermometer: *J. Appl. Metero.,* **1963,** *2,* 156–164.

20. Kaimal, J. C.; Gaynor, J. E.; Another look at sonic thermometry: Boundary-layer meteorology, **1991,** *56,* 401–410.

21. Kimrey, J. O.; **1990,** Potential for ground-water development in central Volusia County, Florida: US Geological Survey Water-Resources Investigations Report 90-4010, 31 p.

22. Knowles, L.; Jr.; **1996,** Estimation of evapotranspiration in the Rainbow Springs and Silver Springs basins in north–central Florida: US Geological Survey Water-Resources Investigations Report 96-4024, 37 p.

23. Lee, X.; Black, T. A.; Atmospheric turbulence within and above a Douglas-fir stand. Part II: Eddy fluxes of sensible heat and water vapor: Boundary-layer meteorology, **1993,** *64,* 369–389.

24. Liu, S.; **1996,** Evapotranspiration from cypress (*Taxodium ascendens*) wetlands and slash pine (*Pinus elliottii*) uplands in north-central Florida: Ph. D.; Dissertation, University of Florida, Gainesville, 258 p.

25. Lowe, P. R.; An approximating polynomial for the computation of saturation vapor pressure: *J. Appl. Metero.,* **1977,** *16(1),* 100–103.

26. McDonald, M. G.; Harbaugh, A. W.; **1984,** A modular three-dimensional finite-difference ground-water flow model: US Geological Survey Open-File Report 83-875, 528 p.

27. Monteith, J. L.; **1965,** Evaporation and environment *in* The state and movement of water in living organisms, Symposium of the Society of Experimental Biology: San Diego, California, (Fogg, G. E.; ed.), Academic Press, New York, 205–234.

28. Monteith, J. L.; Unsworth, M. H.; **1990,** Principles of environmental physics (2d ed.): London, Edward Arnold, 291 p.

29. Moore, C. J. Eddy flux measurements above a pine forest: Quarterly Journal of the Royal Meteorological Society, **1976,** *102,* 913–918.

30. National Oceanic and Atmospheric Administration, Climatological data-annual summary-Florida: **1998,** *102(13),* 21 p.

31. National Oceanic and Atmospheric Administration, Climatological data-annual summary-Florida: **1999,** *103(13),* 21 p.

32. Penman, H. L.; Natural evaporation from open water, bare soil, and grass: Proceedings of the Royal Society of London, Series, A.; **1948,** *193,* 120–146.

33. Phelps, G. G.; **1990**, Geology, hydrology, and water quality of the surficial aquifer system in Volusia County, Florida: US Geological Survey Water-Resources Investigations Report 90-4069, 67 p.

34. Priestley, C. H. B.; Taylor, R. J. On the assessment of surface heat flux and evaporation using largescale parameters: Monthly Weather Review, **1972**, *100*, 81–92.

35. Reifsnyder, W. E.; Radiation geometry in the measurement and interpretation of radiation balance: Agricultural Meteorology, **1967**, *4*, 255–265.

36. Riekerk, H.; Korhnak, L. V.; The hydrology of cypress wetlands in Florida pine flatwoods: Wetlands. **2000**, *20(3)*, 448–460.

37. Rutledge, A. T.; **1985**, Ground-water hydrology of Volusia County, Florida with emphasis on occurrence and movement of brackish water: US Geological Survey Water-Resources Investigations Report 84-4206, 84 p.

38. Schotanus, P.; Nieuwstadt, F. T. M.; de Bruin, H. A. R.; Temperature measurement with a sonic anemometer and its application to heat and moisture fluxes. Boundary-Layer Meteorology, **1983**, *50*, 81–93.

39. Schuepp, P. H.; Leclerc, M. Y.; MacPherson, J. I.; Desjardins, R. L.; Footprint prediction of scalar fluxes from analytical solutions of the diffusion equation. Boundary-Layer Meteorology, **1990**, *50*, 355–373.

40. Simonds, E. P.; McPherson, B. F.; Bush, P.; **1980**, Shallow ground-water conditions and vegetation classification, central Volusia County, Florida: US Geological Survey Water-Resources Investigations Report 80–752, 1 sheet.

41. Stannard, D. I.; Comparison of Penman-Monteith, Shuttleworth-Wallace, modified Priestley-Taylor evapotranspiration models for wildland vegetation in semiarid rangeland. Water Resources Research, **1993**, *29(5)*, 1379–1392.

42. Stannard, D. I.; Interpretation of surface flux measurements in heterogeneous terrain during the Monsoon '90 experiment. Water Resources Research, **1994**, *30(5)*, 1227–1239.

43. Stull, R. B.; An introduction to boundary layer meteorology. Kluwer Academic Publishers, Boston, **1988**, 666 pp.

44. Sumner, D. M.; Evapotranspiration from successional vegetation in a deforested area of the Lake Wales Ridge, Florida. US Geological Survey Water-Resources Investigations Report 96-4244, **1996**, 38 p.

45. Tanner, B. D.; Greene, J. P.; **1989**, Measurement of sensible heat and water vapor fluxes using eddy correlation methods. Final report prepared for US Army Dugway Proving Grounds, Dugway, Utah, 17 p.

46. Tanner, B. D.; Swiatek, E.; Greene, J. P.; **1993**, Density fluctuations and use of the krypton hygrometer in surface flux measurements: Management of irrigation and drainage systems, Irrigation and Drainage Division, American Society of Civil Engineers, July 21–23, 1993, Park City, Utah, 945–952.

47. Tanner, C. B.; Thurtell, G. W.; **1969**, Anemoclinometer measurements of Reynolds stress and heat transport in the atmospheric boundary layer: United States Army Electronics Command, Atmospheric Sciences Laboratory, Fort Huachuca, Arizona, TR ECOM 66-G22-F, Reports Control Symbol OSD-1366, April 1969, 10 p.

48. The Orlando Sentinel, **1998**, Special report—Florida ablaze: Sunday, July 12, 1998, 12.

49. Tibbals, C. H.; **1990**, Hydrology of the Floridan aquifer system in east-central Florida: US Geological Survey Professional Paper 1403-E, 98 p.

50. Twine, T. E.; Kustas, W. P.; Norman, J. M.; Cook, D. R.; Houser, P. R.; Meyers, T. P.; Prueger, J. H.; Starks, P. J. Wesely, M. L.; Correcting eddy-covariance flux underestimates over a grassland: Agricultural and Forest Meteorology, **2000**, *103*, 279–300.

51. US Geological Survey, **1998a**, Water resources data, Florida, water year 1998, v. 1A, northeast Florida surface water. US Geological Survey Water-Data Report FL-98-1A, 408 p.

52. US Geological Survey, **1998b**, Conterminous U.S. AVHRR: US Geological Survey, National Mapping Division, EROS Data Center, 7 compact discs.

53. US Geological Survey, **1999a**, Water resources data, Florida, water year 1999, v. 1A, northeast Florida surface water: US Geological Survey Water-Data Report FL–99–1A, 374 p.
54. US Geological Survey, **1999b**, Conterminous U.S. AVHRR: US Geological Survey, National Mapping Division, EROS Data Center, 7 compact discs.
55. US Geological Survey, **2000**, Water resources data, Florida, water year **2000**, v. 1A, northeast Florida surface water: US Geological Survey Water-Data Report FL-00-1A, 388 p.
56. Volusia County Department of Geographic Information Services, 1996a, Vegetation, Daytona Beach, N. W.; prepared July 29, **1996**, 1 sheet.
57. Volusia County Department of Geographic Information Services, 1996b, Vegetation-Daytona Beach, S. W.; prepared July 29, **1996**, 1 sheet.
58. Webb, E. K.; Pearman, G. I.; Leuning, R.; Correction of flux measurements for density effects due to heat and water vapour transfer. Quarterly Journal of the Royal Meteorological Society, **1980**, *106,* 85–100.
59. Weeks, E. P.; Weaver, H. L.; Campbell, G. S.; Tanner, B. D.; Water use by saltcedar and by replacement vegetation in the Pecos River floodplain between Acme and Artesia, New Mexico. US Geological Survey Professional Paper 491-G, **1987**, 37 pages.

APPENDIX

DERIVATION OF EQ. (24) FOR W_B IN THIS CHAPTER.

The assumptions inherent in the weighting scheme used in Eqs. (23) through (25) can be seen through derivation of Eq. (24) for w b. The latent heat flux measured by the eddy correlation sensors and derived from burned surface covers over a given day of 48 measurements is given by:

$$\lambda E_{bm} = \sum_i g_i \frac{1}{48} \sum_k \lambda E_{bk} \delta_i(\psi_k). \qquad (A1)$$

where, λE_{bm} is daily latent heat flux derived from burned surface covers and measured by the flux sensors, in watts per m^2; g_i is fractional contribution of burned area within burn zone i to the measured latent heat flux when wind direction is from burn zone i; λE_{bk} is latent heat flux from burned surface covers for time step k, in watts per m^2; $\delta_i(\psi_k)$ is a binary function equal to 1 if ψ_k is within burn zone i and otherwize equals 0; and the index i is incremented from zone I to IV, and the index k is incremented from 1 to 48.

By definition, the expression in Eq. (A1) is equal to the second term of the right side of Eq. (23). Setting these two expressions equal and assuming that the high-resolution latent heat flux measurements for burned surfaces are directly proportional to photosynthetically active radiation (PAR), and therefore, that the daily resolution latent heat flux for burned surfaces are directly proportional to average daily PAR:

$$w_b(a\overline{PAR}) = \frac{1}{48} \sum_i g_i \sum_k (aPAR_k) \delta_i(\psi_k). \qquad (A2)$$

where, w_b is the fraction of the measured latent heat flux originating from burned areas, dimensionless; and overbars represent daily average values and the variable a is the constant of proportionality between latent heat flux and PAR.

Solving Eq. (A2), for wb:

$$w_b = \frac{\frac{1}{48}\sum_i g_i \sum_k (aPAR_k)\delta_i(\psi_k)}{a\overline{PAR}} \qquad (A3)$$

$$w_b = \frac{\frac{1}{48}\sum_i g_i \sum_k (aPAR_k)\delta_i(\psi_k)}{a\frac{1}{n}\sum_k PAR_k} \qquad (A4)$$

$$w_b = \frac{\sum_i g_i \sum_k PAR_k \delta_i(\psi_k)}{\sum_k PAR_k} \qquad (A5)$$

Eq. (A5) is identical to Eq. (24). The constant of proportionality a can change from day-to-day as environmental conditions (for example, water level, air temperature, and green leaf density) change and, in fact, as shown in Eq. (A5), w_b is independent of the particular value of the constant. An equivalent expression, equal to $[1 - w_b]$, can be derived for the weight applied to daily latent heat flux from unburned surfaces. The constant of proportionality between unburned latent heat flux and PAR can be different than that between burned latent heat flux and PAR.

It is interesting to note that the use of measured high-resolution λE, rather than PAR, as a means of adjusting the weights for the combination of changing source area composition and diurnal variations in evapotranspiration (ET) (Eq. (25)), produces excessive weighting towards zones with high-ET surface covers. This observation can be illustrated best by an example. Suppose, for a given day, the wind direction were from a lake (high ET) before solar noon and from a desert (near-zero ET) after solar noon. In this case, the appropriate weighting for each surface cover, within an equation of the form of Eq. (23), would be 0.5 and the average, measured ET for the day would be about one-half that of the lake. However, weighting by the fraction of ET measured from each zone would lead to a weight of near 1.0 for the lake zone and 0.0 for the desert zone, leading to a model for lake evaporation that would produce underestimates of true lake evaporation.

PART II: APPLICATIONS

CHAPTER 9

EVAPOTRANSPIRATION FOR PINELANDS IN NEW JERSEY, USA[1]

DAVID M. SUMNER, ROBERT S. NICHOLSON,
and KENNETH L. CLARK

CONTENTS

---

[1]Modified and printed with permission from "Sumner, D. M.; Nicholson, R. S.;Clark, K. L., Measurementand simulationofevapotranspiration at a wetland site in the New Jersey Pinelands: U.S. Geological Survey Scientific Investigations Report 2012–5118, 2012, 1–30. Prepared by the West Trenton Publishing Service Center, U.S. GeologicalSurvey,New Jersey Water ScienceCenter, Mountain View Office Park, 810 Bear Tavern Rd., Suite 206, West Trenton, NJ 08628 USA. Web site: http://nj.usgs.gov/;and prepared in co-operation with the New Jersey Pinelands Commission." © Copyright nj.water.usgs.gov and the New Jersey Pinelands Commission.

9.1 INTRODUCTION

Water budgets are fundamental to the understanding of hydrologic systems. If the various components of the water budget can be quantified, including inflows, outflows, and changes in stored water, then a more complete understanding and evaluation of a hydrologic system becomes possible. In the New Jersey Pinelands area (Fig. 1), wetlands and aquatic habitats are supported by discharge from the Kirkwood – Cohansey aquifer system, and detailed water budgets of the area are needed in order to develop a quantitative understanding of aquifer-wetland-stream interactions at the watershed scale. As described by Rhodehamel [50], groundwater flow in the Pinelands is initiated as aquifer recharge, primarily in upland areas, and follows regional and local subsurface flow paths, with local flowpaths terminating asaquifer discharge to wetlands and streams.Temporal variations in recharge affect aquifer interactions with wetlands and streams, as demonstrated by investigations in the New Jersey Pinelands by Modica [41] and Walker et al. [66].Quantification of aquifer recharge on a seasonal basis, therefore, is needed to understand the dynamics of aquifer interactions with wetlands and streams.

Evapotranspiration (ET) is an important component of the hydrologic budget in the Pinelands, and ET variability exerts considerable control over the amount of water available seasonally for aquifer recharge. On an average annual basis, ET from the Pinelands has been estimated to exceed recent Statewide public water use in New Jersey by more than a factor of two [based on data presented by Rhodehamel [50]; and Hutsonet al. [30]. Inspite ofits significance,the seasonal variability of ET and the relations between ET and other environmental factors in this ecologically important region are poorly quantified. Methods fordirect measurement of ET have been developed [17, 60] and used in a number of settings [44, 55, 58]. Estimating seasonally variable ET rates for specifictime periods at the watershed scale in the Pinelands requires site-specificdata. The U.S. Geological Survey (USGS) conducted a study of ET in the Pinelands, in cooperation with the New Jersey Pinelands Commission, to quantify the temporal variability of ET and to examine relations between ET and environmental variables as part of the Kirkwood-Cohansey Project [48]. During November 10, 2004 through February 20, 2007, ET was monitored above a wetland forest canopy in the McDonalds Branch basin in Burlington County, New Jersey (Fig. 1). Meteorological and eddy-covariance sensors were deployed on a 24.5-meter (m) tower, and ground water levels and soil moisture at the site were also monitored.Three models were evaluated for their utility in simulating ET, and a time series of weekly ET rates was developed for the measurement period.

This chapter presents the results of an ET study at the wet- land forest site in the Pinelands area. The chapter also includes: An explanation ofthe methods used, an analysis ofthe fluxsource area, measured ET rates, relation of these rates to environmental variables, a comparison of wetland ET with ET measured in upland areas in the Pinelands, and a comparison of ET measured during the 27-month monitoring period to ET simulated using selected models. The results of this study can be used in conjunction with information on other components of the water budget to develop an understanding of temporal variations in water exchange among different compartments of the Pinelands hydrologic system. Results can also be extended to formulate approaches for estimating ET in the Pinelands during other time periods using commonly measured environmental variables.Values for ET are shownin figuresand are listed in tables. This chapter can serve as a guideline for similar studies in other locations worldwide.

9.1.1   CURRENT STATUS OF RESEARCH

ET in the New Jersey Pinelands has been the subject of published investigations for more than 115 years. The following summary of previous investigations provides context for the present study. Determination of ET played a key role in early water-supply planning in the region.

Vermeule [64] approximated monthly and annual ET for southern New Jersey watersheds on the basis of calculations made using empirical relations between precipitation and run off in other East Coast watersheds.These calculations were used to estimate the safe yield for Joseph Wharton's 1891 proposal to divert flow from more than 1,191 square kilometers of "the great pine belt" of southern New Jersey and deliver it to the cities of Camden and Philadelphia [13]. Wood [73] quantified interception (the amount ofrain orsnowstored on leaves and branches and eventually evaporated back to the atmosphere), an important component of the hydrologic cycle that contributes to evapotranspiration, in an oak-pine forest in the Pinelands. Anumber of studies were conducted during the 1950s and 1960s to understand the effects of forest management practices on water resources in the Pinelands as part of a research initiative referred to as the "Pine Region Hydrological Research Project" [8]; related research continued into the 1970s. Lull and Axley [39] studied ET in different upland vegetative communities in the Pinelands by measuring soil moisture changes, but significant differences were not found among the communities. Barksdale [6] concluded that ET in the lower Delaware River Basin accounted for about 50 percent of precipitation. Rhodehamel [50] found this 50-percent value applicable to the Pinelands area and estimated the mean annual ET rate in the Pinelands on the basis of the long-term difference between precipitation and runoff to be 572 millimeters per year (mm/yr), or about 50 percent of precipitation. Rhodehamel [51] found reasonable agreement between this ET rate and estimated ET rates from other investigations in the Pinelands and vicinity. Summer ET rates in hardwood-dominated and cedar-dominated wetlands in the Pinelands were estimated from water-table fluctuations, and no differences in the ET rates of these communities were detected [2, 9]. ET rates in lowland shrub communities were found to be lower than those of lowland tree communities [9]. ET rates were shown to be greater in lowland (wetland) areas of the Pinelands, where water is more available to plants, than in upland areas [4]. Ballard [3] examined these differences in an energy fluxcontext and concluded that in wetland tree areas, the net summer loss of groundwater discharge through ET was 250 millimeters (mm).

ET has been estimated as part of water-supply and availability studies in the New Jersey Coastal Plain. Mean ET rates in the major drainage areas of the New Jersey Coastal Plain were estimated by Vowinkel and Foster [65] as the long-term difference between mean precipitation and mean runoff. Estimates of ET (presented as "water loss") for basins partly within the Pinelands ranged from 414 to 653 mm/yr [65]. More detailed examinations of water budgets that include ET in selected drainage areas

that are within the New Jersey Coastal Plain and at least partly within the Pinelands
are presented in a series of reports by Watt and Johnson [67], Johnsson and Barringer
[32], Watt and others [68], Johnson and Watt [31], Watt and others [69], and Gordon
[25]. Although the methods used to estimate ET in these studies vary somewhat, the
ET estimates were all based on the concept of water-budget closure and are, therefore,
consistent with estimates of other water-budget components. The ET estimates from
the previously mentioned series of reports range from 563 to 658 mm/yr. As part of the
Kirkwood – Cohansey Project, the USGS assessed hydrologic conditions in three basins in the Pinelands during 2004–2006, including the McDonalds Branch basin where
the present ET study site is located [66]. Results of the hydrologic assessment of the
McDonalds Branch basin include estimates of water-budget components that can be
used to evaluate the veracity of ET measurements presented in this report. Other recent
investigations have examined the physiological responses of a variety of Pinelands
shrub and tree species to hydrologic stress, fire, and insect defoliation.

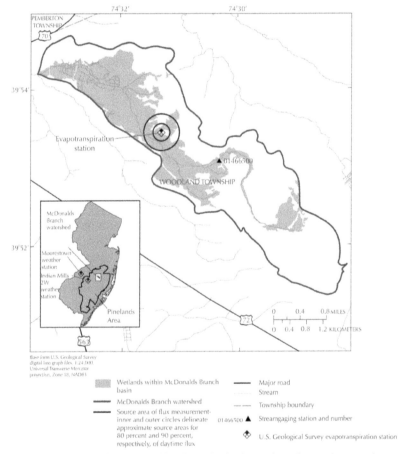

FIGURE 1 Location of McDonalds Branch basin, selected weather stations, and
evapotranspirationstation, Pinelandsarea, New Jersey.

As part of their research on carbon and fire dynamics in the Pinelands, Clark and others [14, 16] measured ET flux using an eddy covariance method at an oak – dominated upland site and two pine-dominated upland sites in the Pinelands. They observed that ET at the oak-dominated upland site was slightly greater in summer and lower in winter than ET at the pitch pine-dominated upland site and that ET averaged 51 to 62 percent of annual precipitation at the sites when they were undisturbed. Additional flux monitoring demonstrated the effects of fire and insect defoliation on ET and water-useefficiency at the three sites; annual ET at one of the defoliated sites was as low as 419 mm/yr, 37 percent of incident precipitation [16]. When all years were considered, maximum seasonal leaf area index at these sites explained 82 and 80 percent of the variation in daily ET during the summer at the oak- and pine – dominated sites, respectively. Results of the study by Clark and others [16] indicate that gypsy moth defoliation disturbance in 2007 may have resulted in a temporary increase in aquifer recharge of approximately 7 percent in upland forests throughout the Pinelands. Schäfer [52] examined changes in stomatal conductance in response to drought, defoliation, and mortality in an upland oak/pine forest in the Pinelands. Drought caused reductions in canopy-level conductance, and corresponding reductions in ET, with the magnitude of the effect varying by species.

9.2 METHODS AND MATERIAL: MEASUREMENT AND SIMULATION OF EVAPOTRANSPIRATION

Evapotranspiration (ET)was measured at a site within McDonalds Branch basin (Fig. 1) using the eddy-covariance method [5, 17, 60]. The site chosen for the ET station is within a pitch-pine lowland stand. Canopy vegetation at the site is dominated by *Pinus rigida* (pitch pine) that reaches a maximum height of about 15 m. Under story vegetation is dominated by *Gaylussacia frondosa* (dangleberry), *Kalmia angustifolia* (sheep laurel), *Eubotrys racemosa* (fetter bush), and *Xerophyllum asphdeloides* (turkey beard). Vegetation types in nearby areas of cedar swamp are dominated by *Chamaecyparis thyroides* (Atlantic white cedar) and *Sphagnum spp* (sphagnum moss); nearby areas of hardwood swamp are dominated by *Acer rubrum* (red maple) and *Nyssa silvatica* (black gum). Depth to the water table at the site fluctuates between about 0.5 to 1.5 meters (m) below land surface although parts of the surrounding area exhibit standing water during wet periods.The site of the ET station is adjacent to one of the vegetation plots used by Laidig and others [35, 36] to develop models for predicting the distribution of wetland vegetation on the basis of hydrologic conditions. Eddy-covariance instrumentation was deployed on a 24.5 m tall Rohn 45G communications, type tower at the site (Fig. 2), and 30-minute data were collected for an 833-day period from November 10, 2004, to February 20, 2007. Other meteorological and hydrologic instrumentation also was deployed on or around the tower to collect data for ET models and to provide ancillary data for the eddy-covariance analysis. The instrumentation used in the study is described in Table 1. Measured values of ET were used to calibrate ET models or for comparison with model results.

9.2.1 MEASUREMENT OF EVAPOTRANSPIRATION

Evapotranspiration can be measured above a forest canopy by using a variety of methods. The method selected for this study is the eddy-covariance method.

9.2.1.1 THE EDDY COVARIANCE METHOD

The eddy-covariance method [17, 60] was used to measure two components of the energy budget of the plant canopy: latent and sensible heat fluxes. Latent heat flux (λE) is the energy removed from the canopy in the liquid-to-vapor phase change of water and is the product of the heat of vaporization of water (l) and the ET rate (E). Sensible heat flux (H) is the heat energy removed from/added to the canopy as a result convective transport along a temperature gradient between the canopy and the air. Both latent and sensible heat fluxes are transported by turbulent eddies in the air that are generated by a combination of frictional and convective forces. The energy available to generate turbulent fluxes of vapor and heat is equal to the net radiation (R_n) less the sum of the heat fluxinto the soil surface (G) and the change in storage (S) of energy in the biomass, air, and any standing water.

The energy involved in fixation of carbon dioxide is usually negligible [10]. Net radiation is the difference between incoming radiation (short- wave solar radiation and long wave atmospheric radiation) and outgoing radiation (reflected short wave and surface-emitted long wave radiation). Energy is transported to and from the base of the canopy by conduction through the soil. Assuming that net horizontal advection of energy is negligible, the energy-budget is described in Eq. (1), for a control volume extending from land surface to a height z_s at which the turbulent fluxes are measured.

$$R_n - G - S = H + \lambda E \tag{1}$$

$$E = \overline{w\rho_v} = \overline{(\overline{w} + w')(\overline{\rho_v} + \rho_v')} \tag{2}$$

$$E = \overline{\overline{w}\overline{\rho_v}} + \overline{\overline{w}\rho_v'} + \overline{w'\overline{\rho_v}} + \overline{w'\rho_v'} \tag{3}$$

$$\overline{w'\rho'}_v = covariance(w, \rho_v) \tag{4}$$

In Eq. (1), the left-hand side represents the available energy and the right-hand side represents the turbulent fluxes; R_n is s net radiation to/from plant canopy, in watts per m²; G is soil heat fluxat land surface, in watts per m²; S is change in storage of energy in the biomass, air, and in any standing water, in watts per m²; H is sensible heat fluxat height z_s above land surface, in watts per m²; λE is latent heat fluxat height z_s above land surface, in watts per m²; and the sign convention is such that R_n and G are positive downwards and H and λE are positive upwards.

The eddy-covariance method is a conceptually simple, one-dimensional approach for measuring the turbulent fluxes of vapor and heat above a surface. For the case of vapor transport above a flat, level landscape, the time-averaged product of measured values of vertical wind speed (w) and vapor density (ρ_v) is the estimated vapor flux (ET rate) during the averaging period, assuming that the net lateral advection of vapor is negligible. Because of the insufficient accuracy of instrumentation available for measurement of actual values of wind speed and vapor density, this procedure generally is performed by monitoring the fluctuations of wind speed and vapor density about their means, rather than monitoring their actual values. This formulation is represented in Eqs. (2)–(4) above and the variables in these equations are defined below:

E Evapotranspiration rate, in grams per m² per second;

w Vertical wind speed, in meters per second;

ρ_v Vapor density, in grams per cubic meter; and

Over-bars and primes indicate means over the averaging period and deviations from means, respectively.

The first term of Eq. (3) is approximately zero because mass-balance considerations dictate that mean vertical wind speed perpendicular to the surface is zero; this conclusion is based on an assumption of constant air density (correction for air-density fluctuations are noted later in this report). The second and third terms are zero based on the definition that the mean fluctuation of a variable is zero. Therefore, it is apparent from Eq. (4) that vertical wind speed and vapor density have to be correlated for a non-zero vapor flux to exist. The turbulent eddies that transport water vapor (and sensible heat) produce fluctuations in both the direction and magnitude of vertical wind speed. The ascending eddies must on average be moister than the descending eddies for ET to occur; that is, upward air movement has to be positively correlated with vapor density, and downward air movement must be negatively correlated with vapor density.

FIGURE 2   Evapotranspiration station at a pitch-pine lowland site in the McDonalds Branch basin, Pinelands area, New Jersey:

A (Left): Tower with instrumentation (Photograph by Anthony S. Navoy, USGS); and

B (Top): Krypton hygrometer (foreground) and sonic anemometer (background) mounted at top of tower at evapotranspiration station [Photograph by Robert S. Nicholson, USGS].

9.2.2 SOURCE AREA OF MEASUREMENTS

The source area for a turbulent flux measurement is defined as the area (up wind of the measurement location) contributing to the measurement. The source area can consist of a single vegetative cover if that cover is adequately extensive.

This condition is met if the given cover extends sufficiently far up wind such that the atmospheric boundary layer has equilibrated with the cover from ground surface to at least the height of the instrumentation. If this condition is not met, the flux measurement is a composite of fluxes from two or more covers within the source area.

Schuepp and others [54] provide an estimate of the source area for turbulent flux measurements and the relative contributions within the source area on the basis of an analytical solution of a one-dimensional (upwind) diffusion equation for a uniform surface cover. In this approach, source area varies with instrument height (z_c), zero displacement height (d), roughness length for momentum (z_m), and atmospheric stability. The instrument height for the turbulent flux measurements in this study was 22.1 m. Campbell and Norman [12] proposed empirical relations for zero displacement height $(d \sim 0.65 \times h)$ and roughness length for momentum $(z_m \sim 0.10 \times h)$, where h is the canopy height. A uniform canopy height of 15 m was assumed in this analysis. The source area estimates were made assuming mildly unstable conditions typical of day time conditions when heat and vapor fluxes are highest; the Obukhov stability length [11] was set equal to [–10 m]. About 80 percent and 90 percent of the source area for the day time turbulent flux measurements were estimated to be within upwind distances of about 205 and 435 m, respectively, as shown in Figs. 1 and 3.

TABLE 1 List of instruments for this research [CSI = Campbell Scientific, Inc.,; HS = Hydrological Services Pvt. Ltd.; REBS = Radiation and Energy Balance Systems, Inc.,; RMY = R. M. Young Inc.,; TE = Texas Electronics, Inc.,]. Note: Negative height is depth below land surface. Number of instruments are listed in brackets; and psi = Pounds per inch^2.

Parameter	Instrument	Height(s) above land surface, meters
Evapotranspiration	CSI eddy correlation system including model CSAT3 3-D sonic anemometer and model KH20 krypton hygrometer (1).	22.1
Air temperature/ Relative humidity	CSI model HMP45 temperature and relative humidity probe (1).	22.1
Solar radiation	LI-COR, Inc., model LI200 pyranometer (1)	22
Net radiation	REBS model Q-7.1 net radiometer (2).	22
Wind speed & direction	RMY model 05305VM wind monitor (1)	22
Photosynthetically active radiation (PAR)	LI-COR, Inc., Model LI-190SB quantum sensor.	35
Soil moisture	CSI model CS615 water content reflectometer (1).	0.0 to [–0.3]

TABLE 1   *(Continued)*

Parameter	Instrument	Height(s) above land surface, meters
Precipitation	HS model TB3 tipping bucket rain gage (1)	14.2
Water level in well	Insitu model miniTroll 5 psi pressure sensor and data logger unit (1).	[−2.0]
Data logging	CSI model 10X data loggers (2);	0 to 1
	12 volt 100 amp-hour deep-cycle batteries (2);	0 to 1 and
	50 watt solar panels (2).	10 to 13

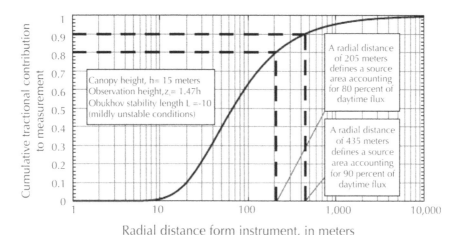

FIGURE 3   Radial extent of source area for day time turbulent flux measurements: Produced using the method of Schuepp and others [54].

The source area during the generally more stable night time conditions could extend considerably further, but turbulent fluxes are relatively small in the absence of sun light. The source area is forested throughout, but includes wetlands with pitch pine and cedar swamps and uplands covered by oak and pine (Fig. 1). Because the measured turbulent fluxes are representative of upwind land covers, the vegetative composition of the source area will change with varying wind direction. For example, wind directions ranging clockwise from south-east to west would provide flux measurements almost exclusively representative of forested wetland; other wind directions provide flux measurements representative of varying mixtures of wetland and upland forests. In this chapter, discrepancies between measured ET and simulated ET from models that are invariant with wind direction are examined as a function of wind direction. This comparison shows the degree to which differences in ET between the two forest communities can be discerned with the available measurements.

9.2.3 INSTRUMENTATION

Instrumentation capable of high-frequency resolution is used in applications of the eddy-covariance method because of the relatively high frequency of the turbulent eddies that transport water vapor. Instrumentation included a three-axis sonic anemometer and a krypton hygrometer to measure variations in wind speed and vapor density, respectively (Fig. 2; Table 2).

The sonic anemometer relies on three pairs of sonic transducers to detect wind-induced changes in the transit time of emitted sound waves and to infer fluctuations in wind speed in three orthogonal directions. The measurement path length between transducer pairs is 10.00 cm (vertical) and 5.8 cm (horizontal); and the transducer path angle from the horizontal is 60 degrees. In contrast to some sonic anemometers used previously [58], the transducers of this improved anemometer are not permanently destroyed by exposure to moisture and thus are suitable for long-term deployment. Operation of the anemometer used in this study ceases when moisture on the transducer disrupts the sonic signal but recommences upon drying of the transducers.

The hygrometer relies on the attenuation of ultraviolet radiation, emitted from a source tube, by water vapor in the air along the 1-cm path to the detector tube. The instrument path line was laterally displaced 10 cm from the midpoint of the sonic-transducer path lines (Fig. 2b). Hygrometer voltage output is proportional to the attenuated radiation signal, and fluctuationsin this signal can be related to fluctuations in vapor density by Beer's Law [71].

Similar to the anemometer, the hygrometer ceases data collection when moisture obscures the windows on the source or detector tubes. Also, the tube windows become "scaled" with exposure to the atmosphere, resulting in a loss of signal strength. The hygrometer is designed such that vapor density fluctuations are accurately measured in spite of variable signal strength; however, if signal strength declines to near-zero values, the fluctuationscannot be discerned. Periodic cleaning of the sensor windows (performed monthly in this study) with a cotton swab and distiled water restores signal strength. Eddy-covariance instrument-sampling frequency was 8 Hertz (Hz) with 30-minute averaging periods. The eddy-covariance instrumentation was placed about 7.1 m above the tree canopy. The 8-Hz data were processed into 30-minute composites and stored in a data logger near ground level.

Flux measurements are made in the constant-flux inertial sublayer, in which lateral variations in vertical flux are negligible, to be representative of the surface cover. Measurements made in the underlying roughness sublayer can reflect individual roughness elements (for example, individual trees or gaps between trees) rather than the composite surface cover [43]. Garratt [23] defines the lower boundary of the inertial sublayer to occur where the difference of the instrument height (z_c) and the zero displacement height (d) is much greater than the roughness length for momentum (z_m). Employing the empirical relations of Campbell and Norman [12] for zero displacement height and roughness length for momentum, and assuming that "much greater than" implies greater by a factor of eight, leads to an instrument height requirement of z_c greater than [1.45h]. A factor of about 1.47 was used in this chapter.

9.2.4   CALCULATION OF TURBULENT FLUXES

Latent heat flux was estimated using Eq. (5), which is a modified form of Eq. (4).

Latent heat flux:

$$\lambda E = \lambda \left((1 + \mu\alpha) \left(\overline{w'\rho'}_v + \frac{\rho_v H}{\rho C_p (T_a + 273.15)} \right) + \frac{F K_o H}{K_w (T_a + 273.15)} \right) \tag{5}$$

Sensible heat:

$$H = \rho C_p \overline{w'T_a'} \tag{6}$$

$$H = \rho C_p \overline{w'T_a'}$$

$$\overline{w'T_a'} = \overline{w'T_s'} - 0.51\overline{(T_a + 273.15)w'q'} \tag{7}$$

The variables in Eq. (5) are defined as: λE is latent heat flux, in watts per m^2; λ is latent heat of vaporization of water, estimated as a function of temperature [57], in joules per gram; μ is ratio of molecular weight of air to molecular weight of water; σ is ratio of vapor density (ρ_v) to air density (ρ); ρ is air density, estimated as a function of air temperature, total air pressure, and vapor pressure [43], in grams per cubic meter; H is sensible heat flux, in watts per m^2; C_p is specific heat capacity of air, estimated as a function of temperature and relative humidity [57], in joules per gram per degree Celsius; T_a is air temperature, in°C; F is a factor that accounts for molecular weights of air and oxygen, and atmospheric abundance of oxygen, equal to 0.229 gram- degree Celsius per joule; K_o is extinction coefficient of hygrometer for oxygen, estimated as [–0.0045] cubic meters per gram per centimeter [61]; K_w is extinction coefficient of hygrometer for water, equal to the manufacturer-calibrated value, in cubic meters per gram per centimeter; and over bars and primes indicate means over the averaging period and deviations from the means, respectively.

The $(1 + \mu\sigma)$ multiplier and the second term of the right-hand side of Eq. (5) account for temperature induced fluctuations in air density [70], and the third term accounts for the sensitivity of the hygrometer to oxygen [60]. Similar to vapor transport, sensible heat can be estimated by using Eq. (6). The sonic anemometer is capable of measuring "sonic" temperature on the basis of the dependence of the speed of sound on this variable [33, 34]. Schotanus and others [53] related the sonic sensible heat, based on measurement of sonic temperature fluctuations, to the true sensible heat given in Eq. (6). Those researchers included a correction for the effect of wind blowing normal to the sonic acoustic path that has been incorporated directly into the anemometer measurement by the manufacturer (E. Swiatek, Campbell Scientific, Inc., written communication, 1998), leading to a simplified form of the Schotanus and others [53] formulation given in Eq. (7). The variables in Eq. (7) are defined as: T_s is a sonic temperature, in °C; and q is a specifichumidity, in grams ofwater vapor per gram of moist air.

Onthe basis of the relation between specific humidity and vapor density, Fleagle and Businger [19] defined specific humidity (q) in Eq. (8), where, R_d is the gas constant for dry air (0.28704 joules per degree Celsius per gram) and P_a is atmospheric pressure, in pascals (assumed to remain constant at 100.5 Kilopascals at top of tower about 58 meters above sea level).

$$q \approx \frac{\rho_v R_d (T_a + 273.15)}{P_a} \tag{8}$$

$$\overline{w'T_a'} = \frac{(T_a + 273.15)}{(T_s + 273.15)} \left(\overline{w'T_s'} - 0.51 R_d \overline{(T_a + 273.15)^2 w' \rho'_v} / P_a \right) \tag{9}$$

Eq. (7) can be expressed in terms of fluctuations in the hygrometer-measured water vapor density rather than fluctuations in specific humidity as shown in Eq. (9) above. Estimation of turbulent fluxes (Eq. (5) and (6)) relies on an accurate measurement of velocity fluctuations perpendicular to the lateral air stream. The study area is relatively flatand level, indicating that the air stream is approximately perpendicular to gravity and the sonic anemometer was oriented with respect to gravity with a bubble level. Measurement of wind speed in three orthogonal directions with the sonic anemometer used in this investigation allowed fora more refined orientation of the collected data with the natural coordinate system through mathematical coordinate rotations. The magnitudes of the coordinate rotations are determined by the components of the wind vector in each 30-minute averaging period. The wind vector is composed of three time-averaged components (u, v, w) in the three coordinate directions (x, y, z). Direction z initially was approximately oriented vertically (with respect to gravity) and the other two directions were arbitrary. Tanner and Thurtell [62] and Baldocchi and others [5] outline a procedure in which measurements made in the initial coordinate system are transformed into values consistent with the natural coordinate system. First, the coordinate system is rotated by an angle η about the z-axis to align u into the x-direction on the x–y plane. Next, rotation by an angle θ is performed about the y-direction to align w along the z-direction. These rotations force v and w equal to zero; therefore, u is pointed directly into the airstream. Athird rotation is sometimes used in complex situations (such as a curving air stream around a mountain) to force equal to zero, although Baldocchi and others [5] indicate that two rotations generally are adequate, and two rotations were used in the current study.

9.2.5   CONSISTENCY OF MEASUREMENTS WITH ENERGY BUDGET

Previous investigators [7, 21, 24, 26, 37, 44, 58, 63] have described a recurring problem with the eddy-covariance method: a frequent discrepancy of the measured latent and sensible heat fluxes with the energy-budget equation (Eq. (1)). The usual case is that measured turbulent fluxes $(H + \lambda E)$ are less than the measured available energy $(R_n - G - S)$. Turbulent fluxes measured above a coniferous forest by Lee and Black [37] accounted for only 83 percent of available energy. Possible explanations for the observed discrepancy in the measured turbulent fluxes include a sensor frequency response that is insufficient to capture high- frequency eddies; an averaging period insufficient to capture low-frequency eddies; drift in the absolute values of anemometer and hygrometer measurements, resulting in statistical nonstationarity within the averaging period; lateral advection of energy; a discrepancy in the measurement points for the wind and vapor density sensors; and overestimation of available energy. Several researchers [24, 26, 44] have shown that the eddy-covariance method obtains better energy-budget closure in windy conditions than during calm conditions.

Twine and others [63] performed an experiment using multiple models of eddy-covariance sensors and net radiometers to measure energy fluxes from a grassland and observed a systematic energy closure discrepancy (turbulent fluxes less than available energy) of 10 to 30 percent. Their conclusion was that eddy-covariance measurements should be adjusted for energy-budget closure. Two common alternatives to adjust turbulent flux measurements for energy-budget closure are to (1) preserve the Bowen ratio, or (2) preserve the measured sensible heat flux. Twine and others [63] indicate a preference for the Bowen ratio approach but state that "the method for obtaining closure appears to be less important than assuring that eddy-covariance measurements are consistent with conservation of energy." The Bowen ratio alternative assumes that the ratio of turbulent fluxes is adequately measured by the eddy-covariance method. The energy-budget Eq. (eq. 1), along with turbulent fluxes (H and λE) measured using the standardeddy-covariance method are used to produce corrected (H_{BREB} and λE_{BREB}) turbulent fluxes in this "Bowen ratio energy-budget variant" of the eddy-covariance method as indicated in Eqs. (10)–(12). On the basis of Eq. (12), this variant fails when the Bowen ratio is close to [−1] and the denominator approaches zero. The second alternative for adjustment of fluxes for energy-budget closure used in this study assumes that the sensible heat flux measured by the standard eddy-covariance method is correct but that latent heat flux is underestimated. Therefore, the corrected latent heat flux is computed as a residual of the energy budget (Eq. (1)). This "residual energy-budget" variant of the eddy-covariance method is shown in Eq. (14).

$$R_n - G - S = H_{BREB} + \lambda E_{BREB} = \lambda E_{BREB}(1 + B) \tag{10}$$

Bowen ratio: $\tag{11}$

$$B = \frac{H}{\lambda E}$$

Rearranging Eq. (10): $\tag{12}$

$$\lambda E_{BREB} = \frac{R_n - G - S}{1 - B}$$

Combine Eqs. (11) and (12); and rearranging: $\tag{13}$

$$H_{BREB} = R_n - G - S - \lambda E_{BREB}$$

"Residual energy- budget" variant: $\tag{14}$

$$\lambda E_{BREB} = R_n - G - S - H$$

Energy generally enters the soil surface and (or) any standing water and is stored in the biomass and air during the day and released at night. Implementation of Eqs.

(10)–(14) was facilitated by using daily composites of terms in these equations and assuming that soil heat flux and changes in energy storage in the biomass, air, and any standing water were negligible at this site over a diurnal cycle. Therefore, although eddy-covariance flux measurements were made at 30-minute resolution, only daily or coarser ET estimates incorporate energy-budget closure considerations. Use of a daily compositing interval of fluxvalues in this study allowed for neglect of those terms of the energy budget that were not measured (soil heat flux and biomass/air/standing water energy storage). However, during periods of rapid temperature changes (for example, cold front passage), the net soil fluxand the net change in energy stored in the biomass/air/standing water over a diurnal cycle may not be negligible.

As mentioned previously, missing 30-minute turbulent flux data can result from scaling of hygrometer windows and from moisture on the anemometer or hygrometer. These data need to be estimated prior to construction of daily composites ofturbulent fluxes. In the present study, regression analysis of measured 30-minute turbulent flux-and net radiation data was used to estimate unmeasured values of 30-minute turbulent fluxes. Additionally, both H and λE were set to zero during periods of rainfall when rainwater on eddy-covariance and net radiation sensors resulted in missing or corrupted data. An assumption of negligible turbulent fluxes during rainfall events was considered reasonable because of the cloudy and, therefore, low net radiation conditions generally prevalent during rain.

9.2.6   SIMULATION OF EVAPOTRANSPIRATION

Several models were evaluated for their utility in estimating ET. The eddy-covariance instrumentation can have extended periods of nonoperation, as discussed previously. However, more robust meteorological and hydrologic instrumentation (sensors for measurement of net radiation, air temperature, relative humidity, solar radiation, wind speed, soil moisture, and water-table depth) provide nearly uninterrupted data. ET models, calibrated to measured turbulent fluxdata and based on continuous meteorological and hydrologic data, can be used to fillgapsin measured data, providing continuous estimates of ET. Additionally ET models can provide insight into the cause-and-effect relation between the environment and ET: Relationships among water-table depth, soil moisture, and ET. Finally, some ET models can provide estimates of maximum (potential or reference) ET under conditions of optimal moisture availability.

Measurement of ET using the eddy-covariance method is resource intensive, so it is practical to use the results of short- term studies to develop and verify ET models that can then be used to estimate ET for other time periods and to understand the cause and effect nature of evaporative processes. The utility of different models is related to the model assumptions or the data used by the models. For example, results of a model that is indifferent to wind direction can be compared with measurements to help determine if ET is related to wind direction. Amodel that requires only temperature data can be used to estimate historical variability in ET using readily available long-term temperature records. This section describes the description of the three ET models explored.

9.2.6.1   PRIESTLEY-TAYLOR EQUATION

Physics-based ET models generally rely on the work of Penman [47], who developed an equation for evaporation from wet surfaces that is based on energy budget and aerodynamic principles. That equation has been applied to estimate ET from well-watered, dense agricultural crops (reference or potential ET). In Penman's equation, the transport of latent and sensible heat fluxes from a "big leaf" to the sensor height is subject to an aerodynamic resistance. The "big leaf" assumption implies that the plant canopy can be conceptualized as a single source of both latent and sensible heat at a given height and temperature. Inherent in the Penman approach is the assumption of a net one-dimensional, vertical transport of vapor and heat from the canopy. The Penman equation is described by Eq. (15). In Eq. (15), the first term is known as the energy term; the second term is known as the aerodynamic term; Δ is slope of the saturation vapor-pressure curve, in kilopascals per degree Celsius; e_s s saturation vapor pressure, in kilopascals; e is vapor pressure, in kilopascals; γ is the psychrometric "constant," equal to approximately 0.067 kilopascals per degree Celsius but varying slightly with atmospheric pressure and temperature; r_h is aerodynamic resistance, in seconds per meter; and other terms are as previously defined.

Penman equation: (15)

$$\lambda E = \frac{\Delta E = \Delta(R_n - S) + \frac{\rho C_p(e_s - e)}{r_a}}{\Delta + \gamma}$$

Priestley and Taylor (1972) for $e_s = e$ [For saturated atmosphere]:   (16)

$$\lambda E = \frac{\Delta(R_n - S)}{\Delta + \gamma}$$

After introducing "Priestley-Taylor coefficient, α" in Eq. (16), we get: (17)

$$\lambda E = \alpha \left[\frac{\Delta(R_n - S)}{\Delta + \gamma} \right]$$

Priestley and Taylor [49] proposed a simplification of the Penman equation for the case of saturated atmosphere ($e_s = e$), for which the aerodynamic term is zero, yielding Eq. (16) above. However, Priestley and Taylor [49] noted that empirical evidence indicates that evaporation from extensive wet surfaces is greater than this amount, presumably because the atmosphere generally does not attain saturation. Therefore, the Priestley-Taylor coefficient, α, was introduced as an empirical correction to the theoretical expression, giving us Eq. (17).

This formulation assumes that the energy and aerodynamic terms of the Penman equation are proportional to each other. The value of α has been estimated to be 1.26, which indicates that under potential ET conditions, where there is no moisture limitation, the aerodynamic term of the Penman equation is about 21 percent of the total latent heat flux. Eichinger and others [18] have shown that the empirical value of α has a theoretical basis; a nearly constant value of α is expected under the existing range of Earth-atmospheric conditions. Previous studies [20, 55, 58] have applied a modified form of the Priestley- Taylor equation. The approach in these studies relaxs the Penman assumption of a free-water surface or a dense, well-watered canopy by allowing α to be less than 1.26 and to vary as a function of environmental factors. The Penman-Monteith equation [42] is a more theoretically rigorous generalization of the Penman equation that also accounts for a relaxation of the Penman assumptions. However, Stannard [55] noted that the modified Priestley-Taylor approach to simulation of observed ET rates was superior to the Penman- Monteith approach for a sparsely vegetated site in the semiarid rangeland of Colorado. Similarly, Sumner [58] found the modified Priestley-Taylor approach performed better than the Penman-Monteith for a site of her baceous, successional vegetation in central Florida. The modified Priestley- Taylor approach was evaluated in the present investigation to simulate daily ET. The selected form of α as a function of environmental variables was determined in this study through trial-and-error exploratory data analysis, and the parameterization of a particular form of α was determined using regression with measured daily values of ET.

9.2.6.2   HARGREAVES EQUATION

The Hargreaves equation [27] is widely used in agricultural studies in the United States and globally to estimate reference ET. Reference ET is defined as the evapotranspiration from an actively growing, well-watered grassoralfalfa vegetative cover of a specific height range [1]. The Hargreaves equation is appealing because of the sparse data requirements; only minimum and maximum daily air temperatures are required, and these are typically measured at most weather stations. The coefficients and form of the equation are empirical and were developed based on a comparison with ET data from precision weighing lysimeters used with grass land covers. The Hargreaves equation is based on an empirical relation between ET and the two most important explanatory variables for this term—incomingsolar radiation and air temperature (Eq. (18)). Another empirical relation (Eq. (19)) is used to relate incoming solar radiation to extraterrestrial radiation and a variable highly correlated with cloud cover (daily temperature range). Combining Eqs. (18) and (19), we get the Hargreaves Eq. (20).

$$ET_S = aR_S(TC + b) \tag{18}$$

$$R_S = K_{RS}R_aTR^{0.5} \tag{19}$$

$$ET_a = aK_{RS}R_a(TC + b)TR^{0.50} \tag{20}$$

In Eqs. (18) to (20), E_a is reference ET, in the same water evaporation units as R_a (for example, millimeters perday); R_s is incoming solar radiation on land surface, the same water evaporation units as ET_a; R_a is extraterrestrial radiation, in the same water evaporation units as ET_a; TC is average daily air temperature, in °C; TR is daily temperature range, in°C; a and b are empirical coefficients [$a = 0.0135$ and $b= 17.8$; and K_{RS} is an empirical coefficient usually estimated as 0.16 for inland areas and 0.19 for coastal areas, respectively. Hargreaves and Samani [28] suggested that the product [aK_{RS}] in Eq. (20) be set equal to 0.0023 for reference conditions. R_a can be estimated using an analytical expression of latitude and day of year [1]. TC and TR

are usually estimated as the average and difference of maximum and minimum daily air temperature, respectively.

In this study, a modifiedform of the Hargreaves equation was considered to allow for the non reference conditions. The modification allows the empirical coefficients [a and b] of Eq. (18) to vary in a regression analysis to best replicate measured daily values of ET and incoming solar radiation. The necessary daily temperature data for these analyze were obtained from measurements at the ET station and also from nearby National Weather Service stations in New Jersey [45].

9.2.6.3 NORTH AMERICAN REGIONAL REANALYSIS

The National Centers for Environmental Prediction (NCEP) – North American Regional Reanalysis (NARR) is an effort to create a long-term set of consistent climate data for North America [40, 46]. NARR is based on a coupled approach to simulation of atmospheric and land-surface processes of energy and mass transfer that assimilates weather data. The period of record of NARR is from 1979 to 2012, and the resolution is 32 kilometers (km) and 3 hours. In the present study, NARR estimates of latent heat flux (at NARR grid centered at 39.75°N. and 74.5°W) were compared to the values measured at the ET station (39.89°N. and 74.52°W).

9.2.7 MEASUREMENT OF ENVIRONMENTAL VARIABLES

Meteorological hydrologic and vegetative data were collected in the study area as ancillary data required by the energy-budget variant of the eddy-covariance method and as independent variables within an ET model. Meteorological variables monitored included net radiation air temperature relative humidity, wind speed, and incoming solar radiation. These data were monitored by data loggers at 15-second intervals, using instrumentation summarized in Table 1; the resulting 30-minute means were stored.

Two net radiometers, deployed at a height of 22 m, provided redundant measurement of net radiation at the ET station. Measured values of net radiation were corrected for wind-speed effects as instructed by the instrument manual. About 90 percent of the source area of the net radiation measurement is contained within a radius of three times the height of the sensor above the canopy [56]. Therefore, the measurement of net radiation had a much smaller source area (radius ofabout 21 m) than did the turbulent flux measurement (about 90 percent of source area within upwind distance of about 450 m during typical daytime conditions). The source area for measured net radiation is composed exclusively of the pitch pine lowlands/cedar swamp typical of forested wetlands in the Pinelands area.

Air temperature and relative humidity were monitored at the ET station at a height of 22.3 m. The slope of the satura- tion vapor pressure curve (a function of air temperature) was computed in the manner of Lowe [38], using the measured air temperature. A propeller-type anemometer to monitor wind speed and direction and an upward-facing pyranometer to measure incoming solar radiation were deployed at heights of 24.7 and 22 m, respectively, at the ET station (Table 1).

Hydrologic variables that were monitored include precipitation, water-table depth, stream discharge, and soil moisture. Precipitation records were obtained from a tipping-bucket rain gage deployed at a height of 14.2 m at the ET station. Soil moisture at a location at the ET station was monitored using a water content reflectometerprobe installed to provide an aver- aged volumetric soil moisture content within the upper 30 cm ofthe soil profile. The measured soil moisture values at this single location are not necessarily representative of the water shedor of the source area of the turbulent flux measurements but are presumed to be correlated with generalized wetting and drying conditions. Soil-moisture measurements were made and recorded on the data logger every 30 minutes (Table 1).

Water levels in a shallow (3-m depth) observation well at the ET station (USGS well number 051604) were measured at 60-minute intervals using a pressure trans-ducer and recorded. The well is situated 5 m from the ET flux tower. Stream discharge of the McDonalds Branch was measured at 15-minute intervals at an upstream site 1.4 km east of the ET station (USGS station number 01466500; Fig. 1). Well-construc-tion data, water-level and stream-discharge data collection methods, water-level data, and stream-discharge data are presented in a report by Walker and others [66].

9.3   RESULTS AND DISCUSSION: EVAPOTRANSPIRATION MEASUREMENT AND SIMULATION

Most (73 percent) of the 30-minute resolution eddy-covariance measurements made during the 833-day study period were acceptable and was used to develop ET mod-els to estimate missing data and discern the effect of environmental variables on ET. Unacceptable measurements resulted from failure of the krypton hygrometer or sonic anemometer and were most extensive (81 percent of missing data) during night- time hours because dew formation at night is common in this humid climate. This diurnal pattern of missing data was less of a problem than it might appear initially because turbulent fluxes are relatively small during the evening to early morning hours when solar radiation is zero or low, so errors associated with gap filling do not translate into substantial errorsin total ET. The environmental variable that provided the best explan-atory value for 30-minute turbulent flux values was net radiation; missing flux data were estimated on the basis oflinear regressions between the turbulent fluxes and net radiation for each month of the study (Eqs. (21) and (22)). This approach reproduced measured 30-minute values of λE and H with r^2 values of 0.80 and 0.84, respectively, and standard errors of 34 and 46 watts per m^2 (W/m^2), respectively. The R_n-to-turbulent flux regression coefficients showed considerable consistency fora given month from year to year (Fig. 4). Although 27 percent of the 30-minute turbulent flux measure-ments were gap filled using Eqs. (21) and (22), the fraction of energy fluxgap filled was small (5% of λE and 7% of H, respectively) because missing data generally oc-curred during periods of low energy flux.

$$\lambda E = a_i R_n + b_i \tag{21}$$

$$H = c_i R_n + d_i \tag{22}$$

In Eqs. (21) and (22): a_i, b_i, c_i, and d_i are coefficients for a given year-month i; a_i and c_i are unitless; b_i and d_i are in watts per m^2. All other variables have been defined before in this chapter.

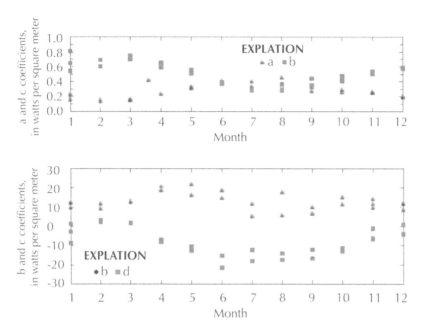

FIGURE 4   Monthly values of regression coefficients used Eqs. (21) and (22).

Examples of measured and regression-simulated energy fluxes are shown for July 1–10, 2006, in Fig. 5. The prominence of daytime over night time turbulent fluxes is evident. Also, a strong correspondence between the diurnal and cloudiness-related variations in net radiation and the turbulent fluxes is apparent. The simple linear relations of Eqs. (21) and (22) also reproduced the variations in measured turbulent fluxes reasonably. The mean, diurnal pattern of turbulent fluxes and net radiation (Fig. 6) indicates that the vast majority of ET occurs during day time, driven by incoming solar radiation. During average daytime conditions, both latent and sensible heat flux are upward. At night, the surface cools below air temperature, producing a reversal in the direction of sensible heat flux. Although the average, night time latent heat flux was slightly upward, dew formation (down ward latent heat flux) commonly occurred.

Daytime wind was predominantly from wetland source areas (Figs. 1 and 7). The two wind arcs most representative of uplands (350° clockwize to 20°and 90° clockwize to 120°) represent only 15 percent of measured wind directions; other wind directions were primarily representative of wetlands, although to varying degrees. From this analysis and the analysis of the source area (Figs. 1 and 3), it is concluded that the source of measured evaporative flux is primarily wetlands.

Energy-budget closure of turbulent fluxes relative to net radiation was examined using weekly, gap-filled composites of the sum of λE and H. As previously discussed, for this analysis, net radiation is equivalent to available energy because it has been assumed that the storage and soil heat flux energy-budget terms are negligible over daily or greater time scales. Energy-budget closure was generally better during relatively windy periods, as indicated in Fig. 8. Measured and gap-filled 30-minute turbulent fluxes accounted for 77 and 83 percent of mean net radiation values of 101 and 100W/ m² for 2005 and 2006, respectively. The relatively large energy-budget discrepancy from early June 2005 to mid-October 2005 led to an exchange of krypton hygrometers on October 19, 2005, and to subsequently improved energy-budget closure.

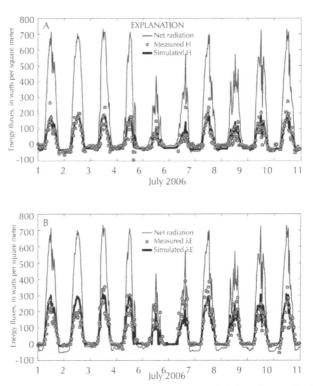

FIGURE 5 Measured and simulated energy fluxes: A, sensible heat flux and B, latent heat flux at the wetland site in the Pinelands area, New Jersey, July 1–10, 2006. (H, sensible heat flux; λE, latent heat flux).

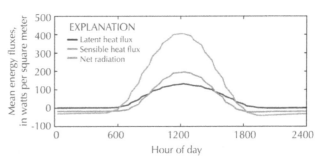

FIGURE 6 Mean diurnal pattern of energy fluxes at the wetland site in the Pinelands area, New Jersey, 2005–06.

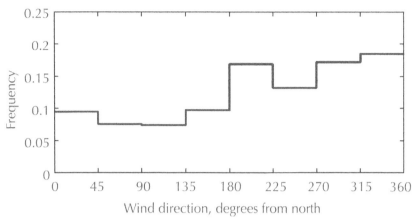

FIGURE 7 Relative frequency of measured wind direction at the wetland site in the Pinelands area, New Jersey, 2005–06.

The hygrometer used in the early part of the study was returned to the manufacturer for evaluation and was diagnosed as exhibiting an unstable voltage output. For this reason, daily valuesof λE computed using either the standard eddy-covariance method or the Bowen ratio variant are considered unreliable for the period June 1, 2005, to October 19, 2005. After the conclusion of the study, the hygrometer used subsequent to October 19, 2005, was returned to the manufacturer for evaluation and was deemed to be stable with a drift in calibration of less than 1 percent. Daily values of ET are shown in Fig. 9 forthe standard eddy-covariance method and the Bowen ratio and residual variants, excluding the standard and Bowen ratio methods for the period in 2005 when hygrometer readings were probably not reliable. For 2006, average ET was 1.66, 2.10, and 2.25 millimeters per day (mm/d) for the standard, Bowen ratio, and residual eddy-covariance methods, respectively; this comparison was restricted to the 343 days when the Bowen ratio was not close to −1 and equation 12 did not have a denominator near zero. As indicated by Twine and others (2,000), energy-budget-closure methods are preferable to the standard eddy-covariance method. The discrepancy in estimated ET between the Bowen ratio and the residual energy – budget methods averaged 7 percent in 2006 with the Bowen ratio variant producing consistently lower ET estimates than the residual variant. The residual variant was selected in this study asthe final method to quantify ET and will be usedfrom here on, primarily because of the loss of data continuity inthe Bowen ratio method associated with the suspected failure of the hygrometer during June–October 2005. However, the 7% over estimation of ET by the residual method relative to the Bowen ratio method can serve as an estimate of possible bias or uncertainty in estimated ET related to the method of energy-budget closure.

Weekly, monthly, and 12-month totals of measured rainfall and ET estimated with the residual energy-budget variant of the eddy-covariance method are summarized in Tables 2 and 3. Annual ET was a remarkably consistent fraction of annual precipitation (0.62 and 0.58 in 2005 and 2006, respectively) and of net radiation (0.62 in 2005 and 2006) as shown in Table 4. An examination of long-term (1902–2011) precipitation data at the nearby National Weather Service Indian Mills weather station indicates that the study period was slightly wetter than mean annual total of 1,173 mm; annual precipitation totals at Indian Mills were 1,325 mm for 2005 and 1,396 mm for 2006, respectively. The Indian Mills precipitation totals are within 30 mm (about 2 percent) of the precipitation measured at the study site during 2005 and 2006 (Table 4). Stream flow measurements during 2005–06 indicate average conditions; mean annual stream flow measured at USGS stream gaging station 01466500 (Fig. 1) during 2005–06 was 0.0593 cubic meters per second (m³/s),which is nearly identical to the long-term mean annual flow of 0.0592 m³/s forthe 1954–2011 period ofrecord.

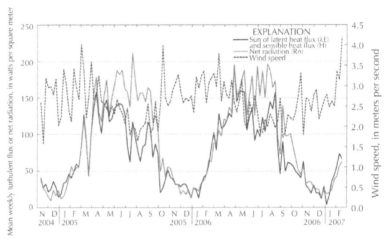

FIGURE 8 Mean weekly values of net radiation, turbulent fluxes, and wind speed at the wetland site in the Pinelands area, New Jersey, November 2004–February 2007.

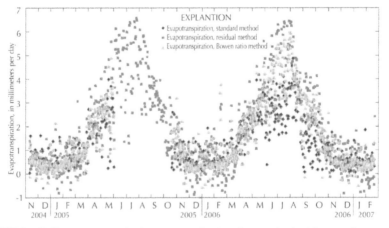

FIGURE 9 Daily evapotranspiration measured using the standard eddy-covariance method, the Bowen ratio method, and the residual energy-budget variant method, at the wetland site in the Pinelands area, New Jersey, November 2004–February 2007.

TABLE 2   Weekly measured rainfall [P, mm] and evapotranspiration [ET, mm] measured using the residual energy-budget variant of the eddy covariance method at the wetland site, pinelands area in New Jersey – USA, during November 2004 to February 2007. Note: Date, month-day-year.

Date	P	ET	Date	P	ET	Date	P	ET
Year 2004								
11-10	37	4	11-17	3	4	11-24	39	3
12-01	33	2	12-08	19	2	12-15	2	0
12-22	20	0	12-29	2	5			
Year 2005								
01-05	38	2	01-12	31	0	01-19	6	2
01-26	11	1	02-02	9	6	02-09	30	5
02-16	14	3	02-23	21	5	03-02	14	6
03-09	2	6	03-16	3	6	03-23	64	5
03-30	44	11	04-06	35	17	04-13	0	17
04-20	10	15	04-27	30	18	05-04	5	17
05-11	0	23	05-18	43	18	05-25	17	25
06-01	34	27	06-08	0	35	06-15	0	32
06-22	51	33	06-29	53	31	07-06	31	31
07-13	47	27	07-20	1	42	07-27	7	34
08-03	29	30	08-10	16	33	08-17	1	31
08-24	1	26	08-31	0	25	09-07	0	24
09-14	8	21	09-21	2	20	09-28	3	17
10-05	92	10	10-12	135	13	10-19	68	8
10-26	1	10	11-02	0	10	11-09	6	7
11-16	70	4	11-23	26	3	11-30	34	2
12-07	17	2	12-14	27	0	12-21	16	3
12-28	67	3						
Year 2006								
01-04	1	4	01-11	31	6	01-18	52	6
01-25	6	5	02-01	17	4	02-08	14	4
02-15	1	6	02-22	2	0	03-01	7	1

03-08	4	10	03-15	0	1	03-22	2	4
03-29	3	13	04-05	24	11	04-12	0	14
04-19	46	15	04-26	1	16	05-03	0	17
05-10	66	19	05-17	9	23	05-24	0	28
05-31	43	22	06-07	26	25	06-14	0	32
06-21	82	27	06-28	28	41	07-05	78	32
07-12	3	37	07-19	13	30	07-26	26	40
08-02	5	37	08-09	0	30	08-16	0	27
08-23	90	19	08-30	127	16	09-06	0	25
09-13	68	21	09-20	15	20	09-27	11	20
10-04	26	16	10-11	76	12	10-18	15	9
10-25	40	8	11-01	5	5	11-08	117	6
11-15	11	6	11-22	44	7	11-29	5	2
12-06	0	4	12-13	4	5	12-20	44	4
12-27	39	3					
Year 2007							
01-03	51	7	01-10	3	4	01-17	7	0
01-24	3	3	01-31	5	0	02-07	0	0
02-14	34	3					

TABLE 3　Monthly and yearly measured rainfall [p, mm] and evapotranspiration [ET, mm] measured using the residual energy-budget variant of the eddy-covariance method at the wetland site, pinelands area in New Jersey – USA, during December 2004 to January 2007.

Year – month	$ET_{monthly}$	12 month moving sum ET	$P_{monthly}$	12 month moving sum P
		mm		
04-Dec	7	--	79	--
05-Jan	7	--	88	--
05-Feb	18	--	60	--
05-Mar	27	--	97	--
05-Apr	66	--	111	--
05-May	94	--	73	--

TABLE 3 *(Continued)*

Year – month	$ET_{monthly}$	12 month moving sum ET	$P_{monthly}$	12 month moving sum P
		mm		
05-Jun	133	--	115	--
05-July	148	--	108	--
05-Aug	134	--	48	--
05-Sep	94	--	13	--
05-Oct	49	--	296	--
05-Nov	27	804	120	1208
05-Dec	8	805	96	1225
06-Jan	22	821	136	1273
06-Feb	14	816	33	1245
06-Mar	24	814	13	1161
06-Apr	58	806	74	1124
06-May	94	807	76	1127
06-Jun	119	794	171	1183
06-Jul	157	802	125	1201
06-Aug	124	792	99	1252
06-Sep	88	786	213	1452
06-Oct	55	792	163	1319
06-Nov	26	790	176	1375
06-Dec	15	797	53	1332
07-Jan	14	789	104	1299

TABLE 4 Summary of selected characteristics measured or simulated annually and dates of spring and fall freezes at the wetland site, pinelands area in New Jersey-USA, during 2005 and 2006.

Characteristic	Units	2005	2006
Measured ET at McDonalds Branch [AET_m]	mm	805	797
Priestly-Taylor PET	mm	1022	1013
Hargreaves reference ET [RET]	mm	986	1008
North American Regional Reanalysis [NARR] actual ET [AET_{NARR}]	mm	831	930
Rainfall [R] at McDonalds Branch	mm	1225	1332

TABLE 4   *(Continued)*

Characteristic	Units	2005	2006
Snow [S] at Indian Mills	mm	749	343
Estimated precipitation [P = R + 0.1*S]	mm	1300	1366
Measured latent heat flux	Watts/m^2	63	62
Incoming solar radiation	Watts/m^2	159	163
Net radiation [R$_n$]	Watts/m^2	101	100
Average Tmin	°C	6.64	7.62
Average Tmax	°C	16.75	17.83
Last spring freeze	---	17 Apr	24 Mar
First fall freeze	--	11 Nov	27 oct
Priestley-Taylor vegetation factor [AET$_m$/PET]	ratio	0.79	0.82
Hargreaves vegetation factor [AET$_m$/RET]	ratio	0.79	0.79
Measured to NARR ET ratio [AET$_m$/AET$_{NARR}$]	ratio	0.97	0.86
ET to P ratio [AET$_m$/P]	ratio	0.62	0.58
Evaporative fraction [Ratio of mean latent Heat to mean net radiation]	ratio	0.62	0.62

Stream flow, water-table altitude, and soil moisture all fluctuated with similar responses to precipitation and no precipitation. Soil moisture and water-table altitude were highly correlated (r^2=0.91). Minimum and maximum ET occurred during December to February and July, respectively (Table 3). Twelve-month ET totals were in a relatively narrow range (786 to 821 mm) over the period of record compared to the range in 12-month rainfall totals (1,124to 1,452 mm). Afirst-orderdata analysis consisting of linear regressions between several environmental variables (incoming solar radiation, air temperature, relative humidity, soil moisture, and net radiation) indicated that net radiation (r^2= 0.72) and air temperature (r^2= 0.73) were the dominant explanatory variables for daily ET. Cross correlation was noted between net radiation and air temperature (r^2= 0.41), and cross correlations are expected between these variables and forest phenological changes, precluding a unique determination of the role of each variable in determining ET. Variationsin the evaporative fraction (ratio oflatent heat flux to net radiation) indicate that air temperature is the strongest explanatory variable in the partitioning of available energy for ET(Fig. 10a). The evaporative fraction wasseemingly unaffected by decreased soil moisture during theApril to September 2005 and the July toAugust 2006 dry periods, until soil moisture fell below a critical threshold of about 0.15, and evaporative fraction decreased. When rains ended the dry period and soil moisture rose above this threshold again, the evaporative fraction recovered (Fig. 10b). A similar dropin evaporative fraction during a dry period in early July 2004 was observed at a nearby upland oak-pine site and attributed to apparent stomatal closure (Kenneth Clark, U.S. Forest Service, written communication, 2010). Schäfer [52] measured a reduction in sap-flux scaled canopy conductance at the upland site during drought conditions in 2006. The observed decrease in the evaporative fraction during the extreme dry periods is an important indication that lower water availability can result in lower rates of ET.

9.3.1 COMPARISON OF MEASURED EVAPOTRANSPIRATION AT WETLAND AND UPLAND SITES

Basin-scale hydrologic analysis requires estimates of ET over large areas, and therefore, an accounting of variability in ET rates across the landscape is needed. Previous investigations have indicated a substantial difference in ET between wetlands and uplands in the Pinelands area. Plot-scale studies using lysimeters concluded that ET from wetland areas in the Pinelands is expected to be greater than ET from upland areas because wetland soils are wetter and water is more readily available for ET [3, 4]. Evaluation of this difference over larger areas is possible by comparing ET measurements collected at the wetland station with those collected at three nearby upland stations operated by the U.S. Forest Service (USFS). The locations of ET measurement stations for the wetland stand and the three upland USFSstands are shownin Fig. 11. The three upland stations are located within 14 km of the wetland station, and fluxes were monitored using eddy-covariance techniques similar to those used at the wetland station.The specific ET measurement methods usedat the three upland sites are described by Clark and others [14–16]. The three upland forest stands are dominated by oak, pine, and mixed oak and pine, respectively. ET was measured at the upland stations during 2005–09, which included periods of disturbance by fireand insect defoliation that had a significant effect in reducing ET. Annual precipitation and ET measured at the three upland stations and the wetland station are listed in Table 4. Clark and others [15] showed that, when averaged across all upland stations for all years of measurement (2005–09), annual ET was 606 mm/yr. The average annual ET measured at the wetland station during 2005–06 (801 mm/yr) is about 32 percent higher than this upland average. When ET at the upland stations is averaged over years without disturbance (685 mm/yr), the average annual wetland ET is about 17 percent higher. As a percentage of precipitation, ET at the wetland station was higher than that of the undisturbed oak and mixed oak/pine upland stations in 2005 and was similar to that of the undisturbed oak and pine upland stations in 2006.

Several factors are likely contributing to differences in ET rates among different stations and among different years. These factors include water availability, dominant plant species, and leaf area. Water availability varies year to year with precipitation and evapotranspiration, and it also varies site to site with depth to the water table and other site conditions. Water availability is less variable in wetlands because the water table is close to or at land surface, whereas in uplands the water-table is deeper. Also, upland soils are more susceptible to drought conditions. Phenological and physiological differences among plant species result in different seasonal patterns of ET and different responses to stress and disturbance in the Pinelands, as described by Clark and others [15].

Leaf area is a function of several site characteristics, including plant species, successional stage, and response to disturbance (including by fireand defoliation). The results presented in Table 5 and Fig. 12 indicate that interannual variability in wetland ETmay be less than that of upland ET because the wetland sites are less susceptible

to periodic drought conditions, disturbance by fire[22],and insect defoliation [29, 72].
Higher ET in wetlands is probably attributable to greater water availability and differences in canopy plant species and leaf area.

A useful approach for validating ET measurements is to use them in a water shed-
scale analysis of the water budget, along with other hydrologic measurements. Ab-
alanced water budget is an indication that the various measurements are internally
consistent. Walker and others [66] describe the water balance in the McDonalds
Branch basin for 2005–2006. The analysis includes an estimate of basin-wide, spatially weighted ET that was based on the respective ET rates for the wetland and upland
sites described previously. The analysis included a land-surface water budget and a
groundwater budget. Both budgets were used to estimate aquifer recharge independently as a residual.The comparability of the independent recharge estimates for the
McDonalds Branch basin indicate that the ET measurements presented in this report
are reasonably consistent with other hydrologic measurements.

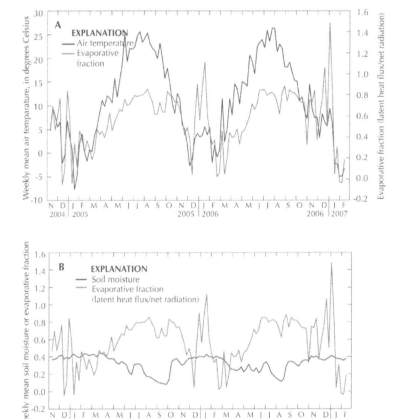

FIGURE 10   Relation of A, weekly mean air temperature and B, weekly mean soil moisture
to evaporative fraction, at the wetland site in the Pinelands area, New Jersey, November
2004-February 2007.

9.3.2   UTILITY OF MODELS TO SIMULATE EVAPOTRANSPIRATION

The utility of three models (Priestley-Taylor, Hargreaves, and North American Regional Reanalysis, NARR) were evaluated through a comparison of model-simulated ET with ET measured at the wetland site using the residual energy-budget variant method. Models that successfully replicated measured ET offer the potential to transfer the results of the site measurements to other locations or to other time periods outside the study period.

9.3.2.1   PRIESTLEY-TAYLOR EQUATION

The environmental variables considered as possible predictors of modified Priestley-Taylor α included soil moisture, solar radiation, air temperature, vapor-pressure deficit, and wind speed. Exploratory data analysis revealed that the greatest explanatory value for the Priestley-Taylor α was related to air temperature, followed by solar radiation, soil moisture, wind speed, and finallyvapor-pressuredeficit. A "Priestley-Taylor α function" composed of a second order polynomial of air temperature (Fig. 13) was optimized to successfully reproduce daily values of latent heat flux measured with the energy- budget residual variant (r^2= 0.90; standard error = 0.65 mm or 19W/m^2;and bias = 0; Table 6).

Comparison of modeled latent heat flux residuals with possible explanatory variables other than air temperature revealed little relation, supporting the use of the simple temperature dependent Priestley-Taylor α function. This function shows an increase in α with air temperature, with the rate of increase decreasing with increasing air temperature. This relation for α probably represents a combination of direct plant stomatal response to air temperature but also a response to phenological changes in the forest plant species that are associated with seasonal temperature changes. A comparison of residuals with vector-averaged daily mean wind direction (Fig. 14) provides a means of evaluating the effects of nonhomogeneous surface covers (wetlands and uplands) within the source area of the latent heat flux measurement. No obvious relation was apparent between ET residuals and wind direction that was consistent with the patterns of wetlands and uplands in the source area, indicating that wind direction, a surrogate for source areas with different vegetation, was not responsible for variability in ET not already accounted for by the temperature relation. Figure 15 shows a good relation between daily measured (energy-budget residual variant) and modified Priestley-Taylor simulated latent heat flux, without noticeable temporal bias. The simulated values oflatent heat flux generally are less erratic in the winter than are the measured values; this phenomenon may be more a consequence of violation of one ofthe assumptions of latent heat flux measurement—negligible changes in canopy heat storage during rapid winter time temperature changes—than an error in the model.

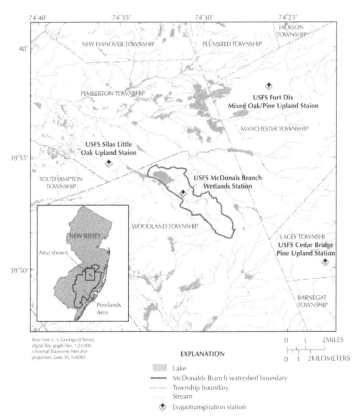

FIGURE 11   Location of the wetland and upland evapotranspiration measurement stations, Pinelands area, New Jersey. (USGS, U.S. Geological Survey; USFS, U.S. Forest Service)

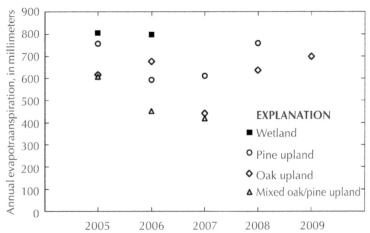

FIGURE 12   Annual evapotranspiration measured at the wetland pine upland, oak upland, and mixed oak/pine upland sites in the Pinelands area, New Jersey, 2005-2009.

TABLE 5 Annual evapotranspiration (ET, mm/year) at the wetland site and three upland forest sites, pinelands area in New Jersey – USA, during 2005-2009. (Data for uplands site from Clark et. al., 2012).

Year	Disturbance, if any	Annual precipitation	Annual ET	Annual ET as a percentage of precipitation
		mm	mm	%
Wetland site, this chapter				
2005	---	1225	805	66
2006	---	1332	797	60
Oak upland				
2005	---	1092	616	56
2006	----	1108	677	61
2007	Complete defoliation	934	442	47
2008	Partial defoliation	936	637	68
2009	---	1173	699	60
Mixed upland				
2005	--	1184	607	51
2006	Burn & defoliation	1163	452	39
2007	Partial defoliation	1135	419	37
Pine upland				
2006	---	1230	757	62
2007	Partial defoliation	1052	593	56
2008	Prescribed burn	1163	611	54
2009	----	1382	759	55

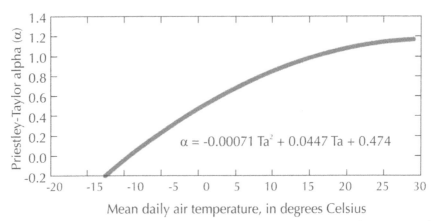

FIGURE 13   Optimized relation between Priestley-Taylor α and mean daily air temperature for the wetland site in the Pinelands area, New Jersey.

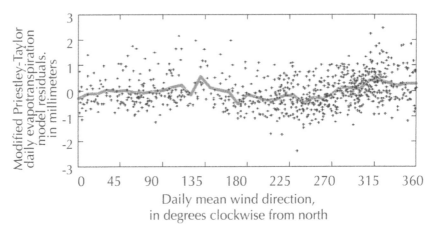

FIGURE 14   Relation of residuals using the modified Priestley-Taylor model to daily mean wind direction, wetland site, Pinelands area, New Jersey, 2004-07. Residual is simulated evapotranspiration minus measured evapotranspiration: The two wind arcs most representative of uplands are 350 degrees clockwise to 20 degrees and 90 degrees clockwise to 120 degrees. The red line is mean residual for a given wind direction.

TABLE 6   Error statistics for evapotranspiration (ET) measured at the wetland site to ET simulated using alternative models, pinelands area in New Jersey – USA.

Alternative model (Parameters are measured in mm)	r^2	RMSE (mm/day)	Bias relative to measured ET (mm/year)
Priestley-Taylor potential evapotranspiration (PET)	0.85	1.08	+216
Hargreaves reference evapotranspiration (RET)	0.87	0.85	+196

TABLE 6   *(Continued)*

Alternative model (Parameters are measured in mm)	r^2	RMSE (mm/day)	Bias relative to measured ET (mm/year)
Modified Priestley – Taylor actual evapotranspiration (AET_{PT})	0.90	0.65	0
Modified Hargreaves actual evapotranspiration (AET_H)	0.89	0.61	0
--- McDonalds Branch wetland site	0.84	0.73	0
--- Indian Mills weather station	0.83	0.76	0
--- Moorestown weather station			
North American Regional Reanalysis (NARR) actual evapotranspiration (AET_{NARR})	0.60	1.21	+80

Note: r^2 = Coefficient of determination with measured actual ET; RMSE = Root mean square error.

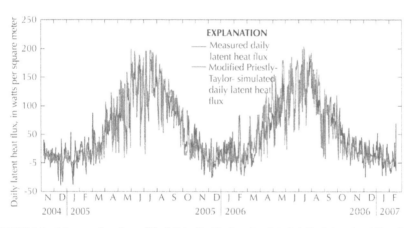

FIGURE 15   Measured and modified Priestly-Taylor-simulated daily latent heat flux for the wetland site, Pinelands area, New Jersey, 2005–06.

The modified Priestley-Taylor model developed in this study is subject to several qualifications. The form of the equation developed for α was empirical rather than physics based and was simply designed to reproduce measured values of ET as accurately as possible. The covariance between environmental variables confounds a unique parameterization. The model was developed for a limited range of environmental conditions, and therefore, extrapolation of the model to conditions not encountered in this study are best done with caution. As noted earlier, deficit soil moisture appeared to play a role in restricting ET during particularly dry periods, but this effect was not considerable enough to be discerned clearly in the identification of the appropriate Priestley-Taylor α function.

Annual measured ET at the wetland site was a relatively constant fraction of potential ET as estimated by the standard Priestley-Taylor method with an α of 1.26 (0.79

and 0.82 in 2005 and 2006, respectively; Table 4). However, potential ET was highly correlated with measured ET (r^2= 0.85), indicating that a constant vegetation factor applied to potential ET also can replicate actual ET rather well at this wetland site, where moisture availability was not often a constraint.

9.3.2.2   HARGREAVES EQUATION

The modified Hargreaves equation performed remarkably well at reproducing measured values of daily ET (Table 6; Fig. 16.) with relatively low error and little temporal bias. Additionally, measurements of daily incoming solar radiation were also well replicated (r^2= 0.70; coefficient of variation = 31 percent; and bias = 3 percent) using the preferred value of K_{RS}= 0.16 for inland areas. The values of a and b identified by the regression analysis to most closely replicate measured daily ET for each source of temperature data are shown in Table 7. As might be expected, use of temperature data from the ET station at the McDonalds Branch site pro vides better explanatory values within the modified Hargreaves equation for ET (r^2= 0.89; standard error = 0.61 mm/d; and bias = 0 mm/yr) than does use of remote temperature data from either of the National Weather Service stations (Indian Mills and Moorestown; Table 6). However, Hargreaves models adjusted for temperature data from either of the remote temperature stations can be considered successful at ET estimation, although the models performed slightly better using values from the closer Indian Mills station (r^2= 0.84; standard error = 0.73 mm/d; and bias = 0 mm/yr) than using values from the more distant Moores town station (r^2= 0.83; standard error = 0.76 mm/d; and bias = 0 mm/yr). The lower values (3.1 to 4.1) of the temperature offset parameter b in the optimized forms of the modified Hargreaves actual ET equation relative to the b value of 17.8 in the standard Hargreaves reference evapotranspiration equation imply that actual ET shows greater sensitivity to temperature at this site than does reference ET. Annual ET was a constant fraction of reference ET, as estimated by the standard Hargreaves method (0.79 in 2005 and 2006; Table 4). However, daily reference ET was highly correlated (r^2 = 0.87) with measured ET, indicating that a constant vegetation factor applied to reference ET can replicate actual ET at the wetland site rather well. Again, no obvious relation was apparent between ET residuals and wind direction that was consistent with the patterns of wetlands and uplands in the source area (Fig. 17).

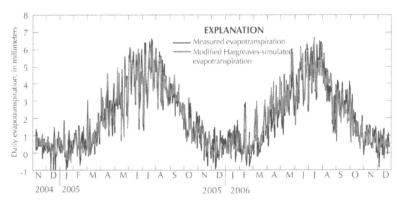

FIGURE 16   Daily measured and modified Hargreaves-simulated evapotranspiration for the wetland site, Pinelands area, New Jersey, 2005–06.

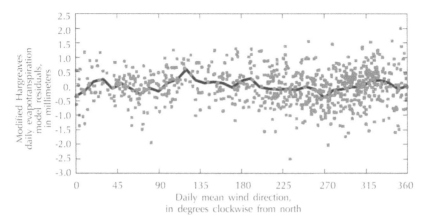

FIGURE 17 Relation of residuals using the modified Hargreaves daily evapotranspiration model to daily mean wind direction, wetland site, Pinelands area, New Jersey, 2004-07. Residual is simulated evapotranspiration minus measured evapotranspiration: The two wind arcs most representative of uplands are 350 degrees clockwise to 20 degrees and 90 degrees clockwise to 120 degrees. The blue line is mean residual for a given wind direction.

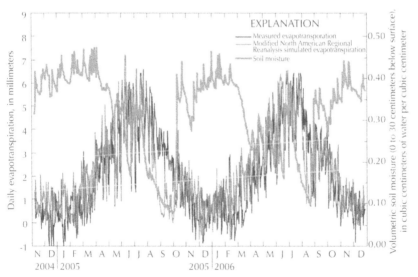

FIGURE 18 Daily measure and modified North American Regional Reanalysis simulated evapotranspiration, and volumetric soil moisture for the wetland site, Pinelands area, New Jersey, 2005-06.

9.3.2.3 NORTH AMERICAN REGIONAL REANALYSIS (NARR)

The North American Regional Reanalysis performed relatively poorly (r^2= 0.60; standard error = 1.21 mm; and bias = 80 mm/yr) at replicating measured values of daily ET relative to the other models considered (Table 6). In particular, NARR showed substantial

under-prediction of measured ET during the dryperiods (Fig. 18) that occurred during
August– October 2005 and in August 2006 and slight over prediction during wetter
periods. Apparently, the NARR algorithms restrict ET as a result of perceived plant
moisture stress during dry periods to a degree that is excessive for the largely wetland environment of the study area. The large spatial resolution of the NARR product
(32-km grid) is intended for a more regional estimate of ET than the relatively small-
scale measurement described in the present study, and this discrepancy in scale can be
expected to account for some of the difference between the two ETestimates.Annual
measured ET totals as a fraction of NARR ET were 0.97 and 0.86 in 2005 and 2006,
respectively (Table 4).

9.3.3   COMPARISON AND LIMITATIONS OF EVAPOTRANSPIRATION MODELS

Of the models investigated in the present study, the modified Hargreaves can be considered the best at replicating the measured daily ET values. The error statistics ofthe modified Hargreaves model were comparable to those of the modified Priestley-Taylor model
and superior to those of the North American Regional Reanalysis product. Additionally, the data requirements ofthe modified Hargreavesmodel (minimum and maximum
daily air temperature) are more easily met than those ofthe modified Priestley-Taylor
model, which requires both mean daily air temperature and the more difficultto obtain
net radiation. The data requirements ofthe modified Hargreaves equation are ideal for
retrospective investigations of ET because they are met by standard National Weather
Service measurements that extend back over a century in parts of the Nation, including at
Moorestown and Indian Mills, New Jersey. Likewise, in the absence of continuing eddy-
covariance measurements, historical or real-time estimates of ET in the study area can be
obtained through use of air-temperature data (available from the National Climatic Data
Center [*http://www.ncdc.noaa.gov/oa/ncdc.html*] or the NJ Weather and Climate Network [*http://climate.rutgers.edu/njwxnet*], for example) and the modified Hargreaves
equation. Sumner and Nicholson [59] presented historical and future probabilistic estimates of ET at the wetland site using a Hargreaves-based approach.

The ET models described in this report are best applied to estimate ET at locations
or during time periods for which the environmental conditions are similar to those
prevailing during the period of record for which these models were calibrated. Use
of the models outside of these environmental conditions introduces additional uncertainty in the ET estimates. For example, under more extreme dry periods than those
that occurred during the study period, ET may be overestimated by models that do not
explicitly account for plant moisture stress.

Based on the results of this research, it can be concluded: Evapotranspiration (ET)
was monitored above a wetland forest canopy in the New Jersey Pinelands during November 10, 2004–February 20, 2007. Meteorological, radiation, and eddy-covariance
flux measurements were made near the top of a 24.5-meter tower; soil moisture and
water- table depth at the site also were monitored. An analysis of the eddy-covariance
sensors' source area and predominant wind directions indicated that the source of
measured ET was primarily pitch pine/cedar wetlands. Three methods were evaluated
for their utility in estimating ET. The standard eddy-covariance method was used to

measure the two turbulent-flux components of the plant-canopy energy budget: latent and sensible heat fluxes. Regression analysis ofmeasured 30-minute turbulent flux and net radiation data was used to estimate missing values of 30-minute turbulent fluxes, which occurred mostly at night and accounted for only 5 percent of the total estimated ET. Two variants of the eddy-covariance method were usedto adjust turbulent flux measurements for daily energy-budget closure; one variant preserves the Bowen ratio (Bowen ratio energy-budget variant), and the other preserves the measured sensible heat flux (residual energy-budget variant). Relations between ET and several environmental variables (incoming solar radiation, air temperature, relative humidity, soil moisture, and net radiation) were explored.

Suspected hygrometer failure during the early part of the measurement period resulted in unreliable ET measurements determined by using the standard eddy-covariance and Bowen ratio energy-budget variant methods.The residual energy- budget variant was selected for use in estimating a time series of daily ET rates for the measurement period. The range of the 12-month ET totals, based on the residual energy-budget variant, is relatively narrow (786 to 821 millimeters (mm)) for the period of record compared to the range of the 12-month rainfall totals (1,124 to 1,452 mm). Minimum and maximum ETvalues were measured during December–February and July, respectively. Net radiation ($r^2= 0.72$) and air temperature ($r^2= 0.73$) were the dominant explanatory variables for daily ET. Air temperature was the dominant control on evaporative fraction with relatively more radiant energy used for ET at higher temperatures. During extended dry periods, soil moisture was shown to limit available energy partitioning into ET. As volumetric soil moisture fell below a threshold of 0.15, the evaporative fraction decreased until rain broke the dry period and the evaporative fraction sharply recovered. This observation indicates that lower water availability can result in lower rates of ET at this wetland site.

Annual ET totals measured at the wetland site were compared with those measured at three nearby upland sites dominated by oak, pine, or mixed oak and pine. A previous investigation by the U.S. Forest Service determined that, when averaged across all upland sites for all years of measurement (2005–2009), annual upland ET was 606 mm/yr. The average annual ET measured at the wetland site during 2005–2006 (801 mm/yr) is about 32 percent higher than the average of that at the upland sites. The average annual ET at the wetland site is about 17 percent higher than ET at the upland sites when averaged over years without disturbance at a particular stand. Factors contributing to differences in ET rates among different sites and among different years include water availability, dominant plant species, and leaf area. Inter-annual variability of wetlands ET may be less than that of uplands ET because the upland sites are more susceptible to periodic drought conditions, disturbance by fire, and insect defoliation.

Three ET models (Priestley-Taylor, modified Hargreaves, and North American Regional Reanalysis) were evaluated to determine their utility in predicting ET at the wetland site using data that may be more readily available in other areas and for other time periods. Of the three models, the modified Hargreaves may be of the most practical use,asit replicated the measured daily ET values reasonably well, and data requirements were relatively easily met.The ET models described in this report are best applied to estimate ET at locations or during time periods for which the environmental conditions are similar to those prevailing during the period of record for which these models were calibrated. Precipitation during the study period at the nearby National-Weather Service Indian Mills weather station was slightly higher than the long- term (1902–2011) annual mean of 1,173 mm, with 1,325 and 1,396 mm of precipitation in 2005 and 2006, respectively.

9.4 SUMMARY

Evapotranspiration (ET) was monitored above a wetland forest canopy dominated by pitch-pine in the New Jersey Pinelands during November 10, 2004–February 20, 2007, using an eddy-covariance method. Twelve-month ET totals ranged from 786 to 821 millimeters (mm). Minimum and maximum ET rates occurred during December–February and in July, respectively. Relations between ET and several environmental variables (incoming solar radiation, air temperature, relative humidity, soil moisture, and net radiation) were explored. Net radiation (r^2= 0.72) and air temperature (r^2= 0.73) were the dominant explanatory variables for daily ET. Air temperature was the dominant control on evaporative fraction with relatively more radiant energy used for ET at higher temperatures. Soil moisture was shown to limit ET during extended dry periods. With volumetric soil moisture below a threshold of about 0.15, the evaporative fraction decreased until rain ended the dry period, and the evaporative fraction sharply recovered. A modified Hargreaves ET model, requiring only easily obtainable daily temperature data, was shown to be effective at simulating measured ET values and has the potential for estimating historical or real-time ET at the wetland site. The average annual ET measured at the wetland site during 2005–2006 (801 mm/yr) is about 32 percent higher than previously reported ET for three nearby upland sites during 2005–2009. Periodic disturbance by fireand insect defoliation at the upland sites reduced ET. When only undisturbed periods were considered, the wetland ET was 17 percent higher than the undisturbed upland ET. Inter-annual variability in wetlands ETmay be lower than that of uplands ET because the upland stands are more susceptible to periodic drought conditions, disturbance by fire,and insect defoliation. Precipitation during the study period at the nearby Indian Mills weather station was slightly higher than the long-term (1902–2011) annual mean of 1,173 millimeters (mm), with 1,325 and 1,396 mm of precipitation in 2005 and 2006, respectively. The methods presented in this chapter can be used for other locations in the world.

KEYWORDS

- **aerodynamic resistance**
- **anemometer**
- **aquifer**
- **atmospheric pressure**
- **big leaf assumption**
- **Bowen ratio**
- **Bowen ratio energy-budget variant (of eddy covariance method)**
- **cypress**

ACKNOWLEDGMENTS

Access to the evapotranspiration (ET) monitoring site was arranged by Christian Bethman, Superintendent of Brendan T. Byrne State Forest, whose assistance in clearing the site and providing road improvements to this previously inaccessible area is gratefully acknowledged. The authors gratefully acknowledge U.S. Geological Survey (USGS) colleagues Amanda Garcia and Amy Swancar for providing helpful comments that greatly improved this report, and William Ellis, Ruth Larkins and Gregory Simpson for their assistance with report preparation and illustrations. Special thanks are extendedto Richard W. Edwards, USGS ElectronicsTechnician,whose technical organization, skill, and resourcefulness ensured a successful ET monitoring campaign at this remote location.

REFERENCES

1. Allen, R. G.; Walter, I. A.; Elliot, R.; Howell, T.; Itenfisu, D.;Jensen, M.; **2005,** The ASCE standardized reference evapotranspiration equation.Reston, Va.; American Society of Civil Engineers, 59 p.
2. Ballard, J. T.; **1971,** Evapotranspiration from lowland vegetation communities of the New Jersey Pine Barrens: New Brunswick, N. J.; Ph.D.; Thesis, Rutgers University.
3. Ballard, J. T.; **1979,** Fluxes of water and energy through the Pine Barrens ecosystems, in Forman, R. T. T.; ed.; Pine Barrens: ecosystem and landscape: New York, Academic Press, revised paperback edition, 133–145.
4. Ballard, J. T.;Buell, M. F.;The role of lowland vegetation communities in the evapotranspiration budget of the New Jersey Pine Barrens: Bulletin of the New Jersey Academy of Science, **1975,** *20(1),* 26–28.
5. Baldocchi, D. D.; Hicks, B. B.;Meyers, T. P.; Measuring biosphere – atmosphere exchanges of biologically related gases with micrometeorological methods. Ecology, **1988,** *69(5),* 1331–1340.
6. Barksdale, H. C.;Ground-water resources in the tri-state region adjacent to the Lower Delaware River: Special Report, New Jersey State Water Policy Commission, State of New Jersey Department of Conservation and Economic Development, Division of Water Policy and Supply, **1958,** *13(10),* 190 p.
7. Bidlake, W. R.; Woodham, W. M.;Lopez, M. A.;Evapotranspiration from areas of native vegetation in west-central Florida: U. S.; Geological Survey Water-Supply Paper2430, **1996,** 35 p.
8. Buell, M. F.;New botanical problems in the New Jersey Pine Barrens: Bulletin of the Torrey Botanical Club, **1955,** v. *82,*no. *3,*p. 237–252.
9. Buell, M. F.;Ballard, J. T.; **1972,** Evapotranspiration from lowland vegetation in the New Jersey Pine Barrens: New Brunswick, N. J.; New Jersey Water Resources Research Institute, Rutgers University (pagination unavailable).
10. Brutsaert, W.; **1982,** Evaporation into the atmosphere–Theory, history, and applications: Boston, Kluwer Academic Publishers, 299 p.
11. Businger, J. A.;Yaglom, A. M.;Introduction to Obukhov's paper "Turbulence in an atmosphere with a non uniform temperature": Boundary-Layer Meteorology, **1971,** v. *2,*p. 3–6.
12. Campbell, G. S.; Norman, J. M.;An introduction to environmental biophysics: New York, Springer, **1998,** 286 p.
13. City of Philadelphia, **1892,** Ninth annual report of the Bureau of Water for the year ending December *31,*1891, Philadelphia, PA.
14. Clark, K. L.; Skowronski, N.;Hom, J.; Invasive insects impact forest carbon dynamics: Global Change Biology, **2010,** *16,* 88–101.
15. Clark, K. L.; Skowronski, N.; Gallagher, M.; Renninger, H.;Schäfer, K.;Effects of invasive insects and fire on forest energy exchange and evapotranspiration in the New Jersey Pinelands: Agric. For. Meteorol.;**2012,** v. 166–167, p. 50–61.

16. Clark, K. L.; Skowronski, N.; Gallagher, M.; Schafer, K. V. R, Renninger, H. J.; **2011,** Effects of invasive insects and fire on forest evapotranspiration and water use efficiency [abs.]: 96th Annual Meeting of the Ecological Society of America, Austin, TX, August 7–12, 2011.

17. Dyer, A. J.;Measurements of evaporation and heat transfer in the lower atmosphere by an automatic eddy-correlation technique: Quarterly Journal of the Royal Meteorological Society, **1961,** v. *87,*p. 401–412.

18. Eichinger, W. E.; Parlange, M. B.;Stricker, H.;On the concept of equilibrium evaporation and the value of the Priestley-Taylor coefficient: Water Resources Research, **1996,** v. *32,*no. *1,*p. 161–164.

19. Fleagle, R. G.;Businger, J. A.; **1980,** An introduction to atmospheric physics: New York, Academic Press, 432 p.

20. Flint, A. L.;Childs, S. W.;Use of the Priestley-Taylor evaporation equation for soil water limited conditions in a small forest clearcut: Agric. For. Meteorol.;**1991,** v. *56,*p. 247–260.

21. Foken, T.;The energy balance closure problem: an over view: Ecol. Appl.;**2008,** v. *18,*no. *6,*p. 1351–1367.

22. Foreman, R. T. T.;Boerner, R. E.;Fire frequency and the pine barrens of New Jersey: Bulletin of the Torrey Botanical Club, **1981,** *108(1),* 34–50.

23. Garratt, J. R.;Surface influence upon vertical profiles in the atmospheric near-surface layer: Quarterly Journal of the Royal Meteorological Society, **1980,** *106,* 803–819.

24. German, E. R.; **2000,** Regional evaluation of evapotranspiration in the Everglades: US Geological Survey Water Resources Investigations Report 00-4217, 48 p.

25. Gordon, A. D.; **2004,** Hydrology of the unconfined Kirkwood-Cohansey aquifer system, Forked River and Cedar, Oyster, Mill, Westecunk, and Tuckerton Creek basins and adjacent basins in the southern Ocean County area, New Jersey, 1998–1999,U. S.; Geological Survey Water-Resources Investigations Report 034337, 5 pp.

26. Goulden, M. L.; Munger, J. W.; Fan, S-M, Daube, B. C.;Wofsy, S. C.;Measurements of carbon sequestration by long-term eddy covariance: methods and a critical evaluation of accuracy: Global Change Biology, **1996,** *2,* 169–182.

27. Hargreaves, G. H.;Allen, R. G.;History and evaluation of Hargreaves evapotranspiration equation: American Society of Civil Engineers Journal of Irrigation and Drainage Engineering, **2003,** *129(1),* 53–63.

28. Hargreaves, G. H.;Samani, Z. A.;Estimating potential evapotranspiration: American Society of Civil Engineers Journal of Irrigation and Drainage Engineering, **1985,** *108(3),* 225–230.

29. Houston, D. R.;Valentine, H. T.;Classifying forest susceptibility to gypsy moth defoliation: US Department of Agriculture, Agriculture Handbook No.542, **1985,** 19 p.

30. Hutson, S. S.; Barber, N. L.; Kenny, J. F.; Linsey, K. S.; Lumia, D. S.; Maupin, M. A.; **2004,** Estimated use of water in the United States in 2000. U. S.; Geological Survey Circular1268,46 p.

31. Johnson, M. L.;Watt, M. K.; **1996,** Hydrology of the unconfined aquifer system, Mullica River Basin, New Jersey, 1991–1992, U. S.; Geological Survey Water-Resources Investigations Report 94-4234, 6 sheets.

32. Johnsson, P. A.;Barringer, J. L.; **1993,** Water quality and hydrogeochemical processes in Mc-Donalds Branch basin, New Jersey Pinelands, 1984–*88,* U. S.; Geological Survey Water-Resources Investigations Report 91–4081, 111 p.

33. Kaimal, J. C.;Businger, J. A.; **1963,** A continuous wave sonic anemometer-thermometer: Journal of Applied Meteo- rology, v. *2,*p. 156–164.

34. Kaimal, J. C.;Gaynor, J. E.;Another look at sonic thermometry: Boundary-Layer Meteorology, **1991,** v. *56,*p. 401–410.

35. Laidig, K. J.; Zampella, R. A.; Brown, A. M.;Procopio, N. A.; Development of vegetation models to predict the potential effect of groundwater withdrawals on forested wetlands: New Lisbon, N. J.; Pinelands Commission, **2010a,** 30 p.

36. Laidig, K. J.; Zampella, R. A.; Brown, A. M.;Procopio, N. A.; Development of vegetation models to predict the potential effect of groundwater withdrawals on forested wetlands: Wetlands, **2010b,** v. *30,*p.489–500.

37. Lee, X.;Black, T. A.;Atmospheric turbulence within and above a Douglas-fir stand, Part II: Eddy fluxes of sen- sible heat and water vapor: Boundary-Layer Meteorology, **1993,** v. *64,*p. 369–389.

38. Lowe, P. R.;An approximating polynomial for the com- putation of saturation vapor pressure: Journal of Applied Meteorology, **1977,** v. *16,*no. *1,*p. 100–103.

39. Lull, H. W.;Axley, J. H.;Forest soil-moisture rela- tions in the coastal plain sands of southern New Jersey: For. Sci.;**1958,** v. *4,*no. *1,*p. 2–18.

40. Mitchell, K.; Ek, M.; Lin, Y.; Mesinger, F.; DiMego, G.; Shaf- ran, P.; Jovic, D.; Ebisuzaki, W.; Shi, W.; Fan, Y.; Janowiak, J.;Schaake, J.;NCEP completes 25-year North American Reanalysis: precipitation assimilation and land surface are two hallmarks: Global Energy and Water Cycle Experiment (GEWEX) News, **2004,** v. *14,*no. *2,*p. 9–12.

41. Modica, E. **1998,** Analytical methods, numerical modeling, and monitoring strategies for evaluating the effects of ground-water withdrawals on unconfined aquifers in the New Jersey Coastal Plain: US Geological Survey Water- Resources Investigations Report 98-4003, 66 p.

42. Monteith, J. L.; **1965,** Evaporation and environment, in Fogg, G. E.; ed.; the state and movement of water in living organ- isms, Symposium of the Society of Experimental Biology, San Diego, California: New York, Academic Press, p. 205–234.

43. Monteith, J. L.;Unsworth, M. H.;*Principles of Environmental Physics*(2d ed.): London, Edward Arnold, **1990,** 291 p.

44. Moore, C. J.;Eddy flux measurements above a pine forest: Quarterly Journal of the Royal Meteorological Society, **1976,** v. *102,* p. 913–918.

45. National Climatic Data Center, **2012,** National Oceanographic and Atmospheric Administration(NOAA) Satellite and Information Service–National Environmental Satellite, Data and Information Service (NESDIS), accessed May22, **2012,** at http: //www.ncdc.noaa.gov/ oa/ncdc.html.

46. North American Regional Reanalysis, **2012,** North American Regional Reanalysis (NARR) Homepage, accessed May 15,**2012,** at http: //www.emc.ncep.noaa.gov/mmb/rreanl/.

47. Penman, H. L.;Natural evaporation from open water, bare soil, and grass: Proceedings of the Royal Society of London, Series, A.; **1948,** v. *193,* p. 120–146.

48. Pinelands Commission, **2003,** Kirkwood-Cohansey Project Work Plan: New Lisbon, N. J. : New Jersey Pinelands Commission, 37 p.

49. Priestley, C. H. B.; and Taylor, R. J.;On the assessment of surface heat flux and evaporation using large-scale parameters: Monthly Weather Review, **1972,** v. *100,* p. 81–92.

50. Rhodehamel, E. C.; **1970,** A hydrologic analysis of the New Jersey Pine Barrens region: New Jersey Division of Water Policy Water Resources Circular *22,*35 p.

51. Rhodehamel, E. C.Hydrology of the Pine Barrens of New Jersey, in Forman, R. T. T.; ed.; Pine Barrens: ecosystem and landscape: New York, Academic Press, revised paperback edition, **1979,** p. 147–167.

52. Schafer, V. R.;Canopy stomatal conductance following drought, disturbance, and death in an upland oak/pine forest of the New Jersey Pine Barrens, USA: Frontiers in Plant Science, **2011,** v. *2,*article *15,*7 p.

53. Schotanus, P.; Nieuwstadt, F. T. M.;de Bruin, H. A. R.;Temperature measurement with a sonic anemometer and its application to heat and moisture fluxes: Boundary- Layer Meteorology, **1983,** v. *50,*p. 81–93.

54. Schuepp, P. H.; Leclerc, M. Y.; MacPherson, J. I.;Desjar- dins, R. L.;Footprint prediction of scalar fluxes from analytical solutions of the diffusion equation: Boundary-Layer Meteorology **1990,** v. *50,*p. 355–373.

55. Stannard, D. I.;Comparison of Penman-Monteith, Shuttleworth-Wallace, and modified Priestley-Taylor evapotranspiration models for wildland vegetation in semiarid rangeland: Water Resources Research, **1993,** v. *29,*no. *5,*p. 1379–1392.

56. Stannard, D. I.;Interpretation of surface flux measurements in heterogeneous terrain during the Monsoon '90 experiment: Water Resources Research, **1994,** v. *30,* no. *5,*p. 1227–1239.
57. Stull, R. B.; **1988,** *An Introduction to Boundary Layer Meteorology.*Boston, Kluwer Academic Publishers, 666 p.
58. Sumner, D. M.; **1996,** Evapotranspiration from successional vegetation in a deforested area of the Lake Wales Ridge, Florida: U. S.; Geological Survey Water-Resources Investigations Report 96-4244, 38 p.
59. Sumner, D. M.;Nicholson, R. S.; **2010,** Estimation of historical and future evapotranspiration: Proceedings of the EWRI of ASCE – 3rd International Perspective on Current & Future State of Water Resources & the Environment, January **2010,** Chennai, India.
60. Tanner, B. D.;Greene, J. P.; **1989,** Measurement of sensible heat and water vapor fluxes using eddy correlation methods: Final report prepared for U. S.; Army Dugway Proving Grounds: Dugway, Utah, 17 p.
61. Tanner, B. D.; Swiatek, E.;Greene, J. P.; **1993,** Density fluctuations and use of the krypton hygrometer in surface flux measurements: Management of irrigation and drainage systems, Irrigation and Drainage Division: Park City, Utah, American Society of Civil Engineers, July 21–23, **1993,** p. 945–952.
62. Tanner, C. B.;Thurtell, G. W.;Anemoclinometer measurements of Reynolds stress and heat transport in the atmospheric boundary layer: Fort Huachuca, Ariz.; United States Army Electronics Command, Atmospheric Sciences Laboratory, TR ECOM 66-G22-F, Reports Control Symbol OSD-1366, April **1969,** 10 p.
63. Twine, T. E.; Kustas, W. P.; Norman, J. M.; Cook, D. R.; Houser, P. R.; Meyers, T. P.; Prueger, J. H.; Starks, P. J.; and Wesely, M. L.;Correcting eddy-covariance flux underestimates over a grassland: Agric. For. Meteorol.;**2000,** v. *103,* p. 279–300.
64. Vermeule, C. C.; **1894,** Report on water-supply, water power, the flow of streams and attendant phenomena: Final report of the State geologist, v. III: Trenton, N. J.; Geological Survey of New Jersey, 352 p.
65. Vowinkel, E. F.;Foster, W. K.; **1981,** Hydrogeologic conditions in the Coastal Plain of New Jersey: U. S.; Geological Survey Open–File Report 81-405, 39 p.
66. Walker, R. L.; Nicholson, R. S.;Storck, D. A.; **2011,** Hydrologic assessment of three drainage basins in the Pinelands of southern New Jersey, 2004–06, U. S.; Geological Survey Scientific Investigations Report 2011-5056, 145 p.
67. Watt, M. K.;Johnson, M. L.; **1992,** Water resources of the unconfined aquifer system of the Great Egg Harbor River Basin, New Jersey, 1989–1990, U. S.; Geological Survey Water-Resources Investigations Report 91-4126, 5 pls.
68. Watt, M. K.; Johnson, M. L.;Lacombe, P. J.; **1994,** Hydrology of the unconfined aquifer system, Toms River, Metedeconk River, and Kettle Creek Basins, New Jersey, 1987–*90,* U. S.; Geological Survey Water-Resources Investigations Report 93-4110, 5 pls.
69. Watt, M. K.; Kane, A. C.; Charles, E. G.; Storck, D. A.; **2003,** Hydrology of the unconfined aquifer system, Rancocas Creek area: Rancocas, Crosswicks, Assunpink, Assiscunk, Blacks, and Crafts Creek Basins, New Jersey: U. S.; Geological Survey Water-Resources Investigations Report 02–4280, 5sheets.
70. Webb, E. K.; Pearman, G. I.;Leuning, R.;Correction of flux measurements for density effects due to heat and water vapor transfer: Quarterly Journal of the Royal Meteorological Society, **1980,** *106,* p. 85–100.
71. Weeks, E. P.; Weaver, H. L.; Campbell, G. S.;Tanner, B. D.; **1987,** Water use by saltcedar and by replacement vegetation in the Pecos River floodplain between Acme and Artesia, New Mexico: US Geological Survey Professional Paper 491-G, 37 p.
72. Whitmire, S. L.;Tobin, P. C.;Persistence of invading gypsy moth populations in the United States: Oecologia,**2006,** *147,* 230–237.
73. Wood, O. M.;The interception of precipitation in an oak- pine forest: Ecology, **1937,** *18(2),* 251–254.

CHAPTER 10

WATER MANAGEMENT IN CITRUS[1]

P. S. SHIRGURE and A. K. SRIVASTAVA

CONTENTS

---

[1] Published with permission and Modified from: P. S. Shirgure, A. K. Srivastava and Shyam Singh, *Water Management in Citrus – A Review.* Agricultural Reviews, **2000,** *21,* 223–230. © Copyright Agricultural Reviews 2000.

10.1 INTRODUCTION

Water is an important natural resource, a basic human need and of vital requirement for all developmental activities. The demand of water is increasing with increase in population and economic activities. Irrigation has been practiced in India since long ago. It is considered as a very important input for agriculture and hence, continuous development has been taking in this field through the centuries. It is a means to mitigate the impact of irregular, uneven and inadequate or wide fluctuations in rainfall from year to year. India's annual rainfall is 117 cm and most of it occurs during monsoon. The irrigation potential in India has risen from 19.5 m-ha in 1950 to 67.89 m-ha in 1985.

The best estimates available indicate that the maximum amount of exploitable irrigation potential by all types of irrigation is 113.5 m-ha. This could be sufficient for 50% of the total cultivable area of the country and 50% of the area, would be left completely dependent on rainfed farming. Conjunctive use of rain and irrigation water offers scope for optimizing water use in areas having problems of surface drainage during, rainy season and water scaricity during the rest of the year.

Citrus is the third largest fruit crop grown in an area of 234570 ha. Nagpur mandarin and acid lime occupies 40% and 25% of the total area under citrus cultivation in the country. The large scale drying of the citrus orchards is mainly due to scarce water resources, frequent drought and lowered water table in mandarin growing areas of Vidarbha (Maharashtra) and Central India [11].

The average yield of these orchards is 7 to 8 t/ha, which is 3 to 4 times less than other citrus producing countries of the world. Citrus plants are more extracting in their demand for irrigation. Direct contact of water with the trunk adversely affects the trees growth. Citrus being an evergreen fruit crop use moisture constantly throughout the year of course at a much slower rate during winter and faster in summer. There is a good amount of research available on irrigation water management of citrus from abroad but a little work has been done under Indian conditions.

There is a need for carrying out the research on estimating the water requirements of the Nagpur mandarin and acid lime under subtropical conditions of the Central India. The use of microirrigation systems is gaining popularity among the citrus growers and it is necessary to standardize the best system for the citrus orchards. The mosisture conservation techniques like mulching and fertigation are also equally important for water and fertilizer conservation point. So, the research in this regard is also required to be carried out for optimizing the productivity and efficient use of inputs including water.

10.1.1 IRRIGATION SCHEDULING AND WATER REQUIREMENT IN CITRUS

The literature on irrigation methods, irrigation systems, scheduling, water requirements and fertigation in citrus in International and under Indian conditions is reviewed. The literature cited related to water requirement and irrigation scheduling in Citrus is reviewed. The growth of 'Valencia' oranges slowed down at 32 cb and 55 cb soil suctions at 30 cm depth in light and medium textured soil, respectively [22]. The preliminary studies on the effect of soil management system on soil moisture in

Sweet orange orchard was initiated by Randhawa et al. [47]. Stolzy et al., [76] found that the treatments irrigated at 20 Kpa tensiometer readings were best as compared to calender schedule. Hashemi and Gerber [19] attempted correlation between actual evapotranspiration (AET) and potential evapotranspiration computed with Penman's model. Koo [25] advized Florida citrus growers to maintain soil moisture at 55 to 65% of field capacity from bloom the young fruit exceeds 1 inch in diameter. Retiz [51] estimated the water requirement of citrus at 40–45 inch/year. Richards and Warnke [53] studied the irrigation systems to lemon and irrigation at 60 cb and extrapolations to 150 cb resulted in no measured differential response in tree growth and fruit yield under coastal conditions. Leyden [28] found that 610 mm irrigation water applied via a drip system at 0, 200, 300 and 400 liters/tree gave the significant difference in total yield and fruit size distribution.

Toledo et al., [79] found that irrigation at 65% field capacity caused drought injury symptoms, excessive defoliation and less water consumption. Best results were obtained with irrigation at 85% filed capacity. Evapotranspiration ranged form 3.78 to 4.42 and 1.46 to 1.3 mm/day for 85% and 65% field capacity irrigations respectively. Kelin [24] compared drip irrigation scheduling according to soil water potential to class A pan evaporation in different horticultural crops using a crop factor and concluded that 12 to 23% water could be conserved by using the irrigation scheduling based on soil water potentials. Moreshet et al., [34] compared the 100% and 40% of soil volume irrigation in 'Shamouti' orange and found that partially irrigated plot was 66% of that of the fully irrigated plot one.

Transpiration from the trees of partially irrigated plots was 72% of that of the fully irrigated plot and the evaporation from the soil surface was 58%. Fruit TSS and acid contents were higher in partially irrigated plots. Smajstrla et al., [71] found that greatest yields were obtained using spray-jet trickle irrigation. Yield increases were not linear with volume of rootzone irrigated but ranged from 39% for the drip irrigation treatments which irrigated 5–10% of the area beneath the tree canopies to 64% for 2-spray jet per tree, which irrigated 50.7% of the areas beneath the tree canopies.

Plessis [38] obtained the highest yields (190 kg/tree) and the largest average fruit size with irrigation at a crop factor of 0.9 on a 3 day cycle, with thin consumption microirrigation gave better results than drip irrigation. Makhija et al., [31] obtained water need for 6 year old Kinnow mandarin varying from 539 to 1276 mm depending upon the level of irrigation with average consumptive use of water in 2 years as 61.5 cm. Smajstrla et al., [73] concluded that the tree growth of young 'Valencia' orange was greatest when irrigations were scheduled at 20 centibar for no-grass and 40 centibar for the grass treatments. Randhawa and Srivastava [48] emphasized on irrigation aspects in Citriculture in India. Autkar et al., [1] studied the distribution of active roots of Nagpur mandarin as it can be useful in planning irrigation nutrition, planting density and drainage management. The root depth and radial extent for trees aged 1–4 years was 7.5–8.0 cm deep and 5–12.5 cm respectively and for 10 years old age tree it was 2–3 m and 80–90 cm. Barbera and Carimin [4] studied the different levels of water stress on yield and quality of lemon tree and found that yield was lower in most stressed plot. The number of flower/m^3 of canopy was higher in most stressed treatment indicating a relationship between severity of stress and flowering response. Mageed et al., [30] carried the research on influence of irrigation and nitrogen on water use and growth of Kinnow mandarin receiving 4 levels of irrigation and three levels of Nitrogen (0, 115 or 230 Kg N./tree). The consumptive use varied from 66.7 cm to132.5 cm. Moreshet et al., [35] studied on water use and yield of a mature 'Shamouti' orange orchard submitted to the root volume restriction and intensive canopy pruning.

Du Plessis [40] with a mature 'Valencia' orange trees and field experiment shown that the water use pattern over the entire season reaching a maximum of 87 lit/day in January. Highest net income was obtained with tensiometer scheduling. He [1989] also demonstrated that 690 liters irrigation when tensiometer-reaching fell to −50kpa gave the highest net income. Use of tensiometer rather than evaporation pan scheduling could save 2000 m^3 water/ha annually. The water requirement of citrus plants varies with species, season and age governed with different climatic conditions. Plant growth retards below certain critical level of available moisture depending upon soil type, climatic factor and plant genetic make up [46]. Autkar et al., [2] also studied the effect of Pan evaporation, canopy size and tree age on daily irrigation water requirement of 1–5, 5–8 and above 8 years old Nagpur mandarin trees over 9 months [October–June] and concluded that the requirement rose with age. Ghadekar et al., [17] estimated that the consumptive use of Nagpur mandarin by modified Penman equation using 40 years air temperature, relative humidity, wind velocity and Solar relation data. Under clean cultivation the water requirement of young, middle age and mature trees was 651.9, 849.0 and 997.3 mm/year, respectively. An equation for daily water use was proposed and it can be used for drip irrigation. Sanehez et al., [57] compared five flood irrigation treatments with daily drip irrigation at 0.475 Epan and concluded that the drip irrigation gave higher yields as compared to flood-irrigated plants. Castel and Buj [8] carried out trials on mature 'Satstuma' average trees grafted on Sour orange rootstocks. Plants were irrigated with 60% of the estimated ET losses from a class A pan and 80% of the control throughout the year. Irrigation treatments affected both yield and fruit quality.

Ray et al., [49] studied the response of young 'Kinnow' mandarin to irrigation. Irrigations were scheduled at −0.05, −0.1, −0.2, −0.4 and −0.8 MPa soil water potential 0.8 IW/CPE ratio and irrigation to replenish estimated crop ET. The water use increased as the frequency of irrigation increased as the frequency of irrigation increased. The highest bio-mass per plant was obtained when irrigation was scheduled at −0.05 MPa soil water potential (SWP] and 18–19 irrigations were required. The best tree growth in terms of trunk diameter, plant height, canopy volume, leaf number and shoot growth was also obtained at −0.05 MPa SWP using 182.4 cm water/tree/annum. He also studied [50] the effect of irrigation on plant water status and stomatal resistance in young Kinnow mandarin and found that the leaf water potential (LWP] and Relative water content [RWC] declined considerably with reduction in soil moisture in rootzone due to differential irrigation schedules. Reduction in RWC was more conspicuous where soil moisture dropped below 11% LWP measurements in early morning hours showed a significant curvilinear relationship with soil water status. Leaf stomatal values were lowest in September and highest in January. Shirgure et al., [60] initiated the irrigation scheduling based on depletion of available water content and fraction of open pan evaporation in acid lime in prebearing stage. He studied [68] the effect of different soil moisture regimes with irrigation scheduling based on available soil moisture depletion and open pan evaporation on soil moisture distribution and evapotranspiration in acid lime and it was concluded that the evapotranspiration varied from 213.6 mm to 875.6 mm in various irrigation schedules. It was also found that the change in soil-moisture distribution in the rootzone of acid lime plants varied from 195.9 mm to 321.3 mm with different irrigation schedules.

10.1.2 IRRIGATION METHODS AND DRIP IRRIGATION SYSTEMS IN CITRUS

The common methods of applying water to the orchards are basin, border strip, furrow, sprinkler and drip irrigation. Ring basin is generally followed in early establishment phase of fruit trees. Micro-irrigation to citrus is common in developed countries. Drip and microjet irrigation has the advantage over surface irrigation methods, for more uniform and complete wetting of the soil surface and adoption on sloppy terrain.

Faton [12] observed better tree growth and yield, less weed growth, evaporation and leaching with 16 gallon water applied through drip to each 4 years old lime trees at two weeks interval compared to 320 gallon water in flood irrigation. Fritz [15] observed that all applied water is transferred directly to rootzone of plants and 20–50% water saving is reported depending on soil and climate. Raciti and Sckderi [44] compared drip irrigation with the basin and found that the fruits under drip system ware more acid and lower maturity ratio. Ronday et al., [55] observed better tree growth and less water consumption in Valencia orange under drip irrigation in sandy soil Sucderi and Raciti [58] compared basin irrigation with different combination of drip irrigation and measured number, weight, quality of fruits in Valencia orange. He also studied miconutrient levels in leaves, annual trunk increments. Drip irrigation gave the higher yields. Simpson [69] found that there is a shift from furrow irrigation and overhead sprinkler irrigation systems to under tree systems like microjets. Slack et al., [70] demonstrated that trickle irrigation on young orange trees used 5,400 liters of water compared to 23,400 liters of water per tree for dragline.

Raciti and Barbargallo [45] found that the yields of lemon were more with localized irrigation amounting to 227.23 q/h and 213.2 q/h for basin irrigation. Ozsan et al., [37] compared furrow, under tree, over tree and drip irrigation in lemons. amounts of water applied were greatest (1,286 mm) with under tree method and least (207 mm) with drip irrigation system. Yield was more with over tree sprinkling and least with furrow. Water use efficiency was high in drip irrigation. Cevik and Yazar [9] demonstrated that a new irrigation system i.e., Bubbler irrigation for the orchards. He observed that under tree sprinkling and drip irrigation had the best pomological effects. Amounts of water applied per tree for over sprinkling, under sprinkling and drip irrigation were 22.01, 17.04 and 10.33 m^3/season. Pyle [43] appraized the use of microirrigation in Citrus especially drip irrigation. Except the higher cost the advantages includes saving in labor, water and power, better orchard uniformity and immediate response to crop need, better soil-water relationships, rooting environment and better yield and quality.

Tash be kov et al., [78] studied different irrigation methods. Drip irrigation and under tree sprinkling produced the highest yield with the least water requirements.

The application rate for drip irrigation of 4 years old lemon trees was 7400 m³/ha annually. Capra and Nicosia [7] studied flooding, sprinkler, and subirrigation with sprays and concluded that the rates of water application affects the rate of growth of fruit diameter. Robinson and Alberts [54] compared under canopy sprinkler and drip irrigation systems in crop like Banana and found that the drip irrigation is superior to under canopy sprinklers. Increased tree growth and yield were recorded in young Valencia orange under drip irrigation method with emitter placed at distance of 1 meter from the trunk [3]. Greive [18] concluded that under tree microsprinklers increased yield by 12% and reduced water application by 9.3% compared to conventional full ground cover. Interligolo and Raciti [23] demonstrated that water saving with subsurface irrigation was 32% over the traditional basin irrigation. The yield was higher but fruit quality was not much different. Marler and Davies [32] studied the effect of microsprinkler irrigation scheduling on growth of young Hamlin orange trees and found that growth was not affected by pattern of irrigation, suggesting that 90% emitters are enough for root system. Zekri and Parsons [80] studied drip, microsprinkler and overhead sprinkler irrigation at two water application rate and found that fruit size and tree canopy area were 9 to 20% greater in the overhead sprinkler treatments. Marler and Davies [33] studied the growth response of microirrigation on growth of young Hamlin orange and found that more than 90% of root dry weight was within 80 cm of the trunk at the end of first growing season.

Rumayor et al., [56] studied three irrigation systems (drip, microsprinkler and flooding) and found that yields were higher for sprinkler-irrigated trees and the fruits were smaller in flood irrigation. Smajstrala [74] researched on microirrigation for citrus production in Florida. Gangwar et al., [16] studied the economics of investment on adoption of drip irrigation system in Nagpur mandarin orchards in Central India and concluded that the drip irrigation system is technically feasible and economically viable with Benefit to Cost ratio as 2.07. Shirgure et al., [64] initiated the research work on evaluation of microirrigation systems in acid lime and a comparison was done with that of basin [ring] method of irrigation. Shirgure et al., [66] studied the effect of dripper 8 liters per hour microjet 300°, microjet 180° and basin irrigation method on water use and growth of acid lime and found that microjet 300° recorded higher growth than rest of the systems. He also studied [67] the efficacy of these microirrigation systems and basin irrigation on fruit quality and soil fertility changes in acid lime.

10.1.3   FERTIGATION IN CITRUS

Fertigation in application of liquid or water-soluble solid fertilizer along with irrigation trough the drip irrigation to the plants. It has many advantages like increasing fertilizer-use efficiency, ensured supply of water and nutrients, labor saving and improvement in yield and quality. It is a very new under Indian conditions but getting popular along with adoption of drip irrigation system.

The research related to injection of fertilizers through the drip irrigation systems was started during 1979 by Smith et al. [75]. Koo [26] appraised the potential advantage of microirrigation systems and its usefulness to fertigation. Bielorai et al., [6] advocated use of fertigation technology in citrus as it resulted in higher production of good quality Shamouti oranges. He compared N. fertigation at 100, 170 and 310 Kg/

ha with broadcast application at 170 kg/ha through irrigation system. Phosphatic and potash fertilizers were given at same rate by conventional method in all the treatments. Average yields for 4 years were 62, 73 and 82 mg/ha with 100, 170 and 310 kg N./ha, through fertigation.

Koo and Smjstrala [27] supplied 15% and 30% of crop N. and K requirements through fertigation and rest through conventional method to Valencia orange. Partial fertigation of N. and K resulted in lower N. contents of leaves. TSS and acid concentration in juice was also reduced but yield was not affected. Haynes [20] discussed the principles of fertilizer use for trickle irrigated crops. Haynes [21] also studied the comparison of fertigation with broadcast applications of urea on levels of available soil nutrients and on growth and yield of trickle irrigated peppers. He found that growth and yields were greatest at the low rate of N. applied as fertigaton or as a combination of broadcast plus fertigation.

Fouche and Bester [14] tried various fertilizer combinations through fertigation on 13 year old Navel oranges. Fertigation was given with a soluble fertilizer 'Triosol' [3: 1: 5] + 350 gm Urea by broadcast, fertigation of N. and K with broadcast of single super phosphate and NPK through broadcast. Highest yields was obtained with fertigation of NPK through Triosol or by complete broadcasting of NPK fertilizers. No significant differences were observed as fruit size, acidity, percent juice content and TSS among treatments. Beridze [5] conducted trial on 5 year old lemon tree and fertilized 150 kg N. + 120 kg P_2O_5 + 90 kg K_2O per hectare as basal dressing. The highest yield of 6.6 ton per hectare was obtained from trees fertilized with basal dressing + 250 kg peat/tree as a mulch + FYM at 25 t/ha. Ferguson [13] studied the fertigation as growth of 'Sunburst' tangerine trees. Two years old citrus *reticulata x C. paradisi cv.* Sunburst was fertilized with 0.66 or 1.32 lb N./tree during 1988–89 and it was 0.52 or 1.05 lb N./tree during 1990. Leaf analysis showed that low to deficient concentrations of N., K, Mn and Zn with both N. treatments. Zekri and Parsons [80] tried micronutrients through fertigation with different sources of various rates. Inorganic forms [NO_3 and SO_4] were ineffective in evaluating micro element levels in oranges. But chelated sources of Fe, Mn, Zn and Cu were very effective and their rates of application were comparable with rates through foliar applications. Neilsen et al., [36] studied that fertigation with calcium ammonium nitrate showed increased vigor and leaf Ca concentration but decreased leaf Mg and Mn compared to trees fertigated with Urea or ammonium nitrate [NH_4NO_3] in apple trees. Fertigation with P increased early tree vigor, leaf and fruit P concentration and decreased leaf Mn.

Syvevtsen and Smith [77] studied the nitrogen uptake efficiency and leaching losses from lysimeter grown trees fertilized at three nitrogen rates. He concluded that Average N. uptake efficiency decreased with increased N. application rates, overall canopy volume and leaf N. concentration increased with N. rate, but there was no effect of N. rate on fibrous root dry weight. In the first 5 years of the experimentation fertigation did not provide a significant enough yield advantage over banded application to warrant the added cost of the fertigation equipment and higher labor requirement. A very little work was done on fertigation in India. The fertigation research in citrus was initiated during 1995 at NRC for Citrus on acid lime. Shirgure et al., [61 and 62] studied the effect of differential doses of Nitrogen fertigation in comparison with band placement of fertilizer application on leaf nutrients, plant growth and fruit quality of acid lime during prebearing stage. The percentage increase in plant height, stock girth and canopy volume was more with 100% N. fertigation followed by 80% N. of recommended dose in acid lime. He also [63and 64] studied that effect of N. fertigation on soil and leaf nutrient build-up and fruit quality of acid lime.

10.2 FUTURE WATER MANAGEMENT STRATEGIES IN CITRUS

a.   Citrus is a very sensitive crop. Any excess or deficit of water even for a short duration adversely affects its growth and productivity. Irrigation scheduling based on scientific principles like available water content, soil water potential and potential crop evapotranspiration is in practice in developed countries. The efficiency of different methods of irrigation scheduling varies with climate, irrigation method and citrus species. Since microirrigation systems are gaining popularity among the farmers due to scarcity of water resources and Govt. subsidy for these systems. A modern system of irrigation will effectively used if it is backed by scientific principles of irrigation application. The water management research pertaining to the citrus is still in a preliminary stage. There is urgent need to evolve efficient irrigation scheduling for citrus crops in different regions of India.

b.   Irrigation scheduling definitely help in maximizing the utilization of water resources and boosting the productivity. Scheduling using tensiometers of various depths, neutron moisture probe, climatological approach like modified Penman equation and water balance approach should be studied. Irrigation scheduling based on canopy temperature and leaf water potential may also be studied for better yield and quality.

c.   Method of irrigation scheduling varies with the irrigation system adopted. It needs to be standardized for both system adopted. It needs to be standardized for both conventional and modern methods like drip, sprinkler, microjet etc. Infiltration rates, water distribution and retention parameters vary greatly with soil composition and structure. Thus study on these aspects will help in formulating the scientific water management.

d.   Another aspect that requires immediate attention is that water requirement and root distribution of fruit crops increases with age. Therefore suitable design needs to be evolved which should enable to irrigate the entire root zone with required quantity of water. A farmer should use the installled system for longer period without many modifications, which incur high cost otherwize.

e.   Citrus growers in Central India give water stress to induce flowering. In absence of any scientific information, farmers apply water stress according to the past experience. Plants are subjected to stress to the extent where complete restoration of vigor may be possible in all the plants. Relationship needs to be established between water stress and flowering on one hand and water stress and plant growth on the other hand. These in turn should be related to soil characteristics. Farmers should have idea about the duration of stress required for different king of soils.

f.   The modern microirrigation systems have one potential advantage of giving soluble fertilizer through irrigation water known on 'fertigation.' It not only saves labors, fertilizer but gives higher yield and better quality. The fertilizers through water are applied to the rootzone, which increases fertilizer use also. A comprehensive research related to different NPK soluble fertilizers, their rates and frequency of application at different growth stages of plants are required to be researched.

g.   Since the water available for irrigation is becoming scarce day by day. The applied water to the tree root zone needs to be conserved for longer period and that is possible with mulching. The material available to the farmers from the farm itself like grass, leaf litter, straw and trashes can be used for mulching. It not only helps in moisture conservation but also in thermal regulation, dizease control and weed control. The research is required on use of organic (grass, straws, leaf litter and trashes) and synthetic (polythene sheets) mulches. The basin area of the citrus trees will be covered with above mulch material and effect may be studied on water saving, growth and yield of the trees. The synthetic mulches are commercially available. But in case of organic mulches around 5 cm thickness of the mulch is required to be maintained uniformly in all the basins. All these strategies mentioned above are definitely make efficient use of water and enhances productivity in citrus.

10.3   SUMMARY

Irrigation management is one of the prime concerns of modern citriculture irrespective of water resource availability. A variety of recommendations have emerged world over on irrigation scheduling based on analysis of meteorological pedigree, evapotranspiration, depletion of available water content, soil and leaf water potential. The review of literature has revealed best promizing results on irrigation scheduling based on depletion pattern of soil available water content. Various microirrigation systems have established their superiority over traditionally used flood irrigation with microjets having little edge over rest of the others. Similarly, fertigation has shown good responses on growth, yield, quality and uniform distribution pattern of applied nutrients with the rootzone compared to band placement on other methods involving localized fertilization. Automated fertigation in citrus orchards is a new concept, which would be the only solitary choice of among many irrigation-monitoring methods in near future.

KEY WORDS

- acid lime
- acidity
- band fertilizer application
- basin irrigation
- black polythene mulch
- canopy volume
- Citrus

REFERENCES

1. Autkar, V. N.; Kolte, S. O.; Bagade, T. R. Distribution of active rooting zones in Nagpur manda-rin and estimates of WR for Vertisoles of Maharashtra. *Ann. Pl. Physiol.* **1988**, *2 (2),* 219–222.

2. Autkar, V. N.; Patel, V. S.; Deshpande, S. L.; Bagade. T. R. Management of Drip irrigation in Nagpur mandarin. *Ann. Pl. Physiol.* **1989**, *3(10),* 74.

3. Azzena, M; Deidda, P.; Dettori. Drip and microsprinkler irrigation for young Valencia orange trees. *Proc. Sixth Intern. Citrus Cong.;* Tel Aviv, Israel, **1988**, *2,* 747–751.

4. Barbera, G.; Carimin, F. Effect of different levels of water stress an yield and quality of lemon tree. *Sixth Intern. Citrus Cong,* Tel Aviv Isreal, **1988**, *2,* 717–722.

5. Beridze, T. R. The effect of organic fertilizers on lemon tree productivity. *Sub tropicheskie Kul'tury,* **1990**, *(3),* 83–86.

6. Bielorai, H.; Deshberg, Erner; Brum, M. The effect of fertigation and partial wetting of the rootzone on production of shamouti orange.*Proc. Int. Soc. of Citriculture,* **1984**, *1,* 118–120.

7. Capra, A.; Nicosia, O. U. D. Irrigation management in citrus orchards. *Irrigazine.* **1987**, *34 (1),* 3–15.

8. Castel, J. R.; Buj, A. Response of Satustinana oranges to high frequency deficit irrigation. *Irrig. Sci.* **1990**, *11 (20),* 121–127.

9. Cevik, B,; Yazar, A. A new irrigation systems for orchards (bubbler irrig.). *Doga Bilim Dergisi,* **1985**, *9(3),* 419–424.

10. Cevik, B.; Kaplankiran, M.; Yurdakul, O. Studies for determining the most efficient irrigation method for growing lemons under Cukurova conditions. *Doga, Tarum ve, Ormaniciuk,* **1987**, *11(1),* 42–43.

11. Dass, H. C. National Research priorities in Citriculture. *Paper Presented at the Citrus Shows cum Seminar, Res, Station, R. F.;* Abhor, Jan 6–7.; **1989**.

12. Faton, J. Drip irrigation at Yuma. *Citrog.* **1970**, *55,* 173–175.

13. Ferguson, J. J.; Davies, F. S.; Bulger, J. M. Fertigation and growth of young 'Sunburst' tangerine trees. *Proc. Fla. Stat. Hort. Sci.* **1990**, *103,* 8–9.

14. Fouche, P. S.; Bester, D. H. The influence of water soluble fertilizer on nutrition and productiv-ity of Navel orange trees under microjet irrigation. *Citrus and Sub-tropical fruit J.* **1986**, *62,* 8–12.

15. Fritz, W. D. **1970,** Citrus cultivation and fertigation. *RUHR, Stickstoff, A. G.; Bochum.*

16. Gangwar, L. S.; Shirgure P. S.; Shyam Singh. Economic viability of investment on adoption of drip irrigation system in Nagpur mandarin orchards. Proc. of National Symp. on Citriculture Nov. 17–19, **1997**, at NRCC, Nagpur, 246.

17. Ghadekar, S. R.; Dixit, S. V.; Patil, V. P. Climatological water requirement of the citrus under Nagpur agroclimatic conditions. *PKV Research J.* **1989**, *13 (2),* 143–148.

18. Grieve, A. M. **1988,** Water use efficiency of microirrigated Citrus. *Proc. 4th Intern. microirriga-tion congress, Oct 23–28, 1988, Albury- Nodonga,* Australia.

19. Hashemi, F.; Gerber, J. F. Estimating evapo transpiration from a citrus orchard with weather data. *Proc. Amer. Soci Hort. Sci.* **1968**, *916,* 173–179.

20. Haynes, R. J. Principles of fertilizer use for trickle irrigated crops. *Fert. Res.* **1985**, *6,* 235–255.

21. Haynes, R. J. Comparison of fertigation with Broad cast applications of Urea-N. On levels of available soil nutrients and on growth and yield of trickle irrigated peppers. *Scientia Hort.* **1988**, *35,* 189–198.

22. Hilgeman, R. H.; Hewland, L. H. Fruit measurement and tensiometers can tell you when to ir-rigate Citrus trees. *Proc. Agri. Arizona.* **1955**, *7,* 10–11.

23. Intrigliolo, F.; Raciti, G. Subsurface system experiments in citrus. *Irrigazione and Drenaggio* **1989**, *36(2),* 25–27.

24. Kelin, I. **1983,** Drip irrigation based on soil matric potentials conserves water in peach and grape. *Hort Science. 18,* 942–944.

25. Koo, R. C. J. Evapotrarspiration and soil moisture distribution as guide to citrus irrigation. *Proc. First Intern. Citrus Symp. Riverside,* **1968**, 269.

26. Koo, R. C. J. Results of Citrus fertigation studies.*Proc. Fla. State Hort. Sci.* **1981,** *93,* 33–36.

27. Koo, R. C. J.; Smjstrala, A. G. Effect of trickle irrigation and fertigation on fruit production and fruit quality of Valencia orange. *Proc. Fla. State Hort. Sci.;* **1984,** *97,* 8–10.

28. Leyden, R. F. Water requirement of grapefruit in Texas. *Proc. Int. Soc. of Citriculture.* **1977,** *3,*1037–1039.

29. Louse Ferguson. Nitrogen fertigation of citrus summery of citrus research.Citrus Research Centre and Agricultural station.*University of California Riverside.* **1990,** 20–22.

30. Mageed, K. J. A.; Sharma, B. B.; Sinha, A. K. Influence of irrigation and nitrogen on water use and growth of Kinnow mandarin. *Indian. J.; Agric. Sci.* **1988,** *58 (6),* 284–86.

31. Makhija, M.; Sharma, B. B.; Sinha, A. K. Estimating water requirements of Kinnow mandarin. *South Indian Hort.* **1986,** *34 (3),* 129.

32. Marler, T. E.; Davies, F. S. Microsprinkler irrigation scheduling and pattern effects on growth of young Hanilin orange trees. *Proc. Fla. State. Hort. Sci.* **1989,** *102,* 57–60.

33. Marler, T. E.; Davies, R. S. Microsprinkler irrigation and growth of young Hamlin' orange trees. J.; *Amer.Soc.Hort.Sci.* **1990,** *115 (1),* 45–51.

34. Moreshet, S.; Cohen, R.; Fuchs, M. Response of mature 'Shomouti' orange trees to irrigation of different soil volumers of similar levels of available water. *Irrig. Sci.* **1983,** *3 (4),* 223–236.

35. Moreshet, S.; Cohen, Y.; Fuchs, M. Water use and yield of a mature Shamouti orange orchard submitted to root volume restriction and intensive canopy pruning. *Proc. Sixth Intern. Citrus Cong. Tel Aviv, Israel,* **1988,** 2739–2746.

36. Nielsen, G. H; P.; Parchomchuck; Wolk W. D.; Lau, O. L.; Growth and mineral composition of Newly planted apple trees following Fertigation with N.; *Amer, P. J. Soc. Hort. Sci.;* **1993,** *118,* 50–53.

37. Ozsan, M.; Tekinel, O, Tuzcu, O.; Cevik, B. Studies on determining the most efficient irrigation method for growing lemons under cukurova conditions. *Doga Bilim Dergisi Dz, Tarim Ve Ormancilik.* **1983,** *7 (1),* 63–69.

38. Plessis, S. F. Du. Irrigation of citrus. *Citrus and Sub tropical Fruit J.* **1985,** *614,* 12.

39. Plessis, S. F. Du. Some factors affecting the fruit growth of citrus. *South African J. Pla. Soil.* **1987,** *4 (1),* 12.

40. Plessis, S. F. Du. Irrigation scheduling. *Proc. Sixth Intern. Citrus Congress,* Tel Aviv, Isreal, **1988,** *988,* 731.

41. Plessis, S. F. Du. Irrigation scheduling of citrus, Research results. *Water and Irrig. Rev.* **1989,** *9 (4),* 4–6.

42. Plessis, S. F. Du.; Laere, H. C.; Van M. E.; Salt accumulation in citrus orchards as influenced by irrigation. *J.; South African Soc. Hort. Sci.* **1991,** *1(1),* 29–32.

43. Pyle, K. R. An appraisal of microirrigation for use in citrus with an emphasis on drip irrigation. *Citrus and Subtropical Fruit J.;* **1985,** *612,* 4–7.

44. Raciti, G.; Scuderi, A. Drip irrigation trial in citrus orchard. *Proc. Int. Soci. Citriculture* **1977,** 31040–1045.

45. Raciti, G.; Barbagallo, A. Localized irrigation in lemon forcing. *Informatore Agravio* **1982,** *38 (41),* 22887–22891.

46. Rajput, R. K. **1989,** Strategies of water management in tropical fruits. *Sixth Binenial workshop AICRP on fruits.* Tirupati, India.

47. Randhawa, Singh, G. S.; Dudani, G. J. P.; Preliminary studies on the effect of soil management systems on soil moisture in sweet orange orchard. *Ind. Jr. Hort.*; **1960,** *7,* 246–249.

48. Randhawa, G. S.; Srivastava, K. C. Citriculture in India. *Hindustan Publishing Co-operation (India).* **1986,** 501.

49. Ray, P. K.; Sharma, B. B. Studies on response of young Kinnow trees to irrigation. *Ind. J.; Hort.* **1990,** *47 (3),* 291–296.

50. Ray, P. K.; Sharma, B. B.; Sinha, A. K. Effect of irrigation on plant status and stomatal resistance in young Kinnow mandarin trees. *South Indian Hort.* **1990,** *38 (3),* 123–128.

51. Retiz, H. J. How much water do Florida citrus trees use? *Citrus Ind.* **1968,** *49 (10),* 4–6.

52. Richards, L. A. Diagnosis and improvement of saline and alkali soils, USDA.; Agri. Handbook No. 60. US Dept. of Agriculture, Washington, D.C., USA. **1954.**

53. Richards, S. J.; Warnke, J. E. Lemon irrigation management under coastal conditions. *Calif. Citrog.* **1968,** *53,* 378–384.

54. Robinson, J. C.; Alberts, A. J.; The influence of undercanopy sprinkler and drip irrigation systems on growth and yield of bananas. *Scientia Hortic.;* **1987,** *32,* 49–66.

55. Rondey, D. R.; Roth, R. L.; Gardner, B. R. **1977,** Citrus response to irrigation methods. Proc. Int. Soc. of Citriculture, *1,* 106–110.

56. Rumayor-Rodriguez, A.; Bravo-Lonzano, A. Effects of three systems and levels of irrigating apple trees. *Scientia Hort.* **1991,** *47,* 67–75.

57. Sanehez Blenco, M. J.; Torrecillas, A.; Leon, A.; Del Amor, F. The effect of different irrigation treatments on yield and quality of verna lemon. *Pl. Soil* **1989,** *120 (2),* 299–302.

58. Scuderi, A.; Raciti, G. Citrus trickle irrigation trials. *Proc. Int. Soc. Citriculture,* **1978,** 244.

59. Shinde, B. N.; Firake, N. N. **1998,** Integrated water management for crop production. Deptt. Of Irrigation and Water management, M. P. K. V.; Rahuri (Maharashtra).

60. Shirgure, P. S.; Marathe, R. A.; Lallan Ram; Shyam Singh. Effect of irrigation scheduling on depth of irrigation, soil moisture distribution and Evapotranspiration in acid lime. *Proc. of National Seminar on Water Management held at, W. T. C.; IARI, New Delhi from* April, 15–17. **1998a.** 30.

61. Shirgure, P. S.; Marathe, R. A.; Lallan Ram; Shyam Singh. Leaf nutrient, growth and fruit quality of acid lime affected by nitrogen fertigation. *Proc. of National Seminar on Water management held at, W. T. C.; IARI, New Delhi from* April, 15–17. **1998b.** 27.

62. Shirgure, P. S.; Srivastava, A. K.; Shyam Singh. Response of fertigation verses band placement of Nitrogen in acid lime. *Presented in Seminar on new horizons in production and postharvest management of tropical and subtropical fruits,* Dec.; 8–9, IARI, New Delhi. **1998c.** 26.

63. Shirgure, P. S.; Lallan Ram, Marathe, R. A.; Yadav, R. P. Effect of Nitrogen fertigation on vegetative growth and leaf nitrogen content of acid lime. *Indian Journal of Soil Conservation* **1999,** *27(1),* 45–49.

64. Shirgure, P. S.; Srivastava A. K.; Shyam Singh. Fruit quality and soil fertility changes in acid lime under drip, microjets and basin irrigation methods. Proc. of International Symposium on Citriculture held at NRCC, Nagpur(India) on 23–27th Nov.; **1999,** 358–366.

65. Shirgure, P. S.; Srivastava A. K.; Shyam Singh. Soil -leaf nutrient build-up and growth response of prebearing acid lime as affected by N. fertigation versus band placement method. Proc. of International Symposium on Citriculture held at NRCC, Nagpur(India) on 23–27th Nov.; **1999,** 551–557.

66. Shirgure, P. S.; Lallan Ram, Shyam Singh, Marathe, R. A.; Yadav, R. P. Water use and growth of acid lime under different irrigation systems. *Indian J. of Agri. Sci.* **2000a.** *70 (2),* 125–127.

67. Shirgure, P. S.; Srivastava A. K.; Shyam Singh. Efficiency of microirrigation systems in relation to growth and nutrient status of acid lime. Micro-irrigation: Eds. Singh, H. P.; Kaushish, S. P.; Murty, T. S.; Jose C.; Samuel. CBIP Publication No. 282 **2000b.** 262–269.

68. Shirgure, P. S.; Marathe, R. A.; Lallan Ram; Shyam Singh. Irrigation scheduling in acid lime as affected by different soil moisture regimes. *Indian J.; Agri. Sci.* **2000c.** *70 (3),* 173–176.

69. Simpson, G. H. Developments in under tree irrigation systems in the Murray valley. *Proc. Int. Soc. Citriculture.* **1978,** 234–235.

70. Slack, J.; Turpin, J. W.; Duncan, J. H.; Mckay, O. L. Trickle irrigation of young citrus on coarse sands. *Proc. Int. Soc. Citriculture,* **1978,** 236–237.

71. Smajstrla, A. G.; Koo, R. C. J.; Weldon, J. H. Effects of volume of the rootzone irrigated on water use and yield of citrus. *J. Am. Soc. of Agri. Eng.;* **1984,** *84,* 2107.

72. Smajstrla, A. G.; Koo, R. C. J. Effects of trickle irrigation methods and amounts of water applied on citrus yields. *Proc. Fla. State Hort. Soc.* **1984,** *97,* 3–7.

73. Smajstrla, A. G.; Parsons, L. R.; Aribi, K.; Yelledis, G. Response of young citrus trees to irrigation. *Proc. Fla. State Hort. Soc.* **1986,** *98,* 25.

74. Smajstrla, A. G. Micro-irrigation for Citrus production in Florida. *HortScience.* **1993,** *28 (4),* 295–298.

75. Smith, M. W.; Kenworthy, A. L.; Bedford, C. L. The response of fruit trees to injections of nitrogen through a trickle irrigation system. *J. Am. Soc. Hort. Sci.* **1979,** *104,* 311–313.

76. Stolzy, L. H.; Taylor, O. C.; Gavber, M. J.; Lambard, P. B. Previous irrigation treatments as factor in subsequent irrigation level studies in orange production. *Proc. Amer. Soc. Hort. Sci.* **1963,** *82,* 119–123.

77. Syvertsen, J. P.; Smith, M. L. Nitrogen uptake efficiency and Leaching losses from Lysimeter grown citrus trees fertilized at three nitrogen rates. *J. Am. Soc. Hort. Sci.* **1996,** *121 (1),* 57–62.

78. Tash be kov, Kh. K.; Saidov, I. I.; Kreidik, B. M. Water requirement of young lemon plantations in central Tajikistan conditions. Tekhnologiya orosheniya; programmirovanie Urozhaya, Moscow, USSR, **1986,** 147–150.

79. Toledo, P.; Rey, E. M.; Cardenas, O. Soil moisture level studies for up to 5 year old 'Olinda Valencia' oranges. *Ciencia y Tecniica en la Agricultura Riego y Drenaje,* **1982,** *5 (1),* 17–30.

80. Zekri, M.; Parsons, L. R. Grapefruit leaf and fruit growth in response to drip, microsprinkler and overhead sprinkler irrigation. *J. Am. Soc. Hort. Sci.* **1989,** *114,* 25–29.

CHAPTER 11

VEGETATION WATER DEMAND AND BASIN WATER AVAILABILITY IN MEXICO[1]

JUDITH RAMOS HERNANDEZ, F. J. GONZÉLEZ, L. MARRUFO, and R. DOMINGUEZ

CONTENTS

[1]Printed with permission and modified from: Ramos, J.; Gonzalez, F. J.; Marrufo, L.; Dominguez, R. **2011,** Semiarid riparian vegetation demand and its influence to compute the Sonora river basin water availability. Chapter 16, 379–395, In: *Evapotranspiration* by LeszekLabedzki (Editor), ISBN 978-953-307-251-7,In-Tech.Available from: http://www.intechopen.com/books/evapotranspiration/semiarid-riparian-vegetation-water-demand-and-its-influence-to-compute-the-sonora-river-basin-water- © Copyright InterTech.com

11.1 INTRODUCTION

During the last decade, the riparian zones have received considerable attention in order to restore and manage them since they have an important role to plan water resources and land, among other functions in the ecosystem. The riparian zones are mainly located in the floodplain where the soil is characterized to be alluvial. However, arid zones channels and their floodplain are transitory and subject to frequent and rapid changes [12, 13], thus there is not a clear distinction from where the channel finishes and the floodplain starts. This condition could be disadvantageous to the vegetation settled in the floodplain, which is subject to the floods' force, the wood debris and eroded sediments carried by. Conversely, this vegetation has the advantage to receive organic matter from the debris and minerals from the eroded soil resulting in a straightforward species capable to survive at adversely conditions [6]. The adaptation of this vegetation generates species drought and flood resistant. In the first case, plants develop long roots in order to access groundwater and be wet. In the second case, plants can afford inundation since the modification of the hydraulic roughness reduces the flood velocity and spread seeds increasing the moisture and nutrients availability in the area [6].

However, the access of the riparian vegetation to groundwater produces high rates of transpiration and a high evaporative demand of the atmosphere. This is because water requirements are bigger in arid and semiarid regions, where rainfall is less and the vapor pressure deficit on the air is large. This implies great water consumption by this vegetation becoming as one of the components to be considered in the groundwater balance. Particularly, in semiarid sites where groundwater is the main water demand source, since water surface resources are highly compromized, temporal and spatially. Scott et al. [27] observed that as groundwater is used to provide water among the different users, there is a depletion of the water levels that affects directly the riparian zones and, in consequence, modify ecology and hydrology of the watershed.

Scott et al. [27] pointed out that in order to compute riparian water requirements it is common to use hydrological models that compute it as the residual discharge resulted after calibrated against known inputs, groundwater levels and discharges. This procedure could over or underestimate the groundwater balance in the basin, thus a better understanding of this component is needed to improve water demand among the users. In order to guarantee a better management of the river basins, Mexico has made different water reforms through the last three decades. Some of these reforms deal with the increasing water overexploitation and, particularly, there were established aquifers management councils supported by a national water law [30].

In order to compute precize riparian water requirements, it is necessary to consider both internal (vegetation growth characteristics) and external conditions (atmosphere, plant and soil characteristics) of the vegetation [8]. The external conditions can be achieved computing the evapotranspiration (ET), which depends of the solar radiation, vapor pressure, wind speed and direction, stage of the plant development, and soils features as soil moisture, among others. Conventional methods can be used to compute ET, but there are also remote sensing methods that have the advantage to consider its spatial and temporal variability. Currently, there are not many studies focused on native riparian zones, researches have mainly carried out analysis for cropped income areas.

This chapter presents the study to establish the water requirement of native vegetation (mesquites and elms) along the main channel and in the floodplain of the Sonora River corresponding to the Pesqueira, Topahue, Ures, La Mesa-Seri and Horcasitas aquifers. To address the water requirements, the ET was estimated using remote sensing techniques based on the energy balance. The study was considered for two hydrological cycles (1996–1997 Spring–Summer and 2002–2003 Autumn–Winter), in particular, for the dry season since the Sonora River tributaries are dry and the flow in the main channel reaches its minimum level. The chapter also presents the importance to consider a temporal and spatial analysis, for determining the water requirements of the riparian vegetation and its impact on the river basin water budget.

11.2 THEORETICAL METHODS FOR EVAPOTRANSPIRATION

Evapotranspiration (ET) is the total moisture lost to the atmosphere from the land surface when the vaporization process starts as function of the input energy received and the vegetation present. Lakshmi and Susskind [18] described ET as a "useful tool," not only because it provides information that can be applied directly in the water budget, but also because ET has a high sensitivity which can be used to define some biophysical parameters. ET has been studied at different scales [21] providing a wide knowledge of how ET is affected or how it can affect the whole system [20]:

1. Macro (e.g., impact of changes in available moisture on cloud formation, radiation budget and precipitation);
2. Meso (e.g., depletion of soil moisture, and therefore, crop water requirements of irrigated lands and partitioning of precipitation); and
3. Micro (e.g., crop water requirements)

In order to assess ET impacts on the water balance, and in consequence, on the land and water management, ET is defined for a specific crop and land condition. Thus, potential ET (ET_p) considers a reference surface as grass with a crop height of 0.12 m, a fixed surface resistance of 70 s.m^{-1} and an albedo of 0.23, whereas crop ET (ET_c) is the rate of ET from disease-free, well fertilized crops, grown in large fields under optimum soil water conditions, and achieving full production under the given climatic conditions. The actual ET (ET_a) refers to the ET from crops grown under management and environmental conditions that differ from the standard conditions [2].

In general, the physics of the ET process is well understood, thus accurate ET values at local level are provided. However, as ET is highly sensitive to various land and atmospheric variables, particularly in their spatially distributed form, it makes regional ET estimations uncertain [7]. This uncertainty increases when the ET contribution from riparian vegetation needs to be considered to determine the water budget [11].

Although some authors have studied the direct and indirect influence of the riparian ET into the water availability in a basin [11], this vegetation is still poorly understood being a noneasy task to explain its hydrogeomorphological influence since it interacts environmentally at different scales [6].

To compute actual riparian evapotranspiration (ET_a) is complex since the vegetation communities are nonuniform linear varying in geometry, altitude and season. Also, the organization and dynamics of the vegetation are strongly related to the channel river and its floodplain, thus the geomorphological process and forms define the pattern for the different aggregation communities maintained by the fluctuations of water discharge (Fig. 1). Goodrich et al. [11] pointed out that this condition limits the application of traditional ET computations since the required fetch conditions are not achieved and in a strict term the definition of potential ET [2] did not apply due to differences in the canopy architecture, available energy, water availability and boundary layers differences between the atmosphere and leaf surfaces, among others. Additionally, only in few cases the crop coefficient has been defined for each riparian type [11].

FIGURE 1   Riparian vegetation along the Sonora River in Mexico.

11.3   TRADITIONAL METHODS FOR EVAPOTRANSPIRATION

The relationship between the different ET definitions (potential, crop and actual) is not always determined easily, but an accurate value is required for different uses and users. Favourably, at regional scales whatever the type of ET, these are not independent of each other. For example, during the crop growing period, water needs to be diverted on to the field to meet the ET_c demands and to compensate losses by seepage and percolation in order to maintain a saturated root zone. Thus, the estimation of ETp is crucial to obtain ET_c rates and moreover to compute ET_a values as the response to different reasons that generate nonstandard conditions such as climate, pest, contamination, water shortage or waterlogging.

In the Sonora River Basin (SRB) and others semidesert basins in Mexico, the Turc equation [28] have been used to estimate the hydrological water balance and to infer the groundwater balance since very few aquifer data is available. Other condition to apply the Turc equation is that the average ET_a is used for longer periods in order to reduce to zero the water retention in the basin. The Turc equation is defined as:

$$if \ \left(\frac{P}{L}\right)^2 > 0.1; \quad ET_{actual} = \frac{P}{\sqrt{0.9 + \left(\frac{P^2}{L^2}\right)}} \tag{1}$$

$$or \ \left(\frac{P}{L}\right)^2 < 0.1; \quad ET_{actual} = P$$

$$L = 300 + 25 * Ta_{prom} + 0.05 * Ta_{prom}^3 \tag{2}$$

where, P is the precipitation and T is air temperature. Sometimes Eq. (2) is modified considering T at the power of 2 instead of 3 without any raison. The constant value of 300 in Eq. (2) corresponds to the base runoff in the basin. Thus, L could increase above 300 as function of the T and ETa is related directly to rainfall. In 1964, the Truc equation was modified by changing the 0.9 coefficient by 1.0 [Pike 1964, 29]. Despite to the simplicity of this method a high uncertainty is associated to it. Other methods based on meteorological data have been used such as Blaney-Criddle, which is a temperaturebased method that can be applied to different climates in a monthly base [11] or the Hargreaves-Samani equation that considers the air temperature and the extraterrestrial radiation [1]. However, these types of methods are recommended only when data is not available. However, when data is available it is recommendable to use the Penman-Montheith (PM) equation [2] but a crop coefficient accurate is necessary to compute ET_c. Gazal et al. [10] applied the PM method to compute the transpiration of cotton/willow forest. The PM equation was inverted in order to asses the seasonal variation in stomatal resistance, results showed that the spatial and temporal heterogeneity of water availability modified the physiology of these riparian vegetation.

11.4   ADVANCED METHODS FOR EVAPOTRANSPIRATION

Although riparian ET_a values have been obtained with good accuracy using traditional methods and compared with available ET measurements for local areas, ET_a values for large areas are still problematic. Also, it has been demonstrated by some authors such as Schultz and Engman [26] that studies based only on conventional field data collection are often limited because they cover a specific area. In addition, the lack of available and reliable data is a big constraint in the application of different methods to compute accurate ET values. As riparian vegetation interacts environmentally at different scales, remote sensing techniques have been shown to be a reliable alternative to estimate ET, since some of the main constraints about suitable and available data can be overcome providing a precize spatial representation. One important advantage is that it provides detailed and independent ET estimations on a pixel-by-pixel basis among other data as mapping soil properties based on the reflectance variations, land use and land cover using the spectral signatures of vegetation, water requirements, monitoring water availability, and detecting some properties like water stress effects. This wide range of alternative data offers new possibilities for managing soil, water and land resources efficiently.

Remote sensing techniques have provided accurate estimations of ET_a at several scales as a function of the spectral resolution of the satellite sensors. ET_a can be estimated using data that describes the conditions in the soil-plant-atmosphere system. Thus, remote sensing provides specific information of some of the parameters involved in the ET estimation such as surface temperature, surface soil moisture, water vapor gradients, surface albedo, vegetative cover and incoming solar radiation.

According to Kite and Droogers [17], methods and methodologies developed to estimate evapotranspiration (ET) using remote sensing data can be classified into four categories: (1) satellite-derived feedback mechanisms, (2) biophysical processes based-model, (3) surface energy balance techniques, and (4) physically based analytical approaches. Qi et al. [24] noticed that the canopy temperature is lower than the air temperature as water evaporates result of the heat extracted from the canopy leaves and from the air near to the canopy surface. This implies a temperature difference and the "Surface Energy Balance Techniques (SEB)" can be applied because they are based on the spatial and temporal variability of the surface temperature (T_s) and air humidity as a reflection of the partitioning of net radiation available to the surface into soil, sensible and latent heat fluxes release to the atmosphere. Thus, the ET is obtained as the residual component of the energy balance equation:

$$\lambda E = (R_n - G - H) \tag{3}$$

where, R_n is the net surface radiation (Wm^{-2}), G is the soil heat flux (Wm^{-2}), H is the sensible heat flux (Wm^{-2}) and λE is the latent heat flux (Wm^{-2}). These methods estimate each parameter of the surface energy balance equation, which are characterized for a diurnal variation and are subject to significant changes from one day to another. To determine these critical variables, these methods first related the flux parameters of the energy balance in terms of variables such as the soil moisture profile, the surface, ground and near-surface air temperatures, and near-surface humidity. The algorithm employed was MEBES (Surface Energy Balance to Measure Evapotranspiration) that was based on the process theory of the original algorithm SEBAL (Surface Energy Balance Algorithm for Land) developed by Bastiaanssen et al. [3, 4] applying some modifications to the data and condition of the region. The algorithm was chosen since it presents major advantages to compute ETa values among other SEB techniques [25] using both remote sensing and meteorological-ground data for local and regional areas.

11.5   SONORA RIVER WATERSHED

This research was conducted in the Sonora River Basin (SRB), that is located in the north-center of the Sonora State in Mexico at the extreme latitude: 28°27' to 30°54' N. and longitude: 110°06' to 111°03' W coordinates. The Sonora River Basin (SRB) is part of the Hydrological Region No. 9 and it has a total surface of 26,010 km^2 almost 14.8% of the Sonora state. The SRB limits at the north with the USA and at the south with the Cortes Sea. The main river is the Sonora with 277 km of longitude, starting at the north in the Cananea City with the union of several streams, which down from the Magallanes, Los Ajos and Bacanuchi Mountain ranges. Then the river flows to the south crossing the Bacoachi, Chinipa, Arizpe, Banámichi, Baviácora, Ures and Hermosillo cities. In the last one the San Miguel de Horcasitas and el Zanjonafluents join the main river; both of them from the same basin. Finally, the Sonora River goes to the Coast but due to the sandy soil it disappears before reach the sea [14]. Figure 2 indicates the divisions of the SRB: the Zanjóna and San Miguel de Horcasitas (8,826 km²), Sonora (11,680.6 km²) and the Coast watersheds (5,503.4 km²).

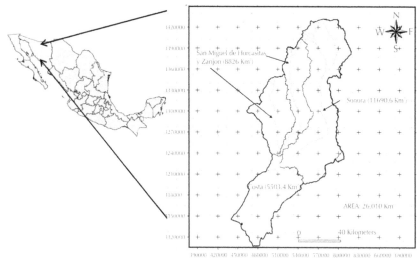

FIGURE 2 The Sonora River Basin and its watersheds: Sonora, Zanjón, San Miguel de Horcasitas and the Coast.

The Sonora River has an erratic behavior and, frequently, the runoff is concentrated into few days during the year. The river flow is controlled by a dam system that includes the Abelardo Rodriguez and Rodolfo Félix Valdés (El Molinito) dams with a total storage capacity of 525 Hm³ [14]. Physiographically, the SRB is divided into two subprovinces [14]:

1. Mountain ranges and North Valleys Subprovincie belonging to the Western Mother Mountain Range Province formed in the mountains by volcanic acid rocks with intrusive igneous outcrops and in the valleys by continental sediments. The mountain system is steep from 1,000 to 2,620 meters above sea level (m.a.s.l.). The weather is dry and semidry and varies as function of the altitude from warm and semiwarm to temperate and semicold.

2. Sonorenses' Mountain range and plain subprovince belongs to the Sonorense' Desert covering 2/3 of the SRB. This subprovince is characterized ingy lower mountain ranges separate by plains being the mountains narrower than the plains; the plains have 80% of the total area. In the central basin intrusive and lavic igneous rocks predominated as well as metamorphic, ancient lime-

stone and Tertiary conglomerates. In the plains alluvial fans are present with smooth slopes from the near mountains. The weather is very dry semi warm and warm.

There is a high climatological variation observed at the SRB as result of the accidental topography with more than 1000 masl in the mountains and less than 100 masl in the coastline. The climate variation produced two severe droughts (1982–1983 and 1997–1998), which have been related with El Niño event [15]. In fact, during 90 s there were three events related to El Niño: 1991–1992, 1993–1994, 1997–1998, and in 2000 s two events more in 2002–2003 and 2004–2005 [22]. The mean annual precipitation is 376 mm with a runoff coefficient of 2.8% being the main use the agriculture, followed by domestic, industrial, livestock and recreational. During the drought season it is necessary the groundwater extraction, which is used in agriculture (93%), domestic and commercial (4.8%), industry (1.5%) and the rest 0.7% in livestock, recreational and others. The intensive use of groundwater in agriculture has generated an overexploitation of the aquifers present in the SRB. The aquifers are characterized according to the subprovinces in the SRB, thus for the Mountain ranges and North Valleys the nonconsolidates material gave a lower infiltration capacity as result of the metamorphic, sedimentary (limestone and conglomerates) and acid extrusive (tuff and riolites) rocks which have a low breaking and porosity reducing the water circulation. Opposite to the Sonorenses' Mountain range and plain subprovince where nonconsolidates materials present a high infiltration capacity since the presence of gravel and sand with different sizes and porosity favoring the intercommunication to the water circulation.

The climatic and topographic variation in the SRB provides a diversity of vegetation: pine woods in the north and mesquites (drought tolerant) in the rest of the basin, except irrigated areas aside the rivers and in the coast. The SRB vegetative coverage is quite similar to the state condition due to the extension of the basin, thus almost 50% of vegetal communities correspond to temperate climates under aridity conditions where drought-deciduous low woodlands of thorny and nonthorny trees and bushes are predominant. The thorny shrubs such as Gobernadora (*Larreatridentata*), Mesquite (*Prosopislaevigata*), Palo Verde (*Parkinsonia aculeate*) and Sangregado (*Jatrophacuneata*) are mainly allocated in hills around the San Miguel de Horcasitas and its affluents, as well as in Bacanuchi, Arizpe and Ures towns (Fig. 3).

a)  b)

FIGURE 3   Mesquite scrubland along the Sonora River and floodplain.

The 20% of the vegetation coverage are other type of mesquites as sarcocaulescent scrubland that is a subtype of the xerophyllous scrubland and elms (*Ulmusamericana*) [19]. These communities are in flat and deep soils, runoff areas, over yermosls, fluvisols or xerosols (Fig. 4).

FIGURE 4   Mesquites as sarcocaulescent scrubland.

Grassland represents 13% of the total vegetation and two types can be found in the area: native and introduced. The first one is represented by bluegrass plants and some herbaceous and shrubs used to livestock. The second one is related to the agriculture practice with only 3% of the total (Fig. 5a). Pine-oak forest and pines with herbaceous secondary vegetation represents 11% of the total vegetation and they can be found in subhumid temperate climates (mountain ranges) (Fig. 5b).

FIGURE 5   (a) Grassland for livestock and (b) pine-oak forest at the SRB.

Finally, introduced vegetation is 6.61% from which more or less 4% are agriculture lands where more than 95% of irrigation and the rest is rainfed. Crops are perennial as Lucerne, walnut and sugarcane and seasonal as ryegrass, barley, fodder oats, wheat, garlic and onion for autumn-winter period and fodder sorghum, beans, maize, and vegetables. The winter crops (i.e., wheat) showed a high vulnerability for prolonged drought seasons, thus farmers seemed to change into drought resistance crops or more economically productive crops, especially once the government stopped subsiding grains.

11.6   METHODS AND MATRIAL

The methodology was divided into three steps: firstly an evaluation of the climate variation, after that a land cover classification was performed in order to identify those areas where the riparian vegetation is present. In this case drought-deciduous low woodlands of thorny and nonthorny trees and bushes were analized, in particular, thorny shrubs such as mesquite (*Prosopislaevigata*) and eml (*Ulmusamericana*). Finally, MEBES was applied to estimated ET_a and analyzed to determine its influence to compute the water availability in the SRB.

11.6.1   CLIMATIC DATA ANALYSIS

The climatological data available include traditional weather stations (CLICOM database) recording precipitation (P) and air temperature (T) (maximum and minimum), and Observatorios weather stations recording P, T, relative humidity, wind speed and isolation hours. Three Observatorios were analyzed: Hermosillo within the SRB, and Nacozari and Empalme at the North and South, respectively, of the neighbor basin at the East.

The CLICOM dataset was updated to 2004 for the Sonora state and it has registered 273 stations. Two filters were applied in order to obtain complete and continuous data for a 20-year period. The results obtained were from 21 stations distributed throughout the entire basin. In order to analyze how the variables related to climate vary in time and space; essential for any study over large areas, data mining techniques were employed. Grouping techniques such as K-means (nonhierarchical) and Ward (hierarchical) methods were evaluated and the result was the basin division into three regions: north, central and south. This definition of homogeneous zones represents the climatic variation in the area accurately.

11.6.2   REMOTE SENSING DATA ANALYSIS

Two hydrological cycles were studied: Spring–Summer, 1996–1997, and Autumn–Winter, 2002–2003. LandSat images were acquired (Table 1) as well as aerial photographs orthorectified.

TABLE 1   Date and path/row of the Landsat satellite images acquired to the USGS.

Sensor	Date of acquisition	Path/row	Day of the year (DOY)
TM	1997(03)04	35/39	63
		35/40	
	1997(05)23	35/39	143
		35/40	
	1997(06)24	35/39	175
		35/40	
ETM+	2003(03)29	35/39	88
		35/40	
	2003(04)30	35/39	120
		35/40	
	2003(01)31	35/39	31
		35/40	

The images were used for both classification of the riparian vegetation and estimation of the riparian ET_a. For classification, it was considered the spectral response of vegetation to the Visible (VIS) and Infrared (IR) electromagnetic spectrum. A high reflectivity is presented in the near IR as consequence of the leaves structure and in the medium IR where the spectral response is affected by the water content in the surface and a low reflectivity observed in the VIS because the chlorophyll absorption. Also, it was observed the dynamics of the stream interact close with the riparian community characterized by the grow form, size, density, and aerial coverage of the plants [9].

The images were submitted to a preprocessing analysis to extract useful information from the images, and also to enhance them, to aid in the visual interpretation, and to correct or restore the images if these were subjected to geometric distortion, blurring or degradation. The preprocessing methods used include enhancement of the image, radiometric, atmospheric and geometric corrections, and the georeference of the scene for a chosen map scale and coordinate system.

In order to estimate the ET_a, the first part involves the determination of the land surface physical parameters from spectral reflectance and radiance. This stage estimates surface albedo (α_o), emissivity (ε_o), surface temperature (T_s), vegetation indices (NDVI and SAVI), fractional vegetation coverage (P_v) and leaf area index (LAI), and the roughness height (or height of the vegetation, z_o). Here ground data are required, however, if the data are not available, it can be replaced using the vegetation indices and considering a standard crop height of 1.0 m. The second stage includes the introduction of meteorological data such as air temperature, humidity, and wind speed at a reference height. The reference height is the measurement height at the weather station (2 m). The last stage includes the estimation of the energy flux parameters and obtains ET as the residual form of the energy balance equation [5].

11.7 RESULTS AND DISCUSSION

11.7.1 CLASSIFICATION

To achieve the classification, vegetation indices were used since they represent a direct relation to the vegetation health linking the biomass and the LAI in a spatial basis. In particular, NDVI was used as well as the soil and wetness indices in the temporal analysis. Additionally, bands 5 and 7 of Landsat were applied to separate areas with high soil moisture, in order to differentiate between irrigated lands and riparian vegetation. The reason to use bands corresponding to the medium IR is to cover a major reflective spectrum allowing the observation of watered overages. Finally, LAI was also monitored, this parameter offers advantages since the geometrical, size, and other conditions for the riparian vegetation is very characteristic.

A supervized classification was performed using the NDVI values for each image available, then other classification was done to a multitemporal image generated with the NDVI values. This classification allowed the identification of temporal homogenous areas. Finally, a multitemporal image was produced using a combination of the vegetation, soil and wetness indices, and the ratio between bands 5 and 7. The number of classes was established into 6, as the main interest was the aggregation of riparian vegetation thus mesquites, elms, paloverde, grass, water and bare soil were selected and the agriculture zones were removed from the images. The error matrix showed a confident level of 12% associated to a mix between paloverde and mesquite, and to a very small surface. Figure 6 illustrates the aggregation of elms, mesquites and some herbaceous vegetation in the Ures and Topahue aquifer.

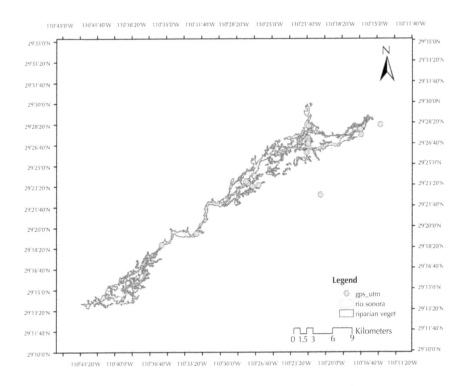

FIGURE 6   Elms and mesquites distribution following the river channel and the floodplain and the gps points used in the supervized classification.

The number of hectares obtained for elms was 68% bigger than the mesquites one within the 2000 ha in the Topahue and Ures region. The mesquite tree is a drought resistance obligates phreatophyte and indicator of the water table, and soil and nitrogen fixed, thus it controls the erosion and the soil fertility. The mesquite tree can be 10 m tall and its roots can go up to 50 m deep reaching the water table, also it has lateral roots that can be extended to 15 m. However, as riparian areas are characterized by the growth form, size, density, and aerial coverage of the plants aboveground, it made elms the predominant species since they are a taller facultative phreatophyte usually with 12 to 15 m in height, although they can reach 20 to 30 m or more (Fig. 7).

FIGURE 7   Elms predominance over mesquites and other thorny and nonthorny shrubs.

As well as mesquites, the roots of elms have the ability to go deep to get water, thus the elms can survive during the drought period. The elms have lateral surface roots that go deep as they need to gather nutrients for the photosynthesis process, approximately from 5 to 20 cm. As elms growth well in moisture and drainage soils, the first precipitation or runoff is enough to obtain the water they need. Elms seeds can be disseminated by wind and water.

The grass area or other secondary herbaceous vegetation was also important for the number of hectares covered (1090 ha). It was observed in the floodplain as a groundcover of the riparian vegetation and their presence is related to the amount of flooding and the radiation allowed by the trees.

Elms LAI values for DOY143 were from 0.7 until 2.0 being the lowest ones during the drought season, whereas for mesquites were from 1.2 to 2.0, these values agreed with Kiniry [16]. LAI for grass or herbaceous vegetation growing around the riparian corridor was between 0.2 and 0.4, and in well irrigated orchards was between 3 and 4. Gazal et al. [10] found that LAI is more sensitive in intermittent streams sites than in perennial ones due to the depth water table, thus plants modified their canopy structure in order to cope with water scarcity as it was observed in the study site.

11.8   ACTUAL EVAPOTRANSPIRATION ESTIMATIONS

Average ET_a values in DOY63 (after winter) showed for elms and mesquites 2.1 mm·d^{-1} and 1.5 mm·d^{-1} for groundcover grass of the riparian vegetation, respectively. It was observed a similar ET_a rate for elms and mesquites and irrigated orchard. At this time the orchard is less irrigated since the fruit development is not impacted thus important water saving is made. For DOY143 and DOY120 during spring-summer where dry conditions are presented having more sunshine hours during the day, the average ET_a values were from 5 to 8 mm·d^{-1} for riparian, and less than 1 mm·d-^{1} for their groundcover grass. Also, ET_a values observed in mature orchard (more than 5 years old) and very well irrigated crops such as Lucerne were quite similar to riparian. The water evaporation was about 28 mm·d^{-1} in these DOYs (Fig. 8).

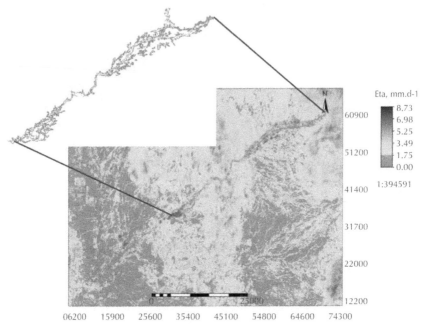

FIGURE 8   Actual evapotranspiration values for DOY143.

A 20 year climatological analysis at the center of the SRB showed that temperature stars to rize from March until June where it reaches its maximum peak (28°C) as well as the wind speed with values of 5.2 ± 0.2 km·h⁻¹, contrary to the precipitation that reach its minimum value in May (11 mm). The monsoon season starts late July and August increasing the relative humidity to $50 \pm 5\%$. As the monsoon rains did not wet the saturated zone for DOY143, the climatic conditions imply a high vapor exchange between the surface and the atmosphere, thus the water necessities of the plants need to be covered to prevent wilting and the only source of water is from the aquifer. Due to the elms are located along the main channel their facultative conditions help them to obtain water at bit deep, thus drought conditions impact less although they are more exposed to radiation. This condition is not the same for mesquites where the main disadvantage is the size competence with elms receiving less energy and being protected from the strong winds. Mesquites as obligate phreatophytes needs to go deep to reach the water table establishing a high soil water dependence and, in consequence, reducing their hydraulic capacity to carried water from the roots to the stomata cavities as Gazal et al. [10] noticed for cottonwood.

11.9   SUMMARY

The riparian vegetation has a very important role in the ecosystems not only because they are a habitat to flora and fauna spices but also because they provide some hydrological control, particularly when flood and drought are present. The vegetation structure is highly related to the geomorphological process of the region being determinant in the valleys where the fluvial soils supply adequate conditions to the development of these plants. Also, the environmental and climatological conditions provide important characteristics since in arid and semiarid zones the main river channel and the flood-plain are undistinguished as result of the erosion and sediments deposition allowing the growth of obligate and facultative phreatophytes as well as herbaceous ground-cover vegetation. This resulted in a hydrogeomorphological influence on the vegetation being severely affected but at the same time benefit by flooding and developing the capacity to survive under drought conditions.

In early stages the hydrological regime and the available energy are the main factors to control ET from riparian corridors, although these conditions change as the vegetation growth. Also, as the dry conditions are a limitation, the plant's capacity to extract water from the aquifer is determinant during their mature lives.

The similitude between the irrigated land and the riparian corridor implies a good water resource, in the first case irrigation is the explanation but in the second the only one possibility is the adaptation of plants to obtain water from the aquifers reaching their roots the water table after the depletion of the soil moisture in the saturated zones.

In the SRB, 95% of irrigation is provided by groundwater extraction with an average volume of 5 $m^3 \cdot s^{-1}$ for a total cropped area of 12,300 ha in 2003. In the Ures and Topahue region the cropped lands were almost 2000 ha in 2003 with a total volume of 1.5 $m^3 \cdot s^{-1}$, as ET_a values were similar to very well irrigated lands, this would imply a volume of 6.6 $m^3 \cdot s^{-1}$ for riparian corridors considering elms, mesquites and grass. This raw estimation pointed out the amount required from the aquifer to cope with irrigation and riparian corridors without consider other uses. This amount needs to be carefully used since there is not still enough data to provide confident recharge values to these aquifers (Ures and Topahue), however, it is clear the high influence of the ET_a to compute water requirements in arid zones. Further work is required in order to compute accurately each component of the water balance paying also attention to the precipitation and interception effect on the riparian corridor since its rapid recovering after the first precipitation during the monsoon months.

KEYWORDS

- aquifer
- arid zone
- drainage
- evaporative demand
- evapotranspiration
- flood plain
- irrigation
- leaf area index
- MEBES
- Mexico
- photosynthesis

- **phreatophytes**
- **remote sensed data**
- **riparianvegetation**
- **Sebal**
- **semi-aridzone**
- **Sonora – Mexico**
- **Sonora river**
- **Sonora riverwatershed**
- **transpiration**
- **water availability**
- **water demand**
- **watershed**

REFERENCES

1. Allen, Jr. L. H. **1999,** Evaporation Responses of Plants and Crops to carbon dioxide and temperature. 37–70 pages. In: Kirkham M. B. (Editor), *Water Use in Crop Production.* The Harworth Press, New York-USA.
2. Allen, R. G.; Pereira, L. S.; Raes, D.; Smith, M. **1998,** *Crop Evapotranspiration. Guidelines for Computing Crop Water Requirements.* Irrigation and Drainage, Paper No. *56,*300 pages. United Nations Food and Agriculture Organization. Rome, Italy.
3. Bastiaanssen, W. G. M.; Menenti, M.; Feddes, R. A.; Holtslag, A. A. M. A remote sensing surface energy balance algorithm for land (SEBAL), Part 1, Formulation. *J. Hydrology,* **1998a,** *212–213,* 198–212.
4. Bastiaanssen, W. G. M.; Pelgrum, H.; Wang, J.; Ma, Y.; Moreno, J.; Roerink, G. J.; van der Wal T. (1998b). The Surface Energy Balance Algorithm for Land (SEBAL), Part 2, Validation. *J. Hydrology, 212/213,* 213–229.
5. Bastiaanssen, W. G. M.; Bandara, K. M. P. S. Evaporative depletion assessments for irrigated watersheds in Sri Lanka. *Irrig. Sci.;* **2001,** *21,* 1–15.
6. Bendix, J.; Hupp, C. Hydrological and geomorphological impacts on riparian plant communities. *Hydrol. Process.;* **2000,** *14,* 2977–2990.
7. Calder, I. R. **1998,** *Water-resource and land-use issues.* SWIM Paper 3. International Water Management Institute. Colombo, Sri Lanka.
8. Ehlers, W.; Goss M. **2003,** *Water Dynamics in Plant Production.* CABI Publishing, Wallingford, UK.
9. Fischenich, J. C.; Copeland R. R. **2001,** Environmental Considerations for Vegetation in Flood Control Channels. Flood Damage Reduction Research Program. US Army Corps of Engineers, Washington D.C., USA
10. Gazal, R. M.; Scott, R. L.; Goodrich, D. C.; Williams, D. G. Controls on transpiration in a semi-arid riparian cottonwood forest. *Agric. For. Meteorol.* **2006,** *137,* 56–67
11. Goodrich, D. C.; Scott, R.; Qi, J.; Goff, B.; Unkrich, C. L.; Moran, M. S.; Williams, D.; Schaeffer, S.; Snyder, K.; MacNish, R.; Maddock, T.; Pool, D.; Chehbouni, A.; Cooper, D. I.; Eichinger, W. E.; Shuttleworth, W. J.; Kerr, Y.; Marsett, R.; Ni, W. Seasonal estimates of riparian evapotranspiration using remote and in situ measurements. *Agric. For. Meteorol.* **2000,** *105,* 281–309

12. Graf, W. L. *Fluvial Processes in Dryland Rivers*. Springer Series in Physical Environment, 3, 1st Edition, Springer-Verlag, New York, **1988a.**
13. Graf, W. L. Definition of flood plains along arid regions rivers. In: Baker, V. R.; Kochel, R. C.; Patton P. C. (eds) *Flood Geomorphology*, John Wiley and Sons, New York, **1988b,** 231–242.
14. National Institute of Statistics, Geography and Informatics (INEGI). **2000,** Synthesis of the Geographic Information of the Sonora State, Mexico.
15. IPCC. **2007,** Climate Change: Synthesis Report, IPCC, Valencia, Spain, November 2007.
16. Kiniry, J. R. **1998,** Biomass accumulation and radiation use efficiency of honey mesquite and eastern red cedar. *Biomass Bioenergy*, 15(6), 467–473.
17. Kite, G.; Droogers P. **2000,** Comparing estimates of actual evapotranspiration from satellites, hydrological models, and field data: A case study from Western Turkey. Research Report No. 42, International Water Management Institute, Colombo, Sri Lanka.
18. Lakshmi, V.; Susskind J. Utilization of satellite data inland surface hydrology: sensitivity and assimilation. *Hydrol. Process*, **2001,** *15,* 877–892.
19. Leon de la Cruz, J. L.; Rebman, J.; Domínguez-Leòn, M.; Domínguez-Cadena, R. The vascular flora and floristicrelationship of the Sierra de la Garganta in Baja California. *Mexican Biodiversity Magazine*, **2008,** *79,* 29–65.
20. Menenti, M. Evaporation. Chapter 8, **2000,** 156–196. In: *Remote Sensing in Hydrology and Water Management* by Gert A.; Schultz and Edwin T.; Engman, Eds. Springer, Heidelberg-Germany.
21. Molden, D. **1997,** Accounting for water use and productivity. International Irrigation Management Institute, SWIM Paper 1, Colombo, Sri Lanka.
22. NOAA. Oceanic Niño Index: Cold and Warm Episodes by Season, Climate Prediction Centre Web page consulted on May, **2008,** http: //www.cpc.ncep.noaa.gov/products/analysis_monitoring/ensos tuff/ensoyears.shtml
23. Pike, J. G. The estimation of annual run-off from metrological data in tropical climate. *J. Hydrology* **1964,** *2,* 116–123.
24. Qi, J.; Moran, M. S.; Goodrich, D. C.; Marsett, R.; Scoot, R.; Chehbouni, A.; Schaeffer, S. Estimation of evapotranspiration over the San Pedro riparian area with remote and in situ measurements. American Meteorological Society Symposium on Hydrology, Phoenix-Arizona – USA, 11–16 January **1998,** Session 1, Paper 1.13.
25. Ramos, J. G.; Cratchley, C. R.; Kay, J. A.; Casterad, M. A.; Martı'nez-Cob, A.; Domı'nguez, R. Evaluation of satellite evapotranspiration estimates using ground-meteorological data available for the Flumen District into the Ebro Valley of Spain, NE, *Agricultural Water Management*, **2009,** *96,* 638–652
26. Schultz, G. A.; Engman, E. T. Remote Sensing in Hydrology and Water Management. Chapter 1, **2000,** 3–14. In: Gert, A.; Schultz and Edwin, T. Engman, Eds. *Remote Sensing in Hydrology and Water Management.* Springer. Heidelberg, Germany.
27. Scott, R. L.; Goodrich, D. C.; Levick, L. R. A GIS-based management tool to quantify riparian vegetation groundwater use. Proc. 1st Interagency Conf. on Research in the Watersheds, edited by Renard, K. G.; S. McElroy, W.; Gburek, E.; Canfield, and Scott, R. L.; Oct. 27–30, Benson-Arizona – USAAZ, **2003,** 222–227.
28. Turc, L. The report of groundwater: relation between precipitation, evapotranspiration and run-off. Agronomic Annals Serie A: **1954,** 491–595
29. Pike, J. G. Estimation of annual run-off from meteorological data in tropical climate. *J.; Hydrol.* **1964,** *2(2),* 116–123
30. Wester, P.; Scott, C. A.; Burton, M. River basin closure and institutional change in Mexico's Lerma-Chapala Basin. *Chapter 8,* **2005,** 125–144 pages. In: Svendsen, M.; (Ed.) *Irrigation and River Basin Management: Options for Governance and Institutions.* Wallingford, UK: CABI Publishing.

CHAPTER 12

WATER CONSERVATION FOR TURFGRASS[1]

BENJAMIN G. WHERLEY

CONTENTS

[1] Printed with permission and modified from: Benjamin Wherley, 2011. *Turfgrassgrowth, Quality and Reflective Heat Load in Response to Deficit Irrigation Practices.* Chapter 18, 419–431 In: *Evapotranspiration* by LeszekLabedzki (Editor), ISBN 978-953-307-251-7, InTech. Available from: http: //www.intechopen. com/books/evapotranspiration/turfgrass-growth-quality-and-reflective-heat-load-in-response-to-deficit-irrigation-practices © Copyright InterTech.com

12.1   INTRODUCTION

Turfgrass irrigation practices have come under intense scrutiny in recent years due to concerns over increasing population growth and diminishing water availability. Municipal water restrictions have become common place and more recently, the U.S. Environmental Protection Agency has developed guidelines that would restrict irrigation and/or amount of turf within the landscape [15]. Thus, turfgrass managers are increasingly faced with the challenge of maintaining acceptable turfgrass quality using less water. Understanding the minimal irrigation requirements and extent of water stress that a particular turfgrassspecies can tolerate while exhibiting acceptable quality is therefore highly valuable information for turfgrass managers and homeowners.

Deficit irrigation is the practice of intentionally under-irrigating of a plant to below its maximum water demand. This practice has been long used in crop production, where it often culminates in overall reductions in growth, development, and yield. Turfgrass systems are perhaps uniquely adapted for deficit irrigation because reductions in shoot growth are perceived to be beneficial, as long as visual and functional quality are not significantly sacrificed. Deficit irrigation has been practiced across a number of species, although the particular level of irrigation needed to maintain acceptable quality appears to vary among species. Using minilysimeters in the field, DaCosta and Huang [2] determined that bentgrass species required $\geq 60\%$ ET_a for maintaining acceptable summer quality, but that irrigating at only 40% ET_a was sufficient for maintaining acceptable quality during fall months. Qian and Engelke [9] found that minimal irrigation requirements for grasses grown along a linear gradient of irrigation ranged from 26% to 68% of Class A pan evaporation (E_p) in a study of five turfgrass species along a linear gradient of irrigation. Feldhake et al. [3] studied deficit irrigation of three turfgrass species grown in lysimeters and determined that irrigation deficits up to ~27% only decreased growth, but greater deficits resulted in significant loss of quality for Kentucky bluegrass and tall fescue.

Zoysiagrass is a warm-season (C_4) turfgrass native to South-east Asia, but has become an increasingly popular turfgrass for use on lawns and golf courses throughout the southern half of the United States and many other tropical, subtropical, and temperate regions of the world [14]. Whereas some turfgrass species are capable of avoiding drought through production of a deep root system, physiological studies indicate that zoysiagrass tolerates drought largely through osmotic adjustment [10]. Very little information is available regarding the minimal irrigation requirements or response of this species to deficit irrigation practices. Therefore, this chapter presents the research on how to determine the response of 'Empire' zoysiagrass to four levels of deficit irrigation and to identify the maximally acceptable irrigation deficit at which acceptable turf quality could be maintained in this species.

12.2   MATERIALS AND METHODS

This study was carried out from August 27 through October 15, 2008 in a greenhouse at the University of Florida campus in Gainesville, FL. 'Empire' zoysiagrass (*Zoysia japonica* Steud.) was grown in pots constructed from 10 cm diameter, 20 cm tall PVC pipes fitted with a flat end cap. A small hole was drilled into the center of each end cap for drainage. The top, open end of the pipes was fitted with a toilet flange for attaching a photosynthesis chambers. Four weeks prior to deficit irrigation studies, zoysiagrass sod pieces (2.5 cm depth) were removed from established, sand-based research plots at the University of Florida G.C. Horn Memorial Turfgrass Research Field Laboratory, Citra, FL, using 10-cm diameter golf cup cutter. The sod was washed free of soil and established atop medium-coarse textured sand in PVC pots. A complete slow-release fertilizer (24–5–11, Turfgro Professional, Sanford, FL) was applied and grasses were grown in the greenhouse for four weeks to fully root in the soil prior to deficit irriga-tion experiments. Greenhouse temperatures during the study period were controlled at 32/22°C (day/night). During this time, grasses were clipped weekly at a height of 6.4 cm. Visual observations confirmed that grass roots had reached container bottoms at the start of the experiment. A full cover of turf was also present on all pots at the start of the experiment, so that water loss was primarily a function of transpiration.

The study was initiated by fully saturating all pots. Following 24-hour period, when drainage had ceased, holes in pot bottoms were plugged to prevent drainage for the duration of the six-week study. At this time, four pots were randomly selected to be fully watered controls; and the initial, well-watered weights of these pots were mea-sured. These pots were kept fully watered throughout the study by adding water daily in an amount equivalent to 100% of ET_a, measured gravimetrically. In order to more rapidly attain the desired water stress levels within the irrigation deficit treatments, drought-stress treatments were allowed to dry down the soil as a result of transpira-tional water loss, as described by Sinclair and Ludlow, 1986 [12].

Four plants were assigned to each of four deficit irrigation treatments (80%, 60%, 40%, and 20% of ET_a), which were initiated by allowing soil water content to decrease below that of fully watered controls prior to the beginning of the experiment.

For example, in the 80% ET_a treatment, plants were allowed to decrease soil water content until the daily transpiration was measured to be 80% of that for well-watered plants. Upon reaching 80% relative transpiration (RT) rate, this stress level was per-manently maintained by replenishing the pots daily with 80% of ET_a. The number of daysof soil dry down required to reach targeted stress levels varied from 2 days for 0.8 RT (80% ET_a) to 6 days for the 0.2 RT (20% ET_a) stress treatment. Aseach of the stress treatmentswererereached, irrigation was returned to pots daily at 100% (controls), 80%, 60%, 40%, or 20% of ET_a. ET rates within the water stress treatments were also measured each afternoon to determine water loss, and corresponded highly to the prescribed irrigation amounts added to the respective treatments (data not shown). As such, ET of deficit treatments maintained a steady state proportional to fully irrigated controls.

Over the course of the six weeks, data were collected including daily ET rates, turfgrass visual quality, clipping dry weights, degree of leaf wilt and firing, photosyn-thetic rates, and reflective heat load in response to deficit irrigation. Turfgrass visual quality was visually estimated on a 1–9 scale, with 6 representing minimally accept-able turf [7]. For measuring clipping dry weights, grasses were clipped to 6.4 cm weekly, with clippings oven dried at 65°C for 72 hours. Percent leaf wilt and firing were determined weekly by visually estimating the percentage of leaves within the 10 cm diameter plug that were either wilted of firing during the afternoon on a clear day.

A portable chamber system modified from that described by Pickering et al., [8], was used to measure canopy photosynthetic rates within the treatments at the conclusion of the study. This involved a portable photosynthesis system (LI-6200, LI-Cor Inc., Lincoln, NE) with an open leaf chamber mounted inside the translucent canopy chamber (Fig. 1). Chambers caused a 20% reduction in incoming photosynthetic active radiation (PAR). Three readings were recorded for each pot on a cloudless day during the afternoon hours, during which PAR levels within chambers always exceeded 1200 μmol m^{-2} s^{-1}. Temperatures inside chambers during the measurement period averaged 37.7 +/– 0.2 C. Canopy net photosynthetic rate (P_n) was expressed as CO_2 uptake per unit turf canopy area (m^{-2}). Reflective heat load from within the treatments was determined with a handheld Crop Trak Mini IR thermometer (Spectrum Technology, Plainfield, IL) by measuring leaf canopy temperatures during the afternoon on a clear day.

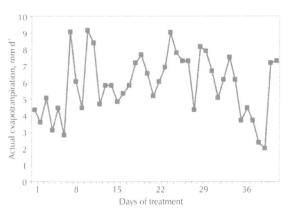

FIGURE 1   Portable chamber system for measuring canopy gas exchange.

12.3   RESULTS AND DISCUSSIONS

The objective of this study was to determine the response of 'Empire' zoysiagrass to four levels of deficit irrigation, provided daily. Actual evapotranspiration (ET_a) and (equivalent to irrigation requirements) for control plants over the 42-day study ranged from 2 to 9 mm d^{-1}, with total irrigation volume applied of 246 mm (Fig. 2). Deficit irrigated treatments received a total of 197 mm (80% ET_a), 157 mm (60% ET_a), 126 mm (40% ET_a), and 101 mm (20% ET_a).

12.3.1   VISUAL QUALITY

Our results demonstrated that irrigating zoysiagrass at > 60% of ET_a was sufficient to sustain acceptable turfgrass quality over the six-week study (Figs. 3 and 9). Conversely, visual quality of 40% ET_a treatments gradually declined over the study, but fell to unacceptable levels after five weeks. Within two weeks of initiating treatments, turf receiving 20% ET_a declined rapidly to unacceptable levels characterized by rapid wilt and significant leaf firing throughout the entire turf canopy. Previous work has shown that irrigating to 80% ET_a twice weekly was necessary to maintain acceptable zoysiagrass quality [4] and irrigating above 73% ET_a three times weekly was necessary for maintaining adequate Kentucky bluegrass quality [3]. Our results indicate that acceptable zoysiagrass quality may be achievable at even greater irrigation deficits if irrigation is supplied with greater frequency.

12.3.2   PHOTOSYNTHESIS AND SHOOT GROWTH

Canopy photosynthetic rates of fully watered (100% ET_a) plants exceeded that of all deficit irrigated treatments and decreasing irrigation amounts resulted in proportional reductions in photosynthetic rates (Fig. 4). Rates of photosynthesis ranged from an average of 6.1 µmol m^{-2} s^{-1} at 100% ET_a to 0.8 µmol m^{-2} s^{-1} at 20% ET$_a$. This is not surprizing, given that dry matter accumulation and transpiration have been shown to be intimately linked to leaf gas exchange through stomata [13].

Shoot growth is a process that is dependent on increases in cell volume, and thus, is generally highly sensitive to water deficit. Turfgrass shoot growth in this study was measured by weekly clipping collections, and was found to be progressively reduced with decreasing irrigation (Fig. 5). Our data show that reducing irrigation to levels of 60% of ET$_a$, while causing little change in overall quality, led to as much as 25% reductions in shoot growth. Increased rates of turfgrass shoot growth have been correlated with higher turfgrass ET in a number of other warm and cool-season turfgrasses [1, 11, 6]. Interestingly, while there was generally a proportional decrease in transpiration and shoot growth as deficits increased; shoot dry weights did not change between the 60% and 80% ET_a treatments. Thus, in terms of shoot growth, transpirational water use efficiency was greatest at 60% ET_a with this species (data not shown). From a practical standpoint, whereas soil water deficits may negatively affect field crop yields, moderate growth reductions in turfgrass systems may actually be desirable, as they could result in less mowing requirements.

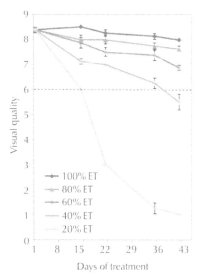

FIGURE 2   Mean evapotranspiration (ET$_a$) rates for 100% ET control plants over the six-week study. Amounts are equivalent to the daily irrigation supplied to 100% ET control plants, with deficit ET treatments receiving a fraction (80, 60, 40, or 20%) of this amount.

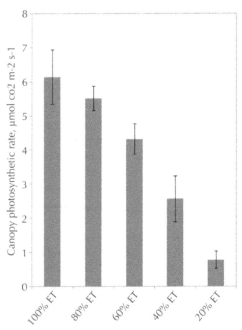

FIGURE 3 Visual quality of turfgrass plants over six-week experiment, based on a 0–9 scale, with 0 = brown, dead turf, 6 = minimally acceptable and 9 = optimal color, density, and uniformity. Error bars denote standard error of the mean (n=4).

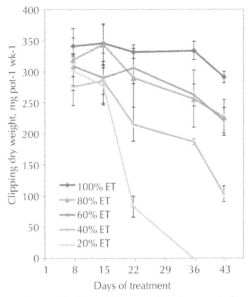

FIGURE 4 Canopy photosynthesis rates measured at the conclusion of the six-week study. Error bars denote standard error of the mean (n=4).

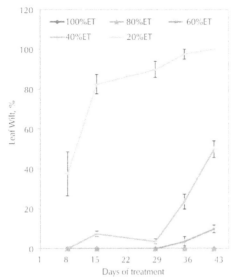

FIGURE 5 Clipping dry weights collected weekly from 81 cm² pots. Error bars denote standard error of the mean (n=4).

12.3.3   LEAF WILT AND FIRING

Leaf wilt and firing were the primary symptoms of plant water stress leading to quality loss in grasses receiving $\leq 60\%$ of ET_a in the study (Figs. 6 and 7). Leaf canopies of 20% ET_a plants declined most rapidly, with half of the canopy wilted within two

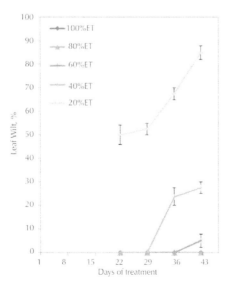

FIGURE 6 Percentage of leaf wilt visible within the 81 cm² turfgrass canopy. Measurements were obtained during mid-afternoon hours on cloudless days. Error bars denote standard error of the mean (n=4).

weeks of initiating treatments, and becoming almost entirely fired by week six. 40% ET_a plants were nearly 50% wilted, with 30% of the canopy firing by week six. Only 5–10% of the canopy in 60% ET_a plants showed signs of wilt or firing; thus, quality was deemed acceptable throughout the study. Zoysiagrass irrigated at ≥80% ET_a never showed signs of wilt during any measurement period, indicating that soil moisture was sufficient to meet transpirational demand.

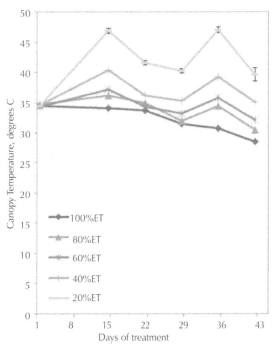

FIGURE 7   Percentage of leaf firing visible within the 81 cm² turfgrass canopy. Measurements were obtained during mid-afternoon hours on cloudless days. Error bars denote standard error of the mean (n=4).

12.3.4   REFLECTIVE HEAT LOAD

A substantial amount of incident solar radiation is converted to latent heat during the transpiration process. Therefore, vegetated surfaces such as turfgrass systems possess significant cooling capacity and play a critical role in heat dissipation within urban areas. It has been reported that green turfs and landscapes can save energy by reducing the energy input required for interior mechanical cooling of adjacent homes and buildings [5]. Decreasing irrigation levels for water conservation conserves resources and maintenance requirements, but it also is likely to substantially impact reflective heat loads. Over the five sampling dates, we observed up to a 16°C increase in canopy temperatures from fully irrigated to 20% ET_a irrigated turf (Fig. 8). Interestingly, this heat load difference was reduced by half when irrigation was only slightly increased to 40% ET_a due to the presence of significantly more green vegetation (Fig. 9). Our data for zoysiagrass are similar to those of Ref. [3], who reported a 1.7°C increase in

Kentucky bluegrass canopy temperature for each 10% reduction in ET. While a nonirrigated control treatment was not used in this study, it seems likely that even greater heat loads would result from turf receiving no irrigation, as no living vegetation would likely be present. From a practical standpoint, these data suggest that large-scale landscape deficit irrigation practices, such as those mandated through municipal water restrictions, could contribute to significantly increased surface temperatures.

FIGURE 8 Canopy temperatures within the irrigation treatments. Measurements were obtained during mid-afternoon hours on cloudless days. Error bars denote standard error of the mean (n=4).

FIGURE 9 Visual appearance of 'Empire' zoysiagrass following six weeks of deficit irrigation, applied daily. (From left to right: 20% ET_a, 40% ET_a, 60% ET_a, 80% ET_a, and 100% ET_a).

12.4 CONCLUSIONS

On the basis of our results, deficit irrigation can be a useful means of conserving water in the management of turfgrass. Irrigating zoysiagrass at up to a 40% deficit (60% ET_a) was sufficient to maintain acceptable turfgrass quality over a six-week period. Zoysiagrass response to the deficit irrigation included a reduction in evapotranspiration and photosynthetic rates, as well as shoot growth reductions. Such effects on shoot growth may be viewed as beneficial, as theywould likely result in fewer mowing requirements. Our results suggest that irrigating this species at >40%irrigation deficits would not be advizable, as significant leaf wilt and firing occurred which negatively affected the appearance of the turf canopy and produced less-than-acceptable visual quality of the turf. While we were able to maintain acceptable quality at these levels with daily watering, it is likely that less frequent irrigation scheduling (2 or 3 days per week) might result in greater loss of quality at similar levels of ET replacement, due to longer periods of water stress encountered. An important consideration with deficit irrigation practices is the increasing reflective heat loads that are generated with diminishing amounts of irrigation. Canopy temperatures in this study progressively increased with greater deficits to a maximum increase of 16° C between 100% and 20% ET_a plants. This is an important consideration that could have large-scale implications on the surface temperatures and human comfort levels within urban environments.

12.5   SUMMARY

Water conservation in the landscape has become critically important in recent years due to diminishing water supplies and population growth. Deficit irrigation, or irrigating to a fraction of a plant's maximal transpiration rate is one-way water may be conserved, but this practice may impact turfgrass quality, performance, and heat load within the landscape. The objective of this study was to determine the response of a commonly used warm-season (C_3) turfgrass 'Empire' zoysiagrass (*Zoysia japonica* Steud.) to deficit irrigation practices. Pots of established zoysiagrass were irrigated at various levels of deficit irrigation including 80%, 60%, 40%, and 20% of actual evapotranspiration (ET_a) of well-watered turf over a six-week period. Over the course of the study, changes in shoot growth, canopy temperatures, turfgrass quality, and rates of photosynthesis were recorded. Results of the study indicated very little change in quality, growth, and photosynthetic rates between plants irrigated at either 100% or 80% of daily ET_a. Supplying turf with only 60% of maximal irrigation produced lower, but still acceptable quality. Deficit irrigation levels of 40% and 20% were not sufficient to maintain acceptable quality and promoted significantly greater reflective heat loads. The results of this study indicate that deficit irrigation can be an important tool for conserving water in managed turfgrass, however, the effect of the practice on reflective heat loads is an important, but often overlooked consideration that could impact the energy balance within the urban environment.

KEYWORDS

- canopy temperatures
- deficit irrigation
- photosynthesis
- shoot growth
- transpiration
- turfgrass
- turfgrass evapotranspiration
- turfgrass irrigation
- turfgrass water conservation
- Turfgrass water requirements
- water use
- Zoysiagrass (Zoysia japonica)

REFERENCES

1. Bowman, D. C.; Macaulay, L. Comparative evapotranspiration rates of tall fescue cultivars. *HortScience* **1991,** *26,* 122–123.
2. DaCosta, M.; Huang, B. Minimum water requirements for creeping, colonial, and velvet bentgrasses under fairway conditions. *Crop Sci.* **2006,** *46,* 81–89.
3. Feldhake, C. M.; Danielson, R. E.; Butler, J. D. Turfgrass evapotranspiration. II.; Responses to deficit irrigation. *Agron. J.* **1984,** *76,* 85–89.
4. Fu, J.; Fry, J.; Huang, B. Minimum water requirements of four turfgrasses in the transition zone. *Hort Science* **2004,** *39,*1740–1744.
5. Johns, D.; Beard, J. B. A quantitative assessment of the benefits from irrigated turf on environmental cooling and energy saving in urban areas. In: *Texas Turfgrass Research.* **1985,** 134–142. Texas Agric. Exp. Stn. PR-4330. College Station.
6. Kim, K. S. Comparative evapotranspiration rates of 13 turfgrasses grown under both nonlimiting soil moisture and progressive water stress conditions. **1983,** 64. M. S. thesis, Texas A & M University, Colege Station, Texas.
7. Morris, K. N.; Shearman, R. C. **2010,** NTEP Evaluation Guidelines. National Turfgrass Evaluation Program. Available at: http: //www.ntep.org/pdf/ratings.pdf Accessed 6 September 2010.
8. Pickering, N. B.; Jones, J. W.; Boote, K. J. Evaluation of the portable chamber technique for measuring canopy gas exchange by crops. Agric. For. Meteorol. **1993,** *63,* 239–254.
9. Qian, Y. L.; Engelke, M. C. Performance of five turfgrasses under linear gradient irrigation. *Hort Science* **1999,** *34(5),* 93–896.
10. Qian, Y. L.; Fry, J. D. Water relations and drought tolerance of four turfgrasses. *J. Am. Soc. Horticultural Sci.* **1997,** *122,* 129–133.
11. Shearman, R. C. Kentucky bluegrass cultivar evapotranspiration rates. *HortScience* **1986,** *21,* 455–457.
12. Sinclair, T. R.; Ludlow, M. M.; Influence of soil water supply on the plant water balance of four tropical grain legumes. Aust. J.; Plant Physiol. **1986,** *13,* 329–341.
13. Sinclair, T. R.; Tanner, C. B.; Bennet, J. M. Water-use efficiency in crop production. BioScience **1984,** *34,* 36–40.
14. Turgeon, A. J. **2002,** Turfgrass Management. Prentice Hall, ISBN 0-13-027823-8, Upper Saddle River, NJ, USA.
15. WaterSense® Single Family New Home Specification. **2009,** http: //www.epa.gov/WaterSense/docs/home_finalspec508.pdf

CHAPTER 13

EVAPOTRANSPIRATION WITH MODIFIED HARGREAVES MODEL: SAUDI ARABIA[1]

MOHAMMAD NABIL ELNESR and ABDURRAHMAN ALI ALAZBA

CONTENTS

[1]M. N. Elnesr, PhD, Assistant Professor, Shaikh Mohammad Alamoudi Chair for Water Research, King Saud University, Kingdom of Saudi Arabia, PO Box: 2460/11451, Riyadh. E-mail:melnesr@ksu.edu.sa or drnesr@gmail.com; A. A. Alazba, Professor in Water Research, King Saud University; and Amin, M. T. Amin, Assistant Professor in Water Research, King Saud University. The authors wish to express thanksto "Shaikh Mohammad Bin Husain Alamoudi" for his kind financial support to the King Saud University [http://www.ksu.edu.sa]. This study is part of the AWC chairactivities in the "Projects and Research" axis.We also thank the Presidency of Meteorology and Environment in Riyadh - KSA for providing the meteorological data. © Copyright InterTech.com

13.1 INTRODUCTION

Estimation of reference evapotranspiration (ET_o) is important for several hydrological and agricultural sciences, like water resources management, crop-water requirements, irrigation scheduling, and land use planning [21, 26]. Though, a good amount of research is done by the scientists to estimate the ET_o accurately, especially for the sensitive fields like irrigation scheduling [9, 25]. Due to its comprehensive theoretical base, the Penman–Monteith method [PM] method is recommended by the United Nations Food and Agricultural Organization (FAO) as the sole method to calculate (ET_o) and for evaluating other ET_o calculation methods as well [3, 19]. The FAO approach to calculate ET_o using the PM method was published in the FAO irrigation and drainage paper number 56 [FAO-56]. The objectives of this chapter were: (1) To investigate whether the ET_o values calculated by PM method are affected significantly by Allen [2] temperature adjustments or not, and (2) To adjust the Hargreaves equation's parameters to use it in prediction of the actual ET_o values from nonstandard weather stations, for all zones of the Kingdom of Saudi Arabia. The "FAO-56 PM" method is shown in Eq. (1) and requires measurements of air temperature, relative humidity, solar radiation, and wind speed.

$$ET_o = \frac{0.408\,\Delta\,(R_n\text{-}G) + \left(\frac{900}{T_a+273}\right)\gamma U_2\left(e_s\text{-}e_a\right)}{\Delta + \gamma(1+0.34\,U_2)} \tag{1}$$

where: Δ is a slope of the vapor pressure curve [kPa.°C^{-1}], R_n is net radiation [MJ.m^{-2}.day^{-1}], G is soil heat flux density [MJ.m^{-2}.day^{-1}], γ is psychrometric constant [kPa.°C^{-1}], T is mean daily air temperature at 2 m height [°C], U_2 is wind speed at 2 m height [m.s^{-1}], e_s is the saturated vapor pressure at average air temperature T_a, and e_a is the actual vapor pressure [kPa]. Eq. (1) applies specifically to a hypothetical reference crop with an assumed crop height of 0.12 m, a fixed surface resistance of 70 sec.m^{-1} and an albedo of 0.23.

The formulas of the equation's parameters are detailed by Allen et al., [3] in chapters 2 and 3. Nonetheless, the FAO-56 method requires climatic data from standard weather stations. Standard FAO56 weather stations are stations having "auniform surface of dense, actively growing vegetation having specified height and surface resistance, not short of soil water, and representing an expanse of at least 100 m of the same or similar vegetation" [6]. Allen [2] reported that data gathered from nonstandard weather station should be adjusted before using it in *ETo* calculation. He suggested a method for this data-adjustment especially for temperature. On the other hand, Jia et al., [18] concluded that nonstandard weather stations yield to nonsignificant bias from the correct *ETo* value, except under high wind speed conditions where the bias is significant.

In fact, the number of weather stations where there are reliable data for these parameters are limited [25] especially for Middle-Eastern and developing countries [10]. Hargreaves [13] developed a temperature based formula to find the *ETo* and it was further modified by Hargreaves and Samani [16]. The modified version is called Hargreaves-85 equation in this chapter and is shown in Eq. (2):

$$ET_o = 0.0135R_s(T_a + 17.8) \tag{2}$$

where, R_s = Global solar radiation, same units as ETo; Ta = Average temperature, °C.

The Eq. (2) was evaluated at several locations [14]. Hargreaves and Samani [15] formula estimates global solar radiation, R_s, when the temperature range (TR) and the extraterrestrial radiation (R_a) are known.

$$R_s = k_r R_a TR^{0.5} \tag{3}$$

In eqs. (3): TR is a temperature range; and kr is an empirical coefficient depending on the station location (Also called coastality value).

The extraterrestrial radiation (Ra) is calculated using procedure described by Allen et al. [3] as shown in Eqs. (4) to (8). Since the Hargreaves' formula requires very few climatic data than the PM-FAO56 method; it urges many investigators to test its reliability compared to PM method. Allen et al., [3]; and Sau et al., [24] concluded the importance of local calibration of the HRG to ensure reliable data. Lee [20] found that the HRG formula provides excellent results for the Korea Peninsula after local calibration. So did Jamshidi et al., [17] for different climates in Iran. Amatya et al., [5] and Trajkovic [27] reported that in humid regions; HRG provides an overestimate to the PM values. Martínez-Cob and Tejero-Juste [22] found that the Hargreaves method overestimated ETo at stations with wind speeds less than 2 m/s in Spain.

Gavilán et al., [12] reported similar results for inland locations with low wind speeds and large daily temperature ranges, and for coastal locations with high wind speeds and small daily temperature ranges. On the other hand, for semiarid environments, the HRG method was found to underestimate the PM values in most stations of Iran [19], Turkey [8], Australia [7], and Jordan [29]. For hyper-arid regions, the HRG equation also underestimates the ETo values for Saudi Arabia as reported by several investigators [1, 4, 23]. The extraterrestrial radiation (Ra) is calculated using procedure described by Allen et. al. [3] as shown in eqs. (4) to (8). The equation /4/ indicates extraterrestrial radiation, R_a, with latitude in radians:

$$R_a = 37.6d_r(\omega_s \sin \varphi \sin \delta + \sin \omega_s \cos \varphi \cos \delta) \tag{4}$$

In eq. (8): d_r, δ, and ω_s are defined in eqs. (5) to (8) below:
d_r is a relative distance from earth to sun (Eq. (5)):

$$d_r = 1 + 0.033 \cos(0.0172J) \tag{5}$$

Solar declination in radians is given by eq. (6):

$$\delta = 0.409 \sin(0.0172J - 1.39) \tag{6}$$

Sunset hour angle in radians is determined by eq. (7), with φ = latitude in radians:

$$\omega_s = \arccos(-\tan\varphi \tan\delta) \tag{7}$$

In eq. (8): J = Julian day, M = Month of the year and D = Day of the month. J ranges from 1 to 366 (366 is for leap year).

$$J = \text{int}\left(\frac{275}{9}M + D - 30\right), \text{ "if } (M > 2): \text{ For leap year, subtract 1; and for not leap year,}$$
subtract 2". (8)

Dew point temperature:

$$T_d = \frac{116.91 + 237.3 \; ln(e_a)}{16.78 - ln(e_a)}$$ (9)

Actual vapor pressure:

$$e_a = 0.005RH_a(e_a[T_n] + e_o[T_x])$$ (10)

Saturation vapor pressure:

$$e_o[T] = 0.611.\exp[(17.27*T)/(T + 237.3)]$$ (11)

Temperature difference between minimum and dew point temperatures:

$$\Delta T = T_n - T_d$$ (12)

Corrected value of maximum temperature, for ΔT= Temperature difference >2:

$$T_x^{\{corr\}} = T_x - 0.5(\Delta T - 2)$$ (13)

13.2 METHODS AND MATERIAL

13.2.1 WEATHER DATA

The most reliable weather dataset in the Saudi Arabia is the database of the Presidency of Meteorology and Environment (PME) in KSA, the official climate agency in the country. Weather stations are equipped with up-to-date monitoring devices and are subjected to regular inspection and replacement for defected devices (personal communication with the PME). This Database represents daily values of 29 meteorological stations distributed spatially over the 13 districts and temporally over a time span since 1980 to the end of 2010 for most of the stations. Details about the weather stations and associated parameters are presented in Table 1. All the PME stations are located inside airports for aviation information services; hence these stations are not standard FAO-56 weather stations, and accordingly Allen [2] corrections of temperature were applied. The recorded climatic factors for all stations are dry bulb and wet bulb temperatures (max., min., and avg.), relative humidity (max., min., and avg.), rainfall, wind speed (average), wind direction, atmospheric pressure (sea and station levels), the cloud cover, and the actual vapor pressure. Most of the listed factors were used to find the FAO-56 *ETo* values.

13.2.2 ESTIMATION OF REFERENCE EVAPOTRANSPIRATION WITH FAO-56 PM METHOD

The procedure is described below to calculate the reference evapotranspiration (*ETo*) using FAO-56 PM method.

Step 1: Calculate the global solar radiation using Eq. (3) and coastality value (kr) for each weather station from Table 1.

Step 2: Calculate the Extraterrestrial radiation, R_a, with Eqs. (4) to (8).

Step 3: The parameters in Eq. (1) were calculated using the procedure described by Allen et al., [3] and summarized by ElNesr and Alazba [10]. The Eqs. (9) to (12) were used.

- Calculate the dew point temperature (Td) with Eq. (9).
- Gather data for Tn, minimum air temperature (°C); Tx, maximum air temperature (°C); RHx, maximum relative humidity (%); RHn, minimum relative humidity [%]; $eo(T)$
- Calculate the actual vapor pressure (e_a) with Eq. (10).
- Calculate the saturation vapor pressure ($eo(T)$) with Eq. (11).
- Calculate the temperature difference between minimum and dew point temperatures with Eq. (12).
- Calculate the ETo values by substituting values of parameters (with Eqs. (9) to (12)) in Eq. (1).

Step 4: The ETo values were calculated for daily basis and then were averaged for monthly basis.

TABLE 1   The Characteristics of weather stations in Kingdom of Saudi Arabia.

Weather station ID	Lat. deg. N	Long. deg. East	Elev. m	Zone	Logged years	Avg. temp. °C	Coastality value Kr*
01 A'Dhahran	26.16	50.10	17	Coastal	30	26.48 ± 7.49	0.184
02 Abha	18.14	42.39	2093	Interior	30	18.60 ± 3.72	0.171
03 Ad Dammam	26.42	50.12	1	Coastal	10	26.72 ± 7.86	0.179
04 Al Ahsa	25.30	49.48	179	Interior	25	27.26 ± 8.27	0.173
05 Al Baha	20.30	41.63	1652	Interior	25	22.83 ± 4.92	0.178
06 Al Jouf	29.47	40.06	671	Interior	30	22.03 ± 8.61	0.177
07 Al Medina	24.33	39.42	636	Interior	30	28.45 ± 7.00	0.178
08 Al Qaisumah	28.32	46.13	358	Interior	30	25.23 ± 9.28	0.173
09 Al Qassim	26.18	43.46	650	Interior	30	24.94 ± 8.26	0.166
10 Al Quraiat	31.50	37.50	560	Interior	5	20.04 ± 7.91	0.167
11 Al Wajh	26.12	36.28	21	Coastal	30	25.00 ± 3.98	0.216
12 Arar	31.00	41.00	600	Interior	30	22.01 ± 9.17	0.173
13 Ar Riyadh Middle	24.63	46.77	624	Interior	30	26.66 ± 8.10	0.181
14 Ar Riyadh North	24.42	46.44	611	Interior	25	25.80 ± 8.12	0.167
15 Ar Ta'if	21.29	40.33	1454	Interior	30	22.92 ± 5.12	0.172
16 Bishan	19.59	42.37	1163	Interior	30	25.69 ± 5.40	0.158
17 Gizan	16.54	42.35	3	Coastal	30	30.22 ± 2.80	0.209
18 Hafr El-Batin	28.20	46.07	360	Interior	20	25.26 ± 9.21	0.172

TABLE 1 *(Continued)*

Weather station ID	Lat. deg. N	Long. deg. East	Elev. m	Zone	Logged years	Avg. temp. °C	Coastality value Kr*
19 Hail	27.26	41.41	1013	Interior	30	22.47 ± 8.20	0.172
20 Jeddah	21.30	39.12	17	Coastal	30	28.23 ± 3.50	0.197
21 Khanis Mushait	18.18	42.48	2057	Interior	30	19.49 ± 3.75	0.168
22 Makkah	21.40	39.85	213	Interior	25	30.78 ± 4.57	0.179
23 Najran	17.37	44.26	1210	Interior	30	25.51 ± 5.54	0.166
24 Rafha	29.38	43.29	447	Interior	30	23.33 ± 9.05	0.169
25 Sharurrah	17.47	47.11	725	Interior	25	28.57 ± 5.81	0.168
26 Tabuk	28.22	36.38	776	Interior	30	21.99 ± 7.53	0.170
27 Turaif	31.41	38.40	818	Interior	30	19.06 ± 8.27	0.173
28 Wadi Al Dawasir	20.5	45.16	652	Interior	25	28.15 ± 7.01	0.168
29 Yenbo	24.09	38.04	6	Coastal	30	27.56 ± 4.72	0.184

*The coastality value data were gathered from El-Nesr et.al. (2011)

13.2.3 REESTIMATION OF REFERENCE EVAPOTRANSPIRATION WITH FAO-56 PM METHOD AND TEMPERATURE CORRECTIONS

Step 5: The *ETo* values with FAO-56 PM method were recalculated after applying Allen [2] temperature corrections and Eqs. (9) to (13). The following procedure was followed for this step:

- Calculate the dew point temperature (*Td*) with Eq. (9).
- Gather data for *Tn,* minimum air temperature (°C); *Tx,* maximum air temperature (°C); *RHx,* maximum relative humidity (%); *RHn,* minimum relative humidity (%); $eo(T)$
- Calculate the actual vapor pressure (e_a] with Eq. (10).
- Calculate the saturation vapor pressure ($eo(T)$) with Eq. (11).
- Calculate the temperature difference between minimum and dew point temperatures with Eq. (12).
- For arid and semiarid regions, calculate corrected value of maximum temperature adjustment using Eq. (13) for ΔT= Temperature difference >2. $T^{(Corr)}$ stands for a corrected value in Eq. (13). Repeat the same procedure for minimum temperature (T_n). Note that no correction is needed for $\Delta T \leq 2$.
- Applying the temperature corrections, the ET_o is recaluclated and is renamed as $ET_o^{(Corr)}$.

13.2.4 ESTIMATION OF REFERENCE EVAPOTRANSPIRATION WITH HARGREAVES – SAMANI METHOD

Reference evapotranspiration [*ETo*] values were estimated using Hargreaves – Samani method [HRG method] and the Eq. (2). In Eq. (2), *Rs* value was calculated using Eq.

(3). The monthly *ETo* values were calculated as the average of the daily-calculated values.

To find a best fit for each station, the linear regression analysis [$y = a + b\,x$, where *a* and *b* are regression coefficients] was used between "the *ETo* values calculated by HRG method" and "both uncorrected and corrected PM values." The resultant Hargreaves equation was compared to the original corrected and uncorrected PM values. The prediction precision was tested with statistical indices: The root mean squared deviation [RMSD: coefficient of variation (RMSD$_{CV}$)]; and the mean percent error (MPE). The procedure is described in Eqs. (14) and (15), where *F* is a forecasted (estimated) value; *A* is an actual (measured) value; *n* is a number of observations; and *i* is a counter.

$$\text{RMSD} = \sqrt{\frac{1}{n}\Sigma_{i=1}^{n}(F_i - A_i)^2} \tag{14}$$

$$MPR = \sqrt{\frac{100}{n}\Sigma_{i=1}^{n}\frac{(F_i - A_i)}{A_i}} \tag{15}$$

13.3   RESULTS AND DISCUSSION

The average monthly values of PM for each station were calculated by three methods; the normal PM method without corrections (PM$^{\{norm\}}$), the corrected PM method (PM$^{\{Corr\}}$) and HRG method. As illustrated in Fig. 1 *a, b*, the HRG method seemed to underestimate the PM$^{\{norm\}}$ method for almost all the 29 studied stations except two stations namely Makkah, and Gizan, in addition to AlWajh station in some months. In Makkah, HRG method overestimated the PM$^{\{norm\}}$ and PM$^{\{Corr\}}$ in all months. Makkah city has special climate (steady hot all the year), and special topography (very rugged mountains and valleys), and even special demographic nature (due to its religious characteristics, it is always occupied by pilgrims and visitors). These special circumstances made this arid city (Makkah) to act like a nonarid city [11]. In humid locations, HRG method always overestimates PM as discussed earlier. For similar reason, Gizan, the semiisland, behaved like humid location and the overestimation occurred. In AlWajh station, HRG method overestimated the PM$^{\{norm\}}$ in summer months (June, July, August, and September) while its underestimation occurred for the rest of the year. This could be due to the weather effects for its coastal location at the top-north of the kingdom's shore on the Red Sea and exposure to the Mediterranean Sea's weather effects as well.

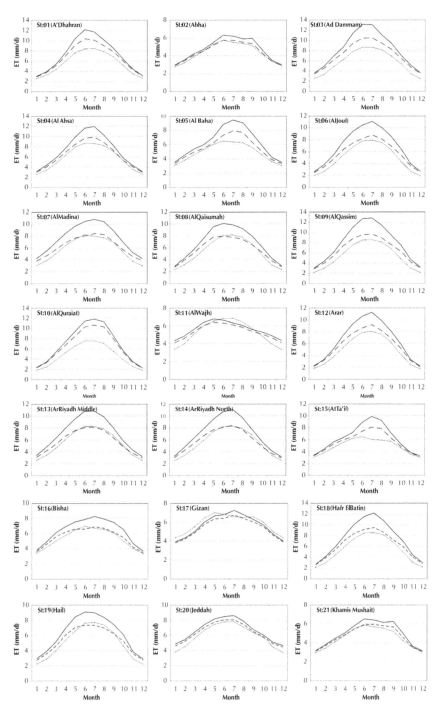

FIGURE 1A Monthly ET_o values of the studied stations calculated by PM$^{\{norm\}}$, PM$^{\{Corr\}}$, and HRG.

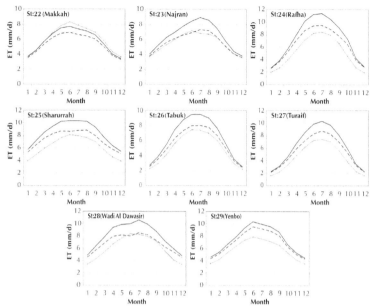

FIGURE 1B Monthly ET_o values of the studied stations calculated by PM$^{\{norm\}}$, PM$^{\{Corr\}}$, and HRG.

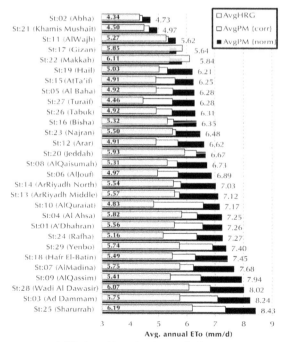

FIGURE 2 Average annual ETo (mm/day) for the studied stations sorted ascending from top to bottom. Numbers outside series indicate PM$^{\{norm\}}$ values, while numbers inside series indicate HRG values.

TABLE 2   Percent of error between the corrected ETo value and the normal ETo value.

Station	Error, %	Station	Error, %
17 Gizan	4.36	11 Al Wajh	5.01
20 Jeddah	5.24	21 Khamis Mushait	6.19
02 Abha	6.37	29 Yenbo	7.01
22 Makkah	8.23	10 Al Quraiat	8.86
01 A' Dhaharan	10.53	15 At Ta'if	11.75
04 Al Ahsa	13.78	05 Al Baha	13.83
27 Turaif	14.28	25 Sharurrah	14.34
24 Rafha	14.41	16 Bisha	14.62
26 Tabuk	15.09	23 Ad Dammam	15.38
12 Arar	16.12	19 Hail	16.37
28 Wadi Al Dawasir	16.77	18 Hafr El-Batin	17.50
08 Al Qaisumah	18.35	06 Al Jouf	18.77
14 Ar Riyadh North	20.95	09 Al Qassim	21.46
07 Al Madina	22.27	13 Ar Riyadh Middle	23.27

TABLE 3   Calculation formulas of PM {Corr} when the PM{norm} is known.

Station ID	Formula, $Y = mX + C$	r^2
01 A'Dhahran	$PM\{corr\} = 0.821 \times PM\{norm\} + 0.610$	0.995
02 Abha	$PM\{corr\} = 0.851 \times PM\{norm\} + 0.420$	0.989
03 Ad Dammam	$PM\{corr\} = 0.744 \times PM\{norm\} + 0.964$	0.990
04 Al Ahsa	$PM\{corr\} = 0.781 \times PM\{norm\} + 0.713$	0.993
05 Al Baha	$PM\{corr\} = 0.758 \times PM\{norm\} + 0.758$	0.987
06 Al Jouf	$PM\{corr\} = 0.734 \times PM\{norm\} + 0.639$	0.994
07 Al Medina	$PM\{corr\} = 0.701 \times PM\{norm\} + 0.849$	0.998
08 Al Qaisumah	$PM\{corr\} = 0.731 \times PM\{norm\} + 0.748$	0.977
09 Al Qassim	$PM\{corr\} = 0.689 \times PM\{norm\} + 1.023$	0.983
10 Al Quraiat	$PM\{corr\} = 0.874 \times PM\{norm\} + 0.319$	0.995
11 Al Wajh	$PM\{corr\} = 0.925 \times PM\{norm\} + 0.155$	0.993
12 Arar	$PM\{corr\} = 0.774 \times PM\{norm\} + 0.566$	0.990
13 Ar Riyadh Middle	$PM\{corr\} = 0.683 \times PM\{norm\} + 0.879$	0.985
14 Ar Riyadh North	$PM\{corr\} = 0.697 \times PM\{norm\} + 0.886$	0.983
15 Ar Ta'if	$PM\{corr\} = 0.735 \times PM\{norm\} + 0.997$	0.978
16 Bishan	$PM\{corr\} = 0.744 \times PM\{norm\} + 0.815$	0.977
17 Gizan	$PM\{corr\} = 0.895 \times PM\{norm\} + 0.357$	0.996
18 Hafr El-Batin	$PM\{corr\} = 0.736 \times PM\{norm\} + 0.811$	0.983
19 Hail	$PM\{corr\} = 0.750 \times PM\{norm\} + 0.663$	0.988
20 Jeddah	$PM\{corr\} = 0.919 \times PM\{norm\} + 0.207$	0.997

TABLE 3 *(Continued)*

Station ID	Formula, Y = mX +C	r^2
21 Khanis Mushait	PM{corr}= 0.858x PM{norm}+ 0.415	0.987
22 Makkah	PM{corr}= 0.850x PM{norm}+ 0.429	0.996
23 Najran	PM{corr}= 0.720x PM{norm}+ 0.954	0.955
24 Rafha	PM{corr}= 0.787x PM{norm}+ 0.633	0.993
25 Sharurrah	PM{corr}= 0.741x PM{norm}+ 1.125	0.956
26 Tabuk	PM{corr}= 0.797x PM{norm}+ 0.451	0.993
27 Turaif	PM{corr}= 0.798x PM{norm}+ 0.483	0.994
28 Wadi Al Dawasir	PM{corr}= 0.698x PM{norm}+ 1.227	0.962
29 Yenbo	PM{corr}= 0.889x PM{norm}+ 0.336	0.995

Y =PM{corr}, X = PM{norm}, m = slope of a line, and C = intercept

For somewhat simpler explanation, Fig. 2 illustrates the annual averages of *ETo* calculated by the three methods for the 29 stations. The overestimation in Gizan and Makkah is clear in Fig. 2, while the partial overestimation appeared for AlWajh station. It appears that Abha station has the least *ETo* in all over the kingdom with an average annual value of 4.73 mm/d. This low value of *ETo*, however, was due to the cold weather of the region due to its high altitude (2093 m above sea level, Table 1). The maximum average annual *ETo*, 8.43 mm/d, was found for Sharurrah station which is the nearest urban zone to the Empty Quarter region (Rub' al Khali). The Empty Quarter region is one of the largest hyper-arid sanddesertsin the world with very high temperatures of about 47°C in normal days and up to 60°C during July and August [25, 28].

For all of the stations except Gizan and Makkah, HRG method resulted in the least value of the *ETo* followed by the PM [Corr] and the PM[norm] methods. This could mean that ET_o values from nonstandard weather stations were always overestimated if calculated from standard AWSs. This could be attributed to the temperature drop in the standard weather stations due to the surrounding vegetation cover. In some stations, however, the overestimating ratio (error) was small, like in Gizan, AlWajh, and Jeddah; where the error was around 5% (Table 2). Table 2 summarizes the error values with eight stations below 10%, between 10 and 15% for similar number of stations, between 15% and 20% for eight stations, and five stations having more than 20% error. The importance of knowing these values was to think twice when dealing with uncorrected values of PM *ETo* in the stations with huge errors, especially for the top five stations of AlJouf, ArRiyadh North, AlQassim, AlMadina, and ArRiadh Middle having error values of 20.95%, 21.46%, 22.27%, 23.27%, and 23.96%, respectively.

In fact, these zones are occupied by massive agricultural projects in addition to "Wadi Al Dawasir" and llllfortllllately some of these agricultural projects probably use the nonstandard weather data to calculate *ETo* resulting up to 25% increase in water usage. This situation clearly demonstrates the importance of using standard weather stations' data in *ETo* calculation especially in these areas, or to use the suggested correction formulas when the **PM{norm)** is calculated from nonstandard weather stations (Table 3).

For most of the small to medium size projects' landowners and agronomists in the KSA, it is not easy to compute the *ETo* by PM equation, however, HRG equation requires less parameters, and minimum efforts to calculate the *ETo*. The HRG equation, however, underestimates the PM value in most regions in the kingdom and hence the correction formulas suggested by Allen et al., [3] were derived for all of the studied stations (Table 4).

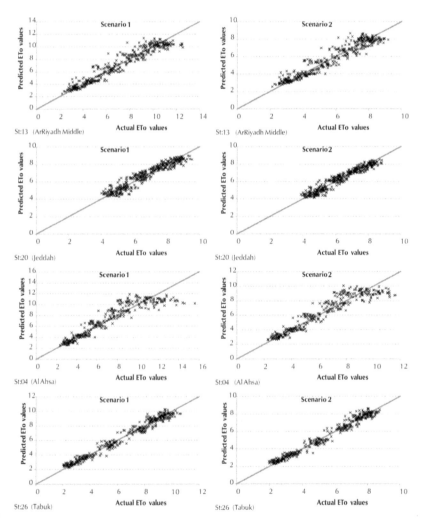

FIGURE 3 Predicted ET_o [*y*] versus actual ET_o [*x*] linear relationships for four of the studied stations in both scenarios: $y = a + b\,x$.

Because of the different scenarios in the agricultural projects, two formulas were derived for each station, one for each scenario. The first scenario's formula is applied when the climatic data (which is used to calculate HRG value) was obtained from the standard AWS. In this case, both PM and HRG were calculated without temperature correction, so HRG value can easily be calculated, and the corresponding PM value was computed through the related formula in Table 4. On the other hand, the second scenario's formula is applied when the climatic data was obtained from nonstandard AWS, so the HRG value was calculated by un-corrected data but the resulting PM value is corrected to the standard conditions.

The coefficient of determination, r^2, for all stations in both scenarios was extremely high and encouraging i.e. an average value of about 0.92 with values ranging from 0.72 to 0.98. Only two stations, AlWajh and Sharurrah, had r^2<0.80, nine stations had 0.80<r^2<0.90, and 18 stations with r^2>0.90. However, the calculated error values (Table 5) indicate an acceptable precision of the derived formulas. The RMSD for 'scenario 1' ranged between 0.36 to 1.19, with an average value of 0.73, while it got better results for 'scenario 2' and ranged from 0.28 to 0.86 with an average value of 0.56. These deviations are reasonable and acceptable, especially when taking into account the excellent MPE values that range from 0.11% to 3.02% with average value of 1.22% for 'scenario 1,' and 0.03% to 2.75% with average value of 1.12% for 'scenario 2'.

To validate the formulas reliability, four stations were selected, one from each geographic zone; 'Tabuk' from the North, 'Arriadh-Middle' from the Middle, 'AlAhsa' from the east, and 'Jeddah' from the South-West. For each station, the measured vs. predicted values of PM *ETo* were plotted, Fig. 3, for both of the studied scenarios. The graphs show the closeness of the points to the 45° line, with some bias for AlAhsa for *ETo* values>12 mm/d while the rest of the stations show very good representation of the efficiency for the predicted formulae. These good values of the validation indices (in Table 5) confirm the reliability of the presented formulae.

TABLE 4 Conversion formulas between HRG and PM values for the studied stations.

Station	Scenario 1*		Scenario 2**	
	Formula, Y = mX + C	r^2	Formula, Y = mX + C	r^2
01 A'Dhahran	PM{norm} = 1.450 x HRG – 0.806	0.948	PM{corr} = 1.203 x HRG – 0.124	0.964
02 Abha	PM{norm} = 1.144 x HRG – 0.241	0.919	PM{corr} = 0.998 x HRG + 0.155	0.935
03 Ad Dammam	PM{norm} = 1.449 x HRG – 0.0.97	0.950	PM{corr} = 1.097 x HRG + 0.784	0.973
04 Al Ahsa	PM{norm} = 1.313 x HRG – 0.395	0.860	PM{corr} = 1.041 x HRG + 0.309	0.882
05 Al Baha	PM{norm} = 1.559 x HRG – 1.396	0.854	PM{corr} = 1.200 x HRG - 0.393	0.870
06 Al Jouf	PM{norm} = 1.335 x HRG + 0.258	0.955	PM{corr} = 0.979 x HRG + 0.833	0.947

TABLE 4   *(Continued)*

Station	Scenario 1*		Scenario 2**	
	Formula, Y = mX + C	r^2	Formula, Y = mX + C	r^2
07 Al Medina	PM{norm} = 1.226 x HRG + 0.630	0.925	PM{corr} = 0.853 x HRG + 1.325	0.901
08 Al Qaisumah	PM{norm} = 1.183 x HRG + 0.447	0.917	PM{corr} = 0.857 x HRG + 1.115	0.880
09 Al Qassim	PM{norm} = 1.419 x HRG + 0.258	0.975	PM{corr} = 0.994 x HRG + 1.112	0.932
10 Al Quraiat	PM{norm} = 1.631 x HRG - 0.712	0.717	PM{corr} = 1.431 x HRG - 0.332	0.978
11 Al Wajh	PM{norm} = 0.637 x HRG + 2.259	0.955	PM{corr} = 0.616 x HRG + 2.099	0.779
12 Arar	PM{norm} = 1.376 x HRG - 0.137	0.955	PM{corr} = 1.065 x HRG + 0.459	0.946
13 Ar Riyadh Middle	PM{norm} = 1.260 x HRG + 0.099	0.919	PM{corr} = 0.860 x HRG + 0.948	0.905
14 Ar Riyadh North	PM{norm} = 1.256 x HRG + 0.082	0.922	PM{corr} = 0.878 x HRG + 0.927	0.913
15 Ar Ta'if	PM{norm} = 1.661 x HRG – 1.916	0.825	PM{corr} = 1.269 x HRG - 0.645	0.870
16 Bishan	PM{norm} = 1.241 x HRG - 0.252	0.875	PM{corr} = 0.927 x HRG + 0.608	0.862
17 Gizan	PM{norm} = 1.110 x HRG – 0.848	0.862	PM{corr} = 1.011 x HRG + 0.505	0.889
18 Hafr El-Batin	PM{norm} = 1.346 x HRG + 0.062	0.932	PM{corr} = 0.992 x HRG + 0.852	0.918
19 Hail	PM{norm} = 1.109 x HRG + 0.643	0.911	PM{corr} = 0.832 x HRG + 1.140	0.903
20 Jeddah	PM{norm} = 0.945 x HRG + 1.074	0.909	PM{corr} = 0.877 x HRG + 1.141	0.925
21 Khanis Mus-hait	PM{norm} = 1.152 x HRG + 0.221	0.881	PM{corr} = 1.011 x HRG + 0.129	0.908
22 Makkah	PM{norm} = 0.906 x HRG + 0.310	0.882	PM{corr} = 0.765 x HRG + 0.723	0.868
23 Najran	PM{norm} = 1.314 x HRG – 0.744	0.829	PM{corr} = 0.949 x HRG + 0.396	0.799
24 Rafha	PM{norm} = 1.314 x HRG + 0.492	0.916	PM{corr} = 1.037 x HRG + 1.005	0.915
25 Sharurrah	PM{norm} = 1.175 x HRG + 1.150	0.762	PM{corr} = 0.885 x HRG + 1.893	0.752
26 Tabuk	PM{norm} = 1.321 x HRG – 0.190	0.970	PM{corr} = 1.057 x HRG + 0.283	0.970
27 Turaif	PM{norm} = 1.358 x HRG + 0.220	0.958	PM{corr} = 1.091 x HRG + 0.628	0.965

Station	Scenario 1*			Scenario 2**		
	Formula, Y = mX + C		r^2	Formula, Y = mX + C		r^2
28 Wadi Al Dawasir	PM{norm} = 1.040 x HRG + 1.708		0.862	PM{corr} = 0.717 x HRG + 2.471		0.810
29 Yenbo	PM{norm} = 1.345 x HRG − 0.316		0.926	PM{corr} = 1.207 x HRG − 0.012		0.940

*Scenario 1: If HRG was calculated from standard agro – climatic weather stations (SAWS); Y = PM{norm}, X = HRG

**Scenario 2: If HRG was calculated from non – standard agro-climatic weather station; Y = PM{corr}, X = HRG.

HRG = Hargreaves method of ET calculation. r^2 = Coefficient of determination.

TABLE 5   Models evaluation parameters of the defined scenarios.

Station	Tested on	Scenario 1*		Scenario 2*	
	months	RMSD	MPE, %	RMSD	MPE, %
01 A'Dhahran	368	0.743	-0.55	0.508	-0.64
02 Abha	369	0.368	-0.60	0.283	-0.45
03 Ad Dammam	131	0.759	-0.32	0.416	-0.39
04 Al Ahsa	307	1.193	-1.62	0.861	-1.43
05 Al Baha	306	0.817	-1.45	0.589	-1.04
06 Al Jouf	368	0.644	-1.46	0.514	-1.36
07 Al Medina	368	0.666	-0.74	0.539	-0.74
08 Al Qaisumah	369	0.775	-1.90	0.690	-2.06
09 Al Qassim	368	1.039	-2.02	0.652	-1.83
10 Al Quraiat	67	0.554	-0.11	0.456	-0.30
11 Al Wajh	367	0.518	-0.83	0.424	-0.62
12 Arar	367	0.692	-1.29	0.592	-1.67
13 Ar Riyadh Middle	369	0.790	-1.31	0.588	-1.37
14 Ar Riyadh North	308	0.779	-1.18	0.579	-1.25
15 Ar Ta'if	365	0.954	-1.35	0.610	-0.89
16 Bishan	368	0.599	-1.00	0.475	-0.86
17 Gizan	368	0.442	-0.54	0.356	-0.40
18 Hafr El-Batin	245	0.870	-1.73	0.712	-1.87
19 Hail	367	0.709	-1.81	0.560	--1.61
20 Jeddah	369	0.439	-0.44	0.368	-0.35
21 Khanis Mushait	367	0.450	-0.81	0.342	-0.54
22 Makkah	367	0.546	-1.00	0.493	-0.97
23 Najran	368	0.764	-1.31	0.610	-1.19
24 Rafha	365	0.950	-3.02	0.756	-2.75

TABLE 5   *(Continued)*

Station	Tested on months	Scenario 1*		Scenario 2*	
		RMSD	MPE, %	RMSD	MPE, %
25 Sharurrah	306	0.992	-1.48	0.769	--1.17
26 Tabuk	369	0.448	-0.95	0.363	-0.86
27 Turaif	367	0.609	-1.56	0.448	-1.43
28 Wadi Al Dawasir	306	0.770	-1.09	0.643	-1.04
29 Yenbo	369	0.602	-0.57	0.483	-0.43
Summary, KSA	-----	**0.730**	**-1.22**	**0.557**	**-1.12**

*Scenario 1: If HRG was calculated from standard agro – climatic weather stations (SAWS); Y = PM{norm}, X = HRG

**Scenario 2: If HRG was calculated from non – standard agro-climatic weather station; Y = PM{corr}, X = HRG. HRG = Hargreaves method of ET estimation.

13.4   SUMMARY

In case of absence of nonstandard AWS, it is recommended to use the temperature correction procedure suggested by Allen [2] for calculating *ETo*. A severe increase of *ETo* values may reult (up to +24%), however, when considering the nonstandard AWS. It is also recommended to use the suggested modified Hargreaves formulas instead of using the FAO56 PM method for small to middle size farms in the cases where the landowners are incapable to deal with FAO56 PM Method. It is also recommended to use the modified Hargreaves formulas when some of the important weather factors are missing like wind-speed and relative humidity. To make it easy, three types of formulas were introduced in this chapter, (1) formulas to convert non standard-to-standard PM value. (2) formulas to convert HRG value to PM value when the HRG values were calculated from nonstandard AWS. (3) same as 2, but when HRG values were calculated from standard AWS.

Due to the lack of standard Agro-climatic Weather Stations (AWS) in Saudi Arabia, simple conversion formulas were derived to find the average daily ET_o value per month for 29 weather stations. The Hargreaves (HRG) formula was used instead of the Penman-Monteith formula due to the ease of calculation of HRG formula, as it almost require only temperature. For each station, two formulas were derived considering both standard and nonstandard AWS. Results show very good statistical representation of the formulas, with average root mean squared deviations of 0.73 and 0.57 and the mean percent errors of 1.22%, and 1.12% while considering the standard and nonstandard AWS, respectively. Results also confirm the importance of temperature correction when using data from nonstandard AWS, especially for the stations in the Middle to Midwest region (e.g. Arriadh, AlQassim, and Almadina); where the errors of using nonstandard AWS data may be up to +24% of the correct *ETo* value, which results in a massive waste of water.

KEYWORDS

- Alfalfa
- Arid regions
- climatic data
- conversion formulae
- evapotranspiration
- extraterrestrial radiation
- FAO56 correction
- Food and Agriculture Organization, FAO
- grass
- Hargreaves
- Hargreaves modified model
- Hargreaves samani
- Makkah
- Penman Monteith
- radiation
- Riyadh
- Saudi Arabia
- temperature correction
- temperature, dew point
- the extraterrestrial radiation
- vapor pressure
- weather correction
- weather station, agroclimatic
- weather station, nonreference
- weather station, reference

REFERENCES

1. Alazba, A.; **2004,** Comparison of the ASCE standardized and other Penman-Monteith type equations for hyper-arid area. ASABE Annual Meeting Paper number 042092 published by American Society of Agricultural and Biological Engineers, St. Joseph, Michigan, USA.
2. Allen, R. G.; Assessing integrity of weather data for use in reference evapotranspiration estimation. *J. Irrig. Drain. Eng.* [ASCE], **1996,** *122(2),* 97–106.
3. Allen, R. G.; Pereira, L. S.; Raes, D.; Smith, M.; **1998,** *Crop Evapotranspiration Guidelines for Computing Crop Water Requirements.* Rome, Italy: United Nations FAO Report #56.
4. Al-Sha'lan, S. A.; Salih, M. A.; Evapotranspiration estimates in extremely arid areas. Journal Irrigation and Drainage Engineering, **1987,** *113(4),* 565–574.
5. Amatya, D. M.; Skaggs, R. W.; Gregory, J. D.; Comparison of methods for estimating ref-et. Journal of Irrigation and Drainage Engineering, **1995,** *121(6),* 427–435.

6. ASCE-EWRI. *The ASCE Standardized Reference Evapotranspiration Equation*. ASCE EWRI Standardization of Reference Evapotranspiration Task Committee Report, **2005,** pp. 180.

7. Azhar, A. H.; Perera, J. C.; **2010,** Evaluation of reference evapotranspiration estimation methods under south east Australian conditions. Journal of Irrigation and Drainage Engineering (ASCE) IR, Paper #1943–4774.0000297.

8. Benli, B.; Bruggeman, A.; Oweis, T.; Üstün, H.; Performance of Penman-Monteith FAO 56 in a semiarid highland environment. Journal of Irrigation and Drainage Engineering, **2010,** *136(11),* 757–765.

9. Doorenbos, J.; Pruitt, W. O.; **1977,** *Crop Water Requirements. FAO Irrigation and Drainage Paper 24*. United Nations Food and Agriculture Organization, Rome.

10. ElNesr, M.; Alazba, A.; Simple statistical equivalents of Penman–Monteith formula's parameters in the absence of nonbasic climatic factors. Arabian Journal of Geosciences, **2010,** 1–11. doi: 10.1007/s12517–010–0231–1

11. ElNesr, M.; Alazba, A.; Amin, M.; **2011,** Estimation of shortwave solar radiation in the Arabian peninsula: A new approach. Unpublished.

12. Gavilán, P.; Lorite, I. J.; Tornero, S.; Berengena, J.; Regional calibration of Hargreaves equation for estimating reference ET in a semiarid environment. Agric. Water Manage. **2006,** *81,* 257–281.

13. Hargreaves, G. H.; Moisture availability and crop production. Trans. ASAE, **1975,** *18(5),* 980–984.

14. Hargreaves, G. H.; Responding to tropical climates. The 1980–1981 Food and Climate Review, Food and Climate Forum, Aspen Institute for Humanistic Studies, Boulder, Colorado. **1981,** 29–33.

15. Hargreaves, G. H.; Samani, Z. A.; Estimating potential evapotranspiration. J. Irrig. Drain. Div.; **1982,** *108(3),* 225–230.

16. Hargreaves, G. H.; Samani, Z. A.; Reference crop evapotranspiration from temperature. Appl. Eng. Agric.; **1985,** *1(2),* 96–99.

17. Jamshidi, H.; Khalili, D.; Zadeh, M. R.; Hosseinipour, E. Z.; Evaluation of Hargreaves equation for ET_o calculations at selected synoptic stations in Iran. World Environmental and Water Resources Congress: Challenges of Change. Section of Planning and Management Council, **2010,** 2791–2801. ASCE, Doi 10.1061/41114(371) 288.

18. Jia, X.; Dukes, M. D.; Jacobs, J. M.; Haley, M.; **2007,** Impact of weather station fetch distance on reference evapotranspiration calculation. ASCE(ID), 243(229).

19. Khoob, A. R.; Comparative study of Hargreaves's and artificial neural network's methodologies in estimating reference evapotranspiration in a semiarid environment. Irrig Sci.; **2008,** *26,* 253–259.

20. Lee, K.; Relative comparison of the local recalibration of the temperaturebased evapotranspiration equation for the Korea Peninsula. Journal of Irrigation and Drainage Engineering, **2010,** *136(9),* 585–594.

21. Mardikis, M. G.; Kalivas, D. P.; Kollias, V. J.; Comparison of interpolation methods for the prediction of reference evapotranspiration—an application in Greece. Water Resources Management, **2005,** *19,* 251–278.

22. Martínez-Cob, A.; Tejero-Juste, M.; A wind-based qualitative calibration of the Hargreaves *ETo* estimation equation in semiarid regions. Agric. Water Manage. **2004,** *64,* 251–264.

23. Salihand, A. M. A.; Sendil, U.; Evapotranspiration under extremely arid climates. J. Irrigation and Drainage Engineering, **1984,** *110(3),* 289–303.

24. Sau, F.; López-Cedrón F. X.; Mínguez, M. I.; **2000,** Reference evapotranspiration: Choice of method. Universidad De Santiago De Compostela.

25. Smith, R. L.; **2001,** Arabian Desert and East Sahero-Arabian xeric shrublands (PA1303). World Wildlife Fund. http: //www.worldwildlife.org/wildworld/profiles/terrestrial/pa/pa1303_full. html.

26. Trajković, S.; Gocić, M.; Comparison of some empirical equations for estimating daily reference evapotranspiration. Facta Universitatis. Series of Arch.itecture And Civil Engineering, **2010,** *8(2),* 163–168.

27. Trajkovic, S.; Hargreaves versus Penman-Monteith under humid conditions. Journal of Irrigation and Drainage Engineering, **2007,** *133(1),* 38–42.

28. Vincent, P.; *Saudi Arabia: An Environmental Overview.* Taylor and Francis. **2008,** pp 141. ISBN 978–0415413879.

29. Weib, M.; Menzel, L.; A global comparison of four potential evapotranspiration equations and their relevance to stream flow modeling in semiarid environments. Adv. Geosci.; **2008,** *18,* 15–23. http://www.adv-geosci.net(18)15/2008/.

CHAPTER 14

EVAPOTRANSPIRATION WITH DISTANT WEATHER STATIONS: SAUDI ARABIA[1]

MOHAMMAD NABIL ELNESR and ABDURRAHMAN ALI ALAZBA

CONTENTS

[1]Mohammad Nabil ElNesr, PhD, Assistant Professor, Shaikh Mohammad Alamoudi Chair for Water Research, College of Agriculture and Food Sciences, King Saud University, Kingdom of Saudi Arabia, POBox: 2460/11451, Riyadh – Saudi Arabia. E-mail: melnesr@ksu.edu.sa or drnesr@gmail.com; Abdurrahman Ali Alazba, PhD, Professor in Water Research, King Saud University; and Amin, M.T. Amin, Assistant Professor in Water Research, King Saud University. The authors wish to express thanks to "Shaikh Mohammad Bin Husain Alamoudi" for his kind financial support to the King Saud University [http://www.ksu.edu.sa]. This study is part of the AWC chair activities in the "Projects and Research" axis. We also thank the Presidency of Meteorology and Environment in Riyadh - KSA for providing the meteorological data. Modified and reprinted from: American Journal of Agricultural and Biological Sciences, 6 (3): 433-439, 2011.   © Copyright American Journal of Agricultural and Biological Sciences

14.1   INTRODUCTION

Determining the evapotranspiration (ET) is the base of many disciplines including the irrigation system design, irrigation scheduling and hydrologic and drainage studies [12]. Perfect determination of ET is a big challenge for investigators especially in arid and hyper-arid regions. Actual crop ET is computed by multiplying reference ET by the crop factor. Reference ET is the summation of evaporation and transpiration produced by a reference crop in specific growth conditions (height, coverage and health). ET value depends on two main factors: the selected reference crop and the climatic data [4]. This chapter discusses the possibility of using the weather data recorded at these stations instead of reference agro-climatic data. Hence, the objective of this chapter is to estimate the ET values for the RWS [Reference weather station] and the NRWS [non reference weather station] at Riyadh – Saudi Arabia; and to compare these ET values with the reference evapotranspiration based on lysimeter observations for reference conditions.

The Kingdom of Saudi Arabia (KSA) is one of the most arid countries in the world and suffers persistent water shortage problems and more than 88% of water consumption in KSA is due to agricultural related activities [10]. Hence, several researches have been performed to assess the ET in KSA. Researchers have estimated the reference ET [2, 9]; and assessed the ET for specified crops [1, 7]; and have determined crop coefficients.

For open field agriculture, the reference ET has traditionally been predicted by using either grass (*ETo*) or alfalfa (*ETr*). Each of these two crops use some conditions to be considered as a reference crop [14]. The selection of either crop as reference crop has been studied by several investigators [6, 11, 13, 14]. It was recommended by the American Society of Civil Engineers Task Committee (ASCE-TC) to use a single equation for both reference crops, each with different constants [5]. They recommended standardizing the equation with two surfaces, the short crop (about 0.12 m height e.g., the clipped cool season grass) and the tall crop (about 0.50 m height e.g., the full cover alfalfa). The heights of crop, however, may vary according to the crop variety and location's geography. When using crop with different height, one should clearly mention the used height beside the ET data.

Climate data are acquired from Weather Stations (WS) whose location is an important consideration for the quality of the data. RWS should be located inside a cropped area (normally with grass) in order to ensure the same environmental conditions for station's gauges as that of the cultivated crops. On the other hand, stations located in these reference conditions usually record less temperature than Non- Reference (NR) weather stations [3]. This was attributed to the cooling effect of the crop. Allen [3] suggested an adjustment for the recorded temperature in NRWS so that the resultant temperatures could be used to give the reference ETo.

In many locations, RWS are not found especially for newly reclaimed desert areas. To perform preliminary studies for an area, one should use the data from nearest station. This situation is probably affected by the distance between the field and the weather station. There are 13 districts in KSA and some of these are larger in size than many countries. Arriyadh District's area, for instance, is 380,000 Km², which is 17%

of the geographical area of KSA. The main weather stations in Arriyadh and other places in the kingdom are situated at airports.

14.2 MATHEMATICAL MODELS FOR ESTIMATING EVAPOTRANSPIRATION

The Eqs. (1–13) describe the selected theoretical models to estimate ET that were used in this chapter. For a complete list of all models, the reader should refer Chapter 4 in this book.

The original Penman-Monteith (PM) equation for determining the evapotranspiration [4] is given in Eq. (1). The Chapters 2 and 3 of the publication by Allen [4] describe in detail all the variables in Eq. (1). The main component of ET calculation is the air Temperature (T). Although it does not appear explicitly in the Eq. (1), yet it is included in most of the parameters: Δ, Rn, c_p, ρ_a, es, ea, λ, γ, r_s, r_a, and G. Allen [3] concluded that correcting the temperature values of the nonreference weather stations (NRWS) to an adjusted value fixes the entire ET equation to give an acceptable value close to the reference weather station (RWS) value.

$$ET_o = \frac{0.408 \, \Delta \, (R_n\text{-}G) + \gamma\left(\frac{900}{T+273}\right)u_2(e_s\text{-}e_a)}{\Delta + \gamma(1+0.34 \, u_2)} \tag{1}$$

where: Δ is a slope of the vapor pressure curve [kPa°C^{-1}], R_n is net radiation (MJm^{-2}day^{-1}), G is soil heat flux density (MJm^{-2}day^{-1}), γ is psychrometric constant (kPa°C^{-1}), T is mean daily air temperature at 2 m height (°C), u_2 is wind speed at 2 m height (m s^{-1}), e_s is the saturated vapor pressure and e_a is the actual vapor pressure (kPa). Eq. (1) applies specifically to a hypothetical reference crop with an assumed crop height of 0.12 m, a fixed surface resistance of 70 sec m^{-1} and an albedo of 0.23.

$$e_a = 0.005(RH_x e_o[T_n] + RH_n e_o[T_x]) \tag{2}$$

where: T_a = Minimum dry bulb air temperature (°C); T_x = Maximum dry bulb air temperature (°C); RH_x = Maximum relative humidity (%); and RH_a = Minimum relative humidity (%).

$$e_a = 0.005 RH_a(e_a[T_n] + e_o[T_x]) \tag{3}$$

$$T_d = \frac{116.91 + 237.3 \, ln(e_a)}{16.78 - \ln(e_a)} \tag{4}$$

$$\Delta T = T_n - T_d \tag{5}$$

$$ET_{HG} = 0.0023 R_a (T_a + 17.8)(T_x - T_n)^{0.5} \tag{6}$$

$$ET_{PM} = c_1 + c_2 ET_{HG} \tag{7}$$

$$R_a = 37.6 d_r (\omega_s \sin \varphi \sin \delta + \sin \omega_s \cos \varphi \cos \delta) \tag{8}$$

In Eq. (8): d_r, δ, and ω_s are defined in eqs. (9) to (11) below:
d_r is a relative distance from earth to sun, given by Eq. (9):

$$d_r = 1 + 0.033 \cos(0.0172 J) \tag{9}$$

Solar declination in radians is given by eq. (10):

$$\delta = 0.409 \sin(0.0172 J - 1.39) \tag{10}$$

Sunset hour angle in radians is determined by eq. (11), with φ = latitude in radians:

$$\omega_s = \arccos(-\tan\varphi \tan\delta) \tag{11}$$

Julian day (J) is calculated with eq. (1), with M = Month of the year and D = Day of the month. J ranges from 1 to 366 (366 is for leap year).

$$J = \mathrm{int} \left(\frac{275}{9} M + D - 30 \right),$$ "if ($M > 2$): For leap year, subtract 1; and for not leap year, subtract 2". (12)

Allen's method is summarized in Eqs. (1–5). The actual vapor pressure is calculated with Eqs. (2) and (3). The dew point temperature is determined with Eq. (4), in the absence of measured values. The temperature difference is determined with Eq. (5), where T_n is a minimum temperature. For arid and semiarid environments, if $\Delta T > 2$ then the maximum temperature (T_x) is adjusted to: $T_x^{(corr)} = (T_x - 0.5 (\Delta T - 2))$, where $(corr)$ = Refer to the corrected value. Finally doing the same for T_n: if $\Delta T \leq 2$, then no correction is needed.

For sites with limited weather data, Allen[4] suggested using a modified version of the Hargreaves equation (HG) as an alternative method for determining ET. He also suggested calibrating the HG equation (Eq. (6)) using linear regression equation (Eq. (7)) using the corresponding values from the PM Eq. (1). The units are: MJ-m^{-2} for extraterrestrial radiation in Eq. (8); MJ-m^{-1}-h^{-2} for ET_{HG} in Eq. (6); and mm-day^{-1}. The parameters in Eq. (6) were estimated with Eqs. (8–12).

14.3   WEATHER DATA SOURCES

Two types of data were used in this study: weather data to estimate the ET value; the field data to determine ET experimentally. For weather data, two weather stations (Campbell and Davis) at the educational farm of the King Saud University were selected as agro-climatic RWS [reference weather station]. Also two weather stations at Old Riyadh airport and King Khaled airport (Fig. 1) were used as domestic NRWS (nonreference weather stations). Reference field data was obtained from Al-Amoud et al. [1] based on five years project of ET evaluation through lysimeter studies in 9 zones throughout the country.

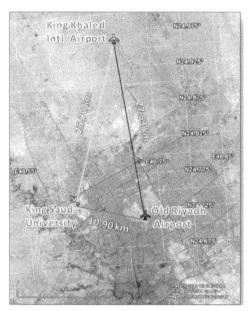

FIGURE 1   Location and distances between the weather stations in this research.

All of the weather data were recorded on daily bases while the field data was recorded on monthly bases. Hence, daily ET values for all the studied weather stations were calculated and later, the data was summarized as average monthly ET values. The recorded dataset varies from station to station. For airport's weather stations, complete records from 1985 to 2009 were obtained. For Campbell and Davis weather stations, the records were from 1993–2006. This research was limited to the least-size dataset, i.e., Campbell's dataset for an appropriate comparison. The databases of the studied weather stations were not so coincident. For all stations, the commonly available data parameters included: the dry bulb temperature (max., min. and avg.), relative humidity (max., min. and avg.), rainfall and wind speed (average). In addition to the common variables for airports stations, the wet bulb temperature (max., min. and avg.), the atmospheric pressure at sea level and at station level and the actual vapor pressure were also recorded. While for Campbell station, the actual vapor pressure and solar radiation are recorded. Finally, for Davis station, the only addition to common parameters was the solar radiation. This information is summarized in Table 1. Solar Radiation (Rs) and vapor pressure (ea) are essential parameters for computing ET. If not recorded at the weather station, these parameters were calculated using Eqs. (2)–(5) and Eqs. (8)–(12).

As mentioned above, the field data were obtained from Al-Amoud et al. [1]. The five years project used Alfalfa cultivated in weighing lysimeters located at Riyadh and at 8 more locations in the KSA. The daily and monthly values of irrigation, drainage, precipitation and water consumption were recorded for these stations. However, only the monthly results were available in the published research.

14.4   METHODS AND MATERIAL

For each of the four stations mentioned in Table 1, weather data were observed as daily records. Using the raw data, the researchers calculated ET_{PMg}, ET_{PMa} and ET_{HG}; where the suffixes PMg and PMa stand for Penman Monteith formula for 0.12 m grass reference crop and 0.25 m alfalfa reference crop, respectively.

The entire calculations were repeated after applying Allen [3] corrections to the temperature data but only for non agro-climatic stations. To simplify data representation and discussion, the symbol and numerical value were assigned to each data source as the shown in Table 2. Since the published data by Al-Amoud et al. [1] were on monthly bases and Allen [4] suggested calibrating Hargreaves formula using monthly data, subsequently, the research team converted the daily calculated data to the data on monthly bases.

14.5   RESULTS AND DISCUSSION

The ET data for the six datasets are presented graphically in Figure 2. The charts are denoted by letters 'a' to 'f' for Campbell reference WS, Davis reference WS, old Riyadh airport corrected dataset, old Riyadh airport raw dataset, King Khalid airport corrected dataset and King Khalid airport raw dataset, respectively. Four ET values were plotted for each dataset: measured ET, {Px(L)}; grass based PM evapotranspiration{g}; alfalfa based PM evapotranspiration{a}; and Hargreaves method ET.

All of the calculated data groups were compared with the measured dataset and the correlation coefficient for each data group pair was calculated. The correlation coefficients are illustrated in Figure 3. Then, the regression coefficients, mentioned in Eq. (7), were calculated between HG and PM formulas (See Table 3). Finally, the ET ratio between alfalfa and grass was calculated and compared to the value of 1.15 that has been reported by Doorenbos and Pruitt [8]. Table 4 shows the regression coefficients for the linear relationships between ET alfalfa and ET grass ET in this study.

TABLE 1   The climatic parameters that were recorded at the weather stations.

Parameter	Weather station			
	Campbell	Davis	Riyadh old airport	King Khaled Int. airport
Dry bulb temp.	yes	yes	yes	yes
Wet bulb, relative	no	no	yes	yes
Wind humidity	yes	yes	yes	yes
Solar speed	yes	yes	yes	yes
Vapor radiation	yes	no	no	no
Atm. Pressure	yes	yes	no	no
Commutative sea level	no	no	yes	yes
Commutative at station	no	no	yes	yes
Rainfall	yes	yes	yes	yes

TABLE 2 The data source for the research and identification.

Variable	Location and name						
	Data is measured: Educational Farm, KSU			Data is calculated Airport			
	Project data	Campbell	Davis	Riyadh old airport		King Khaled Int. airport	
	0	1	2	3	4	5	6
Reference data	no	yes	yes	no	no	no	no
Corrected data	No need	No need	No need	yes	no	yes	no
Symbols	Px(L)	Cs(a,g,H)	Ds(a,g,H)	Oc(a,g,H)	On(a,g,H)	Kc(a,g,H)	Kn(a,g,H)
Longitude: 24N	44'12.24"	44'12.24"	44'12.24"	42'35.46"	42'35.46"	57'27.00"	57'
Latitude: 46E	37' 14.90"	37' 14.90"	37' 14.90"	43' 30.54"	43' 30.54"	41' 55.54"	27.00"

Symbols: A = Alfaalfa; C = Corrected, C = Campbell; D = Davis; g = grass; H = Hargreaves;
K = King Khaled airport; L = Lysimeters; n = Normal; O = Old airport; P = Project data; s = Reference;
X = Experimental.

TABLE 3 Linear regression analysis for ET values between Hargreaves [HS] and Penman-Monteith [PM] based on equations in this chapter.

Grass			Alfalfa		
	Equation	r^2		Equation	r^2
Cs	PMg = 1.280 x HG + 0.324	0.998	Cs	PMa = 1.608 x HG + 0.526	0.999
Ds	PMg = 1.563 x HG – 0.122	0.959	Ds	PMa = 1.993 x HG – 0.046	0.975
Kc	PMg = 1.189 x HG – 0.140	0.979	Kc	PMa = 1.481 x HG – 0.140	0.987
Kn	PMg = 1.296 x HG + 0.013	0.986	Kn	PMa = 1.619 x HG + 0.109	0.991
Oc	PMg = 1.214 x HG – 0.211	0.981	Oc	PMa = 1.517 x HG – 0.243	0.988
On	PMg = 1.298 x HG – 0.021	0.991	On	PMa = 1.624 x HG + 0.055	0.994

TABLE 4 The regression coefficients for the linear relationships between ET-alfalfa, ET-grass and ET in this study: Y = mX + C.

	Cs	Ds	Kc	Kn	Oc	On
Slope, m	1.255	1.269	1.243	1.247	1.246	1.249
Intercept, C	0.125	0.158	0.053	0.107	0.040	0.092
r^2	0.999	0.999	0.999	0.999	0.999	0.999

Campbell and Davis weather stations are located at the educational farm of the King Saud University; the lysimeters' experiment of Al-Amoud et al. [1] was held at the same location. Figures 2a and 2b show the results of measured and calculated ET by the three mentioned methods. Both stations show underestimation of Hargreaves formula and overestimation of PM alfalfa, 'a,' calculations. It is strange that the grass ET, 'g,' almost coincides with the measured alfalfa data. This may be attributed to some lack of precision either in the field measurements devices or to some calibration errors of the weather stations. The data at old airport (Figs. 2c and 2d) behaves differently and the closer values to the measured alfalfa ET are the calculated alfalfa values, especially at months 1–3 and 9–12. The situation is different for King Khalid airport's station (Figs. 2e and 2f) as the raw data appear to give fuzzy trend dissimilar to the measured data, Fig. 2f. After applying the data correction, the shape of the curve improved dramatically, Fig. 2e. This is confirmed in Fig. 3, which shows the Correlation Coefficient (CC) between measured data versus each data group. Although all the values of CC are more than 0.9, which is a very good value, however, the 'Kn' dataset is the worst representation of actual state. On the other hand, it strangely appears that the corrected values of King Khalid's airport (K) are the top most accurate representatives of the measured data. This is probably due to the geographic condition of the 'K' airport, which is outside of the city and almost surrounded with desert lands, in addition to the long distance between the Educational Farm stations (EF) and the 'K' airport (about 25.7 km), as shown in Figure 1. The Old airport (O) is near (10.9 km), in fact almost in the middle of the city and surrounded by buildings, roads and some green areas. The correction of the 'O' data improves the 'g' and 'a' data groups, while it worsens the 'H' data group, as shown in Fig. 3.

From Fig. 3, it can be concluded that the PM calculations improve dramatically after applying the Allen [3] correcting algorithm to the data, while for HG formula, applying the corrections improves the accuracy for 'K' station but worsens it for 'O' station.

In general, Hargreaves equation gives very satisfactory results of ET for Riyadh city and the equation can be used trustfully especially in the absence of some climatic factors like wind speed and radiation. However, we applied the linear correction equations and found some excellent fitted equations, as listed in Table 2. All the equations are excellently fitted with minimum value of coefficient of determination (r^2) of 0.974. It can be approximated that $PM^g = (1.30 \times HG - 0.05)$ for grass reference crop, while $PM_a = (1.64 \times HG + 0.1)$ for alfalfa reference crop. For more accurate values each station should be calibrated, as shown in Table 3. The ratio between alfalfa ET and grass ET, (ETalfalfa/ETgrass, or ET_r/ET_o), is always taken as 1.15 for arid regions, as recommended by Doorenbos and Pruitt [8]. This value has been used by many researches [2] when studying the Saudi Arabia ET, We evaluated this value for each of the studied data groups and a linear relationship between ETalfalfa and ETgrass was obtained, as shown in Table 4. The slope of ET_r/ET_o is almost 1.25 for all stations and using less value may result in some bias in data.

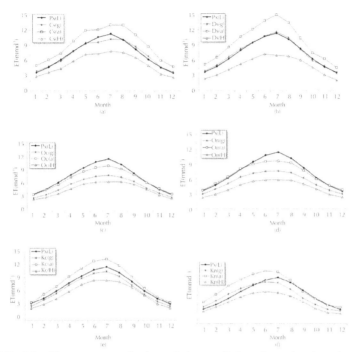

FIGURE 2 Monthly evapotranspiration of the studied region, showing three datasets in each chart compared to measuredevapotranspiration, where, *a*: Alfalfa; *c*: Corrected; C: Campbell; D: Davis; g: Grass; H: Hargreaves; K: King Khalid airport; L: Lysimeters; n: Normal; O: Old airport; P: Project data; s: Reference; *x*: Experimental.

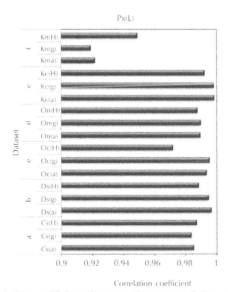

FIGURE 3 The correlation coefficients between the measured data and the calculated data for all data groups in the six datasets.

14.6   SUMMARY

Due to the easiness of finding nonreference agro- climatic WS than the agro-climatic ones in the newly reclaimed areas, their weather data had to be corrected through simple procedure. Two reference and two nonreference WS were taken in Riyadh city and corrections were applied to nonreference WS only. We calculated the evapotranspiration using Penman Monteith and Hargreaves formulas. PM was calculated for two reference crops, i.e., alfalfa and grass. Calculated data were compared with measured data. Results show an admirable enhancement in data accuracy after applying the data correction to the nonreference stations. The simple ET formula of Hargreaves underestimates the actual ET. The situation changes after applying the simple linear fitting equation to the resulted values. The ratio between alfalfa and grass ET was found to be 1.25 for Riyadh area. It is concluded to use the temperature correction method when using nonreference stations. Hargreaves formula is recommended to be used after applying the suggested fit in this study, especially when the wind speed and radiation data are missing.

Reference agro-climatic Weather Stations (WS) are rarely found in newly reclaimed areas. The usage of weather data from nonreference WS may lead to inaccurate estimations of Evapotranspiration (ET), especially if the nonreference stations are distant from the reclaimed location. Weather data from four WS located at Riyadh were used to calculate ET by using Penman Monteith (PM) and Hargreaves equations. PM equation was applied with both alfalfa and grass reference crops. Calculations were done with and without temperature correction for nonreference weather stations. All calculations were compared with measured lysimeter data and corrections in Hargreaves formula were suggested.

Results: (1) Weather data from nonreference WS can be used safely to calculate ET only when temperature corrections are applied. (2) Hargreaves formula underestimates ET at all locations in the study area. By applying the simple linear correction to the data, highly acceptable results are obtained. (3) The ET ratio between alfalfa and grass in Riyadh is 1.25. The study concludes that temperature correction for nonreference WS is essential to ensure acceptable ET calculations. Usage of Hargreaves formula is recommended with the corrections suggested in the study due to its simplicity.

KEYWORDS

- Alfalfa
- Arid Regions
- climatic data
- climatic stations integrity
- data correction
- dew point temperature
- ET
- evapotranspiration

- **extraterrestrial radiation**
- **grass**
- **Hargreaves method**
- **Hargreaves modified method**
- **Kingdom of Saudi Arabia**
- **Lysimeter**
- **nonreference weather station**
- **Penman-Monteith method**
- **reference crop**
- **Regression analysis**
- **Riyadh, Saudi Arabia**
- **Saudi Arabia**
- **vapor pressure**
- **weather station**
- **weather station, nonreference**
- **weather station, reference**

REFERENCES

1. Al-Amoud, A.; Al-Takhis, A.; Awad, F.; Al AbdelKader and A. Al-Mushelih, *A guide for Evaluating Crop Water Requirements in the Kingdom of Saudi Arabia.* King Abdulaziz City for Science and Technology, Abdulaziz – KSA, **2010,** 104.
2. Al-Ghobari, H. M.; Estimation of reference evapotranspiration for southern region of Saudi Arabia. Irrig. Sci.; **2000,** *19,* 81–86.
3. Allen, R. G.; Assessing integrity of weather data for reference evapotranspiration estimation. J.; Irrig. Drainage Eng.; **1996,** *122,* 97–106.
4. Allen, R. G.; *Crop Evapotranspiration: Guidelines for Computing Crop Water Requirements.* First edition by Food and Agriculture Organization of the United Nations, Rome. **1998,** 300.
5. Allen, R. G.; *The ASCE Standardized Reference Evapotranspiration Equation.* ASCE Publications, Reston, VA. **2005,** 216.
6. Allen, R. G.; Walter, I. A.; Elliott, R.; Mecham, B.; Jensen, M. E.; **2000,** Issues, requirements and challenges in selecting and specifying a standardized ET equation. CMIS.
7. Al-Omran, A. M.; Mohammed, F. S.; Al-Ghobari, H. M.; Alazba, A. A.; Determination of evapotranspiration of tomato and squash using lysimeters in central Saudi Arabia. Intl. Agric. Engr. J.; **2004,** *13,* 27–36.
8. Doorenbos, J.; Pruitt, W.; *Guidelines for Predicting Crop Water Requirements.* Food and Agriculture Organization of the United Nations, Rome. **1977,** 144.
9. ElNesr, M.; Alazba A.; Abu-Zreig, M.; Analysis of evapotranspiration variability and trends in the Arabian Peninsula. Am. J. Environ. Sci.; **2010,** *6,* 535–547.
10. Faruqui, N. I.; Biswas, A. K.; Bino, M. J.; Water Management in Islam. United Nations University Press, Tokyo. **2001,** 149.
11. Howell, T. A.; Evett, S. R.; Schneider, A. D.; Dusek, D. A.; Copeland, K. S.; Irrigated fescue grass ET compared with calculated reference grass ET.; Proceedings of the Decennial Symposium, Nov. 14–16, American Society of Agricultural Engineers, Phoenix, AZ, **2000,** 228–242.

12. Irmak, S.; Haman, D. Z.; **2003,** Evapotranspiration: Potential or Reference. University of Florida.

13. Wright, J. L.; Derivation of Alfalfa and Grass Reference Evapotranspiration. In: *Evapotranspiration and Irrigation Scheduling*, Camp, C. R.; Sadler, E. J.; (Eds). American Society of Agricultural Engineers, Michigan. **1996,** 133–140.

14. Wright, J. L.; Allen, R. G.; Howell, T. A.; Conversion between evapotranspiration REFERENCES and methods. Proceedings of the 4th Decennial Symposium, NWISRL Publisher, USA- AZ-Phoenix. **2000,** 251–259.

CHAPTER 15

ACTUAL EVAPOTRANSPIRATION USING SATELLITE IMAGE IN THAILAND[1]

PREEYAPHORN KOSA

CONTENTS

[1]Printed with permission and modified from: P. Kosa, 2011. The effect of temperature on actual evapotranspiration based on Landsat-5TM satellite imagery. Chapter 9, pages 209-229. In: *Evapotranspiration* by Leszek Labedzki [Editor], ISBN 978-953-307-251-7, InTech. Available from: <http: //www.intechopen.com/books/evapotranspiration/the-effect-of-temperature-on-actual-evapotranspiration-based-on-landsat-5-tm-imagery> © Copyright InterTech.com

15.1 INTRODUCTION

In the water cycle, evapotranspiration is one of the most important components, but it is one of the most difficult to measure and monitor. Evapotranspiration relates to the exchange of energy in the atmosphere, ground surface, and root zone. Some elements of calculated evapotranspiration can be measured by weather stations, while others are estimated from empirical equations. Then, the calculated evapotranspiration has some inaccuracy. To improve upon this problem, the combination of meteorological data and remote sensing observersions are an alternative evapotranspiration [14, 18] On the other hand, temperature is normally measured in a number of weather stations. Since temperature relates to many weather data, temperature can imply the characteristic of other weather data. For example, low temperature is included high humidity but low evaporation can be occurred in the condition of low temperature. Moreover, the variable spatial resolution is the characteristic of both actual evapotranspiration and temperature. At present, satellite images are used for studying the Earth's surface and it bolsters spatial resolution. This chapter discusses effects of temperature on actual evapotranspiration using satellite image.

Evapotranspiration occurs from evaporation and transpiration, and it can be obtained from weather data and satellite images. Evaporation is the primary process of water transfer in the hydrological cycle. The water is transformed into vapor and transported into the sky. Evaporation can be classified into potential evaporation and actual evaporation. The potential evaporation is defined as the amount of evaporation that would occur if a sufficient water source were available. On the other hand, the actual evaporation is the amount of water which is evaporated a normal day. The potential evaporation is the maximum value of the actual evaporation. Transpiration is included by the vaporization of liquid water contained in plant tissues and the vapor removal to the atmosphere [14, 18]. Evapotranspiration is normally computed from the Penman-Monteith FAO 56 equation using weather data. This equation is affected by principal weather parameters such as radiation, air temperature, humidity and wind speed. These parameters can be measured by weather station and computed by the equation of FAO irrigation and Drainage Paper No. 56 [2]. In addition to the Penman-Monteith FAO 56 equation, evapotranspiration can be estimated from the concept of energy balance. The main components of the energy balance equation are sensible heat flux, latent heat flux, soil heat flux, and net radiation. These elements relate with incoming and outgoing radiation in the atmosphere, ground surface, and root zone. They are estimated from remote sensing data and weather data [4, 5, 26, 27].

Remotely sensed data are used for studying the Earth's surface. Current technology allows continuous acquisition of data, regular revisit capabilities [resulting in up-to-date information], broad regional coverage, good spectral resolution [including infra-red bands], good spatial resolution, ability to manipulate/enhance digital data, ability to combine satellite digital data with other digital data, cost-effective data, map-accurate data, possibility of stereo viewing, and large archives of historical data. Remote sensing helps to record data from remote locations. Satellite data provides timely and detailed information about the Earth's surface, especially in relation to the management of our renewable and nonrenewable resources. The advantages of satellite data for many fields include for example; assessment and monitoring of vegetation types and their status, soil surveys, mineral exploration, map making and revision, production of thematic maps, water resources planning and monitoring, urban planning, agricultural property planning and management, crop yield assessment, and natural disaster assessment.

The Surface Energy Balance Algorithm for Land or SEBAL [7, 8] is an image processing model to calculate evapotranspiration by using satellite imagery and weather data with the concept of energy balance at the land surface, so evapotranspiration for each pixel is calculated. Satellite images that can be used in SEBAL are LANDSAT, NOAA-AVHRR, MODIS, and ASTER. The advantages of SEBAL are easily applicable because of minimal collateral data needed, applicable to various climates because of physical concepts, no need for land use classification, no need to involve data demanding hydrology, and is a suitable method for all visible, near-infrared and thermal radiometers. However, the disadvantages of SEBAL include cloud-free conditions being required and that surface roughness is poorly described, so that SEBAL is only suitable for flat terrain [1, 3, 16, 17, 20, 25].

15.2   LANDSAT 5™

Landsat 5™ has a unique and necessary role in the realm of earth observing satellite orbits, because no other satellite of the earth observing system matches with the synoptic coverage, high spatial resolution, spectral range and radiometric calibration of Landsat's system. Landsat 5™ satellite imagery Path/Row: 127/48 was mainly input data. The spatial resolution of Landsat 5™ satellite imagery is 30 m and the swath width covered by the image is 185 km. The details of Landsat 5™ characteristics are shown in Tables 1 and 2 [12]. At the ground stations, the Landsat ground system consists of a spacecraft control center, ground stations for uplinking commands and receiving data, a data handling facility and a data archive, which were developed by the Goddard Space Flight Center, Greenbelt, MD, in conjunction with the U.S. Geological Survey (USGS) EROS Data Center (EDC), Sioux Falls, SD. These facilities will receive, process, archive, and distribute ETM+ data to users. Raw ETM+ data can be transmitted to the EROS Data Center by the ground system within 24 hours of its reception [12, 22].

TABLE 1   Landsat 5™ mission specifications.

Characteristics	Specifications
Swath width	185 km
Repeat coverage interval	16 days [233 orbits]
Altitude	705 km
Quantization	Best 8 of 9 bits
On-board data storage	~ 375 Gb [solid state]
Inclination	Sun-synchronous, 98.2 degrees
Equatorial crossing	Descending node; 10: 00 am +/–15 min
Launch vehicle	Delta ll
Launch date	April, 1999

TABLE 2   Landsat 5™ characteristics.

Channel	Spectral range [microns]	Ground resolution [m]
1 [Visible and near infrared: VNIR]	0.450–0.515	30
2 [Visible and near infrared: VNIR]	0.525–0.605	30
3 [Visible and near infrared: VNIR]	0.630–0.690	30
4 [Visible and near infrared: VNIR]	0.750–0.900	30
5 [Short wavelength infrared: SWIR]	1.550–1.750	30
6 [Thermal long wavelength infrared: LWIR]	10.40–12.50	60
7 [Short wavelength infrared: SWIR]	2.090–2.350	30
Pan [Visible and near infrared: VNIR]	0.520–0.900	15

This Landsat 5™ image covers Sri Songkhram subriver basin whch is located in the lower part of Mekhong river basin, in the north-east of Thailand. The time period of this study is from November 2006 to January 2007, belonging to the dry season. Also, These are conditions of clear sky. Moreover, there is the cultivation during this time so there are both evaporation and transpiration that is evapotranspiration.

15.3   RADIATION

Extraterrestrial radiation (R_a) is the solar radiation received at the top of the earth's atmosphere on a horizontal surface. The values of extraterrestrial radiation depend on seasons change, the position of the sun, and the length of the day. Therefore, the extraterrestrial radiation is a function of latitude, the date and time of day. The solar constant is the radiation striking a surface perpendicular to the sun's rays at the top of the earth's atmosphere and it is some 0.082 MJ m^{-2} min^{-1}. If the position of the sun is directly overhead, the incidence angle of extraterrestrial radiation is zero. In this case, extraterrestrial radiation is some 0.082 MJ m^{-2} min^{-1}.

Solar or shortwave radiation (R_s) is the amount of radiation penetrating from the atmosphere to a horizontal plane. The sun emits energy by electromagnetic waves that include short wavelengths so solar radiation is referred to as shortwave radiation. In the atmosphere, radiation is absorbed, scattered, or reflected by gases, clouds, and dust. For a cloudless day, the solar radiation is about 75% of the extraterrestrial radiation, while it is about 25% of the extraterrestrial radiation on a cloudy day. The solar radiation, which is known as global radiation, is a summation of direct shortwave radiation from the sun and a diffuse sky radiation from all upward angles.

Relative shortwave radiation (R_s/R_{so}) is a relationship between shortwave radiation [R_s] and clear-sky solar radiation (R_{so}). The shortwave radiation is solar radiation that actually reaches to the earth's surface in a given time, while clear-sky solar radiation is solar radiation that reaches to the same area with a clear-sky condition. The relative shortwave radiation is affected by the cloudiness of the atmosphere. On a cloudy day, the ratio is smaller than on a cloudless day. The range of this ratio is between 0.33 [cloudy condition] to 1.00 [cloudless condition].

Relative sunshine duration (n/N.) shows the cloudiness in the atmosphere. It is the relationship between the actual duration of sunshine (n) and the maximum possible duration of sunshine, or daylight hours (N.). For the cloudless condition, n is equal to N., while n and n/N. are nearly zero for the cloudy condition. The maximum possible duration of sunshine, or daylight hours (N.), depends on the position of the sun, so it is a function of latitude and date. The daily values of N. throughout a year differ with latitude.

Albedo (α) is a relationship between reflected radiation and total incoming radiation. It varies with both the characteristics of surface and the angle of incidence, or the slope of ground surface. Albedo can be more than 0.95 for freshly fallen snow, and it is smaller than 0.05 for wet bare soil. The range of albedo for green vegetation is about 0.20–0.25 and albedo for the green grass reference crop is 0.23.

Net solar radiation (R_{ns}) is the fraction of the solar radiation that is reflected from the ground surface. It can be calculated by Eq. (1):

$$R_{ns} = (1 - \alpha) R_s \tag{1}$$

Net long wave radiation (R_{nl}) is the difference in value between outgoing and incoming long wave radiation. The longwave radiation is solar radiation absorbed by the earth and turned to heat energy. Since the temperature of the earth is less than the sun, so the earth emits longer wavelengths. Terrestrial radiation is referred to as longwave radiation. The emitted long wave radiation ($R_{1,up}$) is absorbed by the atmosphere or lost into space. The long wave radiation received by the atmosphere ($R_{1,down}$) increases its temperature. Therefore, the earth's surface both emits and receives longwave radiation. The value of outgoing longwave radiation is normally more than incoming longwave radiation, so the net longwave radiation is used to present the energy loss.

Net radiation (R_n) is the difference in value between incoming and outgoing radiation of both short and long wavelengths. It is the balance among energy absorbed, reflected, and emitted by the earth's surface. The net radiation is also the difference in value between the incoming net shortwave (R_{ns}) and the net outgoing long wave (R_{nl}) radiation. It is a positive value during daytime, while it is a negative value during nighttime. For the total daily value, it is a positive value except for the condition of high latitude.

Soil heat flux (G) is energy that is used in heating the soil. It is a positive value under the condition of warming soil and negative value under the condition of cooling soil. The soil heat flux is very small when compares with net radiation but it cannot be ignored.

15.4 TEMPERATURE ESTIMATIONS

Normally, Landsat 5™ satellite image data is in the form of Digital Number (DN) so it is necessary to convert from Digital Number to Radiances for all bands in Landsat 5™ imagery. The conversion equation is shown below:

$$L_\lambda = \frac{(LMAX_\lambda - LMIN_\lambda)}{(QCALMAX - QCALMIN)} \times (QCAL - QCALMIN) + LMIN_\lambda \tag{2}$$

where, L_λ is spectral radiance at the sensor aperture in watts/(meter squared *ster* μm), QCAL is the quantized calibrated pixel value in DN, $LMIN_\lambda$ is the spectral radiance that is scaled to QCALMIN in watts/(meter squared *ster* μm), $LMAX_\lambda$

is the spectral radiance that is scaled to QCALMAX in watts/(meter squared *ster* μ m), QCALMIN is the minimum quantized calibrated pixel value (corresponding to $LMIN_\lambda$) in DN and QCALMAX is the maximum quantized calibrated pixel value (corresponding to $LMAX_\lambda$) in DN.

$LMAX_\lambda$ and $LMIN_\lambda$ are the spectral radiances for each band at digital number 1 and 255 (i.e., QCALMIN, QCALMAX), respectively. QCAL or DN, $LMAX_\lambda$ and $LMIN_\lambda$ are input data. These elements are values in header file information. Thereafter a thermal band or band 6 imagery is converted to the effective at satellite temperature (T_{bb}) calculated by Eq. (3):

$$T_{bb} = \frac{K_2}{\ln\left(\dfrac{K_1}{L_6}+1\right)} \tag{3}$$

For thermal band, calibration constants, K_1 and K_2, are 666.09 watts/(meter squared *ster*μm) and 1282.71 Kelvin, respectively. L_6 is the spectral radiance for band 6 in watts/(meter squared *ster*μm). Then, surface temperature (T_s) is computed by Eq. (4).

$$T_s = \frac{T_{bb}}{\varepsilon_o^{0.25}} \tag{4}$$

where, ε_o is surface emissivity.

15.5 ESTIMATIONS OF ACTUAL EVAPOTRANSPIRATION

SEBAL is a tool to estimate actual evapotranspiration for flat areas with the most accuracy and confidence. Satellite image and weather data are used in the SEBAL model to calculate actual evapotranspiration by using a surface energy balance at the land surface. SEBAL evaluates an instantaneous actual evapotranspiration flux for the image time, because the satellite image provides information for the overpass time only. The actual evapotranspiration flux can be calculated for each pixel of the image as a residual of the surface energy budget equation. SEBAL needs both shortwave and thermal bands. The required ground-based data is wind speed. The SEBAL energy balance calculates actual evapotranspiration for each pixel for the time of the satellite image, so the results are instantaneous actual evapotranspiration. To obtain actual evapotranspiration using the conception of SEBAL, following equations are applied [6–10].

Firstly, data in the format of radiance is converted to reflectance for all bands. For this converting, thermal band (band 6) is not considered. In practice, band 6 is converted, but it is a dummy in this file. The Eq. (5) is used to convert radiance to reflectanc.

$$\rho_\lambda = \frac{\pi \times L_\lambda}{ESUN_\lambda \times \cos\theta \times d_r} \tag{5}$$

where $\rho\lambda$ is unitless planetary reflectance, L_λ is calculated from Eq. (1), $ESUN_\lambda$ is mean solar exoatmospheric irradiances from Table 3, θ is solar zenith angle in degrees, and d_r is the Earth-Sun distance in astronomical that can be obtained from Eq. (6):

$$d_r = 1 + 0.033 \cos\left(DOY \frac{2\pi}{365}\right) \tag{6}$$

where, DOY (or J.) is number of day in one year for example DOY for January 1 is 1 while DOY for December 31 is 365.

$$\cos\theta = \cos(90 - \beta) \tag{7}$$

where, β is sun elevation angle in degree and $\cos\theta$ is in degree. Albedo for the top of atmosphere (α_{toa}) can be considered from Eq. (9). The weighting coefficient is calculated from Eq. (10).

$$[Radians] = \frac{\pi}{180} \times [decimal\ deg\,rees] \tag{8}$$

$$\alpha_{toa} = \sum(\omega_\lambda \times \rho_\lambda) \tag{9}$$

$$\omega_\lambda = \frac{ESUN_\lambda}{\sum ESUN_\lambda} \tag{10}$$

where, ω_λ is weighting coefficient, which is constant value. Solar spectral irradiances values are shown in Table 3.

TABLE 3   Solar Spectral Irradiances.

Band	Solar Spectral Irradiances watts/[meter squared *ster*µm]
1	1969.00
2	1840.00
3	1551.00
4	1044.00
5	225.70
7	82.07
Pan	1368.00

Surface albedo (α) can be estimated using Eq. (11):

$$\alpha = \frac{\alpha_{toa} - \alpha_{path\_radiance}}{\tau_{sw}^2} \tag{11}$$

$$\tau_{sw} = 0.75 + 2 \times 10^{-5} \times z \tag{12}$$

In Eqs. (11)–(16): α_{toa} is calculated from Eq. (9) and $\alpha_{path\_radiance} \approx 0.03$; z is an eleva-tion of area which is defined from Digital Elevation Map (DEM); G_{sc} is solar constant value of 1367 W/m²; ρ_3 and ρ_4 are reflectance value in red and near-infrared bans (band 3 and 4), respectively; L is constant for SAVI (L= 0.5, when an area have no information for L). L = 0.5 is suitable for this practice. Incoming solar radiation ($R_{s\downarrow}$) is estimated in spreadsheet using Eq. (13). The vegetation indices can be considered from Eqs. (14)–(16).

$$R_{s\downarrow} = G_{sc} \times \cos\theta \times d_r \times \tau_{sw} \tag{13}$$

$$NDVI = \frac{\rho_4 - \rho_3}{\rho_4 + \rho_3} \tag{14}$$

$$SAVI = \frac{(1+L)(\rho_4 - \rho_3)}{L + \rho_4 + \rho_3} \tag{15}$$

$$LAI = -\frac{\ln\left(\frac{0.69 - SAVI}{0.59}\right)}{0.91} \tag{16}$$

Surface emissivity (ε_0) can be estimated using Eq. (17):

$$\varepsilon_o = 1.009 + 0.047 \times \ln(NDVI) \tag{17}$$

Outgoing longwave radiation ($R_{L\uparrow}$) can be calculated by following equation:

$$R_{L\uparrow} = \varepsilon_o \sigma T_s^4 \tag{18}$$

where, σ = 5.67 × 10⁻⁸ W/(m²·K⁴). For the selection of "anchor pixel", SEBAL pro-cess uses two "anchor" pixels to fix boundary condition for the energy balance. (a) "Cold" pixel: a wet, well-irrigated crop surface with full cover ($T_s \cong T_{air}$). In cold pixel, sensible heat flux (H) is usually zero so cold pixel should be selected from wa-ter area. (b) "Hot" pixel should be located in a dry and bare agricultural field where one can assume there is no evapotranspiration taking place, and should have a surface albedo similar to other dry and bare field in the area of interest. It should have a LAI in the range of 0 to 0.4. After the temperatures of both cold and hot pixel are defined, these values are used for calculation in the next step. Incoming longwave radiation ($R_{L\downarrow}$) is computed in spreadsheet using Eqs. (19) and (20):

$$R_{L\downarrow} = \varepsilon_a \times \sigma \times T_{cold}^4 \tag{19}$$

$$\varepsilon_a = 0.85 \times (-\ln\tau_{sw})^{0.09} \tag{20}$$

where ε_a is an atmospheric emissivity. Net surface radiation flux (R_n) can be com-puted by Eq. (21). The soil heat flux (G) is given by Eq. (22).

$$R_n = R_{s\downarrow} - \alpha R_{s\downarrow} + R_{L\downarrow} - R_{L\uparrow} - (1 - \varepsilon_o) R_{L\downarrow} \tag{21}$$

$$\frac{G}{R_n} = \frac{T_s}{\alpha} \left(0.0038\alpha + 0.0074\alpha^2 \right) \left(1 - 0.98 NDVI^4 \right) \tag{22}$$

After the parameters in Eqs. (1–22) computed, now we can estimate sensible heat flux, latent heat flux, evaporative fraction, and 24-hour actual evaporatranspiration as shown below. Sensible heat flux (H) is the flux of heat from the earth's surface to the atmosphere (Eq. (23)). It is not associated with phase changes of water.

$$H = \frac{\rho \times c_p \times dT}{r_{ah}} \tag{23}$$

where, ρ is air density (kg/m³), c_p is air specific heat (1004 J./kg/K), dT (K) is the temperature difference $(T_1 - T_2)$ between two heights (z_1 and z_2), and r_{ah} is the aerodynamic resistance to heat transport (s/m). The estimation of sensible heat flux is most tedius job, because the temperature difference and aerodynamic resistance to heat transport are unknown for the sensible heat flux calculations. To find these unknowns, SEBAL first calculates the sensible heat flux at extreme dry and wet locations. They are manually identified by the user on the image. The aerodynamic resistance to heat transport is computed from the lower integration constant for r_{ah} (z_1 = 0.1 m.) and the upper integration constant for r_{ah} (z_2 = 2 m.).

For a dry pixel or hot pixel, it should be located in a dry and bare agricultural field where one can assume that there is no evapotranspiration taking place. The wet pixel or cold pixel will include a surface temperature equal to air temperature. The sensible heat flux for the cold pixel is usually zero. The linear relationship is developed between the temperature difference ($dT = a + bT_s$) and the surface temperature (T_s), and the regression coefficients (a and b) are defined from the two (dT, T_s) pairs applicable to the hot and cold pixels. Then sensible heat flux can be computed for every pixel that has the condition of free convection. Next, the values of friction velocity (u*) are estimated from the wind speed at the blending height, a value of 200 m will be used. Thereafter, the condition of mixed convection is applied, and the pixel-dependent aerodynamic resistance to heat transfer, r_{ah}, is calculated by using the Monin-Obukhov hypothesis. The new temperature difference is calculated. Finally the processes from the calculation of sensible heat flux to the temperature difference are repeated until the aerodynamic resistance to heat transfer and temperature difference are stable values. To compute sensible heat flux, following procedure is considered:

Step 1 Friction velocity (u*) can be computed using Eq. (24).

$$u* = \frac{k u_x}{\ln\left(\dfrac{z_x}{z_{om}} \right)} \tag{24}$$

The calculation of the friction velocity requires a wind speed measurement (u_x) at a known height (z_x) in the time of the satellite image. k is a constant (0.41). Then, u_x and z_x are know, but z_{om} is unknown. z_{om} can be calculated in many ways: from $z_{om} = 0.12 \times$ height of vegetation (h) for agricultural area, from a land-use map, or from a function of NDVI and surface albedo. At weather station, u_x, z_x, z_{om}, and $u*$ can be determined.

Step 2: Wind speed at a height 200 m above the weather station can be computed using Eq. (25).

$$u_{200} = \frac{u * \ln\left(\dfrac{200}{z_{om}}\right)}{k} \tag{25}$$

Step 3: The friction velocity for each pixel is calculated using the wind speed at a height 200 m (u_{200}) that is assumed to be constant for all pixels of the image because it is defined as occurring at a "blending height" unaffected by surface features. From Eq. (26), z_{om} is unknown to calculate the friction velocity, so z_{om} need to be estimated.

$$u* = \frac{ku_{200}}{\ln\left(\dfrac{200}{z_{om}}\right)} \tag{26}$$

where, z_{om} is the particular momentum roughness length of each pixel. The z_{om} for each pixel can be computed by two methods: using a land-use map or using NDVI and surface albedo data (z_{om} is calculated in spread sheet). For this pattern, a land-use map is not available, and then NDVI and surface albedo data are used. In the method using NDVI and surface albedo, z_{om} is computed from the following equation:

$$z_{om} = \exp\left[\left(a \times \frac{NDVI}{\alpha}\right) + b\right] \tag{27}$$

where, a and b are correlation constants derived from a plot of $\ln(z_{om})$ vs $\frac{NDVI}{\alpha}$ for two or more sample pixels representing specific vegetation types. To determine a and b, a series of sample pixels representing vegetation types and conditions of interest are selected and the associated values for NDVI and surface albedo are obtained.

Typical surface albedo values are 0.17 to 0.22 for rice field and 0.15 to 0.20 deciduous forest, respectively.

Step 4: Aerodynamic resistance to heat transport (r_{ah}) is computed as following equation:

$$r_{ah} = \frac{\ln\left(\dfrac{z_2}{z_1}\right)}{u * \times k} \tag{28}$$

where, z_1 and z_2 are 0.1 m and 2 m, respectively.

Step 5: Near surface temperature difference (dT) for each pixel is calculated using Eq. (29). A linear relationship is assumed between T_s and dT (Eq. (30)).

$$dT = T_s - T_a \tag{29}$$

$$dT = b + aT_s \tag{30}$$

In Eqs. (29) and (30): T_a is unknown; a and b are the correlation coefficients. To define coefficients a and b, SEBAL uses the "anchor" pixel when a value for sensible heat flux (H) can be reliably estimated.

a. At the "Cold" pixel:

$$H_{cold} = R_n - G - LE_{cold} \tag{31}$$

$$dT_{cold} = \frac{H_{cold} \times r_{ah}}{\rho \times c_p} \tag{32}$$

If "Cold" pixel is chosen from a body of water, then $H_{cold} = 0$ and Eq. (31) reduces to:

$$LE_{cold} = R_n - G \tag{33}$$

$$dT_{cold} = 0 \tag{34}$$

b. At the "Hot" pixel:

$$H_{hot} = R_n - G - LE_{hot} \text{ and } LE_{hot} = 0 \tag{35}$$

Therefore,

$$H_{hot} = R_n - G \tag{36}$$

$$dT_{hot} = \frac{H_{hot} \times r_{ah}}{\rho \times c_p} \tag{37}$$

Therefore, $dT_{cold} = 0$ from Step 5a and $dT_{hot} = \frac{H_{hot} \times r_{ah}}{\rho \times c_p}$ from Step 5b.

Step 6: The sensible heat flux is calculated in this step, called initial sensible heat flux.

Step 7: Monin-Obukhov theory in an iterative process is applied in SEBAL to account for the buoyancy effects, which are generated by surface heating. The Monin-Obukhov length (L) is used to define the stability conditions of the atmosphere in the iterative process)this is not the same "L" as used in the SAVI computation). It is a function of the heat and momentum fluxes and is computed as follows:

$$L = -\frac{\rho c_p u *^3 T_s}{kgH} \tag{38}$$

or

$$L = -\frac{1 \times 1004 \times u *^3 T_s}{0.41 \times 9.81 H} \tag{39}$$

Step 8: The values of the stability corrections for momentum (ψ_m) and and heat transport (ψ_h) are computed as follows. These values depend on the conditions of atmosphere.

    a.   If $L < 0$ [unstable atmospheric condition]:

$$\psi_{h(200m)} = 2\ln\left(\frac{1+x_{(200m)}}{2}\right) + \ln\left(\frac{1+x_{(200m)}^2}{2}\right) - 2ARCTAN\left(x_{(200m)}\right) + 0.5\pi \quad (40)$$

$$\psi_{h(2m)} = 2\ln\left(\frac{1+x_{(2m)}^2}{2}\right) \quad (41)$$

$$\psi_{h(0.1m)} = 2\ln\left(\frac{1+x_{(0.1m)}^2}{2}\right) \quad (42)$$

In Eqs. (40)–(42):

$$x_{(200m)} = \left(1-16\frac{200}{L}\right)^{0.25} \quad (43)$$

$$x_{(2m)} = \left(1-16\frac{2}{L}\right)^{0.25} \quad (44)$$

$$x_{(0.1m)} = \left(1-16\frac{0.1}{L}\right)^{0.25} \quad (45)$$

    b.   If $L > 0$ (stable atomspheric condition):

$$\psi_{h(200)} = -5\left(\frac{2}{L}\right) \quad (46)$$

$$\psi_{h(2m)} = -5\left(\frac{2}{L}\right) \quad (47)$$

$$\psi_{h(0.1m)} = -5\left(\frac{0.1}{L}\right) \quad (48)$$

    c.   If $L = 0$ (neutral condition): ψ_m and $\psi_h = 0$

Step 9: The friction velocity ($u*$), which is a corrected value, is now computed for each successive iteration as follows:

$$u* = \frac{u_{200}k}{\ln\left(\dfrac{200}{z_{om}}\right) - \psi_{m(200m)}} \quad (49)$$

where, u_{200} is in m/s and k is 0.41.

Step 10: The aerodynamic resistance to heat transport (r_{ah}), which is a corrected value, is now computed during each iteration as follows:

$$r_{ah} = \frac{\ln\left(\frac{z_2}{z_1}\right) - \psi_{h(z_2)} + \psi_{h(z_1)}}{u* \times k} \quad (50)$$

Step 11: Repeat the steps 5 to 10 until the successive values for dT_{hot} and r_{ah} at the hot pixel have stabilized.

Step 12: The latent energy of evaporation (LE) was computed using following equation:

$$LE = R_n - G - H \quad (51)$$

After the latent energy of evaporation has been computed, then the evaporative fraction (Λ) is obtained using Eq. (52). The evaporative fraction at each pixel of a satellite image can be estimated using the 24-hour evaporatranspiration for the day of the image. The evaporative fraction is assumed to be a constant value over the full 24-hour period.

$$\lambda = \frac{LE}{R_n - G} = \frac{LE}{LE + H} \quad (52)$$

The 24-hour actual evaporatranspiration is estimated using Eq. (53):

$$ET_{24} = \frac{86400\Lambda\left(R_{n24} - G_{24}\right)}{\lambda} \quad (53)$$

where, R_{n24} is daily net radiation, G_{24} is daily soil heat flux, 86,400 is the number of seconds in a 24-hour period, and λ is the latent heat of vaporization (J./kg). The 24-hour actual evaporatranspiration, ET_{24}, can be expressed in mm/day. Since energy, on average, is stored in the soil during the daytime and released into the air at night, G_{24} is very small for the combined vegetative and soil surface, so it can be assumed as zero at the soil surface [24]. Therefore, Eq. (53) can be rewritten as:

$$ET_{24} = \frac{86400\Lambda R_{n24}}{\lambda} \quad (54)$$

Steps 1 to 11 are summarized in Fig. 1. Thereafter, to determine the relationship between the temperature and actual evapotranspiration, the maximum and minimum values of actual evapotranspiration in each temperature were ignored. The selected actual evapotranspiration in each temperature was averaged. Finally, the relationship between temperature and actual evapotranspiration was determined using a polynomial equation.

15.5.1 VALIDATION OF TEMPERATURE

For the validation of temperature computed by using Landsat satellite image, it was compared with recorded temperature. On the other hand, to validate actual evapotranspiration calculated by SEBAL, recorded pan evaporation was used. Pan evaporation is the amount of water evaporated during a period [mm/day] with an unlimited supply of water (potential evaporation). It is a function of surface and air temperatures, insolation, and wind, all of which affect water-vapor concentrations immediately above the evaporating surface [11]. On the other hand, actual evapotranspiration is a function of temperature, wind, humidity and net radiation. It can be concluded that there is a relationship between the pan evaporation and actual evapotranspiration. [11, 13, 15] indicate that the decreasing trend detected in the pan evaporation and actual evapotranspiration can be attributed to the significant decreasing trends in the net radiation and in the wind speed. Also, it can be attributed to the significant increasing trend in the air temperature.

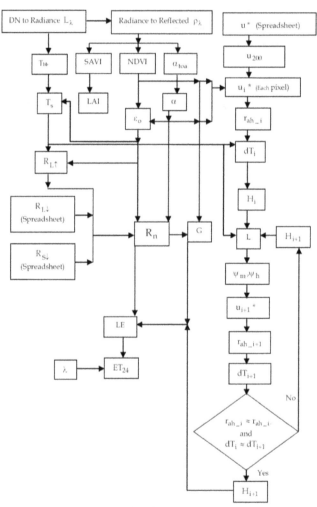

FIGURE 1   The surface energy balance algorithm for land process.

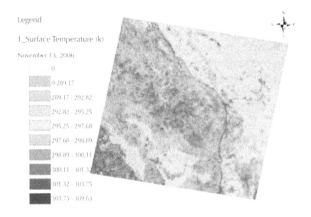

FIGURE 2   The surface temperature on November 13, 2006.

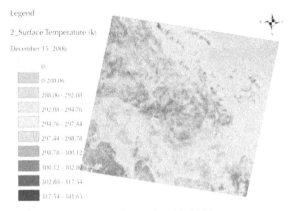

FIGURE 3   The surface temperature on December 15, 2006.

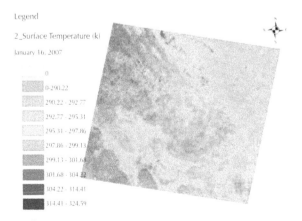

FIGURE 4   The surface temperature on January 16, 2007.

15.6 RESULTS AND DISCUSSION

Since there are three main parts for the calculation of this study, there are three main parts of the result that are the spatial temperature, the spatial actual evapotranspiration, and the relation between temperature and actual evapotranspiration. The results for each part are presented as follows:

15.6.1 SPATIAL TEMPERATURE

The spatial distributions of temperature calculated by using Landsat 5™ satellite images are presented from Figs. 2–4 [19]. These figures can be presented that the mean temperatures from Figure 2 to Figure 4 are 297.34, 295.74 and 296.25°K, respectively. On the one hand, the minimum temperatures from these three figures are 283.21, 278.93 and 284.02°K, respectively while the maximum temperatures from these three figures are 308.62, 313.63 and 310.65°K, respectively.

15.6.2 SPATIAL ACTUAL EVAPOTRANSPIRATION

The spatial distributions of actual evapotranspiration calculated by using Landsat 5™ satellite images and SEBAL are presented from Figs. 5–7 [19]. These figures cab be presented that the mean actual evapotranspiration from these three figures are 3.67, 4.50 and 4.26 mm, respectively.

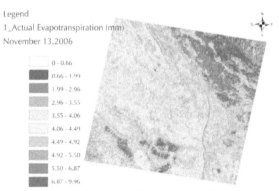

FIGURE 5   The actual evapotranspiration on November 13, 2006.

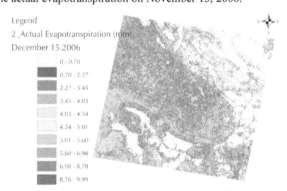

FIGURE 6   The actual evapotranspiration on December 15, 2006.

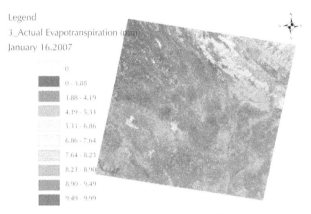

FIGURE 7 The actual evapotranspiration on January 16, 2007.

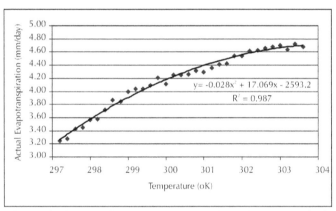

FIGURE 8 The relation between the temperature (°K) and actual evapotranspiration (mm/day).

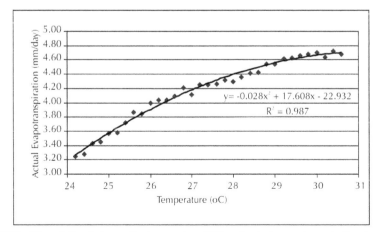

FIGURE 9 The relation between the temperature (°C) and actual evapotranspiration (mm/day).

15.6.3   RELATION BETWEEN TEMPERATURE AND ACTUAL EVAPOTRANSPIRATION

After the spatial temperature and spatial actual evapotranspiration were calculated as shown in Figs. 2–7. The relationships between the temperature and actual evapotranspiration are shown in Figs. 8 and 9 [19]. In the regression Eqs. (55) for Fig. 8 and Eq. (56) for Fig. 9, y is the actual evapotranspiration [mm/day] and x is the temperature in the unit of °K and °C, respectively.

$$y = -0.028x^2 + 17.069x - 2593.2 \ [R^2 = 0.987] \text{ and} \tag{55}$$

$$y = -0.028x^2 + 1.7608x - 22.932 \ [R^2 = 0.987] \tag{56}$$

15.7   SUMMARY

The results in this chapter indicate that the mean temperature and the mean actual evapotranspiration are 296.44°K (23.44°C) and 4.14 mm, respectively. The relationship between the temperature (°K and °C) and actual evapotranspiration (mm/day) is in the format of the polynomial equation. For the temperature in Kelvin, an equation is $y = -0.028x^2 + 17.069x - 2593.2$ and for the temperature in Celsius, an equation is $y = -0.028x^2 + 1.7608x - 22.932$.

The spatial temperature and spatial actual evapotranspiration during November 2006 to January 2007 that are the result are in the condition of both dry season and clear sky. Also, the relationship between temperature and actual evapotranspiration is a result. These results are useful for irrigation project and water management. For spatial temperature, it presents temperature for each area. Since temperature relates to many weather data, temperature can imply the characteristic of other weather data. For example, high temperature is included low humidity but high evaporation can be occurred in the condition of high temperature. For spatial actual evapotranspiration, it can be used to present daily actual evapotranspiration. This daily actual evapotranspiration leads to the planning of water management in each area. Then, it is easy to manage water for irrigation. For the relationship between temperature and actual evapotranspiration, it can be used to estimate actual evapotranspiration when temperature is not unknown.

KEYWORDS

- **Landsat 5™**
- **SEBAL**
- **spatial actual evapotranspiration**
- **spatial temperature**
- **Thailand**

REFERENCES

1. Abou E.; Tanton. T.; **2003,** Real time crop coefficient from SEBAL method for estimating the evapotranspiration. Proceedings of SPIE, 5232.

2. Allen, R. G.; Pereira, L. S.; Dirk, R.; Smith, M. **1998,** *Crop evapotranspiration-Guidelines for computing crop water requirements.* FAO Irrigation and drainage paper 56.

3. Allen, R. G.; Morse, A.; Tasumi, M.; Bastiaanssen, W.; Kramber W.; Anderson H. Landsat Thematic Mapper for evapotranspiration via the SEBAL process for water rights management and hydrology water balances, SEBAL, Available Source: http: //www.agu.org/meetings/sm01/sm01_pdf/sm01_B42C. pdf, August 8, **2001.**

4. Amin, M. S. M.; Nabi A.; Mansor, S.; **1997,** Use of satellite data to estimate areal evapotranspiration from a tropical watershed. Asian Conference on Remote Sensing, 20–24 October 1997, GIS Development Pvt. Ltd.; Malaysia.

5. Andrew, N. F.; Schmugge, T. J.; Kustas, W. P.; Estimating evapotranspiration over El Reno – Oklahoma with ASTER imagery. EDP Sciences, **2002,** *22,* 105–106.

6. Bastiaanssen, W. G. M.; Meneti, M.; Feddes, R. A.; Holtslag, A. A. M.; A remote sensing surface energy balance algorithm for land (SEBAL) 1 Formulation. Journal of Hydrology, **1998a,** *212–213,* 198–212.

7. Bastiaanssen, W. G. M.; Meneti, M.; Feddes R. A.; Holtslag, A. A. M.; A remote sensing surface energy balance algorithm for land (SEBAL) 1 Validation. Journal of Hydrology, **1998b,** *212–213,* 213–229.

8. Bastiaanssen, W. G. M. SEBAL-based sensible and latent heat fluxes in the irrigated Gediz Basin, Turkey. Journal of Hydrology, **2000,** *229,* 87–100.

9. Bastiaanssen, W. G. M.; Ahmad, M.; Chemin, Y.; Satellite surveillance of evaporative depletion across the Indus Basin. Water Resources Research, **2002,** *38,* 12.

10. Chemin, Y.; Ahmad, M.; Estimating evaporation using the surface energy balance algorithm for land (SEBAL). In: *A Manual for NOAA-AVHRR in Pakistan.* International Water Management Institute *(IWMI).* Report No. R-102. **2000.**

11. Chong-yu, X. U.; Lebing G.; Jlang T.; Deliang C. Decreasing reference evapotranspiration in a warming climate-a case of Changjiang (Yangtze) river catchment during 1970–2000. Advances Atmospheric Sciences, **2006,** *23,* 115–131.

12. Farr, R. **1999,** Landsat 7 Project, Policy, and History, Landsat. Available at: http: //landsat.gsfc. nasa.gov/main/project.html, October 21, 2003.

13. Grismer, M. E.; Orang, M.; Snyder, R.; Matyac R.; Pan evaporation to reference evapotranspiration conversion methods. Journal of Irrigation and Drainage Engineering, **2002,** *128(3),* 180–184.

14. Hongjie, X.; Jan H.; Shirley K.; Eric S. **2002,** Comparison of evapotranspiration estimates from the surface energy balance algorithm (SEBAL) and flux tower data, middle Rio Grande Basin, Evapotranspiration. Available at: http: //www.nmt.edu/~hjxie/sebal-agu.htm, November 29, 2002.

15. Humphreys, E.; Meyer, W. S.; Prathapar S. A.; Smith, D. J.; Estimation of evapotranspiration from rice in southern New South Wales: A review. Aust. J. Exp. Agric.; **1994,** 34(7), 1013–1020.

16. Ines A. V. M. **2002,** Improved crop production integration GIS and genetic Algorithm. Thesis for Doctorate Degree by Asian Institute of Technology, Thailand.

17. Jacob, F.; Olioso, A.; Gu, X. F.; Hanocq, J. F.; Hautecoeur O.; Leroy, M. Mapping surface fluxes using visible-near infrared and thermal infrared data with the SEBAL algorithm. In: *Physics and Chemistry of the Earth.* **2003.**

18. Kalluri S.; Gilruth P.; Bergman P.; Plante, R. **2003,** Impacts of NASA's remote sensing data on policy and decision making at state and local agencies in the United State. Evapotranspiration. Available at:
< http: //earth-outlook.east.hitc.com: 1500/05_07_11.00_kallyri.pdf>, August 5, 2003.

19. Kosa, P. Air Temperature and Actual Evapotranspiration Correlation Using Landsat 5™ Satellite Imagery. *Kasetsart J. Nat. Sci.,* **2009,** *43,* 605–611.

20. Lal, M.; Chemin Y.; Chandrapala L.; **2001,** Variability of soil moisture in the Walawe river basin: A case study in Sri Lanka using low-resolution satellite data. Asian Conference on Remote Sensing, 5–9 November. GIS Development Pvt. Ltd.; Singapore.

21. Liang, S.; Fang H.; Chen M.; Atmospheric correction of Landsat ETM+ Land Surface Imagery-Part I: Methods. IEEE Transactions on Geoscience and Remote Sensing, **2001,** *39(11),* 2490–2498.

22. Liu, J. G. Evaluation of Landsat-7 ETM+ panchromatic band for image fusion with multispectral bands. Natural Resources Research, **2000,** *9(4),* 269–276.

23. Marco, A. F. C. Reference evapotranspiration based on class a pan evaporation. Scientia Agricola Journal, **2002,** *59(3),* 417–420.

24. Morse, A.; Tasumi, M.; Allen R. G.; Kramber, J. W.; **2000,** Application of the SEBAL methodology for estimating consumptive use of water and stremflow depletion in the Bear river basin of Idaho through remote sensing. Final Report submitted to The Raytheon System Company – Earth Obseration System Data and Information System Project.

25. Timmermans, W. J.; Meijerink A. M. J.; Lubczynski, M. W.; **2001,** Satellite derived actual evapotranspiration and groundwater modeling, Botswana. Proceedings of a symposium held at Santa Fe – New Mexico – USA, April 2001, IAHS Publ, No. 267.

26. Wei, Y.; Sado. K. **1994,** Estimation of areal evapotranspiration using Landsat™ data Alone. Asian Conference on Remote Sensing, 17–23 November 1994, GIS Development Pvt. Ltd.; Bangalore, India.

27. Xihua, Y.; Zhou, Q.; Melville, M. Estimating local sugarcane evapotranspiration using Landsat™ image and a VITT concept. *Int. J. Remote Sens.* **1997,** *18(2),* 453–459.

CHAPTER 16

SENSOR BASED IRRIGATION SCHEDULING[1]

B. KEITH BELLINGHAM

CONTENTS

[1]Author is thankful to Stevens Water Monitoring Systems, Inc. for the support. Printed with permission from: <http://www.stevenswater.com/articles/irrigationscheduling.aspx> © Copyright Stevenswater.com

16.1   INTRODUCTION

In the western United States, irrigation accounts for about 80% of the water consumed [8]. Concerns about changes in land use due to growing populations, climate change, and the protection of aquatic habitats are driving a need to conserve water. Optimization of irrigation will not only benefit the environment, but also benefit local economies. Over irrigation may lead to dangerous increases in the total maximum daily loads (TMDL) of temperature, nitrates, and salinity in natural waters [6]. Nitrate fertilizers leached out of the soils get transported to natural waters causing eutrophication and other aquatic impairments. Run off from over irrigation may affect water quality parameters such as pH, total suspended solids (TSS), and dissolved oxygen [18]. Other negative impacts associated with over irrigation include wastes of water and energy, and reduced crop yields. The negative impacts associated with under irrigation are more intuitive. Under irrigation may reduce crop yields which will reduce profit margins. This chapter discusses a soil water balance model incorporated into a data acquisition system that is a power tool for scheduling and optimizing irrigation. A case study for blueberries is presented.

Advancements in computer microprocessors, memory and software development tools has improved data acquisition methods and made data acquisition system integration more reliable and more cost effective. The soil water balance model incorporates inputs of soil moisture, water application and evapotranspiration (ET). The soil moisture data acquisition system retrieves the input parameters via telemetry and populates software that accommodates the soil water balance model. The soil data acquisition software integrated with a soil water balance model is commercially available from Stevens Water Monitoring Systems, Inc.

16.2   SOIL MOISTURE BUDGET

To begin our discussion about soil moisture budgets (Fig. 2), we first describe the components and the hydrological conditions of soil. In general, inorganic soil is composed of mixes of sands, silts and clays. Sands, silts and clays differ not only by particle size distribution, but also in the atomic arrangement and charge distribution at the molecular level [9]. Soil geomorphology is the process by which sands and silts chemically and physically transform into clays as the soil ages [2]. The soil textural class is determined by the gravimetric percentage of sand silt and clay. Figure 1 shows the soil texture classifications based on gravimetric percentage.

Sands, silts, clays and organics represent the solid particle composition of soil while air and water fill the pore spaces between the solid particles. When soil is completely saturated with water, the porosity will be equal to the volumetric soil moisture content [16]. The amount of organics in soil will affect the bulk density and the porosity. Some organic soils may have porosities of over 90%, but in general, most inorganic agricultural loams will have a porosity of near 50%. The pores can be nearly microscopic (micropores) or visible with the naked eye (macropores) [3].

The hydrologic properties of soil play an important role in a crop's ability to transpire water with their root systems. Knowledge of volumetric soil moisture content (θ, m^3 m^{-3}) is an important input into the soil water balance model. Permanent wilting point (θ_{PW}) is the soil moisture level at which plants can no longer adsorb water from the soil. Plant transpiration and direct evaporation will decrease the moisture level in soil to a point below θ_{PW} and, in some cases, down to near dryness (Fig. 3).

Field capacity (θ_{FC}) is defined as the threshold point at which the soil pore water will be influenced by gravity. Above field capacity, the gravitational force will overcome the capillary forces suspending the moisture in the pores of the soil allowing for down movement of water in the soil column. Below θ_{FC}, there will be a net upward movement of water driven by ET. Field capacity and permanent wilting point are heavily influenced by soil textural classes, particularly clay content [10]. Clays interact with water in ways uniquely different from sand, silt and organics. Clays will have a physical and chemical affinity for water due to the negative charge distribution and the planner molecular lattice. The positive portion of the water molecule will be oriented toward the negatively charged clay lattice and the oxygen's lone electron pair will be pointed outwards [7].

Positively charged cations will also be influenced by the negative charged distribution of clay [9]. Figure 4 shows two cations of different valance states (Ca^{++} and Na^{+}) chemically influenced by clay at the molecular level. Figure 4 also shows the charge distribution of the water molecule. The available water capacity (θ_{AC}) of soil is the water that is available to a plant. It represents the range of soil moisture values that lie above permanent wilting point and below the field capacity.

$$\theta_{PW} < \theta ac < \theta_{FC} \tag{1}$$

Table 1 shows the typical values for permanent wilting point and field capacity for common soil textural classes [10]. Plants are able to uptake water from soil if the soil moisture is above permanent wilting point. As the soil moisture approaches permanent wilting point, the plant will become increasingly stressed as the soil pore water becomes depleted. The point below field capacity where plants become stressed is called the maximum allowable depletion (MAD).

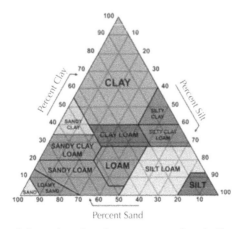

FIGURE 1   Soil textural classes based on the percentage of sand, silt, and clay.

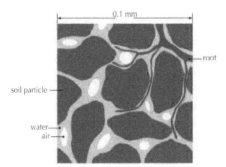

FIGURE 2 Unsaturated soil is composed of solid particles, organic material and pores. The pore space will contain air and water.

FIGURE 3 Soil moisture: Saturation, field capacity and permanent wilting point.

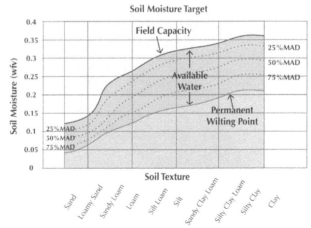

FIGURE 4 Planner clay lattice: a negative charge distribution (left); cation influence (center); and Dipole moment of a water molecule (extreme right).

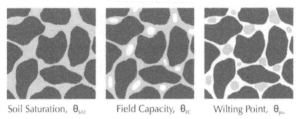

FIGURE 5 The relationship between soil textural classes and the hydrological thresholds θ_{PW}, θ_{AC}, and θ_{FC}. The 25%, 50% and 75% MAD levels are displayed in the available water capacity region.

The MAD value is expressed as a percent of the available water capacity. Table 2 shows typical MAD values for a few selected crops. Figure 5 shows the soil field capacities and the permanent wilting points for common soil textural classes. The dark shaded region in Fig. 5 is the available water capacity showing 25%, 50% and 75% MADs. As shown in Fig. 5, the field capacity and the permanent wilting point will increase with the percentage of clay. With specific knowledge of field capacity, soil textural class and the maximum allowable depletion, a soil moisture target can be determined for irrigation optimization [4]. The soil moisture target is the range of soil moistures that lie above the MAD but below the field capacity. Below the MAD value the crop will still have the ability to receive water from the soil, however the crop will become stressed after a period of time. If the crop becomes stressed due to the lack of water, the plant will have a reduced yield and become more susceptible to pathogens. If the soil moisture gets above field capacity, water will be transported downward by gravity potentially wasting water and leaching nutrients. Upper soil moisture target for the soils in the root zone will be the field capacity. The lower soil moisture target is determined by the MAD, θ_{FC}, and θ_{PW}:

$$\text{Lower Soil Moisture Target} = \theta_{FC} - (\theta_{FC} - \theta_{PW}) \times \text{MAD} \qquad (2)$$

For example, green beans with a MAD of 50% have a root zone depth of 18 inches. If the green beans are growing in a silt loam, the field capacity will be 0.3 water fraction by volume (wfv) and the permanent wilting point will be 0.15 wfv. Using Eq. (2), the lower soil moisture target will be 0.23 wfv. In this example, the soil moisture target for the green beans will lie between 0.23 wfv and 0.3 wfv from 5 inches to 18 inches deep adjacent to the root ball. It is important to note that the values in Table 1 are typical values and could vary slightly with bulk density of soil, mineralogy and organic content. Similarly, the MAD values in Table 2 are typical values and may vary by species, age of crop, region and soil chemistry.

TABLE 1 Field capacity and permanent wilting point for common soil textural classes.

Soil Texture	Field capacity	Permanent wilting point
Sand	0.12	0.04
Loamy Sand	0.14	0.06
Sandy Loam	0.23	0.1
Loam	0.26	0.12
Silt Loam	0.3	0.15
Silt	0.32	0.165
Sandy Clay Loam	0.33	0.175
Silty Clay Loam	0.34	0.19
SiltyClay	0.36	0.21
Clay	0.36	0.21

TABLE 2   Maximum allowable depletion and effective root zone depth for selected crops [12].

Crop	Maximum allowable depletion (MAD)	Effective root Depth, inches
Grass	50%	7
Table beet	50%	18
Sweet Corn	50%	24
Strawberry	50%	12
Winter Squash	60%	36
Peppermint	35%	24
Potatoes	35%	35
Orchard Apples	75%	36
Leafy Green	40%	18
Cucumber	50%	24
Green Beans	50%	18
Cauliflower	40%	18
Carrot	50%	18
Blue Berries	50%	18

TABLE 3   Typical values for sprinkler efficiencies for various sprinkler systems [12].

Irrigation System	Sprinkler efficiency (*Ef*)	Sprinkler efficiency (sprinkler spacing over 40 x40 feet)
Solid set	0.70	0.63
Hand move or side roll	0.80	0.74
Pivot or linear move	0.90	0.81
Offset managed hand move	0.90	0.81

16.3   WATER APPLICATION

While soil moisture data provides information about the root zone, the measured application of water can be used concurrently with the soil moisture values to provide a more complete suite of tools for the irrigator. The measured application of water (D) is the amount of water applied to the crops with sprinklers, plus the amount of natural precipitation measured in inches/day. It is the total depth of water received by the crop.

16.3.1   SPRINKLER EFFICIENCY

In order to effectively use the application of water in a water budget model, a high sprinkler efficiency (*Ef*) is required. Sprinkler efficiency (*Ef*) is the measure of uniformity of water application. Ponding of irrigation water, and uneven application of water over the field is the result of poor sprinkler efficiency. Soil moisture data and rain gauge data are less meaningful if the monitoring site receives more or less water than the rest of the irrigation regime. Sprinkler efficiency is determined by placing catch cans or a set of containers of uniform size in the field. The catch cans can be placed in grid or uniformly distributed among the crops. After running the sprinklers for a length of time, the amount of water in the catch cans is measured. The sprinkler efficiency is expressed as a fraction and an *Ef* value of 1 is perfect uniformity. There are a number of methods for calculating *Ef*. The most common method for determining *Ef* involves averaging the lower 25% of the measured catchment of catch cans divided by the mean. An *Ef* value greater than 0.8 is preferred. Table 3 shows typical *Ef* values for several different types of sprinkler systems.

16.3.2   EVAPOTRANSPIRATION

An important factor for quantifying the water budget is the evapotranspiration rate (*ET*). Evapotranspiration is the water that is transpired out of the soil by the plant plus the amount of water lost to evaporation [1]. *ET* represents the rate of water consumed by the plant and lost by direct evaporation. The factors that affect the *ET* rate include wind, temperature, relative humidity, and solar radiation. The units for *ET* are inches/day. Based on the Penman Monteith model for *ET* estimations, *ET* is not measured directly for an individual crop, but rather it is determined from a standard reference grass and then adjusted for different crops and plants with a crop coefficient [1]. The evapotranspiration for a reference grass is referred to as the potential evapotranspiration (*ET°*). Potential evapotranspiration values will vary regionally and seasonally and are available in the literature. If literature values for *ET°* are not available or if the irrigator wishes to have a real time *ET* measurements, *ET* data acquisition systems are commercially available. ET data acquisition systems consist of weather sensors, telemetry and software that can retrieve the weather sensor inputs and perform the Penman Monteith model calculations. While an *ET* data acquisition system could potentially provide accurate real time *ET°* values, these systems are very expensive and do not necessarily represent microclimates. Because *ET°* is the *ET* for a standard reference grass, a crop coefficient (*Kc*) is necessary to determine the *ET* for the crop of interest. With information about sprinkler efficiency, crop coefficient and potential evapotranspiration, the water consumption (*ET"*) for a specific crop (in inches per day) are calculated from the Eq. (3). Typically, *Kc* values will range from 0.75 to 1.25 depending on species of the plant, the growth stage of the plant, and vary regionally. In practice, *ET°* and *Kc* values can be obtained from a local government crop extension or a local crop advisor.

$$ET" = [ET° \times Kc] / Ef \tag{3}$$

16.3.3   APPLIED WATER SCHEDULING

In general, the water application (*D*) in inches/day should be roughly equal to the system water loss (*ET"*) due to ET and sprinkler uniformity. The water loss calculated by Eq. (3) can be compared to the applied water measured with a rain gauge to set an irrigation target.

$$D \approx ET" \tag{4}$$

It is difficult to keep $D \approx ET"$ on an hourly or daily basis due to factors such as pivot lap speed and soil infiltration rates. Eq. (4) should define a water application target on a weekly basis. In general, depending on the crop and the irrigation system, crops should be irrigated 3 to 7 times a week and net weekly sum of the daily *D* values should be roughly equal to the net weekly sum of the daily *ET"* values. Figure 5 demonstrates a weekly water application target. In Fig. 5, there are three irrigation events and an *ET"* rate of 0.26 inches per day. Based on an ET'' rate of 0.26 inches per day and the E_p by the end of the week, 1.80 inches of water was consumed and approximately 1.80 inches would need to be applied. The application rate in Fig. 5 is 0.3 inches per hour for 2 hours.

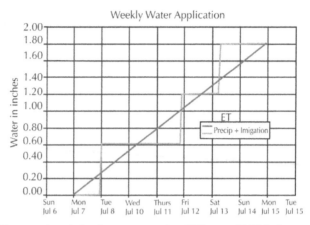

FIGURE 6 There are three irrigation events, and an ET'' rate of 0.26 inches per day. $D \approx ET''$ after the three irrigation event at the end of the week during the July, 2008.

TABLE 4 Typical Infiltration rates based on soil texture.

Soil texture	Typical infiltration rate inches/hour
Sand	1.5 or more
Sandy Loam 1 to 1.5	
Loam	0.5 to 1
Clay Loam	0.25 to 0.5
Clay	0.05 to 0.25

To minimize the water loss due to direct evaporation, the irrigation events take place between sunset and sunrise. It is important to irrigate at a rate that is less than the infiltration rate of the soil. Runoff and ponding may occur if the rate of application exceeds infiltration rate of the soil. Table 4 provides infiltration rates of soils based on soil textural class [2]. The infiltration of water into soil will vary with texture, but it will also depend on soil moisture, vegetation, bulk density and soil geomorphology among other factors. Soil infiltration rates can be determined from tests and area soil surveys data.

16.4   DATA ACQUISITION

Data acquisition systems are the most effective tool for identifying and reaching soil moisture and water application targets for irrigation optimization. A data acquisition system with the water budgeting method was constructed and is commercially available from Stevens Water Monitoring Systems, Inc.. The Stevens Agricultural Monitoring (SAM) Package integrates the input from sensors, displays the data from the remote field locations and integrates the water balance method described in the previous section. The SAM package includes rain gauges, the Stevens Hydra Probe Soil Sensor, a

Stevens DL3000 data logger, telemetry and the software program. Described below is the engineering that collects field data (soil moisture and precipitation) and the software program that acquires the data from the data loggers through the telemetry. The data is either exported to the internet or is imported into the SAM software where it can be used to make informed decisions about irrigation scheduling.

16.4.1   SOIL MOISTURE DATA COLLECTION

The soil moisture is collected using the Stevens Hydra Probe. The Hydra Probe is the soil sensor used in the USDA's Soil Climate Analysis Network (SCAN) and NOAA's Climate Reference Network (CRN). The Hydra Probe uses electromagnetic waves to measure both the real and imaginary dielectric permittivity [5]. The real component of the dielectric permittivity represents the energy storage based on the high rotational dipole moment of water compared to that of dry soil [14]. The measured real dielectric permittivity (εr) is used to accurately calculate the soil moisture in water fraction by volume (θ) in most soils [11] with the calibration equation:

$$\theta = A\sqrt{\varepsilon_r + B} \tag{5}$$

where, A is 0.109 and B is equal to −0.179. The Hydra Probe is digital and Eq. (5) is written into the firmware of the probe. The digital communication between the Hydra Probe and the data logger is the standard communication format Serial Data Interface at 1200 Baud (SDI-12). The advantages of SDI-12 include connecting many sensors on a single serial addressable bus and cable lengths up to 1000 feet from the sensor to the data logger. Multiple digital sensors are "daisy chained" together and the longer cable lengths provide flexibility in the architecture of the system in the field. Up to 4 or more SDI-12 soil moisture profiles can be installed up to 1000 feet away from the data logger reducing the cost by using common data loggers and telemetry.

16.4.2   RAIN DATA COLLECTION

The precipitation and the irrigation from sprinklers are measured together with a tipping bucket rain gauge. A tipping bucket is a 6 to 10 inch in diameter cylinder with a screen at the top facing end and a drain out the bottom. Inside of the bucket is a dual sided tray that is located under a funnel. The tray will tip over and drain after receiving 0.01 inches of rain. After tipping, the other half of the tray will fill with water, tip and drain after receiving another 0.01 inches of water. Every time the tipping bucket's tray tips (0.01 inch of rain), an electrical pulse is sent to the DL3000 data logger. The data logger counts the tips and calculates the depth of rainfall over time. It is important that the tipping bucket remain level and is placed in a location that will receive a representative application of water from the sprinklers.

If an irrigation method is used that does not include the use of sprinklers such as furrow or drip irrigation, the method described in Fig. 5 and Eq. (4) will not be as applicable. In this case, one or no rain gauge would be used in the data acquisition package.

16.4.3   DATA LOGGER AND FIELD STATION

The Stevens Data Logic 3000 (DL3000) data collection platform resides inside a weather proof fiber glass enclosure located in the field. The cable from each SDI-12

Hydra Probe enters the enclosure by running through bulkhead bushings located on the bottom of the enclosure. The Hydra Probe power, ground and SDI-12 communication wires are "daisy chained" together with a multiplex inside the enclosure. A single SDI-12 communication wire runs from the multiplexer to the DL3000's SDI-12 communication port. The DL3000 will log data on a set time interval typically every 30 minutes, and will hold up to 2 Gigabytes of data. The wire from the tipping bucket also runs into the enclosure through a bulkhead and is wired into the DL3000's pulse port. The data logger has a wireless RS232 communication radio attached. A coaxial cable runs from the radio out of the enclosure through the bulkhead to an Omni directional antenna. Also contained in the field enclosure is a 9 Amp/hour 12 volt DC battery, and charge regulator for the solar panel power supply. Figure 9 describes a field station with a subsurface soil moisture monitoring profile.

16.4.4 WIRELESS TELEMETRY

After the data from the sensors is received by the data logger, the data is transmitted from the field to the base station computer via radio. The frequency and type of radio would depend on the distance from the field to the base station computer. The radio communication between the field and the base station is usually line of sight. Large obstacles such as buildings, mountains and trees will impede the radio signal and prevent the signal from reaching its destination. If there is a large obstacle in the way, a repeater station could be installed, however repeater stations will increase the overall cost of the system. Radio communication always takes place between two or more radios. The radio at the base station is called the server or master radio and the radios in the field are call client or slave radios. The master radio is connected to the base station computer and a directional Omni antenna. Each radio has a Media Access Control (MAC) address written into the radio's firmware, identifying it. When the master radio needs communication with a specific radio, the master radio will address the radio with the MAC address. Radios will only respond to their specific MAC address from the master radio. In a network of radios, the master radio will communicate with each slave radio one by one and retrieve the sensor data from each logger individually.

Distance from the field site to the base station is the main factor determining the most appropriate radio and frequency. In most agriculture applications, 900 MHz Spread Spectrum radio with a 5 miles line of sight range is the most common. While satellite communication is common in the water resources industry, it is less common at the farm level due to licensing and hardware costs. Table 5 lists the different kinds of telemetry solutions, the ranges and the frequencies.

TABLE 5 Summary of telemetry options and ranges.

Radio	Range	Frequency
Blue Tooth	100 m	2,400 to 2,483.5 MHz
Spread Spectrum	5 miles	902 to 928 MHz
Wi-Fi	100 m	2.4 GHz
VHF	30 miles	30 to 300 MHz
UHF	30 miles	300 to 1,000 MHz

TABLE 5   *(Continued)*

Radio	Range	Frequency
Wi-Max	30 miles	2.3 to 3.5 GHz
Cellular Modem	Cell Coverage	824.01 to 848.97 MHz
Geosynchronous Satellite	1/3 the of Earth	401.7010 to 402.0985 MHz
Low Earth Orbiting Satellite	Global Coverage	148 to 150.05 MHz

16.4.5   SOIL PROFILE

Soil moisture probes at different depths in the soil column are referred to as a soil profile. Depending on the root zone depth, the typical soil profile consists of four soil sensors. One probe in the top soil (2 to 4 inches) two probes in the root zone (6 to 30 inches) and one probe below the root zone (36 inches). The Hydra Probe in the top soil will experience the greatest moisture fluctuation because it will be the most influenced by ET and downward flow. The top soil may reach saturation or reach a soil moisture value over the field capacity thus conducting water downward into the root zone of the crop. The lower soil moisture target for the two Hydra Probes in the root zone however are calculated from the MAD, θ_{FC} and θ_{PW} in Eq. (2) and the upper soil moisture target in the root zone will be the soil's field capacity. The soil sensor below the root zone should stay below field capacity. If the soil moisture below the root zone reaches values above field capacity, there will be downward conductance of water.

The soil profile should be placed in a location that will most represent the irrigated area. Soil moisture can be highly variable spatially [17]. The factors that affect soil moisture variability are slope, vegetation type, bulk density, soil type, microclimate, and other variables. An irrigation regime represents an area that is homogenous enough that the soil moisture variability will be low and the soil moisture data will represent the entire irrigation regime. There should be at least one soil profile for every irrigation regime. Irrigation regimes are determined by crop type, crop age, soil type, slope, and irrigation method. If the irrigation regimes are less than 1000 feet apart, it may reduce cost to tie multiple soil profiles into one data logger. By tying multiple profiles into a single data logger, the irrigator can save on the number of solar panels, batteries, radios, data loggers and other necessary accessories.

16.4.6   DATA ACQUISITION SOFTWARE

The central user interface of the data acquisition package is the software. The Stevens Agricultural Monitoring (SAM) Software is commercial available and can be subsidized by some energy and water conservation grants. The SAM software runs on a computer that is connected to the master radio. A master radio is not necessary if the system has a field cellular modem or satellite transceiver. The SAM Software acquires the sensor data in the field from a polling sequence. The polling sequence runs at a user specified time interval, which is usually every 15 or 30 minutes. Communication begins with a serial command from the software to the data logger to take a current reading from all of the sensors. The SAM sends the command to the master with instructions to use a specific slave radio. The data logger becomes active after receiving the command and takes a current reading from all of the sensors that are connected to it. Next the data logger sends a comma delimitated string of sensor data back to the SAM software through the slave and master radio. The SAM software parses the data and populates the tables and graphical displays in the software.

The irrigator can then view the real time data and make decisions about when to irrigate based on the soil moisture targets and the rate of water consumption by the crop from the ET. Other features in the software include battery voltages for power management. In the SAM Software, a display of MAD, θ_{FC}, θ_{PW} and the lower soil moisture limit based on the calculations from Eq. (2) are superimposed onto the real time soil moisture data. The real time soil moisture, superimposed onto the targets, is shown on a screen similar to Fig. 8.

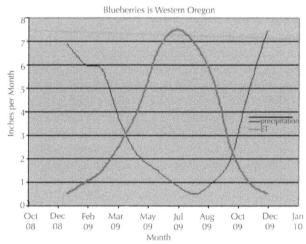

FIGURE 7 Typical values for monthly ET and precipitation for blue berries in western Oregon.

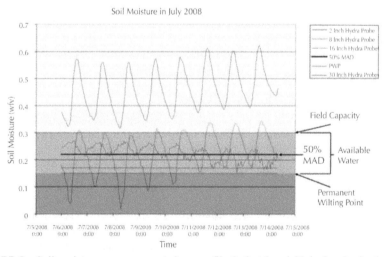

FIGURE 8 Soil moisture measurements in a profile 2, 8, 16 and 30 inches in depth. Daily irrigation events with subsequent decrease in soil moisture from a high ET rate.

At the beginning of the irrigation season, the irrigator can manually input the weekly ET values or the values from Eq. (3) into the SAM setup page. A real time display similar to Fig. 6 is displayed. With real time displays of the real time data superimposed onto the targets in a graphical representation will allow the irrigator to easily interpret the data.

16.4.7   SAM DATA ACQUISITION POLLING SEQUENCE FOR STATION 1

The flow chart below describes the process by which the SAM [Stevens Agriculture Monitoring] software communicates with the field stations. Figure 9 shows a diagram of a field station. The SAM Software will poll data from each station in consecutive order starting with the first field station. After retrieving the data from one field station the software will move on to the next field station.

1. The Polling Sequence initiates on a fixed time interval.
2. The Acquisition command "Take Current Readings Data Logger 1" along with a command to the master radio to communicate with radio 1 with its MAC address. These two commands are sent by the software out the serial port of the computer.
3. With an RS232 or USB connection to the computer, the Master Radio receives the "Take Current Readings Data Logger 1" message and transmits this message to slave radio 1 as commanded by the SAM software.
4. Slave radio 1 receives the "Take Current Readings Data Logger 1" and passes the message to the data logger via a RS232 cable.
5. Data Logger 1 receives the command "Take Current Readings Data Logger 1" from the slave radio and one by one collects the current data readings from each sensor that is connected to it.
6. Data Logger 1 sends a comma delimited data string back to the SAM software through the radios and serial ports.
7. The SAM software receives the data string, parses the data, and populates the graphical displays and tables in the software viewable by the user.
8. After the SAM software receives the data from data logger 1, it repeats steps 1 through 7 for data logger 2 and slave radio 2.

16.4.8   BLUEBERRY FARM IN WASHINGTON COUNTY OREGON: CASE STUDY

A SAM Soil Moisture data acquisition package complete with telemetry and software was installed on a 200 acre blueberry farm in Washington County, Oregon. The soil unit is Woodburn Silt Loam with less than 3% slope and the soil taxonomic description is Typic Plinthoxeralf. There are two irrigation regimes based on the age of the crop. Two stations, one in each irrigation regime, were installed with 4 Hydra Probe soil sensors, a tipping bucket rain gauge, and an air temperature sensor. Soils data for this location and most locations in the United States are provide for free by the US Department of Agriculture's Web Soil Survey Program, <http://websoilsurvey.nrcs.usda.gov/app/WebSoilSurvey.aspx>.

Figure 7 shows the annual precipitation and ET rate for blueberries in Washington County Oregon [12]. The ET exceeds precipitation from April to October and this generally defines the irrigation season.

Each station is located 1 mile away from the computer with the master radio; therefore, this network uses spread spectrum radios. The stations each have one soil profile consisting of 4 Hydra Probes at various depths (2", 8" 16" and 30"). The SDI-12 Hydra Probe Soil Sensors are wired into a multiplexer which is connected to the Stevens Data Logger. Each station is power with a solar panel and the enclosure houses the battery, multiplexer, charge regulator and radio. The radio antennas are mounted to the same mast as the tipping bucket. Figure 9 illustrates one of the field stations with the soil profile.

Using Tables 1 and 2, the permanent wilting point is 0.15 the field capacity is 0.3 and the MAD is 50%. The lower soil moisture target as calculated from Eq. (2) is 0.22.

Figure 8 shows thesoil moisture for a warm week in July 2008. The region of the chart for, wfv = 0.3 to 0.7, represents soil moisture levels over field capacity, there gion for, wfv = 0.15 to 0.30, shows the range ofsoil moistures available to the crop (available water capacity) and the region for, wfv = 0.0 to 0.15, is below permanent wilting point. The two inch deepsoil moisture values fluctuate the most for downward conductivity and ET and stays above field capacity. This is typical because if the top 2 inches of the soils tayed below field capacity then the root zone would not receive the water. The 8 inch soil moisture values fluctuate wide lydueto ET and there is a 4 hour lagtime between the 2 and 8 inch soil moisture probes from the downward movement time of the wetting front. During extremelyhot days, itis not uncommon to have the soil moisture values briefly drop below permanent wilting point between irrigation cycles. The 16 inch soil moisture mirrors the 8 inch values with a4 hour latency from the soil moisture values above it and the rise and fall of soil moisture values with the irrigationevents. The 30 inch deep soil moisture probe below the root zone is remaining constant about 0.10 wfv indicating that water is not peculating downward to the water table.

The solid set sprinklers rotator (with an efficiency of 0.90) apply water daily. For the month of July ET (=ET° × Kc) is 0.25 inches per day. Using Eq. (3), the daily water consumption will be 0.28 inches. A weekly display similar to Fig. 6 is displayed in the software which will allow the irrigator to meet the soil moisture and water application targets.

It was concluded: As the demand for water increases, along with the need to protect aquatic habitats, water conservation practices for irrigation need to be effective and affordable. Precision irrigation will optimize irrigation by minimizing the waste of water, and energy, while maximizing crop yields. The most effective method for determining the water demands of crops is the based on the real time monitoring of soil moisture, and direct water application used in conjunction with the information about soil hydrological properties and evapotranspiration. The Stevens Agriculture Monitoring data acquisition system wirelessly acquires rain and soil data from the field and integrates the data into water management tools. The water management tools use information about evapotranspiration, soil and the crop to set specific irrigation targets. These irrigation targets will help the irrigator optimize the amount of water used on a weekly basis. Optimization of irrigation water will increase crop yields while conserving water resources.

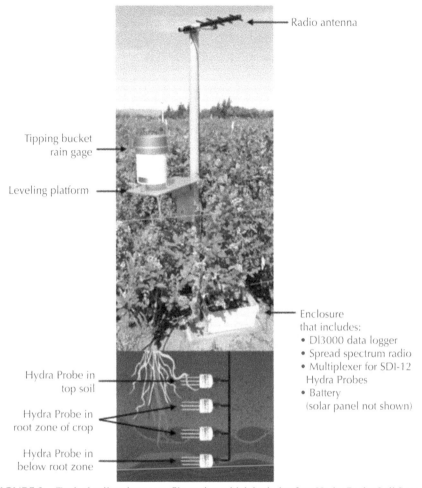

Radio antenna

Tipping bucket
rain gage

Leveling platform

Enclosure
that includes:
• Dl3000 data logger
• Spread spectrum radio
• Multiplexer for SDI-12
  Hydra Probes
• Battery
  (solar panel not shown)

Hydra Probe in
top soil

Hydra Probe in
root zone of crop

Hydra Probe in
below root zone

FIGURE 9 Typical soil moisture profile station which includes four Hydra Probe Soil Sensors, Stevens DL3000 data logger, radio, antenna and accessories.

16.5 SUMMARY

The water requirements of crops are dependent on ET, soil chemistry, and the MAD. Direct measurements of root zone soil moisture, water application along with published ET values and soil textures, can be used in a soil water balance model that can significantly optimize irrigation efficiency. Over the past five years, advancements in computer microprocessors, memory, and software development tools has improved data acquisition methods and made data acquisition system integration more reliable and more cost effective. This chapter presents an irrigation scheduling method based on a volumetric soil moisture balance model and data acquisition.

KEY WORDS

- antenna
- available water capacity
- Baud rate
- blue berry
- Bluetooth
- capillary forces
- catch can
- cellular modem
- clay
- cloud based server
- computer software
- crop
- crop coefficient
- data acquisition
- data collection platform
- data logger
- data polling
- dielectric permittivity
- dipole moment
- enclosure
- eutrophication
- evapotranspiration (ET)
- field capacity (FC)
- geosynchronous satellite communication
- Grants
- green beans
- hydra probe soil sensor
- imaginary dielectric permittivity
- internet
- irrigation
- irrigation optimization
- irrigation regime
- irrigation scheduling
- irrigation system

- loam
- MAC address
- master radio
- Maximum Allowable Depletion (MAD)
- microclimate
- mineralogy
- natural waters
- omni directional antenna
- pathogens
- Penman-Monteith ET method
- Permanent Wilting Point (PWP)
- ponding
- porosity
- radio frequency
- rain gage
- real dielectric permittivity
- root zone
- RS232 communication
- salinity
- sand
- SDI-12 communication
- silt
- slave radio
- soil bulk density
- soil chemistry
- soil climate analyzes network
- soil geomorphology
- soil infiltration
- soil macro pores
- soil micro pores
- soil moisture
- soil moisture budget
- soil moisture sensor
- soil particle size
- soil saturation

- **soil sensor**
- **soil survey**
- **soil textural class**
- **solar panels**
- **spatial variability**
- **spread spectrum radio**
- **sprinkler**
- **sprinkler efficiency**
- **Stevens Water Monitoring Systems**
- **telemetry**
- **tipping bucket**
- **total maximum daily load**
- **total suspended solids**
- **unsaturated soil**
- **volumetric soil moisture content**
- **water molecule**
- **web soil survey**
- **wetting front**
- **wireless telemetry**

REFERENCES

1. Allen, R. G.; Pereira, L. S.; D.; Raes, M.; Smith, **1998**, *Crop Evapotranspiration Guidelines for Computing Crop Water Requirements*. FAO Irrigation and Drainage Paper 56. Food and Agriculture Organization of the United Nations.
2. Birkeland, P. W.; **1999**, *Soils and Geomorphology*. Third Edition. Oxford University Press.
3. Brady, N. C.; **1974**, *The Nature and Properties of Soils*. Eighth Edition. Macmillan Publishing Co., Inc.
4. Brouwer, C.; **1988**, *Irrigation Water Management: Irrigation Methods*. Training Manual Number 5. Food and Agriculture Organization of the United Nations-Land and Water Development Division.
5. Campbell, J. E. Dielectric properties and influence of conductivity in soils at one to fifty Megahertz. *Soil Sci. Soc. Am. J.* **1990**, *54*, 332–341.
6. Chapman, D.; **1994**, *Water Quality Assessments*. World Health Organization.
7. Grim, E. G.; **1968**, *Clay Mineralogy*. 2nd Edition. McGraw-Hill Co.
8. Hutson, S. S.; **2000**, *Estimated Use of Water in the United States in 2000*, USGS Circular 1268, US Geological Survey.
9. McBride, M. B.; **1994**, *Environmental Chemistry of Soils*. Oxford University Press.
10. Rowell, D. L.; **1994**, *Soil Science Methods and Applications*. John Wiley & Son Inc.
11. Seyfried, M. S.; Grant, L. E.; E. Du and K.; Humes, **2005**, Dielectric loss and Calibration of the Hydra Probe Soil Water Sensor. Vadose Zone Journal *4*, 1070–1079.
12. Smesrud, J. M. . Hess, J. Selker, **1997**, *Western Oregon Irrigation Guide*. Oregon State University.

13. Stevens Water Monitoring Systems, Inc.; <www.stevenswater.com>
14. Topp, G. C.; Davis, J. L.; Annan, A. P.; **1980,** Electromagnetic Determination of Soil Water Content: Measurement in Coaxial Transmission Line. Water Resources Research *16,* 574–582.
15. US Department of Agriculture, -NRCS, *Cooperative Web Soil Survey* <http: //websoilsurvey. nrcs.usda.gov/app/WebSoilSurvey.aspx>
16. Warrick, A. W.; **2003,** *Soil Water Dynamics.* Oxford University Press.
17. Western, A. W.; S.; Zhou, Grayson, R. B.; T. A. McMahon, G.; Bloschl, Wilson, D. J. Spatial Correlation of Soil Moisture in Small Catchments and Its Relationship to Dominant Spatial Hydrological Processes. *J. Hydrology* **2004,** *286,* 113–134.
18. Winter, T. C. **2002,** *Ground Water and Surface water A Single Resource.* USGS Circular 1139, by US Geological Survey.

CHAPTER 17

SNOW DATA AND WATER RESOURCES[1]

B. KEITH BELLINGHAM

CONTENTS

[1]Author is thankful to Stevens Water for the technical support. © Copyright Stevenswater.com

17.1 INTRODUCTION

There are many complex political, social, environmental, and scientific challenges surrounding water resources in the western United States [9]. Water is scarce and does not occur when and where people need it the most. According to the US Geological Survey (USGS), about 80% of fresh water used in the western United States is appropriated for irrigation [10]. Over the past 150 years, society has diverted streams, built wells and created reservoirs to better distribute water to where it is needed the most. Over the past 30 years, the impact to the environment of water withdrawals from natural sources has become more evident with increasing demand. Some of the environmental challenges associated with water withdrawals from streams include maintaining enough flow to support the habitats, increase temperature, and nutrient loading [3]. For more than 100 years, knowing and predicting stream flow has been very important for water managers, not only to provide enough water for human consumption, but to protect the aquatic habitats in our natural water ways as well. United States Irrigation Companies such as Stevens Water provide instrumentation that is critical to the prediction of water resource availability in the western United States and throughout the world. This chapter presents critical role of the USDA SNOTEL NETWORK in Protecting Water Resources in the Western United States of America [13].

17.2 HISTORY OF USDA SNOTEL NETWORK

Much of the water in the western United States comes from the winter snowpack in the mountainous regions. The snowpack in the mountains of western USA can range from nothing or very little up to 30 or 40 feet of snow in the high Cascades [9]. In 1906, Dr. James Church, Hydrologist at the University of Nevada [13], began to document the relationship between winter snowpack in the mountains and stream flow throughout the year for certain watersheds (Fig. 1). Dr. Church enhanced the existing Russian technology for measuring snow water equivalent (SWE). Shortly after Dr. Church developed these snow measurement techniques, the United States Department of Agriculture (USDA) began to construct "Snow Courses" in the mountainous areas of the west so that hydrologists could make stream flow predictions from snow data [5].

These snow courses were areas free of trees where the snow survey staff could take manual measurements of the snowpack. About that same time, the USGS began installing stream gauge stations so that stream data could be compared to the snow data. In 1911, these USGS stream gauge stations began using mechanical chart recorders an innovative new technology for automatically measuring water level developed by J. C. Stevens, one of the founders of Leupold and Stevens, which later became Stevens Water Monitoring Systems [12].

Beginning in the 1980 s, the USDA's snow courses became more sophisticated, adding an array of weather sensors, data loggers and telemetry systems. These snow course telemetry sites were named SNOTEL. Today, the US Department of Agriculture – National Resources Conservation Service (USDA-NRCS), manages and operates over 600 (and growing) SNOTEL stations [13]. Using http://www.wcc.nrcs.usda. gov/snow/, the hourly data is now displayed on the internet for every station The data from SNOTEL is of high quality, and SNOTEL is known the world over for having the one of best quality control protocols of any environmental network.

17.3   SNOW COURSE AND SNOW WATER EQUIVALENT

For the traditional stream flow prediction models, the parameter Snow Water Equivalent (SWE) was needed. SWE is the amount of water contained within a core of snowpack. This is a manual measurement where a technician would push a preweighed cylindrical tube into the snow. The tube is then weighed to get the weight of the snow. From this weight, they are able to determine the amount of water equivalent of the snow. The density of snow can change with temperature and precipitation throughout the year. The same depth of snowpack can yield different water amounts depending on the density. With the SWE measurement, apples to apples comparisons can be made with snow data across regions and time [5]. Some locations have ground truthing markers for determining snow depth visually. Sometimes ground truthing is performed from an air craft.

17.4   SNOTEL COMPONENTS

While this manual SWE measurement method is still used on most snow courses several times a year, SNOTEL has automated ways for collecting information about snowpack. Each SNOTEL site is equipped with a radar sensor that can provide snow depth, and a precipitation gauge that that measures the total amount of precipitation (both solid and liquid) using a pressure transducer inside of a collector. These total precipitation gauges are made of metal and stand 10 to 15 feet tall and are often called "rockets" because they look like rockets. Another device at a SNOTEL is that measure SWE is a snow pillow. A snow pillow is a big bladder filled with a nontoxic liquid antifreeze solution. As the snowpack builds on a snow pillow throughout the winter, the antifreeze is displaced up a stand pipe. From the pressure of antifreeze in the stand pipe, a SWE is calculated. SNOTEL sites also collect air temperature, wind speed and direction, relative humidity, barometric pressure, and solar radiation data [13].

FIGURE 1   Dr. James Church in 1906: Picture compliments of the USDA Natural Resources Conservation Service [13]

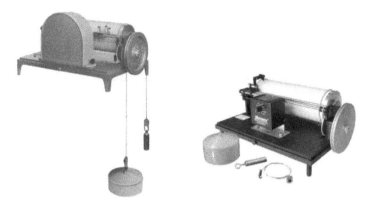

FIGURE 2   Left: A Stevens Type F Chart Recorder from the 1960's, which was used in many USGS Stream Gaging sites. Right: The current production model of the Stevens Type F Chart Recorder [12].

FIGURE 3   Typical SNOTEL site: Picture complements of the USDA Nation Resources Conservation Service [13].

17.5   ESTIMATIONS OF STREAM FLOW FORECASTS

The stream flow predictions from the USDA's SNOTEL data are derived from the statistical relationship between the SWE on April 1st and the stream flow throughout the summer (Figs. 2–4). The snow courses are standardized plots of ground where a transect of SWE measurements can be taken. These SWE measurements need to be collected consistently year after year so that the hydrologic trends can be statistically quantified.

Based on many years of historical (antecedent) data between the snow courses, SNOTEL and the USGS stream flow data, a mathematical algorithm can be generated from a matrix regression method to correlate the data so that a stream flow prediction can be generated [7]. The comparison between the stream flow prediction and the actual flow is called "skill." The closer the skill is to 1.0, the closer the prediction was to the actual stream flow. The skill is the same as a Pearson correlation coefficient or r squared value used in statistics. Many stream flow forecasts provided by SNOTEL have a skill (r^2) of 0.9 or greater [7].

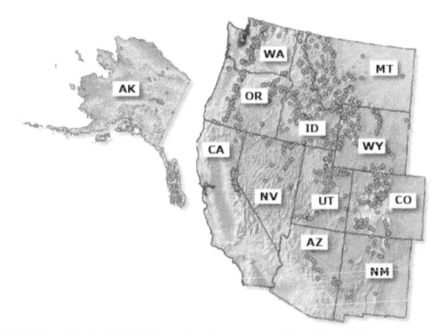

FIGURE 4   USDA's SNOTEL Network within the western United States [13].

17.5.1 HOW DOES STREAM FORECASTING WORK?

The stream flow forecasts or Surface Water Supply Index (SWSI) is calculated for specific points along streams and is SNOTEL's main publically available product. Stream flow forecasts are available for over 750 locations in the western US. The SWSI is given as a distribution of probabilities that accommodates a wide number of applications depending on the user's interest.

TABLE 1 Example of a Stream Flow Forecast Table (USDA 13).

Forecast Pt	Chance of Exceeding					
Forecast	90%	70%	50%	30%	10%	30 Yr. Avg.
Period	(1000AF)	(1000AF)	(1000AF)	(% Avg.) (1000AF)	(1000AF)	(1000AF)
Emma at Lake Outlet: 1000 AF = 1000 acre – feet of water						
APR–JUL 285	345	385	65	425	485	590

The Table 1 is an example of a typical stream flow forecast from April to July in units of 1000 acre feet of water. The 50% exceedance forecast provides the percent of average. In this example, the out flow of Emma Lake is 65% of average for April to July of this particular year. An irrigation district that withdraws water directly downstream of Emma Lake outlet would plan for the 70% exceedance. There is a 70% chance that there will be at least 345,000 acre feet of water flowing out of the lake from April through July and a 30% chance that there would be less than 345,000 acre feet of water. Because in this example, the 50% exceedance forecast only shows 65% of average, the irrigation district would use caution and use the 90% exceedance forecast to be on the safe side. There will be a 10% chance that there will only be 285,000 acre feet of water flowing out of the lake from April to July.

A hydrologist at the US Highway Department Planning for Flood Conditions, for example, would use this forecast differently. There is only a 10% chance that there will be more than 485,000 acre feet draining out of the lake for this time period. Like the irrigation district, the highway department can look at the USGS stream gauge station to get actual daily rates. In this example, the forecast is 65% of average, so the highway department should not be concerned about flooding, while the irrigation district needs to use more caution with water allocations.

17.5.2 STREAM FLOW AND SOIL MOISTURE

Traditionally, stream flow forecast models primarily use the antecedent SWE values in their calculations and ignore soil moisture. Soil represents a huge water reservoir and can introduce errors into the forecasts. Excluding runoff, water from snow melt and precipitation will first percolate through the soil before entering the water table or saturated ground water zone. Once the water is in the saturated zone, it will travel down gradient and eventually discharge into a stream or lake (Fig. 5). The vadose zone is the soil above the water table and represents a hydrological regime that can hold large amounts of water.

Once water enters the vadose zone, it will generally only move in one dimension, up and down [16]. In unsaturated soils, water will migrate upward because of evaporation and the uptake of water by plants and trees. This upward movement of water is called evapotranspiration [1]. Evapotranspiration is the primary mechanism responsible for removing water from soil. During the winter months with a snowpack, evapotranspiration is almost zero, and the soil moisture values stay relatively constant.

The downward movement of water in soil obeys an entirely different set of rules then it would in saturated conditions. Water will suspend itself and adhere to soil particles. This attraction between water and soil particles is called capillary force. As the soil moisture increases, gravity will pull the water downward. The point at which the gravitational influence exceeds the capillary influence is called field capacity. Above field capacity, water will be conducted downward through the soil and will discharge into the water table. If the soil moisture stays below field capacity, water can only travel upward due to evapotranspiration [16].

Field capacity is an important parameter in determining the fate and direction of water transport in soil (Fig. 6). The field capacity of a particular soil is largely a function of the soil texture. In general, the more clay-rich the soil is the more water it will hold on to.

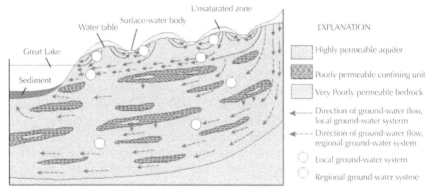

FIGURE 5   Groundwater flow diagram taken from the USGS Report 00–4008 [8].

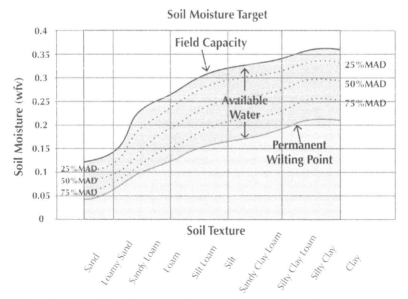

FIGURE 6   The availability of water at different soil moisture levels for various types of soil [2].

Soil moisture targets are used also in irrigation scheduling for optimization of water. Assuming the soil stays frozen, the soil moisture value before the winter snow arrives will generally be the same soil moisture value in the spring when the snow begins to melt because evapotranspiration will be negligible in the winter [6]. If there is a dry fall, and if winter arrives quickly, the soil moisture under the snowpack will be low. When the snow melts in the spring in western USA, much of the water will be retained by the soil, and not as much water will reach the streams. There can be a below-average stream flow even if there is an above-average snowpack because the soils dried out the previous fall. Conversely, if there is a rainy autumn, the soil will already be at field capacity when the snow melts in the spring, and all of the water from the snowpack will enter the water table pushing an equal amount of water out into the streams. A wet fall can cause flooding in the spring even if there is a below average snowpack.

The transport of water from snow pack is very complex and can be driven by not only air temperature, but by thermal fluxes in the soil and surrounding vegetation as well as sublimation (solid to vapor phase). Because soil moisture was recognized as a major factor of the hydrology of a watershed starting in the late 1990s, STOTEL began installing high quality soil sensors at SNOTEL sites. The soil sensor selected to go into SNOTEL sites was the Stevens Hydra Probe. Now there are almost 300 sites nationwide that are equipped with Stevens Hydra Probes.

Even though many SNOTEL sites are equipped with Hydra Probe Soil Sensors (Fig. 7), the official stream flow forecasts do not yet include soil moisture data in their models. Recently, the SNOTEL office in Utah began evaluating soil moisture under snowpack. They suggest a correction to the forecasts based on a parameter called soil moisture deficit (Vaughan 14). The soil moisture deficit is the difference between the current soil moisture and the soil's field capacity and represents the amount of water that can enter into the soil before migrating downward to the water table. The soil moisture deficit can then be used to adjust the chance of exceedance forecasts. While this technique shows promize in improving forecasts by incorporating soil moisture data, it is not yet used system wide. More new SNOTEL sites equipped with Hydra Probes are scheduled to be installed.

17.6   SNOTEL AND ENVIRONMENTAL ISSUES

"The Clean Water Act's 303d stream listing" is a list of streams that do not meet the water quality standards required by law. Streams on the 303d list are said to be "impaired." While streams in the US no longer catch on fire from excessive pollution, they can still become impaired from storm water runoff and from agricultural/irrigation land uses.

As water is withdrawn from streams for irrigation purposes, the temperature may increase, as well as the salinity and nutrient level. This loading of temperature, nutrients and salinity causes algal blooms and threatens the aquatic ecozystems [3]. The demand for water from urban areas and for irrigated crops is increasing every year. The stream forecasts provided by the NRCS offer part of a potential solution for water conservation in western US.

Climate change is also a concern for many water managers and stake holders. Climate change can of course affect the quantity of the snowpack, but, it also impacts the timing of the melt. Timing of the availability of the water dictates the management and the operation of reservoirs, which in turn affects power generation.

The SNOTEL Network also provides high quality data for scientists modeling climate change. SNOTEL offers uniform, hourly data in real time for hundreds of watersheds in mountainous regions, and it is all available to the public for free.

The quality control policies for SNOTEL are mimicked by other network and mesonets in the US and in Europe such as the NRCS's Soil Climate Analysis Network (SCAN), NOAA's Climate Reference Network and the Kentucky Mesonet. Some of the policies include strategic placement of sensors to capture the particular characteristic of the site. For example, air temperature sensors and precipitation gauges are placed high enough off the ground so that they do not become buried in snow. The precipitation gauges use elaborate wind shields that help diminish the effects of wind on precipitation measurements. Each station is also ground truthed several times a year by highly trained staff that can collect measurements as well as perform maintenance on the stations, ensuring accurate readings year-round.

The data is also evaluated with special modeling software called "Parameter-Elevation Regression on Independent Slopes Model (PRISM)" developed at Oregon State University [4]. PRISM statistically examines the data from SNOTEL sites and can calculate the variability of the data. PRISM can make regional prediction of the parameters. For example, if an air temperature gauge is suspected of malfunctioning, PRISM software can take temperature data from other nearby SNOTEL sites and other sources and calculate what the expected temperature would be at the suspect site. If necessary, the data archives can be edited to provide the highest quality data possible.

17.6.1   WATER RIGHTS IN THE WESTERN UNITED STATES

Water rights in the western United States are exceedingly complex and vary from region to region. In general, water rights are based on Prior Appropriation which means whoever got there first has the senior water rights.

The availability of water determined where people settled in the west in the 1800s. Reservations were created for the Indian tribes which often time included large amounts of land high in the mountains that hold a snowpack much of the year. In 1908 the US Supreme Court ruled that the Indian tribes own the water rights on their reservations. Because snow occurs up gradient from the irrigated lands, as the snow melts, it travels though several if not many jurisdictions before returning to the natural water body. Figure 8 shows a water withdraw head works diverting water from a river by an irrigation district that will sell water for irrigation purposes. Figure 9 shows a gaging station on an irrigation lateral that is logging water level and flow [Stevens Water 12].

Often times, the first water jurisdiction is snowpack on Indian Reservations, however many tribes have chosen not to exercize their water rights. Structures that divert state or tribe owned natural waters are built and paid for by the US Bureau of Reclamation. The US Bureau of Reclamation will then sell the water to irrigation districts, municipalities, and other water users that hold or service holders of water rights. Under the US Department of the Interior, the US Bureau of Reclamation is the nation's largest wholesale supplier of water and the nation's second largest producer of hydroelectric power. Its facilities also provide substantial flood control, recreation, and fish and wildlife benefits.

When water passes though jurisdictions, the stakeholders and the holders of the water rights do not always come to full agreement on the pricing, usages, Total Maximum Daily Loads, and ownership of the water. These disagreements can erupt into a "water war" where SNOTEL and USGS data are often times used in litigation. In 1969, the National Environmental Policy Act (NEPA) was established to help settle disputes while using sound environmental assessments of the particular issues.

FIGURE 7   Installation of a Stevens Hydra Probe into native soil at a SNOTEL site [2].

FIGURE 8   An example of irrigation district river diversion in Oregon [12].

FIGURE 9 Measurement of water flow in an irrigation ditch, using the Stevens AxSys Flowmeter [12].

17.7 THE SNOW SCIENCE COMMUNITY

The Western Snow Conference started in the early twentieth century as the American Geophysical Union's committee on snow hydrology under the chairman Dr. Church. In the 1930s, a special conference was held highlighting the advances in stream forecasting from snow surveys (Fig. 10). Since the 1930s, the Western Snow Conference has been held every year [15].

To this day, the yearly Western Snow Conference continues to promote snow science and stream forecast modeling. At the yearly conference there are technical exhibits, a forum of talks and a peer reviewed journal. The organization is divided into four suborganizations based on region. Some of the topics at this year's Western Snow Conference included effects of dust on the melt rate of snow, water management, the challenges of hydrological modeling, the bark beetle and other invasive species, and prediction of landslide hazards based on SNOTEL data.

For almost 100 years, snow surveys in the western US have helped resolve some fundamental issues regarding the availability of water. With increasing water demand, climate change, and changes in land use, we will need to rely more and more on monitoring and modeling technology to help protect valuable water resources. SNOTEL, in cooperation with the USGS, Bureau of Reclamation, NOAA and other agencies, is providing products such as real time data and forecasts free of charge to help water resource managers meet the new challenges facing water in the west.

To learn more about SNOTEL, the reader is advized to visit the US Department of Agriculture NRCS SNOTEL site at: <http://www.wcc.nrcs.usda.gov/snow/>.

FIGURE 10   Western Snow Conference visit to the Columbia Ice Sheet in 2009 [15].

17.8 SUMMARY

The US Department of Agriculture, Natural Resources Conservation Service operates about 600 remote snow monitor stations (SNOTEL) in the mountains of the western United States. About 30% currently have soil moisture sensors and the number of sites that have soil moisture sensor increases every year. In the western United States, the majority of the water used for irrigation originates as snow pack. Data from SNOTEL sites along with hydrological models are used to provide water availability forecasts to whole sales water users, local governments, and the general public. Adding addition sensors such as soil moisture sensors and can improve forecast accuracies. The snow science community gathers once a year at the Western Snow Conference to promote the advances in measurement technology, hydrological model and snow science.

KEY WORDS

- **Acre foot**
- **American Geophysical Union**
- **available water capacity**
- **clean water act**
- **climate change**
- **data acquisition**

- data logger
- evapotranspiration (ET)
- exceedance forecast
- field capacity
- Field Capacity (FC)
- flood
- head works
- hydra probe soil sensor
- hydroelectric power
- hydrological model
- impaired stream
- irrigation district
- irrigation scheduling
- Maximum Allowable Depletion (MAD)
- National Environmental Policy Act (NEPA)
- Natural waters,
- Permanent Wilting Point (PWP)
- prior appropriation
- radar sensor
- rain gage
- rocket gauge
- SNOTEL
- snow course
- snow depth
- snow pack
- snow pillow
- snow survey
- Snow Water Equivalent (SWE)
- soil climate analyzes network
- soil moisture
- soil moisture sensor
- soil saturation
- Stevens Water Monitoring Systems
- stream gaging
- Surface Water Supply Index (SWSI)

- **telemetry**
- **tipping bucket**
- **total maximum daily loads**
- **type F recorder**
- **US Bureau of Reclamation**
- **US Department of Agriculture (USDA)**
- **US Geological Survey (USGS)**
- **vadose zone**
- **water allocation**
- **water forecast**
- **water resources**
- **water rights**
- **water war**
- **water withdraws**
- **Western Snow Conference**
- **wetting front**

REFERENCES

1. Allen, R. G.; Pereira, L. S.; D.; Raes, M.; Smith, **1998**, *Crop Evapotranspiration Guidelines for Computing Crop Water Requirements*. FAO Irrigation and Drainage Paper 56. FAO of the United Nations.
2. Bellingham, B. K.; **2013**, *Comprehensive Stevens Hydra Probe Manual*.
3. Chapman, D.; **1994**, *Water Quality Assessments*. World Health Organization.
4. Daly, C.; Redmond, K.; Gibson, W.; Doggett, M.; Smith, J.; Taylor, G.; Pasteris, P.; Johnson, G.; **2005**, Opportunities for Improvements in the Quality Control of Climate Observations. 15th AMS Conference on Applied Climatology, Amer. Meteorology Soc.
5. *Field Office Guide to Climatic Data US Department of Agriculture Online Manual*. <http://www.wcc.nrcs.usda.gov/climate/foguide.html>
6. Flint, A. L.; Flint, L. E.; **2006**, Modeling Soil Moisture Processes and Recharge under a Melting Snowpack. TOUGH Symposium 2006, Proceedings.
7. Garen, D. C.; **1992**, Improved Techniques in Regression-Based Stream flow Volume Forecasting. J. of Water Resources Planning and Management, *118(6)*, November.
8. Grannemann, N. G.; Hunt, R. J. Nicholas, J. R.; Reilly, T. E.; Winter, T. C.; The Importance of Ground Water in the Great Lakes Region. USGS Water Resources Investigations Report 00-4008 by USGS.
9. Gottfried, G. J. Neary, D. G.; Ffolloitt, P. F.; **2002**, Snowpack-Runoff relationships for Mid-Elevation Snowpacks on Workman Creek Watersheds of Central Arizona, USDA Research Paper RMRS-RP–33.
10. Hutson, S. S.; **2000**, Estimated Use of Water in the United States in 2000, USGS Circular 1268, by US Geological Survey.
11. Serreze, M. C.; Clark, M. P.; Armstrong, R. L.; Characteristics of western United States snowpack from snowpack telemetry (SNOTEL) data. Water Recourses Research, **1999**, *35(7)*, 2145–2160, July.
12. Stevens Water Monitoring Systems, Inc. <www.stevenswater.com>

13. US Department of Agriculture Natural Resources Conservation Service Snow Survey. http: // www.wcc.nrcs.usda.gov/snow/
14. Vaughan, K. R. **2010,** Relationship Between Climate Conditions and Soil Properties at SCAN SNOTEL Sites in Utah. Western Snow Conference Proceedings.
15. Western Snow Conference <http: //www.westernsnowconference.org/>
16. Warrick, A. W. **2003,** *Soil Water Dynamics*. Oxford University Press.

PART III: WATER MANAGEMENT IN THE TROPICS

CHAPTER 18

HISTORICAL OVERVIEW OF EVAPOTRANSPIRATION IN PUERTO RICO[1]

ERIC W. HARMSEN

CONTENTS

---

[1] Modified and reprinted with permission from: "Harmsen, E. W., Fifty years of crop evapotranspiration studies in Puerto Rico. *Journal of Soil and Water Conservation,* **2003**, 58*(4)*, 214–223." © Copyright 2003 [<http://www.swcs.org/>] by the Soil and Water Conservation Society. © Copyright Soil and Water Conservation Society

18.1 INTRODUCTION

The countries within the humid tropics contain almost one-third of the total world population [1]. Within this region there are many small islands that rely on agriculture to feed their populations, and whose water resources are subject to an ever-increasing risk. Therefore, it is imperative at this time to better understand the hydrology of small tropical islands, with the goal of improving water conservation practices. Irrigation being among the largest consumers of water in society, special emphasis should be placed on developing techniques for estimating crop water requirements. In support of this goal, this paper reviews the majority of crop evapotranspiration studiesconducted during the last 60 years in Puerto Rico. It is hoped that this information may be useful not only in Puerto Rico but within other areas of the humid tropics as well.

This chapter includes topics such as, evapotranspiration reference materials, consumptive use obtained from field studies, studies predicting consumptive use, studies predicting reference evapotranspiration, studies comparing the Penman-Monteith equation with other evapotranspiration methods, estimating climatic parameters for use with the Penman-Monteith equation, use of pan evaporation data for estimating consumptive use, peak evapotranspiration estimates for irrigation system design and use of satellite remote sensing for estimation of evapotranspiration. The chapter also includes a list of priorities for future research.

18.1.1 SETTING

Puerto Rico is an island between the Caribbean Sea and the North Atlantic Ocean, east of the Dominican Republic. It is located at geographic coordinates: 18° 15' N, 66° 30' W and has an area of 9,104 square kilometers, slightly less than three times the size of Rhode Island. The highest elevation in Puerto Rico is 1,338 m above mean sea level at Cerro de Punta, PR [2]. Many of the islands of the West Indies archipelago share similar characteristics with respect to climate. This is due in part to the influence of the trade winds and the islands' mountainous topography [3]. Generally, for these islands, rainfall is greatest in the north-east and interior mountain areas. Due to orographic effects, the leeward side may be quite dry and even semiarid, as in the case of south-west PR. The rainy season tends to be from June to November and the dry season from December to May. For a given location, air temperature variations throughout the year are small; however, air temperatures are highly correlated with elevation. An exception to this occurs within interior mountain valleys where warm air can become trapped. In some cases average temperatures within interior mountain valleys may be higher than coastal areas at lower elevations [4].

18.1.2 CROP EVAPOTRANSPIRATION

Crop evapotranspiration (ET_c) is defined as the combination of evaporation from soil and plant surfaces, and transpiration from plant leaves. Evaporation is the process whereby liquid water is converted to water vapor and removed from the evaporating surface [5]. Transpiration is the vaporization of liquid water contained in plant tissues and its subsequent removal to the atmosphere. Crops predominately evaporate water through small openings in their leaves called stomata. Evapotranspiration can be expressed in units of mmd^{-1} (in.d^{-1}), or as an energy flux in units of MJm^{-2}d^{-1} (calft^{-2}d^{-1}).

Evapotranspiration is often the largest component of the hydrologic cycle after rainfall. Under arid conditions, potential evapotranspiration can easily exceed rainfall. The following water balance equation illustrates the relationship between hydrologic variables at any location in the field:

$$S_2 = P + IRR - DP - RO - ET_c + S_1 \tag{1}$$

where, P is precipitation, IRR is irrigation, DP is deep percolation, ET_c is crop evapotranspiration, RO is surface runoff, and S is the amount of water in the soil profile, equal to $\theta_v z$, where θ_v is the average volumetric soil moisture content and z is the root zone depth. All the terms in Eq. 1 have units of mm (in.). The subscripts 1 and 2 represent the beginning and end of the period under consideration. Equation 1 can be used over periods ranging from hours to many months or even years. A common approach for estimating evapotranspiration involves rearranging equation(1) and solving for ET_c:

$$ET_c = (S_1 - S_2) + (P + IRR) - (DP - RO) \tag{2}$$

In Eq. (2), the water balance has been separated into three components: stored water = $(S_1 - S_2)$, water input = $(P + IRR)$, and water losses = $(DP - RO)$.

Other general methods used to estimate evapotranspiration include the energy balance, microclimatological, and weighing and nonweighing lysimeter methods [6]. In Puerto Rico, all of these methods have been used except for the weighing lysimeter method. Determination of evapotranspiration with a weighing lysimeter has been considered to be the most accurate of all methods. However, the major disadvantages of the method are its high cost and nonportability. Crop evapotranspiration can also be estimated as the product of the reference evapotranspiration and crop coefficient:

$$ET_c = K_c ET_o \tag{3}$$

where, ET_c is crop evapotranspiration (mm), K_c is the crop coefficient (dimensionless) and ET_o is reference evapotranspiration (mm).

In much of the crop water use literature published in Puerto Rico, evapotranspiration has been referred to as consumptive use (CU). Therefore, in this review, ET_c and CU will be used interchangeably. The crop coefficient (K_c) accounts for the effects of characteristics that distinguish the field crop from a reference crop [5]. Reference evapotranspiration has been defined as the evapotranspiration from an extended surface of 0.08 to 0.15 m (3.2 in. to 5.9 in.) tall, green grass cover of uniform height, actively growing, completely shading the ground and not short of water [7]. Alternatively, reference evapotranspiration has been defined as the "upper limit or maximum evapotranspiration that occurs under given climatic conditions with a field having a well-watered agricultural crop with an aerodynamically rough surface, such as alfalfa with 12 in. to 18 in. of top growth" [8].

Numerous mathematical equations have been developed for computing ET_o. One such expression, having global validity and recommended by the United Nations Food and Agriculture Organization (FAO), is the Penman-Monteith equation [5] given below:

$$ET_0 = \frac{0.408\Delta\left(R_n - G\right) + \gamma\left(\dfrac{900}{T+273}\right)u_2\left(e_s - e_a\right)}{\Delta + \gamma\left(1 + 0.34u_2\right)} \tag{4}$$

where, Δ is the slope of the vapor pressure curve (kPa°C^{-1}), R_n is net radiation (MJ m^{-2}d^{-1}), G is the soil heat flux density (MJ m^{-2} d^{-1}), γ is the psychrometric constant (kPa^{-1}), T is mean daily air temperature at 2 m height (°C), u_2 is wind speed at 2-m height, e_s is the saturated vapor pressure (kPa^{-1}) and e_a is the actual vapor pressure (kPa^{-1}). Eq. (4) applies specifically to a hypothetical reference crop with an assumed crop height of 0.12 m, a fixed surface resistance of 70 sec m^{-1} and an albedo of 0.23. Note that the derivation of Eq. (4) was based on SI units and therefore nonSI units should not be used.

18.2 HISTORICAL REVIEW OF EVAPOTRANSPIRATION IN PUERTO RICO

This literature review surveys the efforts to either measure or estimate evapotranspiration in Puerto Rico. The majority of the literature indicates that before the late 1980s, Goyal or others in collaboration with him have made almost all of the currently available estimates of crop consumptive use and reference evapotranspiration. These studies are detailed in a compilation document called *Irrigation Research and Extension Progress in Puerto Rico* [9]. Throughout the 1990s, there appears to have been a cessation in evapotranspiration research in Puerto Rico, except in the use of pan evaporation-estimated CU for managing irrigation application amounts. Since 2000, some studies have been initiated involving comparisons between the Penman-Monteith method and other calculation methods [10–12]. Also, one recent study has attempted to standardize and validate climate parameter estimation procedures for use with the Penman-Monteith method in Puerto Rico [13].

18.2.1 EVAPOTRANSPIRATION REFERENCE MATERIALS

There is a chapter on evapotranspiration in the book *Management of Drip/Trickle or Micro Irrigation* published by the Apple Academic Press Inc., [14]. The document provides basic definitions and descriptions of the following methods: three variations of the Penman method, three variations of the SCS Blaney-Criddle method, two Hargreaves methods, Jensen-Haise, Stephens-Stewart, Priestley-Taylor, Thornthwaite, Linacre, Makkink, pan evaporation, water balance, radiation and regression methods. The document also covers local calibration techniques and crop coefficients. Much of this material is also contained in a 40-five page extension document called *Evapotranspiración* by Goyal and González [6].

18.2.2 CONSUMPTIVE USE OBTAINED FROM FIELD STUDIES

The following five studies are significant because they represent a limited dataset in which actual field measurements were made to determine CU.

Fuhriman and Smith [15] conducted a study for the Aguirre area of southern Puerto Rico on the CU of sugar cane under differing irrigation treatments. At the time of this study, between 1949 and 1950, sugarcane was the dominant crop in Puerto Rico, grown on 95 percent of the irrigated land. The experiment involved field studies using nonweighing lysimeters. Consumptive use was estimated using the water balance method, which accounted for rainfall, irrigation, percolation below the root zone and changes in soil moisture. Daily water requirements varied between 2.5 and 4.6 mm d^{-1} (0.1 and 0.18 in. d^{-1}). Soil moisture was estimated using tensiometers and nylon resistance blocks. Soil evaporation measured on several of the lysimeters was subtracted from CU data on the sugar cane-covered plotsto determine plant transpiration rate.

Vázquez [16] determined CU using the water balance method for guinea grass, para grass and guinea grass-kudzu and para grass-kudzu mixtures at Lajas, Puerto Rico. Soil moisture content was estimated by means of tensiometers, gypsum resistance blocks and by gravimetric analysis of soil samples. The study was conducted from March 1959 through June of 1960. Sixty-day CU values were determined for the two grasses and two grass mixtures over the 15-month study period. Annual water useby the guinea grass and Para grass was 1,494 mm (58.8 in.) and 1,466 mm (57.7 in.), respectively. This was the only study found reporting CU for grasses.

Vázquez [17] used similar experimental methods to determine the CU of sugar cane at Lajas, Puerto Rico. This study included use of a neutron probe method for soil moisture determination. The study was conducted from April 1965 through May 1966. Consumptive use was determined for various irrigation treatments. Average daily consumptive use varied between 2.5 and 4 mm d^{-1} (0.1 and 0.16 in. d^{-1}). The study concluded that about 250 mm (9.84in. d^{-1}) of water consumption could be saved under soil and climatic conditions similar to those existing at Lajas, if the crop is irrigated frequently during the early part of the growing season and no further irrigation is applied after 5 months prior to harvest.

Abruña et al. [18] conducted a study of water use by plantains at the Gurabo Substation in Puerto Rico. Consumptive use was determined by a water balance procedure based on soil moisture measurements and rainfall data. The study was conducted from September 1976 through April 1977. For irrigations scheduled when available soil moisture was depleted by 20%, the average daily evapotranspiration was estimated to be 2.9 mmd^{-1} (0.11 in. d^{-1}), and was found to be equivalent to 79% of pan evaporation. The minimum and maximum daily CU were 1.5 and 3.6 mm d^{-1} (0.05 and 0.14 in. d^{-1}), respectively, for September and December.

Ravalo and Goyal [19] estimated water requirements for rice during December through May of 1973 at the Lajas Substation using a hydrologic balance approach, accounting for farm inflow from the irrigation canal, effective rainfall, evapotranspiration, deep percolation, seepage through borders, border overflow, depth of flooding and water for saturating soil. Daily water requirementsvaried between 6.9 and 11.8 mmd^{-1} (0.27 and 0.47 in. d^{-1}). It is interesting to note that in none of the above studies was there an attempt to calculate crop coefficients from field data. The crop coefficient is calculated from rearrangement of Eq. (3) (i.e., $K_c = ET_c/ET_o$). In future studies that derive crop coefficients in Puerto Rico data from these studies should be evaluated.

18.2.3 STUDIES PREDICTING CONSUMPTIVE USE

Monthly water consumption by 15 different vegetable crops for two locations in Puerto Rico (Fortuna and Isabela) was calculated by Goyal [20, 21]. The SCS Blaney-

Criddle approach was used based on monthly percentage of annual daylight, mean air temperature and a humid area factor for Puerto Rico. The SCS Blaney-Criddle equation is given below:

$$CU = (K_{crop} K_t p \, T \, H)/100 \tag{5}$$

where, CU is the monthly water consumptive use (in. mo^{-1}), K_t is a climatic coefficient related to mean air temperature, p is a monthly percentage of annual daylight hours (percent), T is mean air temperature (°F), H is a humid area factor, and K_{crop} is a crop growth coefficient reflecting the growth stage. It should be noted that the crop growth coefficients used in the SCS Blaney-Criddle method are not equivalent to the evapotranspiration crop coefficient used in Eq. (1) [22].

The purpose of the study was to estimate monthly CU and net and total gross irrigation requirements. The net irrigation requirement was based on the monthly CU minus the monthly effective rainfall as determined by the Soil Conservation method [23]. The gross irrigation requirement was calculated by dividing the monthly consumptive use by irrigation efficiency (80% for drip, 60% for sprinkler and 40% for surface irrigation). This study was valuable because it included a large number of crops and surveyed both humid and semiarid locations in Puerto Rico. The applicability of the data was further enhanced by the fact that each crop was evaluated for various lengths of growing season, and for season starting dates at 15-day intervals throughout the year (e.g., Jan. 1, Jan.15, Feb. 1, Feb. 15, etc.).

Using the SCS Blaney-Criddle method Goyal [24] estimated monthly CU for papaya at seven Agricultural Experimental Stations located at Adjuntas, Corozal, Fortuna, Gurabo, Isabela, Lajas and Mayagüez. Average daily CU varied between 2.8 mmd^{-1} (0.11 in. d^{-1}) at Adjuntas and 3.7 mmd^{-1} (0.15 in. d^{-1}) at Fortuna. In this report, the author cautioned that the CU estimates have not been compared with experimental data, and that such experimental data are obtained with lysimeter studies, which are not available for Puerto Rico.

Goyal and González-Fuentes [25] used the SCS Blaney-Criddle method to estimate monthly CU for sugar cane at four locations in Puerto Rico: Fortuna, Gurabo, Isabela and Lajas. Monthly CU was minimum in April and maximum in August at all four sites. Average daily CU varied between 4.1 and 4.4 mmd^{-1} (0.16 and 0.17 in. d^{-1}). Consumptive use estimates from this study compared reasonably well with the range for sugar cane determined by Fuhriman and Smith [15] for a field project in the Aguirre area ("normally irrigated" treatment).

Several other studies were conducted using the SCS Blaney-Criddle method involving sorghum at two locations [26], plantain at seven locations [27], and bell and cubanelle peppers at two locations [28]. In the plantain study, CU data were estimated for the Gurabo Experiment Station, which can be compared with the September and December 1977 results (1.5 and 3.6 mm d^{-1} [0.06 and 0.14 in. d^{-1}], respectively) from the water balance study by Abruña et al. [18]. For these same months Goyal and González [27] derived values of 3.8 and 5.1 mm (0.15 and 0.2 in. d^{-1}), respectively. Assuming the field study data is accurate, the error in the Blaney-Criddle-estimated minimum and maximum CU were 30 and 60 percent, respectively. It should be noted that Goyal and González [27] based their calculations on mean monthly average climate data, whereas actual monthly data for the period were used by Abruña et al. [18]. Although these data were not available for comparison, inspection of the 1977 annual pan evaporation for the Gurabo Experimental Station [29] indicates only a 6.4 percent deviation below the long-term average annual pan evaporation. Since pan evaporation is correlated with evapotranspiration, the 30 and 60 percent differences between the field and estimated data do not appear to be reasonable.

González-Fuentes and Goyal [30] estimated CU for sweet corn grown at the Fortuna Experiment Station using the Hargreaves-Samani method to calculate reference evapotranspiration, which was then multiplied by crop coefficients applicable to sweet corn. The data were used to determine net irrigation requirements between December 10 and March 10, 1986. This was the only study found in which the Hargreaves-Samani method was used in combination with a crop coefficient to estimate CU. Minimum and maximum CU were 2 and 7 mm d^{-1} (0.08 and 0.28 in d^{-1}) during December and March, respectively.

18.2.4 STUDIES PREDICTING REFERENCE EVAPOTRANSPIRATION

This section summarizes several studies in which reference evapotranspiration (ET_o) was estimated. It should be noted that a shift in the kind of information is being reported. That is, reference evapotranspiration (ET_o) instead of consumptive use (CU). The advantage of reporting ET_o is that it leaves one free to calculate CU for any crop. The disadvantage is that crop coefficients are not yet available for numerous crops in Puerto Rico.

Goyal [31] used the Hargreaves-Samani method to estimate monthly ET_o for Central Aguirre, Fortuna and Lajas substations, and for Magueyes Island located on the south coast of Puerto Rico. The Hargreaves-Samani equation for reference evapotranspiration is given by [32]:

$$ET_o = [0.0023 \ R_a][(T + 17.8) \ (T_{max} - T_{min})]^{0.5} \qquad (6)$$

where, ET_o (mm d^{-1}) is the reference evapotranspiration, R_a (mm d^{-1}) is the extraterrestrial radiation, T is the mean daily average temperature (°C), and T_{min} and T_{max} are the mean daily minimum and maximum temperatures (°C), respectively. Daily estimated values of ET_o varied between 3.68 to 5.37 mmd^{-1} (0.15 to 0.21 in d^{-1}). The minimum estimated value of ET_o occurred in December and the maximum occurred in July. The same procedure was applied for Vieques Island, PR, with estimated monthly ET_o ranging between 3.29 mm d^{-1} (0.13 in. d^{-1}) in December to 4.94 mmd^{-1} (0.19 in. d^{-1}) in July [33].

Goyal et al. [34] used the Hargreaves-Samani method (Eq. (6)) to estimate ET_o for thirty-four locations within Puerto Rico. Average monthly minimum and maximum air temperatures were based on long-term measured data. Average monthly values of the extraterrestrial radiation (R_a) were based on the average latitude of Puerto Rico. They also developed a regression analysis in which several monthly climatic factors (mean daily minimum, maximum and average air temperature) were correlated with surface elevation in Puerto Rico.

18.2.5   STUDIES COMPARING EVAPOTRANSPIRATION METHODS IN PUERTO RICO

In 1990, a committee of the United Nations Food and Agriculture Organization [35, 36] recommended the Penman-Monteith method as the single approach to be used for calculating reference evapotranspiration (ET_o). This recommendation was based on comprehensive studies, which compared numerous calculation methods with weighing lysimeter data [37, 38]. The study of the American Society of Civil Engineers [37] found the Penman-Monteith method to produce superior results relative to all other methods (including the SCS Blaney-Criddle and Hargreaves-Samani methods).

Harmsen et al. [10] reported large differences between the SCS Blaney-Criddle method (estimates obtained from 20] and the Penman-Monteith method in a study that compared seasonal consumptive use for pumpkin and onion at two locations in Puerto Rico. The Penman-Monteith approach used crop coefficients as determined by the FAO procedure [5]. Average monthly minimum and maximum air temperatures and wind speed for the Experiment Stations were obtained from the National Oceanographic and Atmospheric Administration (NOAA) Climatological Data Sheets. Long-term average monthly humidity and radiation were not available for the Experiment Stations; therefore these data were estimated by procedures recommended by the FAO [5]. Crop growth stage durations, needed to construct the crop coefficient curves, were based on crop growth curve data presented by Goyal [20]. The maximum observed differences in the estimated seasonal consumptive use were on the order of 100 mm per season (3.94 in d^{-1}). The study concluded that large potential differences could be expected between the SCS Blaney-Criddle and the Penman-Monteith methods, with underestimations some months and overestimation in other months. Figure 1 shows differences (DELTA) in the seasonal consumptive use (CU) estimates between the SCS Blaney-Criddle (SCS BC) and Penman-Monteith (PM) methods for pumpkin at Fortuna, PR. In this example, the SCS Blaney-Criddle method overestimated the seasonal CU by approximately 90 mm (2.54 in.) for crop seasons beginning in June through August, and under estimated by approximately 60 mm (2.36 in.) for crop seasons beginning in December and January.

Harmsen et al. [12] compared the Penman-Monteithand Hargreaves-Samani-methods for estimating ET_o at 30-four locations in Puerto Rico (Fig. 2). The Hargreaves-Samani results were obtained from Goyal et al. [34]. Generally, the two methods were in reasonable agreement ($r^2 = 0.86$). Mean monthly climate data needed as input to the Penman-Monteith method were estimated using procedures derived for Puerto Rico (see next section). Some of the observed differences in the two methods may be because equation 6 does not account for the effects of wind and humidity as does the Penman-Monteith method. According to Allen et al. [5], Eq. 6 tends to under-predict under high wind conditions and overpredict under conditions of high relative humidity. Differences in the two methods might also be attributable to the fact that Goyal [34] used an average single value for latitude when estimating extraterrestrial radiation (R_a).

The maximum positive and negative differences were observed at the Juncos 1E and Aguirre stations, respectively. If the Penman-Monteith method is taken as the best estimator of ET_o, then it can be stated that the Hargreaves-Samani method overestimated

ET_o at Juncos 1E and underestimated ET_o at Aguirre. Juncos 1E is considered to have a humid climate, while Aguirre is considered to be semiarid. The maximum underestimate of -0.75 mmd^{-1} (0.03 in. d^{-1}) occurred during May at Aguirre (semiarid) was equal to a -13% error, and the maximum overestimate of 0.92 mmd^{-1} (0.04 in. d^{-1}) occurred during December at Juncos 1E (humid) was equal to a 28% error. These results are consistent with the findings of the ASCE study [37], which found the Hargreaves-Samani method to underestimate on average by 9% in arid regions and overestimate on average by 25% in humid regions.

The assumption that the Penman-Monteith method is the best estimator of reference evapotranspiration may be inappropriate under nonreference conditions [5]. Reference conditions imply that the weather data are measured above an extensive grass crop that is actively evapotranspiring, or in a well-watered environment with healthy vegetation.

A summary of studies in Puerto Rico that have measured or predicted consumptive use, and/or have predicted reference evapotranspiration is presented in Table 1. Five studies have been conducted in which CU has been measured, prior to the year 2000 computational approaches used were limited to only two methods, CU has been estimated for a relatively large number of crops, and that reference evapotranspiration has been estimated for a large number of locations within Puerto Rico.

18.2.6 ESTIMATING CLIMATIC PARAMETERS FOR USE WITH THE PENMAN-MONTEITH EQUATION

Harmsen et al. [12] evaluated climate parameter estimation procedures for use with the Penman-Monteith equation in Puerto Rico. With only site latitude, surface elevation and specification of the site's NOAA Climate Division, they were able to estimate all other needed inputs to the Penman-Monteith method. Mean monthly minimum and maximum air temperatures were estimated from surface elevation data. Dew point temperature was estimated from the minimum temperature plus or minus a temperature correction factor. Temperature correction factors and average wind speeds were associated with six climatic divisions in Puerto Rico. Solar radiation was estimated from a simple equation for island settings (elevation < 100 m [328 ft]) or by the Hargreaves' radiation equation (elevations > 100 m [328 ft]) based on air temperature differences.

Comparisons of reference evapotranspiration using the climate parameter estimation procedures and measured climate parameters showed good agreement ($r^2 = 0.93$) for four locations in Puerto Rico. Recently Harmsen and González [39] developed a Windows-based computer program for implementing the climate estimation procedures. The computer program includes a database of information for the 15 vegetable crops analyzed by Goyal [20].

18.2.7 USE OF PAN EVAPORATION DATA FOR ESTIMATING CONSUMPTIVE USE

A number of studies have been performed to determine optimal irrigation rates based on pan evaporation data in Puerto Rico during the 1990s [40, tanier]; [41, banana's under mountain conditions]; [42, bananas under semiarid conditions]; [43, plantains under semiarid conditions]; [44, watermelon under semiarid conditions]; [45, sweet pepper]. The pan evaporation method estimates crop water requirement from the following equation:

$$\text{CWR} = [K_c K_p E_{pan}] - [P] \tag{7}$$

where, CWR is crop water requirement (mm d^{-1}), K_c is the crop coefficient, K_p is the pan coefficient, E_{pan} is pan evaporation (mm) and P is precipitation (mm). The product $K_p E_{pan}$ is an estimate of ET$_o$. An alternative estimate of CWR, recommended by the University of California – Davis Vegetable Research and Information Center [46] and promoted by the UPR Agricultural Experiment Station [47], is given below:

$$\text{CWR} = K_c[(K_p E_{pan}) - (ER)] \tag{8}$$

where, ER is effective rainfall (mm). Eq. (8) is an empirical equation and there is no physical basis, which the author is aware of, for multiplying K_c times ER.

Most of the studies have recommended applying water to plants at a rate equal to 1 or greater times the pan-estimated CU rate. Because this approach is easy and inexpensive, these studies represent valuable contributions to agricultural production in the tropics. However, problems may result from this approach owing to the inherent differences in water loss from an open water surface and a crop [5]. Another potential limitation is that only a single value of crop coefficient is commonly used, and by definition, the crop coefficient varies throughout the season. Although recommended irrigation application rates by this method may maximize crop yields, the method may also result in the overapplication of water, contributing to the degradation of groundwater resources from leaching of agricultural chemicals.

According to Allen et al. [5], estimates of evapotranspiration from pan data are generally recommended for periods of 10 days or longer. However, in Puerto Rico Eqs. (7) and (8) are usually applied for periods of 2 to 4 days. The shorter-than-recommended periods used in Puerto Rico may be due to the widespread use of microirrigation systems, in which smaller and more frequent applications of water are common.

In Puerto Rico, the K_p values commonly used were derived from a study by González and Goyal [48]. These data were developed based on the ratio of long-term average pan evaporation and the estimated evapotranspiration using the SCS Blaney-Criddle method [23]. Because of inherent errors associated with the SCS Blaney-Criddle method and observed long-term changes in pan evaporation, the recommended K_p values may also be somewhat in error. The FAO currently recommends using the ratio of pan evaporation divided by the Penman-Monteith-estimated reference evapotranspiration for determining pan coefficients [5].

Harmsen et al. [49] evaluated historical pan evaporation data to determine if statistically significant increasing or decreasing trends existed for data from seven UPR Experimental Stations. Significant decreasing pan evaporation was observed at Lajas and Río Piedras. Significant increasing pan evaporation was observed at Gurabo and Adjuntas. No significant trends were observed at Fortuna, Isabela and Corozal. A significant difference was found to exist between the mean K_p calculated with pan evaporation data from 1960–1980 and 1981–2000. An updated table of monthly average pan coefficients was provided that could be used to estimate ET_o for the seven Agricultural Experiment Stations.

TABLE 1 Summary of studies that measured or predicted consumptive use (ETc), and/or predicted reference evapotranspiration (ETo) in Puerto Rico.

No.	Method	Number of locations	Crop	References
	Measured Consumptive Use			
1	Water balance	1	Sugar Cane	15
2	Water balance	1	Guinea Grass Para Grass and Guinea Grass-Kudzu Para Grass-Kudzu Mixtures	16
3	Water balance	1	Sugar Cane	17
4	Water balance	1	Plantain	18
5	Water balance	1	Rice	19
	Predicted Consumptive Use ($K_c ET_o$)			
6	SCS Blaney-Criddle	2	Green Beans Cabbage, Carrots Cucumber, Egg Plant Lettuce, Melons Okra, Potatoes Sweet Peppers Sweet Potato Tomotoes	20
6	SCS Blaney-Criddle	2	Onion, Pumpkin	10
7	SCS Blaney-Criddle	7	Papaya	24
8	SCS Blaney-Criddle	4	Sugar Cane	25
9	SCS Blaney-Criddle	2	Sorhgam	26
10	SCS Blaney-Criddle	7	Plantain	27

TABLE 1   *(Continued)*

No.	Method	Number of locations	Crop	References
11	SCS	2	Bell pepper	28
	Blaney-Criddle		sweet pepper	
12	Hargreaves	1	Sweet Corn	30
	-Samani			
	with crop			
	coefficient			
	Predicted Reference Evapotranspiration (ET$_o$)			
13	Hargreaves-	4	N/A	21
	Samani			
14	Hargreaves-	1	N/A	33
	Samani			
15	Hargreaves Samani	34	N/A	11, 12, 34

TABLE 2   Comparison of peak evapotranspiration estimates determined by three different methods for six vegetable crops at three locations in Puerto Rico [12].

Crop	Peak Evapotranspiration (mm d^{-1})		
	SCS Irrigation Guide for Caribbean Area [50]	SCS Blaney-Criddle Method[20]	Penman-Monteith Method[1]
	Fortuna		
Cabbage	4.1	5.3	6.1
Eggplant	4.1	5.3	6.1
Cucumbers	4.1	5.1	5.8
Melons	4.1	4.8	5.8
Sweet			
potatoes	5.3	6.4	6.7
Tomatoes	5.3	5.8	6.7
	Isabela		
Cabbage	4.1	5.1	5.7
Eggplant	4.1	5.3	5.7
Cucumbers	4.1	4.6	5.4
Melons	4.1	4.6	5.4
Sweet Potatoes	5.3	6.1	6.2
Tomatoes	5.3	5.6	6.2
	Aibonito		
Cabbage	4.1	NA[2]	5.5
Eggplant	4.1	NA	5.5

TABLE 2   *(Continued)*

Crop	Peak Evapotranspiration (mm d⁻¹)		
	SCS Irrigation Guide for Caribbean Area [50]	SCS Blaney-Criddle Method[20]	Penman-Monteith Method[1]
Cucumbers	4.1	NA	5.3
Melons	4.1	NA	5.3
Sweet Potatoes	5.3	NA	6.0
Tomatoes	5.3	NA	6.0

[1]Input to the Penman−Monteith equation for reference evapotranspiration were determined using the method Harmsen et al. [11]. Crop coefficients for the mature growth stage were obtained from [5].
[2]NA = Data not available.

18.2.8   PEAK EVAPOTRANSPIRATION

Design of irrigation systems requires knowledge of the peak evapotranspiration (ET_{peak}) rate. The Natural Resource Conservation Service (NRCS) has published values of ET_{peak} for various crops grown in Puerto Rico in its Irrigation Guide [50]. This document is a widely used source of peak evapotranspiration values in Puerto Rico.

Harmsen et al. [11] compared ET_{peak} estimated using three methods for six vegetable crops for three locations in Puerto Rico. The three methods were the SCS irrigation Guide [50], the SCS Blaney-Criddle method [20] and the Penman-Monteith method. A comparison of the peak ET data is presented in Table 2. It should be noted that the SCS Irrigation guide recommends a single value of ET_{peak} for the entire Island for a given crop. Peak ET for the SCS Blaney-Criddle method was obtained by using the maximum monthly consumptive use divided by the number of days in the month. The SCS Blaney-Criddle-estimates of ET_{peak} were not available for the Aibonito location. Input data for Penman-Monteith-estimates of reference evapotranspiration were determined using procedures described by Harmsen et al. [11]. Estimates of ET_{peak} were based on the maximum daily reference evapotranspiration (ET_o) times the published value of the crop coefficient (K_c) for the mid-season growth stage. The crop coefficients were obtained from Allen et al. [5].

For the three methods, estimates of ET_{peak} ranked lowest to highest were: SCS Irrigation Guide, the SCS Blaney-Criddle method and the Penman-Monteith method, respectively. The implications of these results are important because designers of irrigation systems in Puerto Rico may be under-designing systems at this time (assuming the Penman-Monteith method is the most accurate method). Normally, an under-designed drip irrigation system will be corrected by operating the system longer; for example, a system could be operated for eight hours instead of six hours. However, if the system was designed to run more hours per day (e.g., 22 hours, which is the maximum recommended by the American Society of Agricultural Engineers, [51]), then increasing the operating time may not be an option.

FORTUNA-PUMPKIN

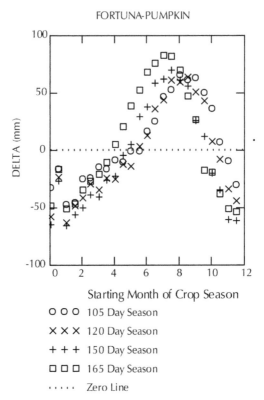

Starting Month of Crop Season

O O O 105 Day Season

X X X 120 Day Season

+ + + 150 Day Season

□ □ □ 165 Day Season

· · · · · Zero Line

FIGURE 1 Differences (DELTA) in the seasonal consumptive use (CU) estimates between the SCS Blaney-Criddle (SCS BC) and Penman-Monteith (PM) methods. [DELTA = {[SCS Blaney-Criddle] minus [Penman-Monteith]}, [11].

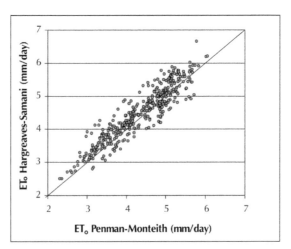

FIGURE 2 Comparison of reference evapotranspiration estimated by the Penman-Monteith (P-M) and Hargreaves-Samani (H-S) Methods for 30-four locations in Puerto Rico [12].

18.2.9 USE OF REMOTE SENSED CLIMATIC DATA FOR EVAPOTRANSPIRATION

Harmsen et al. [52, 53] developed an algorithm for estimating reference evapotranspiration in Puerto Rico and Hispañola. Harmsen et al. [52] compared estimated reference evapotranspiration values for Puerto Rico using the Hargreaves [54, 55], Penman-Monteith [5] and Priestly Taylor [56] methods, and obtained comparable results. As an example, Fig. 3 shows the estimated reference evapotranspiration using the original radiation-based Hargreaves reference evapotranspiration formula for June 29, 2010. These reference evapotranspiration methods rely heavily on solar radiation as input. Solar radiation was derived from a physical model for estimating incident solar radiation at the ground surface from GOES satellite visible and albedo data (1-km spatial resolution). The method was first proposed by Gautier et al. [57–60].

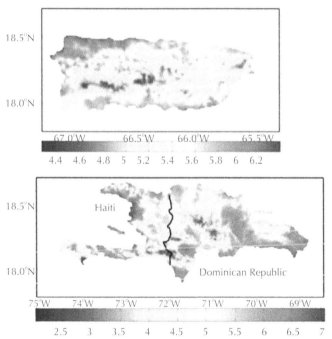

FIGURE 3   Estimated reference evapotranspiration for Puerto Rico (A) and Hispañola on June 29, 2010 [53].

Harmsen et al. [53] modified the GOES satellite-based algorithm for Puerto Rico to include estimation of the water and energy budget, and is referred to as GOES-PRWEB.GOES-PRWEB uses an energy balance approach similar to Yunhao et al. [61] for estimating actual evapotranspiration, which is then incorporated into a water balance calculation. Twenty four-hour rainfall is obtained from NOAA's Advanced Hydrologic Prediction Service (AHPS) website (http://water.weather.gov/precip/). Runoff is estimated using the Curve Number method of the USDA Natural Resource Conservation Service [62, 63]. Average daily wind speed for Puerto Rico is obtained from the National Weather Service's National Digital Forecast Database (http://www.nws.noaa.gov/ndfd/). Although the wind speed is a model forecast, it is the best source of spatially distributed wind speed over the island. Minimum, average, maximum and dew point air temperatures are obtained from a lapse rate approach calibrated for Puerto Rico by Goyal et al. [34] with regression equations relating average air temperature with surface elevation. These temperatures are adjusted daily with a nudging technique, using forecast temperature data from the NDFD. Figure 4 shows an example of the annual water budget components for 2010 and Fig. 5 shows the average annual crop water stress factor for Puerto Rico produced by GOES-PRWEB, using the method of Allen et al. [5]. A lower value of the crop stress factor indicates more stress caused by drier conditions. GOES-PRWEB is an operational algorithm that produced daily values of 25 hydro-climate variables for Puerto Rico. Daily results and monthly and annual averages/totals can be obtained at the website: http://pragwater.com.

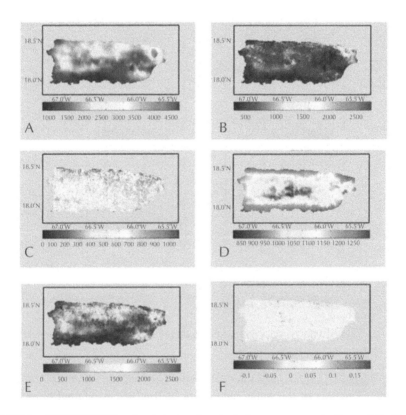

FIGURE 4 GOES-PRWEB-estimated annual water balance components (in mm) for Puerto Rico for 2010: a) Rainfall, b) Surface Runoff, c) Actual ET, d) Reference ET, e) Aquifer Recharge, and f) Change in Soil Moisture Content (fraction).

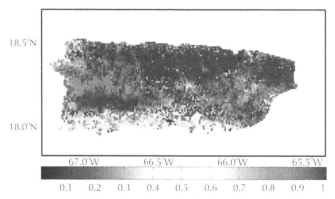

FIGURE 5 Average annual crop stress factor (K_s) for Puerto Rico for 2010.

18.3 RECOMMENDATIONS FOR FUTURE EVAPOTRANSPIRATION RESEARCH

Several areas of evapotranspiration research are needed in Puerto Rico at this time. Some of the suggested research will be of direct benefit to the Caribbean Region as well as in Puerto Rico, and may be of benefit to other tropical regions of the world.

Development of crop coefficients (K_c). To date, only one study has produces crop coefficients in the island for [64–66]. Although crop coefficients derived in other parts of the world can be used to provide approximate estimates of evapotranspiration, K_c in fact depends upon the crop cultivar and other local conditions. Perhaps more importantly, crop coefficients are not available in the literature for many of the local crops grown in Puerto Rico. The NRCS in Puerto Rico has given high priority to developing crop coefficients for the following cultivars [67]: Acerola, Tannier, Guanabana, Malanga, Parcha, Star Grass, Recao, Pangola Grass, Ñame, and Buffel Grass.

Water balance and nonweighing lysimeter studies performed during the 1960s and 70s in Puerto Rico did not determine crop coefficients. A study should be performed to reevaluate these data for this purpose. The crops considered in these studies included: sugar cane, guinea grass, para grass, guinea grass-kudzu and para grass-kuduz mixtures, plantains, rice and pumpkin.

Validation of the pan evaporation method for scheduling irrigation. This method has become popular in Puerto Rico in recent years because its ease of use. However, the method may result in over or under-application of water relative to the crop water requirements when a single value of the crop coefficient (K_c) for the entire season is used. It is preferable to use data from an evapotranspiration crop coefficient curve, which takes into account the development of the crop. Allen et al. [5] provides an excellent discussion on the construction of crop coefficient curves. Another potential source of error is the use of published values of K_p derived from the study of González and Goyal [48], which estimated evapotranspiration using the SCS Blaney-Criddle method, based on pre1980 climate data. Pan coefficients have been updated for Puerto Rico using the Penman-Monteith reference evapotranspiration [49].

Harmsen and Torres-Justinian [11] proposed procedures for estimating input to the Penman-Monteith method in Puerto Rico. A geographic information system (GIS) could be developed for Puerto Rico, incorporating the estimation procedures. This GIS consumptive use system could be made available on the internet. In combination with the GIS consumptive use system, a rainfall and soil database for Puerto Rico could be included which would permit irrigation planning by means of a water balance analysis [53].

Instead of estimating ET_c using a crop coefficient and reference evapotranspiration (Eq. (4)), ET_c can be determined using a "one-step" method as:

$$ET_c = f(r_a, r_b) \qquad (9)$$

where, $f(r_a, r_b)$ represents the general form of the Penman-Monteith equation, which is a function of the aerodynamic resistance (sec m^{-1}) and the bulk surface resistance (sec m^{-1}). For convenience, the other parameters/variables used in the Penman-Monteith equation are not shown. At this time, the parameters r_a and r_s are not easily obtained. However, through research efforts it is hoped that in the not too distant future it will become routine to apply Eq. 9 directly. Harmsen et al. [53] have used the one-step approach in Puerto Rico in the GOES-PRWEB water and energy balance algorithm.

The use of remote sensing techniques for estimating evapotranspiration. Satellite data can be used to estimate evapotranspiration over relatively large areas. Examples of this technique include the Surface Energy Balance Algorithm for Land (SEBAL) [68]; or by estimating biophysical processes using remotely sensed data [69, 70]. Numerous remote sensing projects are already being conducted at the University of Puerto Rico and collaborations exist between the Agricultural and Biosystems Engineering Department and NOAA's Cooperative Remote Sensing Science and Technology Center. Harmsen [52, 53] has developed a GOES satellite-based water and energy balance algorithm for Puerto Rico, which estimates reference and actual evapotranspiration.

Expert Consultation on the revision of FAO methodologies for crop water requirements [36] has suggested the following areas of research (which could be supported in Puerto Rico):

1. To evaluate the effect of advective conditions on crop resistance factors (i.e., r_a and r_b);
2. To make a systematic effort reviewing various research results, in developing sound values for crop resistance factors for a range of crops;
3. To review the effect on crop resistance factors of reduced evapotranspiration under soil moisture stress and adverse growth conditions.

18.4 SUMMARY

Currently, the water resources of Puerto Rico are being threatened by population pressure, development, pollution, and potentially adverse changes in the climate. Accurate determination of evapotranspiration is essential in managing water resources and practicing water conservation in Puerto Rico. In support of this goal, a review is presented covering the majority of research on crop water use and evapotranspiration estimation methods used in Puerto Rico during the last 60 years. Specifically the review considers consumptive use determined from field water balance studies, the pan evaporation method, meteorological methods and satellite remote sensing. Several studies considered the estimation of climate parameter data needed as input to the reference evapotranspiration calculation methods. Recommendations for research priorities are provided.

KEYWORDS

- aerodynamic resistance Blaney-Criddle
- Caribbean region
- climate
- consumptive use
- crop coefficient
- crop surface resistance
- crop water requirement
- crop water use
- drip irrigation
- evapotranspiration
- Hargreaves
- Hargreaves-samani
- irrigation requirement
- micro irrigation
- pan coefficient
- pan evaporation
- Priestly-Taylor
- peak evapotranspiration
- Penman-Monteith
- Puerto Rico
- reference evapotranspiration
- remote sensed data
- soil moisture
- surface resistance
- trickle irrigation
- tropical climate
- tropical region
- vapor pressure deficit

REFERENCES

1. Bonell, M.; Hufschmidt, M. M.; Gladwell J. S.; (editors), *Hydrology and Water Management in the Humid Tropics*. Cambridge University Press. **1993**, 620 pages.
2. US Central Intelligence Agency, **2002**, The World Fact Book 2002 (http: //www.cia.gov/cia/publications/factbook/geos/rq.html)
3. Kent, R. B. **2002**, West Indies. Microsoft® Encarta® Online Encyclopedia 2002.
4. Capiel, M.; Calvesbert, R. J.; **1976**, On the climate of Puerto Rico and its agricultural water balance. *J. Agric. Univ. P.R.* LX(2): 139–153.
5. Allen, R. G.; Pereira, L. S.; Dirk Raes; Smith, M.; **1998**, *Crop Evapotranspiration Guidelines for Computing Crop Water Requirements*. FAO Irrigation and Drainage Paper 56, Food and Agriculture Organization of the United Nations, Rome.
6. Goyal, M. R.; González-Fuentes, E. A.; **1990**, "Evapotranspiration" in thebook: Manejo de Riego por Goteo, Universidad de Puerto Rico Recinto Mayaguez, Colegio de ciencias Agricolas, Servico de ExtensionAgricola. Edited by Megh R.; Goyal. SEA IA 80, ISBN 0-9621805-2-3, Library of Congress Catalog No. 89–50060.
7. Doorenbos, J.; Pruitt, W. O.; **1977**, *Guidelines for Predicting Crop Water Requirements*. FAO Irrigation and Drainage Paper 24, Revized. United Nations, Rome.
8. Jensen, M. E.; Robb, D. C. N.; Franzoy, C. E.; **1970**, Scheduling irrigations using climate-crop-soil data. Proc. Am. Soc. Civ. Engr.; *J. Irrig. And Drain. Div.* 96(IR1): 25–38.
9. Goyal, M. R. (editor) **1989**, Irrigation Research and Extension Progress in Puerto Rico. Prepared for the First Congress on "Irrigation in Puerto Rico," March 8th, 1989, Ponce, Pr. UPR-Ag. Experiment Station.
10. Harmsen, E. W.; Caldero J.; Goyal, M. R.; **2001**, Consumptive Water Use Estimates for Pumpkin and Onion at Two Locations in Puerto Rico. Proceedings of the Sixth Caribbean Islands Water Resources Congress. Editor: Walter F.; Silva Araya. University of Puerto Rico, Mayagüez, PR 00680.
11. Harmsen, E. W.; Torres-Justiniano, S.; **2001**, Estimating Island-Wide Reference Evapotranspirations for Puerto Rico Using the Penman-Monteith Method. ASAE Paper 01–2174. 2001, American Society of Agricultural Engineering Annual International Meeting.Sacramento Convention Center, Sacramento, California, USA.; July 30-August 1, 2001.
12. Harmsen, E. W.; Goyal M. R.; Torres-Justiniano, S.; EstimatingEvapotranspiration in Puerto Rico. Puerto Rico. *J. Agric. Univ. P.R.* **2002**, *86(1–2),* 35–54
13. Harmsen, E. W.; Torres-Justiniano, S.; **2001**, Evaluation of Prediction Methods for Estimating Climate Data to be Used with the Penman-Monteith Equation in Puerto Rico. ASAE Paper 01–2048. 2001, American Society of Agricultural Engineering Annual International Meeting. Sacramento Convention Center, Sacramento, California, USA.; July 30-August *1,*2001.
14. Goyal, M. R.; **2012**, Evapotranspiration. Chapter *2,*pages, In: Management of Drip/Trickle or Micro Irrigation edited by Megh R. Goyal. Apple Academic Press Inc.; pages 422.
15. Fuhriman, D. K. Smith, R. M.; Conservation and consumptive use of water with Sugar Cane under irriation in the south coastal area of Puerto Rico. J. of Agric. University of Puerto Rico, **1951,** *35(1),* 1–45.
16. Vázquez, Roberto, Effects of Irrigation and nitrogen levels on the yields of Guinea Grass, Para Grass, and Guinea Grass-Kudzu and Para Grass-Kudzu Mixtures in Lajas Valley.J.; Agric. Univ. Vol, XLLIX, P.R. No. 4, **1965**, 389–412.
17. Vázquez, Roberto, **1970**, Water Requirements of Sugarcane under irrigation in Lajas Valley, Puerto Rico.Bulletin 224, University of Puerto Rico Agricultural Experiment, Rio Piedras, PR.
18. Abruña, F.; Vicente-Chandler, J.; Irizarry H.; Silva, S. Evapotranspiration with plantains and the effect of frequency of irrigation on yields. J. Agric. U.P.R. **1979**, *64(2),* 204–10.
19. Ravalo, E. J. Goyal, M. R.; Water Requirement of rice in Lajas Valley, Puerto Rico. Dimension, **1988,** *8(2),* Jan-Feb-Mar.

20. Goyal, M. R.; **1989,** Estimation of Monthly Water Consumption by Selected Vegetable Crops in the Semiarid and Humid Regions of Puerto Rico. AES Monograph 1989–90, June, Agricultural Experiment Station, University of Puerto Rico Rio Piedras, PR.

21. Goyal, M. R.; González, E. A.; **1988,** Water Requirements for Vegetable Production in Puerto Rico. ASCE Symposium on Irrigation and Drainage, July.

22. Burman, R. D.; Nixon, P. R.; Wright, J. L.; Pruitt, W. O.; **1981,** Water requirements. Chapter 6, In: *Design and Operation of Farm Irrigation Systems* edited by Marvin E.; Jensen. American Society of Agricultural Engineers, St. Joseph, Michigan.

23. SCS, **1970,** *Irrigation Water Requirements.* Technical Release No. 21. USDA Soil Conservation Service, Engineering Division. (SCS Chapter 2, 1993)

24. Goyal, M. R.; **1989,** Research Note: Monthly consumptive use of papaya at seven regional sites in Puerto Rico. *J. Agric. U.P.R.; 73(2),* 00

25. Goyal, M. R. González-Fuentes, E. A.; **1989,** Estimating Water Consumptive Use by sugarcane at four regional sites in Puerto Rico. *J. Agric. U.P.R.;* 73(1)

26. Goyal, M. R.; González, E. A.; Requisitos de riego para sorgo en las costas Sur *y* Norte de Puerto Rico. J. Agric. University of Puerto Rico, **1988,** *72(4),* 585–98.

27. Goyal, M. R.; González, E. A.; Requisitos de riego para plátano en siete regions ecológicas de Puerto Rico. J.; Agric. University of Puerto Rico, **1988c,** *72(4),* 599–608.

28. Goyal, M. R.; González, E. A.; Seasonal consumptive use of water by bell and Cuanelle peppers in semiarid and humid coastal sites in Puerto Rico. *J. Agric. U.P.R.;* **1988d,** *72(4),* 609–14.

29. González, E.; Goyal, M. R.; Pan coefficients for Puerto Rico [Spanish]. *J. Agric. Univ. P.R.* **1989,** *73(1).*

30. González-Fuentes E. A.; Goyal, M. R.; Irrigation requirement estimations for sweet corn (Zea mays cv. Suresweet) in the south coast of Puerto Rico [Spanish]. J. of Agric. University of Puerto Rico, **1988,** *72(2),* 277–84.

31. Goyal, M. R. Research Note.Potential evapotranspiration for the south coast of Puerto Rico with the Hargreaves-Samani technique. *J. Agric. U.P.R.;* **1988a,** *72(1),* 57–63.

32. Hargreaves, G. H.; Samani, Z. A.; Reference crop evapotranspiration from temperature. *J. Appl. Eng. Agric.* **1985,** *1(2),* 96–99.

33. Goyal, M. R. Research note: Potential Evapotranspiration for Vieques Island, Puerto Rico, With the Hargreaves and Samani Model. *J. Agric. U.P.R.;* **1988b,** *72(1),* 177–78.

34. Goyal, M. R.; Gonzalez E. A.; Chao de Baez, C. Temperature versus elevation relationships for Puerto Rico, J. Agric. U.P.R.; **1988,** *72(3),* 440–67.

35. FAO, United Nations, **1990,** Expert consultation Italy, on revision of FAO methodologies for crop water requirements, 28–31 May. Annex V; Rome.

36. Smith, M.; Allen, R.; Monteith, J. L.; Perrier, A.; Santos Pereira, L.; Segeren. A.; Report on the Expert Consultation on Revision of FAO Methodologies for Crop Water Requiements. Held in FAO.; Rome, Italy, **1990,** 28–31 May, 1990.

37. Jensen, M. E.; Burman, R. D.; Allen, R. G.; **1990,** *Evapotranspiration and irrigation water requirements.* ASCE Manuals and Reports on Engineering Practice No. 70. 332.

38. Choisnel, E.; De Villele, O.; Lacroze, F.; **1992,** A uniform approach to calculate the Uneapproacheuniformisée du calcul de l' évapotranspirationpotentielle pour l'ensemble des pays de la CommunautéEuropéenne." Com. Commun. Européennes, EUR 14223 FR, Luxembourg, 176.

39. Harmsen, E. W.; Gonzaléz, A.; **2002,** Puerto Rico EvapoTranspiration Estimation Computer Program PR-ET Version 1.0 USER'S MANUAL.; University of Puerto Rico Experiment Station-Rio Piedras. Grant SP-347. August 2002, 44.<www.uprm.edu/abe/PRAGWATER>

40. Goenaga, R. **1994,** Growth, nutrient uptake and yield of tanier (Xanthosoma spp.) grown under semiarid conditions. *J. Agric. Univ. P.R.* 78(3–4)-87–99.

41. Goenaga, R.; Irizarry, H.; Yield of Banana grown with supplemental drip irrigation on an Ultisol. Expl. Agric.; **1998,** *34,* 439–448.

42. Goenaga, R.; Irizarry, H.; Yield Performance of Banana Irrigated with Fractions of Class A pan Evaporation in a Semiarid Environment. Agron. J.; **1995,** *97,* March –April.

43. Goenaga, R.; Irizarry H.; Gonzalez, E.; Water requirement of plantains (Musa acuminata X Musa balbisiana AAB) grown under semiarid conditions. Trop. Agric. (Trinidad), **1993**, *70(1)*, January.

44. Santana Vargas, J. **2000**, Wetting patterns under subsurface microirrigation [Spanish].Thesis for Master of Science (), University of Puerto Rico – Mayaguez Campus.

45. Harmsen, E. W.; Colón Trinidad, J.; Arcelay, C.; Sarmiento Esparra, E.; Evaluation of the Pan Evaporation Method for Scheduling Irrigation on an Oxisol in Puerto Rico. Proceedings of the Caribbean Food Crops Society, Thirty Eighth Annual Meeting, 2002. Martinique. **2002b**, Vol. 38.

46. Hart, T. K.; Hochmuth, G. J.; Fertility Management of Drip Irrigated Vegetables. Davis, U. C.; Vegetable Research and Information Center. **1998**, 10 pp.

47. Rivera-Martínez, L. **2002**, Revision of Technological Practices for Eggplant and Sweet Peppers [Spanish].AgriculturalExperimentStation, University of Puerto Rico – MayaguezCampus, Río Piedras, PR.

48. Goyal, M. R.; González-Fuentes, E. A.; **1989a**. Evapotranspiration [Spanish]. University of Puerto Rico – Cooperative Agricultural Extension Service. *14*, IA-72.45 pp.

49. Harmsen, E. W.; Gonzaléz A.; Winter. J. A.; Re-Evaluation of Pan Evaporation Coefficients at Seven Locations in Puerto Rico, By Agric, E.W. Univ. P. R. **2004**, *88(3–4)*, 109–122.

50. SCS, **1969**, *Technical guide for Caribbean Area*, Section IV-Practice Standards and Specifications for Irrigation System, Sprinkler. Code 443. US Department of Agriculture Soil Conservation Service.

51. ASAE, **1999**, *Design and Installation of Microirrigation System*. American Society of Agricultural Engineers, ASAE EP405.1 DEC99.

52. Harmsen, E. W.; Mecikalski, J.; Cardona-Soto, M. J.; Rojas Gonzalez A.; Vasquez, R.; Estimating daily evapotranspiration in Puerto Rico using satellite remote sensing. WSEAS Transactions on Environment and Development. **2009**, *6(5)*, 456–465.

53. Harmsen, E. W.; Mecikalski, J.; Mercado A.; Tosado Cruz, P.; **2010**, Estimating evapotranspiration in the Caribbean Region using satellite remote sensing. Proceedings of the AWRA Summer Specialty Conference, Tropical Hydrology and Sustainable Water Resources in a Changing Climate. San Juan, Puerto Rico. August 30-September *1,*2010.

54. Hargreaves, G. H.; Moisture availability and crop production. Transactions of the ASAE, **1975**, *18(5)*, 980–984.

55. Hargreaves, G. H.; Samani, Z. A.; Estimating potential evapotranspiration. Journal of Irrigation and Drainage Division, Proceedings of the ASCE, **1982**, *108(IR3)*, 223–230.

56. Priestly, C. H. B.; Taylor, R. J.; On the assessment of surface heat flux and evaporation using large scale parameters. Mon. Weath. Rev.; **1972**, *100*, 81–92.

57. Gautier, C.; Diak, G. R.; Masse, S.; A simple physical model to estimate incident solar radiation at the surface from GOES satellite data. J.; Appl. Meteor.; **1980**, *19*, 1007–1012.

58. Diak, G. R.; Bland, W. L.; Mecikalski, J.; *1996*, A note on first estimates of surface insolation from GOES-8 visible satellite data. Agric. For. Meteorol.; *82*, 219–226.

59. Otkin, A. J.; Anerso, M. C.; Mecikalski J.; Diak, G. R. R.; Validation of GOES-Based Insolation Estimates Using Data from the US Climate Reference Network. J. Hydrometeorology, **2005**, *6*, 460–475.

60. Sumner, D. M.; Pathak, C. S.; Mecikalski, J. R.; Paech, S. fJ.; Wu, Q.; Sangoyomi, T.; **2008**, Calibration of GOES-derived Solar Radiation Data Using Network of Surface Measurements in Florida, USA.; Proceedings of the ASCE World Environmental and Water Resources Congress **2008**, Ahupua'a.

61. Yunhao C.; Xiaobing L.; Peijun, S.; Estimation of regional evapotranspiration over North-west China by using remotely sensed data. Journal of Geophysical Sciences, **2001**, *11(2)*, 140–148.

62. US Soil Conservation Service, A Method for Estimating Volume and Rate of Runoff in Small Watersheds.SCS-TP-149, Washington, D.C. **1973**.

63. Fangmeier, D. D.; Elliot, W. J.; Workman, S. R.; Huffman, R. L.; Schwab, G. O.; *Soil and Water Conservation Engineering*, Fifth Edition. John Wiley and Sons. **2005**, pp. 528.

64. Ramírez-Builes, Víctor H. **2007,** Plant-Water Relationships for Several Common Bean Genotypes (Phaseolus vulgaris L.) with and Without Drought Stress Conditions. MS Thesis. Department of Agronomy and Soils, University of Puerto Rico – Mayaguez, Puerto Rico, December 2007.

65. Ramirez Builes, V. H.; Porch, T. G.; Harmsen, E. W.; Genotypic Differences in Water Use Efficiency of Common Bean under Drought Stress. Agron. J. **2011,** *103,* 1206–1215.

66. Ramírez-Builes, V. H.; Harmsen, E. W.; **2011,** Water Vapor Flux in Agroecosystems: Methods and Models Review, published in the book Evapotranspiration edited by Leszek Labedzki, ISBN: 978-953-307-251-7, InTech, Publishing.

67. Martínez, J. **2000,** Personal communication. Meeting with the author and the representatives of the NRCS at the Juan Diaz, PR, Field Office. May, 2000.

68. Bastiaanssen, W. G. W.; **2000,** SEBAL-based sensible and latent heat fluxes in the irrigated Gediz Basin, Turkey. *J. Hydrology, 229,* 87–100.

69. Choudhury, B. J.; DiGirolamo, N. E. A biophysical process-based of global land surface evaporation using satellite and ancillary data. I.; Model description and comparison with observations. *J. Hydrology,* **1998,** *205,* 164–185.

70. Kite, G.; Droogers, P. *Comparing Estimates of Actual Evapotranspiration from Satellites, Hydrologic Models, and Field Data: A Case Study from Western Turkey.* International Water Management Institute, Research Report 42. **2000,** 41.

CHAPTER 19

REFERENCE EVAPOTRANSPIRATION FOR COLOMBIAN COFFEE[1]

VICTOR H. RAMIREZ

CONTENTS

---

[1] Special thanks to the UNISARC – Colombia for providing financial support for this project; to the technical staff at the Agroclimatology Section of Cenicafé for providing the meteorological data. © Copyright Café Columbia

19.1 INTRODUCTION

The appropriate estimation of evapotranspiration (ET) is necessary for the crop water requirement estimations, for climatic characterization, management of the water resources, and in the irrigation scheduling [23]. At the biological level, the knowing the ET helps to understand the magnitude of the gas interchanges between the eco and agro ecosystems with the atmosphere.

The calculation of the crop water requirement need an appropriate selection of method or model for the estimation. The direct measuring can be done with field lysimeters, which measure mass balance variables like rainfall, percolation, runoff and soil moisture changes and by default estimate the crop water requirement or the ET [11, 19], also the ET estimation can be done using micrometeorological measurements, which are base on the energy balance equation, or methods based on gradients of air temperature and moisture, or the eddy covariance technique [19]. All these direct methods in practice present some difficulties to be use, basically by his high cost of acquisition and keeping.

The United Nations Food and Agriculture Organization (FAO) recommend the method known like "the two steps" for the crop evapotranspiration (ET_c) estimation. This method is useful for condition without limitation of water and pest [2, 6]. The "two steps" method consist in the estimation of the reference evapotranspiration (ET_0) times the crop coefficient (K_c) which should be estimated in field using mass balance, energy balance, temperature and humidity gradient or eddy covariance methods. The FAO approach is very usefully because is possible make water requirement estimation by phenological phases or discriminate water uses by crops, using common meteorological information that is the most available in our media.

At global level the use of the FAO approach has increasing the knowledge of the crop water uses, the knowledge of the water consumes of the different land covers, the precision of the climate and hydrological studies. At agronomical level, the FAO approach is very useful in the estimation of the available water in the root zone, all this indicate a need a good estimates of the reference evapotranspiration (ET_0). The first step in the appropriated ET_0 estimation is the local validation or calibration of the ET_0 models [2]. In countries like Colombia, where the studies that measure directly the crop evapotranspiration (ET_c) are scarce, the FAO approach in an important alternative.

Few are the studies that evaluate the ET models in Colombia. In the case of the Colombian coffee zone, can be mentioned the Jaramillo [12] and Jaramillo [13] studies that compared the class A pan evaporation with equations base on weather information. Subsequently Jaramillo [14]. calculated the ET_0 variation with altitude in several locations of the Colombian Andes discriminating the Cauca and Magdalena river watershed, in that study he compare the relationship between the ET_0 estimated by the Penman-Monteith (P-M) model and the class A pan evaporation. Giraldo et al. [9]. compared several ET_0 methods in the North of Santander coffee zone (Francisco Romero station), in that work they compare the reference P-M model with the Turc, Linagre, Hargreaves, Jensen-Haise and Garcia and Lopez modified by Jaramillo, indicating not statistical differences between the Garcia and Lopez modified by Jaramillo and the reference Penman-Monteith model. In that study they do not include direct measures with lysimeter or other micrometeorological methods. Barco et al. [5], made an a macro scale estimation of the evaporation in Colombia, using several methods like the Turc, Morton, Penman, Holdridge and Budyko, in that study they made an quantitative and qualitative analysis of these models, but do not include a comparison with direct field measures.

The reference P-M model is the most useful method because include most climate variables that other models, and specially because include in his calculation the effect of the several factor in the ET like the energy availability (R_{n-}G), de water vapor pressure deficit (VPD) and the wind speed. The precision of the ET_o estimation by the P-M model, depends of the data quality. For the specific case of the Colombia coffee zone, exist a potential limitation for the ET_o estimation using the P-M model, because some direct meteorological measurement are not available, like the net radiation (R_n), soil heat fluxes (G), wind speed at 2 m level, and vapor pressure deficit (VPD). In the specific case of the R_n and G, we use empirical relations base on sunshine measurements. Similar situation has been reported by other author in other locations [17, 27].

The FAO-56 paper recommends the P-M model when is possible use the solar radiation, wind speed, temperature and air humidity information, or when the empirical models for the solar radiation has been previously calibrated, and then recommends the Hargreaves model when only exist air temperature information (maximum and minimum).

This chapter evaluates the reference evapotranspiration (ET_o) models recommended by the FAO and compares these estimations with the Garcia and Lopez modified model and direct field measurements using lysimeter.

19.2   MATERIALS AND METHODS

19.2.1   LOCATION

This study was conducted in the Campus of the University of Santa Rosa de Cabal (Risaralda-Colombia), located in the western slope of the Cauca river watershed at 04° 55' North, 75° 38' West, at 1,600 meters a.s.l. The climate characteristics during the study are listed in Table 1. The dominant soils are derivate for volcanic ash, classifieds as Andisols [26], the main characteristics of this soils are: depth of the A horizon higher than 20 cm, high levels of organic matter, lower content of exchangeable bases, acid, with high infiltrations rates and high water retention capacity. The dominant crops in this area are the coffee and grass.

TABLE 1   Climatic conditions during the research period. September 2008 to September 2009.

Variable	Unit	Value Range	Mean Value
Sunshine	hours	0.0 – 10.0	4.0
Global radiation	W m^{-2}	101 - 297	173.1
Mean temperature	°C	17.3 – 23.2	20.5
Mean Relative humidity	%	56.3 – 95.3	74.0

19.2.2 THE PENMAN-MONTEITH MODEL

For this study we use the P-M model for the ET_o estimation recommended by the FAO-56 paper [2] and standardized by the American Society of the Civil Engineers-ASCE [1]. The model calculations were based at daily level for a reference crop with 0.12 m height:

$$ET_o = \frac{0.408(R_n - G) + \gamma \dfrac{900}{T + 273} u_2 (e_s - e_a)}{\Delta + \gamma(1 + 0.34 u_2)} \quad (1)$$

where, R_n is the net radiation (MJ m^{-2} día^{-1}), G is the soil heat fluxes (MJ m^{-2} día^{-1}) calculated as 0.1 of the R_n, g is the psychometric constant (kPa°C^{-1}), e_s is the saturated vapor pressure, e_a is the actual vapor pressure (kPa), D is the slope of the vapor pressure curve (kPa°C^{-1}), u$_2$ is the wind speed at 2 m, T is the air temperature (°C) at 2 m level. For this study the e_a, e_s, Δ y g were calculated how is describe by the FAO-56 paper [2]. The mean air temperature was calculated using the average between the maximum and minimum. The net radiation (R_n) was calculated using the equation for solar radiation budget as follow:

$$R_n = \left[R_a \left(0,26 + 0,506 \frac{n}{N} \right)(1 - \alpha) \right] - \left[\lambda T^4 \left(0,56 - 0,079 \sqrt{e_a} \right)\left(0,1 + 0,9 \frac{n}{N} \right) \right] \quad (2)$$

where, R_a is the astronomic solar radiation (MJ m^{-2} d^{-1}) and calculated following the model presented by Allen et al. [2]; n is the sunshine (h) measure by the heliograph, N. is the astronomic sunshine (h), λT^4 is the Stefan-Boltzman constant (W m^{-2}), e_a is the actual vapor pressure (mb). The a and b values are the Ángstrom-Prescott coefficients, we use a=0.26 and b=0.56 calculated by Gómez and Guzmán [10]. for the central coffee zone, α is the albedo we use 0.23 for wet grass used for the reference conditions. The solar radiation budget used in this study is different at the recommend by FAO-56 paper, due that we use the Ángstrom-Prescott model for the calculation.

19.2.3 THE HARGREAVES MODEL

When some of the climate information for the P-M model estimation is not available, the FAO-56 paper [2] recommend the Hargreaves model:

$$ET_o = 0.0023 \left(T_{mean} + 17.8 \right)\left(T_{max} - T_{min} \right)^{0.5} R_a \quad (3)$$

where, R_a is the astronomical radiantion in mm per day.

Modified model of García and López [8] proposing a model for the reference evapotranspiration calculation in Venezuela, that was later modified by Jaramillo [12] for the conditions of the central coffee zone (with elevation between 1,000–2,000 m), this model use the mean air temperature (T_{mean},°C) and the mean relative humidity (R.H$_{mean}$, %) at daily level:

$$ET_o = 1.22 * 10^n \left[\left(1 - 0.01 * R.H_{mean} \right) + \left(0.2 * T_{mean} \right) - 1.80 \right] \tag{4}$$

19.2.4   THE LYSIMETER

We used in this study a drainage type lysimeter, with 0.255 m² of collecting area and 0.60 m of depth. For the lysimeter installation, the soil was excavated at 0.70 m depth, removing the soil in three layers with the aim to reduce the soil disruption by the excavation. One polyethylene container was placed in the hole with the installation to collect the drainage water. The drainage system was composed by 0.1 m of fine gravel and 0.1 m of large gravel, the system for the drainage water conduction was connected in the bottom of the lysimeter with a PVC tube connection to other small collector tank of 12 liters of capacity. Above the drainage system the soil was stockpiled in two layer of 0.10 m each one in the inverse order of excavation, and then a grass layer was installed (Fig. 1). The lysimeter was installed at side at weather station. Once installed the grass cover, after the lysimeter installation we leave during 15 for the grass installation inside and outside of the lysimeter.

FIGURE 1   A. Cross section of the drainage lysimeter. (B) Top view of the drainage system 0.1 m of fine gravel + 0.1 m of large gravel. (c) Lysimeter with soil layers stockpiled. (D) Lysimeter with the grass cover.

The field capacity of the soil was 0.81 cm³ cm⁻³, the wilting point 0.48 cm³ cm⁻³, the bulk density 0.7 g cm⁻³, and the available water at 30 cm of 46.2 mm. Before start the evapotranspiration reading we applied irrigation at the lysimeter at field capacity to reduce the influence of the soil moisture changes in the water balance calculation, the irrigation was applied the day before to allow the free drainage. The ET_o estimates were made daily after the lysimeter reached the field capacity. Days without rain fall we applied a known irrigation.

19.2.5   THE REFERENCE EVAPOTRANSPIRATION (ET$_o$) CALCULATIONS

$$ET_{o(i-1)} = \{I_{i-1} - D_i + [R_{13:00(i-1)} + R_{18:00(i-1)} + R_{07:00\,(i)}]\} \qquad (5)$$

where, ET$_{o(i-1)}$ is the reference ET for the previous day in mm d^{-1}; I$_{i-1}$ is the irrigation applied in the previous day (mm), D$_i$ is the drainage in the day i (mm), R$_{13:00(i-1)}$+R$_{18:00(i-1)}$ are the rainfall measured at the previous day at 13:00 and 18:00 hours, and R$_{07:00\,(i)}$ is the rainfall measured at the day i at the 07:00 hours that correspond at the night rainfall at the previous day. The lysimeter had a small trench 0.1 m wide and 0.15 m depth to prevent the entry of the runoff from the adjacent area (Fig. 1).

19.2.6   MEASUREMENT OF METEOROLOGICAL VARIABLES

Since, September 2008 to September 2009, the meteorological information was recorded as follow. The minimum temperature was measure in an alcohol thermometer, the maximum temperature in a mercury thermometer with strangulation, the relative humidity was measure in a thermo-hygrograph with bi metallic sensor and hair bundle for the air temperature and humidity respectively, these instruments were placed inside of a shelter at 2-m height of the ground. The sunshine was measure in a heliograph type Campbell Stokes. The rainfall was measured three time per day (07:00–13:00 and 17:00 hours) using a rain gage with 200 cm^2 of collecting area and a pluviograph type Hellman. All the instruments were placed at side of the lysimeter area following the specification of the World Meteorological Organization (WMO), and the weather station was operated by the National Coffee Research Center (Cenicafé-Colombia).

19.2.7   EVALUATION FOR THE REFERENCE CONDITIONS

According with the FAO-56 paper [2], the references equations for the ET_o computing is based on requirements that weather data be measured in environmental conditions that correspond to the definition of reference evapotranspiration, that means that the weather variables are measured above extensive grass cover crop that is actively evaporating with constant leaf area, surface resistance and height, or in an environment with healthy vegetation not short of water, this mean that do not exist additional energy source for evaporation [3]; proposed the difference between the minimum temperature and the dew point temperature to evaluate the reference conditions. If the difference is higher than +3°C, it is not considered as a "reference day" [15]. The dew point temperature was computing using the Tetens equation as follows:

$$T_{pr} = \frac{237.3\,Ln\left[\dfrac{e_a}{4.584}\right]}{12.27 - Ln\left[\dfrac{e_a}{4.584}\right]} \qquad (6)$$

where, e_a is in mm of Hg.

19.2.9   DATA ANALYSIS

For the models comparison, we used the hypothesis test with a T-student as a statistical test and a linear regression analysis especially the slope of the line to identify subestimation of overestimation, similarly how was used in similar studies by [4, 16, 20, 21, 24].

19.3   RESULTS AND DISCUSSION

During the measured time, the differences between the minimum air temperature and dew point temperature was negative, indicating that all measurements were under references conditions [2], with difference up to 11°C.

19.3.1   COMPARISONS OF ET_O METHODS

For wet tropical Andean zone like the include in this study, the Penman-Monteith (P-M) model without wind conditions and the Garcia and Lopez Modified, showing similar ET_o values that those measured in the lysimeter (Table 2). The Hargreaves model how is recommended by the FAO-56 paper, Eq. (3), overestimated the ET_o values respect to the Lysimeter and the Garcia and Lopez Modified. Similar results were reported in the coffee zone of Brazil by Souza et al. [25], respect to the Hargreaves model.

When increasing the wind speed in the P-M model the ET_o is overestimated (Table 2). This situation is associated at the fact that in this study the wind speed was not measured. As general term at meteorological level, when the wind speed increase the vapor pressure deficit increase and the air temperature decrease, increasing the ET_o rates. For these reason, is not recommended assume an arbitrary value for the wind speed when this in not measured because the ET_o is overestimated. By other hand, the results indicating that the model of the Garcia and Lopez modified by Jaramillo is appropriated for the ET_o estimations without wind conditions. The initial Hargreaves model overestimated the ET_o values respect to the lysimeter measures and the P-M model without wind (Table 2), we propose a modification at the Hargreaves model as follows:

$$ET_o = 0.0018\left(T_{media} + 17.8\right)\left(T_{máxima} - T_{min\,ima}\right)^{0.5} R_a \qquad (7)$$

Once the Hargreaves model was modified (Eq. (7)) we do not observed statistical differences between modified model and the lysimeter measures and the P-M model without wind (Table 3), how the results indicate modification was appropriated for the local conditions.

TABLE 2 ET_o models comparisons for the Colombian coffee zone.

Relation	b	S.E	T-test[i]
$ET_{o\text{-}P\text{-}M\text{-}(0,0\,m\,s^{-1})}/ET_{o\text{-}Lysimeter}$	1.10	0.100	True
$ET_{o\text{-}Garcia\_lópez\_mod}/ET_{o\text{-}Lysimeter}$	1.02	0.110	True
$ET_{o\text{-}Hargreaves}/ET_{o\text{-}Lysimeter}$	1.53	0.186	False
$ET_{o\text{-}P\text{-}M\text{-}(2,0\,m\,s^{-1})}/ET_{o\text{-}Lysimeter}$	1.62	0.149	False
$ET_{o\text{-}P\text{-}M\text{-}(4,0\,m\,s^{-1})}/ET_{o\text{-}Lysimeter}$	2.17	0.200	False
$ET_{o\text{-}Garca\_lópez\_mod}/ET_{o\text{-}Hargreaves}$	0.71	0.008	False

NOTE: b = ratio between $ET_{o\,method1}$ to $ET_{o\,method2}$;
S. E. = the standard error.

TABLE 3 Statistical relationship for the ET_o estimated with the *Hargreaves* model (modified), the lysimeter measurements, and *Penman-Monteith* estimation without wind and daily level.

Relation	b	S.E	T-test[7]
$ET_{o-Lysimeter}/ET_{o-Hargreaves-modificated}$	1.14	0.117	True
$ET_{o-P-M\,(0.0\,m.s)}/ET_{o-Hargreaves-modificated}$	1.03	0.016	True

NOTE: b = ratio between $ET_{o\,method1}$ to $ET_{o\,method2}$.; S. E. = the standard error.

19.3.2 RESTRICTIONS FOR APPLICATION OF THE PENMAN-MONTEITH MODEL IN THE COLOMBIAN COFFEE ZONE

The P-M model is highly susceptible at the wind speed changes. In general, most of the weather stations installed in the Colombian coffee zone, do not measure the wind speed and directions at 2-m height, for this reason is not appropriated the ET_o calculation with the P-M model if the wind speed at 2-m height is not measured.

Only the central Colombian coffee zone has the Ángstrom- Prescott coefficients adjusted used in the equation 2 for the net solar radiation estimation base on sunshine readings, and is necessary derivative these coefficients for other locations for get more precize the calculations.

When we compared the solar radiation estimation using the Ángstrom-Prescott relations using the FAO-56 coefficients and those derivative by [10], the difference in the estimation is low with a mean error of 0.53 MJ m^{-2} d^{-1}, equal to 0.22 mm d^{-1} indicating that the difference in the estimation is in the long wave radiation estimation, as is described as follow. How the difference in the short radiation is low, we analyzed the long wave radiation using two models, Eqs. (8) and (9).

$$R_{nl} = \lambda T^4 \left[0.56 - 0.079\sqrt{e_a} \right] \left[0.1 + 0.9\frac{n}{N} \right] \tag{8}$$

The long wave radiation model used for cloudy locations using sunshine readings is the Penman model, Eq. (8), and, the FAO-56 recommended model for the ET_o estimating is presented in the Eq. (9).

$$R_{nl} = \lambda T^4 \left[0.34 - 0.14\sqrt{e_a} \right] \left[1.35\frac{R_s}{R_{so}} - 0.35 \right] \tag{9}$$

where, R_{so} is the solar radiation for clear sky conditions, Eq. (10), Rs is the solar radiation using the Ángstrom- Prescott model first part of the Eq. (2).

$$R_{so} = \left(0.75 + 2x10^{-5} \, Altitude \right) R_a \tag{10}$$

where, The altitude is in meters and Ra is the astronomical radiation.

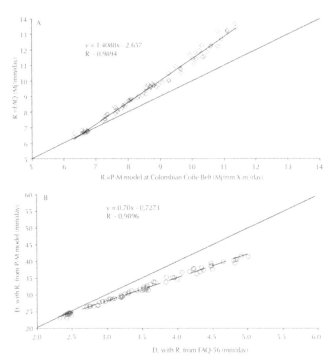

FIGURE 2 (A) Net radiation relationship estimated with the P-M model used in the Colombian coffee zone and the FAO-56 model. (B) ET_o relationship derived from the P-M model and FAO – 56 solar radiation models.

When compared the ET_o estimating with the long wave radiation models from the equations 7 and 8, we can see that the long wave radiation model proposed by FAO-56 overestimate the net radiation up to 2,5 MJ m^{-2} d^{-1} (Fig. 2a) and also the ET_o (Fig. 2b), with difference in the estimation up to 0,9 mm d^{-1}; similar results were reported in China by Liu et al. [17], who used uncalibrated FAO-56 coefficients for the radiation model, reporting variations between 3% to 15% daily ET_o calculation. For these reason is necessary to conduct more field research for the Colombian coffee zone that permit the development of the appropriated coefficients for the long wave radiation estimation and increase the precision of the ET_o calculation.

Faccioli et al. [7] compared the ET_o calculations using the FAO-56 model for net radiation using sunshine readings and direct solar radiation measurements in the Brazilian Coffee zone of Vicosa-Mina Gerais, and they finding that under cloud conditions the ET_o was over estimated with the FAO-56 model.

19.4 CONCLUSIONS

The Garcia and Lopez Modified model and the Hargreaves Modified model area usefully for the ET_o estimation for Colombian Coffee zone without wind or when only the air temperature and relative humidity data are available. The use of sunshine data in the Angstrom-Prescott solar radiation model can be use in the net radiation estimation for the ET_o estimation in the P-M model, the potential source of variability of the estimation could be direct related with the long wave radiation estimation that affect the precision of the net radiation calculation (R_n) and clearly the ET_o estimation. It is necessary conduct studies that permit the long wave radiation coefficients estimation (R_{nl}) in the Colombian coffee zone with the aim to make more precize the ET_o estimation from the P-M model. Also, is necessary study the short wave radiation coefficient for other location of the Colombian coffee zone, especially in that zone were presumably the relationship R_a/R_g differs substantially from those estimated by Gómez and Guzmán [10].

19.5 SUMMARY

The reference evapotranspiration (ET_o) is an important variable for hydrological studies, crop water requirements estimations, climatic zonification and water resources management. The FAO recommends the Penman-Monteith (P-M) and/or the Hargreaves models as the worldwide useful for the ET_o calculation. The objective of this work was to test the performance of these models in one place of the Colombian Coffee Belt, and identify limitations and proposes modifications. The ET_o calculation were compared with daily lysimeter measurements. The principal disadvantages of the P-M model were: the lack of calibrated coefficient for the long wave radiation estimation (R_{nl}), which affected seriously the net radiation estimation and finally the ET_o, highly sensitivity at the wind speed changes, that make it inappropriate for locations without this data. The Hargreaves model, as FAO proposed, overestimate the ETo, which made necessary a modification. The ET_o estimation for this location was most sensible to atmospheric vapor and air temperature than the available energy in the atmosphere $(R_n–G)$.

KEY WORDS

- **astronomical radiation**
- **atmospheric humidity**
- **crop coefficients**
- **crop evapotranspiration**
- **crop water requirements**
- **Drainage Lysimeters**
- **evapotranspiration**
- **FAO-56 approach**
- **Garcia and Lopez**
- **global radiation**
- **Hargreaves**
- **Hargreaves, Garcia and Lopez**

- long wave solar radiation
- lysimeter
- maximum air temperature
- mean air temperature
- minimum air temperature
- net radiation
- Penman-Monteith
- radiation
- reference evapotranspiration
- short wave solar radiation
- temperature

REFERENCES

1. Allen, G. R.; Walter, I. A.; Elliot, R. L.; Howell, T. A.; Itenfizu, D.; Jensen, M. E.; Snyder, R. L.; **2005,** The ASCE standardized reference evapotranspiration equation. Committee of standardization of reference evapotranspiration of the environmental and Water Resources Institute of the American Society of Civil Engineers.
2. Allen, G. R.; Pereira, S. L.; Raes D.; Smith, M.; Crop evapotranspiration: Guidelines for computing crop water requirements. Food and Agricultural Organization of the United Nations (FAO). Publication No. 56. Rome. **1998,** 300 pp.
3. Allen, R. G. Assessing integrity of weather data for reference evapotranspiration estimation. Journal of Irrigation and Drainage Engineering **1996,** *122(2),* 97–106.
4. Alves, L.; Pereira, L. S.; Modeling surface resistance from climatic variables? Agric. Water Manage, **2000,** *42,* 371–385.
5. Barco, J.; A.; Cuartas, O.; Mesa, G.; Poveda, Velez, I. J.; Mantilla, R.; Hoyos, A.; Mejia, F. J.; Botero B.; Montoya, M.; Estimation of evaporation in Colombia [in Spanish]. Avances Recursos Hidráulicos **2000,** *7,* 43–51.
6. Doorembos, J.; Pruitt, W. O.; *Guidelines for predicting crop water requirements.* Food and Agricultural Organization of the United Nations (FAO). Publication No 24. Rome. **1977,** 300p.
7. Faccioli, G. G.; L. Souza, O. C.; Mudrik, A. S.; Mantovani, E. C.; Effects on the solar radiation balance using measured solar radiation and estimated sunshine in the reference evapotranspiration by the Penman-Monteith method [in Portuguese]. **2003,** 560–568. In: II Symposium of Coffee Orchards. Brazil – Viçosa- MG.
8. Garcia, B. J.; Lopez, D. J.; Equation for the potential evapotranspiration estimation adapted for the tropic [in Spanish]. Agronomía Tropical **1970,** *20(5),* 335–345.
9. Giraldo, J. A.; Lince, L. A.; Cuartas, A. F.; Gonzalez. H.; Evaluation of the empirical equations for the potential evapotranspiration estimation [in Spanish]. Fitotecnia **2008,** *141,* 2.
10. Gómez, G. L.; Guzmán, M. O.; Empirical relationship between global solar radiation and the sunshine in the area of cenicafe, Chinchiná, Caldas. [in Spanish]. Cenicafé **1995,***46(4),* 205–218.
11. Howell, T. A. **2004,** Lysimetry. 379–386 pages. In: Hillel, D. (ed.), Encyclopedia of Soils in the Environment. Elsevier Press, Oxford, UK.
12. Jaramillo, R. A. Comparison between pan class A evaporation and estimated evaporation by equations [in Spanish]. Cenicafé **1977,** *28(2),* 67–72.
13. Jaramillo, R. A. Relationship between the evaporation and climate variables [in Spanish]. Cenicafé **1989,** *40(3),* 67–72.

14. Jaramillo, R. A. Reference evapotranspiration in the Andean Colombian zone [in Spanish]. Cenicafé **2006**, *57(4),* 288–298.

15. Jia, X.; Martin, E. C.; Slack, D. C.; Temperature adjustment for reference evapotranspiration calculation in Central Arizona. Journal of Irrigation and Drainage Engineering **2005**, *130(5),* 384–390.

16. Kjelgaard, J. K.; Stockle, C. O.; Villar Mir, J. M.; Evans, R. G.; Campbell, G. S.; Evaluation methods to estimate corn evapotranspiration from short-time interval weather data. Trans. ASAE. **1994**, *37(6),* 1825–1833.

17. Liu, X.; X.; Mei, Y. Li, Q.; Wang, Y.; Zhang, Porter, J. R.; Variation in reference evapotranspiration caused by Ángstrom- Prescott coefficient: Locally calibrated versus the FAO recommended method. Agric. Water Manag. **2009**, *96(7),* 1137–1145.

18. Malone, R. W.; Stewardson, D. J.; Bonta, J. V.; Nelsen T.; Calibration and quality control of the Coshocton weighing lysimeters. Transction of ASAE **1999**, *42(3),* 701–712.

19. Meyers, P. T.; Baldocchi, D. D.; **2005,** Current micrometeorological flux methodologies with applications in agriculture. 381–396. In: Micrometeorology in Agric. Sys. Agronomy Monograph No. 47. American Society of Agronomy, Crop Sci. Society of America, Soil Sci. Society of America, Madison-WI-USA.

20. Ortega-Farias, S. O.; Olioso, A.; Antonioletti, R.; Brisson N.; Evaluation of the Penman- Monteith model for estimating soybean evapotranspiration. Irrig. Sci. **2004**, *23,* 1–9.

21. Ortega-Farias, S. O.; Olioso, A.; Fuentes, S.; Valdes H.; Latent heat flux over a furrow irrigated tomato crop using Penman-Monteith equation with a variable surfaces canopy resistance. Agric. Water. Manage. **2006**, *82,* 421–432.

22. Prenger, J. L.; Fynn, R. P.; Hansen, R. C.; A comparison of four evapotranspiration models in a greenhouse environment. Trans. ASAE. **2002**, *45(6),* 1779–1778.

23. Ramirez-Builes, V. H.; Harmsen, E. W.; Water vapor flux in agroecosystems: methods and models review. In: Labedski, L. (ed.), Evapotranspiration. INTECH Open Access Publisher. **2011,** 3–48.

24. Rana, G.; Katerji, N.; Mastrorilli, M.; El Moujabber, M. A model for predicting actual evapotranspiration under soil water stress in a mediterranean region. Theor. Appl. Climatol. **1997**, *56,* 45–55.

25. Souza, L. O. C.; Faccioli, G. G.; Mudrik, A. S.; Mantovani, E. C.; Comparison of the reference evapotranspiration estimated by the Penman-Monteith and Hargreaves-Samani models with the software SISDA. (in Portuguese). **2003**, 506–511. In: II Simpósio de Pesquisa dos Cafés do Brazil, Viçosa (Brazil).

26. Suárez, V. S. **1998,** Physical characterization of the Risaralda's soils related with the use, management and conservation [in Spanish]. Avances Técnicos-Cenicafé. No 257.

27. Yoder, R. E.; Odhiuambo, L. O.; Wright, W. C.; Effects of the water pressure deficit and the net-irradiance calculation methods on accuracy of standardized Penman-Monteith equation in a humid climate. Journal of Irrigation and Drainage Engineering **2005**, *131(3),* 228–237.

CHAPTER 20

WATER MANAGEMENT FOR AGRONOMIC CROPS IN TRINIDAD[1]

MEGH R. GOYAL

CONTENTS

[1]Modified and printed with permission from: "Goyal, Megh R., 2005. Water consumption by selected crops and climatology: Case study in Trinidad. Corporación Universitaria Santa Rosa de Cabal [UNISARC], Colombia. Boletín: Investigaciones de UNISARC, 4 (1st January): 1–18." © Copyright RevistaInvestigaciones de UNISARC Columbia

20.1 INTRODUCTION

Irrigation provides plants with sufficient water to prevent yield-reducing stress. The frequency of irrigation and quantity of water depend on local climatic conditions, crop species, stage of plant growth and soil-plant-moisture characteristics. The need for irrigation can be determined in several ways that do not require data on rate of evapotranspiration (ET). Visual indicators such as plant color or leaf wilting and an early drop can be used [1]. However, this information appears too late to prevent reduction in crop yield or quality. Other methods of irrigation scheduling include determination of plant water stress, soil moisture status and soil water potential [2]. Methods of estimating crop water requirements using ET in combination with soil characteristics are useful in determining not only when to irrigate but also the quantity of water needed [1]. Estimates of ET have not been used in Trinidad even though necessary climatic data are available. Trinidad's water supply is dwindling because of luxury consumption of ground water resource (e.g., sprinkler irrigation of golf courses). There is an increasing demand of ground water for domestic, municipal and industrial uses. Water quality is declining as well. Thus, water is a limiting factor in Trinidad's goal for self-sufficiency in agriculture. Intelligent use of water will prevent sea water from entering into aquifers. Irrigation of crops in the tropics and on these soils requires appropriate working principles for the effective use of all resources peculiar to the local conditions. Adequate water supply for the entire growing season is essential for the optimum production of crops. The crop water requirements are often provided by both rainfall and irrigation. In places where sufficient rainfall is received throughout the growing period, irrigation is minimal. For good water management and irrigation planning, it is necessary to know the water consumption of crops grown in the project area. Hackbart [7] developed a computer program to estimate net irrigation requirements for various crops in which he has combined information from a modified Blaney-Criddle model [1, 4] together with the USDA-SCS Technical Release No. 21 of USDA-SCS [10]. Irrigation water requirements for vegetables and other agronomic crops have been computed by Rogers et al. [9] for Florida and by Goyal and González-Fuentes [3, 4] for Puerto Rico. They have estimated net irrigation requirements (NIR) with mean monthly rainfall data and with 20% rainfall probability (dry years) data. Methods presented in this chapter to estimate PET and total water consumption can be employed in other countries in the world to develop local data bases. This chapter indicates: (1) Estimation of potential evapotranspiration (PET) with Hargreaves-Samani and modified Blaney-Criddle methods; and (2) Estimation of total water consumption for agronomic crops at five locations in Trinidad in the Caribbean region.

20.2 METHODS AND MATERIALS

The five weather stations were the University of the West Indies, Piarco Airport, Hollis, Navet and Penal [Fig. 1]. The potential evapotranspiration (PET) for these locations was estimated with available climatic data and the Eqs. (1) and (2):

$$PETHS = 0.0023 \times Ra \times (T + 17.8) \times (Tmax - Tmin)^{0.5} \tag{1}$$

$$PETBC = Kt \times H \times P \times (0.46 \times T + 8.128) \tag{2}$$

where, PETHS = potential evapotranspiration with Hargreaves-Samani method [8], mm/day; PETBC = PET with modified Blaney-Criddle model [2, 4–6], mm/day; Ra = extra-terrestrial radiation, mm/day; T = average temperature, °C; TMAX = maximum mean temperature, °C; Tmin = minimum mean temperature, °C; Kt = (0.03114 × T + 0.5222) = temperature coefficient; H = humid area factor of 0.8 for Trinidad; P = monthly percentage of total daylight hours. With values of Kt and H for Trinidad, Eq. (2) reduces to:

$$PETBC = (0.024912 \times T + 0.41776) \times (0.46 \times T + 8.128) \times P \qquad (3)$$

$$= K1 \times K2 \times P \qquad (4)$$

where, $$K1 = (0.41776 + 0.24912 \times T) \qquad (5)$$

$$K2 = (8.128 + 0.46 \times T) \qquad (6)$$

K1 versus T and K2 versus T curves are shown in Fig. 2. K1 and K2 are temperature dependent coefficients [7]. For each monthly value of T, K1 and K2 were interpolated. With monthly K1, K2 and P in Eq. (4), PETBC was determined [7, 10]. Monthly water consumption was estimated for banana, cabbage, cucumber, dry onion, eggplant, grain corn, honeydew, lettuce, plantain, potato, pumpkin, snap bean, wet and dry season rice, snap bean, sugarcane, sweet pepper, sweet potato, tomato and watermelon. Monthly water consumption values (CUm, inches per month) were estimated with climatic data from Trinidad, Hackbart's computer model, USDA-SCS Technical Release No. 21 and the following equations:

$$CUm = Kc \times Kt \times P \times TF \times H \qquad (7)$$

$$CU = CUml + CUm_2 + — + CUm_i \qquad (8)$$

$$NIR = CU - ER \qquad (9)$$

where, CUm = Monthly water consumption for the first, —, last month (inches/month); CU = Total water consumption during the season; Kc = Crop growth coefficient; TF = Mean air temperature, °F; ER= Effective rainfall, inches, (Calculated using Technical Release No. 21); NIR = Net Irrigation for normal years, inches. Monthly water consumption values (CUm) were summed to obtain seasonal water consumption (CU). A net irrigation requirement (NIR) was estimated with Eq. (9).

FIGURE 1 Location of hydrometric stations in Trinidad.

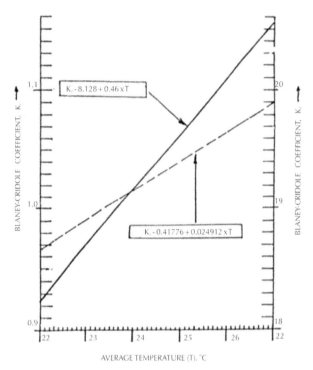

FIGURE 2   Blaney Criddle coefficients, K1 and K2, versus average temperature in Trinidad using USDA – SCS Technical Bulletin No. 21.

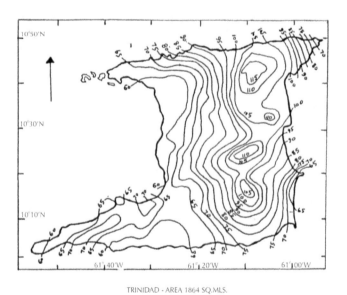

TRINIDAD - AREA 1864 SQ.MLS.
Mean Isohyetal Map

FIGURE 3   Mean annual distribution of rainfall (inches) in Trinidad.

20.3 RESULTS AND DISCUSSIONS

20.3.1 CLIMATOLOGY

All results are shown in Tables 1–4. Mean monthly temperature was highest in May at the University of the West Indies and Piarco Airport; and in August at Hollis, Navet and Penal. Mean monthly temperature was lowest in January at the University of the West Indies, Piarco, Navet and Penal; and in December at Hollis.

TABLE 1 Seasonal irrigation requirement and water consumption for selected crops in Trinidad, using modified Blaney – Criddle method.

Crop		Total CU and NIR, mm/season				
		Location				
		Univ. of W. I.	Piarco	Hollis	Navet	Penal
Banana	CU	1527	1505	1302	1468	1512
	ER	594	611	981	808	577
	NIR	933	894	321	660	935
Cabbage	CU	295	291	257	290	288
	ER	92	100	144	138	95
	NIR	203	191	113	152	193
Cantaloupe	CU	250	946	218	246	246
	ER	91	98	148	138	95
	NIR	159	148	70	108	151
Cucumber	CU	180	177	156	177	177
	ER	78	81	110	110	79
	NIR	102	96	45	67	98
Eggplant	CU	416	410	358	295	402
	ER	105	114	173	70	108
	NIR	311	296	185	225	294
Lettuce	CU	196	194	171	193	193
	ER	79	82	110	108	80
	NIR	177	112	61	85	113
Onion	CU	325	320	282	318	317
(dry)	ER	95	105	162	144	98
	NIR	230	217	120	174	219
Pepper	CU	422	417	364	412	409
(transplanted)	ER	105	114	175	159	109
	NIR	317	303	189	253	300
Plantain	CU	1527	1506	1302	1469	1513
	ER	594	611	982	809	577
	NIR	933	895	320	660	936

TABLE 1   *(Continued)*

Crop		Total CU and NIR, mm/season				
		Location				
		Univ. of W. I.	Piarco	Hollis	Navet	Penal
Potato	**CU**	483	474	416	474	475
	ER NIR	164	169	251	241	169
		319	305	165	233	306
Pumpkin	**CU**	383	378	330	374	371
	ER	103	113	171	155	107
	NIR	280	265	159	219	264

Crop		Total CU and NIR, mm/season				
		Location				
		Univ. of W. I.	Piarco	Hollis	Navet	Penal
Rice	CU	539	532	464	526	522
(dry)	ER	126	135	229	189	131
IR8	NIR	413	397	235	337	391
Rice	CU	550	537	464	532	555
(wet)	ER	263	262	411	353	256
IR8	NIR	287	275	53	179	299
Rice	CU	411	404	356	402	400
(dry)	ER	110	119	205	167	113
IR747-B26	NIR	301	285	151	235	287
Rice	CU	440	435	377	418	438
(wet)	ER	240	250	342	309	223
IR747-B26	NIR	200	185	35	109	215
Snap	CU	293	289	254	287	286
bean	ER	92	100	142	137	95
	NIR	201	189	112	150	191
Sugar-	CU	1559	1535	1330	1503	1541
cane	ER	585	595	945	795	568
	NIR	974	940	385	708	973
Sweet	CU	301	296	262	295	294
corn	ER	93	101	151	141	96
	NIR	208	195	111	154	198
Sweet	CU	467	586	507	575	572
potato	ER	107	136	225	188	132
	NIR	360	450	282	385	440

TABLE 1 *(Continued)*

Crop		Total CU and NIR, mm/season				
		Location				
		Univ. of W. I.	Piarco	Hollis	Navet	Penal
Tomato	CU	441	434	380	430	428
	ER	107	116	182	164	111
	NIR	334	318	198	266	317
Water-	CU	386	381	334	378	246
melon	ER	104	113	174	159	95
	NIR	282	268	160	219	151

1/ mm/season CU = water consumption; ER = effective rainfall; NIR = net irrigation requirement = [CU – ER].
Note: Univ. of West Indies is located at St. Augustine (Fig. 1).

TABLE 2 Factors that affect irrigation requirements at five locations in Trinidad.

Description	Location				
	Univ. W. I.	Piarco	Hollis	Navet	Penal
Geographical Factors					
Weather Station no.	NA	9:32	2:3	3:8	7:7
Latitude	10°58'N	10°35'N	10°41'N	10°24'N	10°10'N
Longitude	61°23'W	61°21'W	61°11'W	61°15'W	61°27'W
Elevation above sealevel, mm	15	11	150	122	12
Soil Factors					
Soil type	Clay loam	Fine sand	Clay loam	Clay loam	Clay loam
AWC depth, in	3.6	3.6	3.8	3.6	3.0
Allowable depletion	50%	50%	50%	50%	50%
Crop Factors					
Crops	Planting	Last Harvest	Crops	Planting	Last Harvest
Banana	Jan 01	Dec 31	Rice IR_8, dry	Dec 01	Apr 11
Cabbage	Dec 01	Feb 28	IR_8, wet	Sep 01	Jan 02
Cantaloupe	Dec 01	Feb 15			
Cucumber	Dec 01	Jan 31	IR747 B2– 6dry	Dec 01	Mar 08

TABLE 2 *(Continued)*

Description	Location				
	Univ. W. I.	Piarco	Hollis	Navet	Penal
Eggplant	Jan 01	Mar 31	IR 747B2–6 wet	Jun 01	Sep 04
Lettuce	Dec 01	Jan 31	Snap bean	Dec 01	Feb 28
Onion	Dec 01	Feb 28	Sugar cane	Jun 01	May 31
Pepper	Dec 01	Mar 31	Sweet corn	Dec 01	Feb 28
Plantain	Jan 01	Dec 31	Sweet potato	Dec 01	Apr 30
Potato	Nov 01	Mar 03	Tomato	Dec 01	Mar 31
Pumpkin	Dec 01	Mar 31	Water melon	Dec 01	Mar 31

NA = Not Available.

TABLE 3 Percentage day light hours at five locations in Trinidad and Tobago.

Location	Mean day light hours percentage											
	Month											
	Jan	Feb	Mar	Apri	May	Jun	Jul	Aug	Sep	Oct	Nov	Dec
1. Univ. of the West Indies	8.68	8.89	9.95	8.73	8.73	7.39	8.33	7.98	7.39	7.98	7.84	8.10
2. PiarcoAirport	9.00	8.79	9.55	9.10	9.12	7.15	8.16	8.19	7.69	7.79	7.56	7.89
3. Hollis	7.95	7.96	10.5	7.64	8.82	7.20	8.99	9.81	8.52	7.67	7.24	7.72
4. Navet	9.36	8.96	8.76	8.98	8.39	8.31	8.22	8.33	7.73	7.56	7.66	7.75
5. Penal	8.20	9.46	8.57	9.74	9.28	7.53	8.12	8.66	7.52	7.09	6.68	8.29

TABLE 4 Potential evapotranspiration (PET) in mm/day with modified Blaney – Criddle (ETBC) and Hargreaves-Samani (ETHS) models at five locations in Trinidad.

Location	Potential evapotranspiration, mm/day												
	Month												
	PET	Jan	Feb	Mar	Apr	May	Jun	Jul	Aug	Sep	Oct	Nov	Dec
1. University of the West Indies	ETHS	3.82	4.15	4.58	4.52	4.52	3.73	4.30	4.01	4.56	4.33	3.92	3.60
	ETBC	5.67	6.55	6.75	6.19	6.28	5.37	5.83	5.61	5.47	5.69	5.67	5.46
	Epan	ND	ND	ND	ND	ND	ND	ND	ND	ND	ND	ND	ND
2. Piarco Airport	ETHS	3.87	4.24	4.66	4.78	4.53	3.69	4.23	4.40	4.46	4.21	3.89	3.59
	ETBC	5.80	6.37	6.46	6.61	6.53	5.14	5.66	5.69	5.59	5.41	5.32	5.17
	Epan	5.00	6.00	6.80	7.00	6.70	5.30	5.50	5.40	5.40	5.00	4.60	4.50

TABLE 4   *(Continued)*

Location							Month						
	PET	Jan	Feb	Mar	Apr	May	Jun	Jul	Aug	Sep	Oct	Nov	Dec
3. Hollis	ETHS	3.55	3.91	4.32	4.49	4.43	3.92	4.54	4.68	4.72	4.44	4.09	3.67
	ETBC	4.71	5.14	6.11	4.77	5.43	4.51	5.48	6.12	5.50	4.65	4.49	4.46
	Epan	3.10	3.60	4.00	4.30	3.50	3.40	3.50	4.80	3.60	3.50	3.00	2.90
4. Navet	ETHS	4.26	4.21	4.41	4.53	4.47	3.99	4.43	4.49	4.75	4.15	3.97	3.60
	ETBC	5.96	6.50	5.74	6.29	5.77	5.82	5.57	5.57	5.42	5.17	5.37	5.16
	Epan	4.00	4.80	5.00	5.70	5.30	4.40	4.30	4.90	4.40	4.10	4.00	4.10
5. Penal	ETHS	4.26	4.75	5.19	5.34	5.13	4.29	4.39	4.56	4.69	4.43	4.05	3.82
	ETBC	5.25	6.72	5.61	6.83	6.45	5.41	5.64	6.22	5.58	5.09	4.83	5.45
	Epan	4.10	4.90	5.30	5.90	5.20	4.70	4.20	4.50	3.90	3.80	3.00	3.20

Note: EPAN = Class A pan evaporation; ND = No data available; PET = Potential evapotranspiration.

Average monthly temperature range (°C) was 24.8 to 26.9° at the University of the West Indies; 24.6 to 26.7° at Piarco Airport; 22.3 to 23.9° at Hollis; 24.3 to 25.9 at Navet; and 24.3 to 26.8° at Penal. Mean annual distribution of rainfall (inches/year) in Trinidad is shown in Fig. 3. West and south coasts are driest, and central Trinidad is wettest. In Trinidad, January through April is dry season and May through December is wet season. March is the driest month. Wettest month is August at the University of the West Indies; July at Piarco and Hollis; June at Navet and Penal. Mean monthly rainfall range (mm/month) was 32.8 to 241.6 at the University of the West Indies; 35.8 to 263.8 at Piarco; 90 to 411 at Hollis; 56 to 343 at Navet; 35 to 228 at Penal. Mean monthly relative humidity (%) ranges from 71 to 87 during the year. The wind speed ranking was Piarco > Navet > Hollis > Penal. Mean monthly daylight hour percentage ranged between 7.39 to 9.95 at University of the West Indies, 7.15 to 9.55 at Piarco, 7.20 to 10.48 at Hollis, 7.56 to 9.36 at Navet, 6.68 to 9.74 at Penal (Table 3).

20.3.2   POTENTIAL EVAPOTRANSPIRATION WITH HARGREAVES-SAMANI AND MODIFIED BLANEY – CRIDDLE METHODS

Monthly PET (mm/day) with Blaney-Criddle (ETBC), Hargreaves and Samani (ETHS) methods and class A pan evaporation (Epan, mm/day) at five locations in Trinidad are shown in Table 4. Epan is minimum in December at Piarco and Hollis; and in November at Navet and Penal. Epan (mm/day) ranged between 4.5 to 7.0 at Piarco, 2.9 to 4.8 at Hollis, 4.0 to 5.7 at Navet, 3.0 to 5.9 at Penal during the year. At all locations during the year, ETBC is greater than ETHS. At Hollis, Navet and Penal, monthly ETBC was always higher than Epan. At Piarco, monthly ETHS were lower than Epan. ETHS range (mm/day) was 3.60 to 4.58 at the University of the West Indies; 3.59 to 4.75 at Navet; 4.05 to 5.34 at Penal. ETBC range (mm/day) was 5.37 to 6.75 at the University of the West Indies; 5.14 to 6.61 at Piarco; 4.46 to 6.12 at Hollis; 5.16 to 6.50 at Navet; and 4.83 to 6.83 at Penal.

20.3.3 SEASONAL NET IRRIGATION REQUIREMENTS

Table 1 gives the total water consumption (CU) and seasonal net irrigation require-ments for normal years (NIR) at five locations in Trinidad. Among these five locations, CU range (mm/season) was 1302 to 1527 for banana; 257 to 295 for cabbage; 156 to 180 for cucumber; 282 to 325 for dry onion; 295 to 416 for eggplant; 262 to 301 for grain corn; 218 to 250 for honeydew; 171 to 196 for lettuce; 1302 to 1527 for plantain; 403 to 483 for potato; 330 to 383 for pumpkin; 400 to 539 for dry season rice; 377 to 555 for wet season rice; 254 to 293 for snap bean; 1330 to 1559 for sugarcane; 476 to 586 for sweet potato; 380 to 441 for tomato; 364 to 422 for transplanted sweet pepper; and 246 to 386 for watermelon, respectively. Higher values belong to the University of the West Indies, and lower values belong to the Hollis location.

Average daily CU range (mm/day) was 3.6 to 4.2 for banana; 2.9 to 3.3 for cab-bage; 2.5 to 2.9 for cucumber; 3.1 to 3.6 for dry onion; 3.2 to 4.5 for eggplant; 2.9 to 3.4 for grain corn; 2.8 to 3.2 for honeydew; 2.8 to 3.2 for lettuce; 3.6 to 4.2 for plantain; 2.7 to 3.2 for potato; 5.3 to 6.2 for pumpkin; 3.0 to 4.2 for rice dry season; 3.6 to 4.5 for wet season rice; 2.8 to 3.3 for snap bean; 3.6 to 4.3 for sugarcane; 5.9 to 6.8 for sweet pepper; 3.1 to 3.9 for sweet potato; 3.1 to 3.6 for tomato; and 2.0 to 3.2 for watermelon. In Trinidad net irrigation requirement (NIR, mm/season) ranged between 321 and 935 for banana; 113 to 203 for cabbage; 45 to 102 for cucumber; 120 to 230 for dry onion; 185 to 311 for eggplant; 111 to 208 for grain corn; 70 to 159 for honeydew; 61 to 117 for lettuce; 320 to 936 for plantain; 165 to 319 for potato; 159 to 280 for pumpkin; 151 to 413 for dry season rice; 35 to 299 for wet season rice; 112 to 201 for snap bean; 385 to 974 for sugarcane; 282 to 450 for sweet potato; 198 to 334 for tomato; 189 to 317 for transplanted sweet pepper; and 151 to 282 for watermelon. Average daily NIR (mm/day) range was 0.9 to 2.6 for banana; 1.3 to 2.3 for cabbage; 0.7 to 1.6 for cucumber; 1.3 to 2.6 for dry onion; 2.0 to 3.3 for eggplant; 1.2 to 2.3 for grain corn; 0.9 to 2.1 for honeydew; 1.0 to 1.9 for lettuce; 0.9 to 2.6 for plantain; 2.7 to 4.5 for pumpkin; 1.1 to 3.1 for dry season rice; 0.3 to 2.4 for wet season rice; 1.2 to 2.2 for snap bean; 1.1 to 2.7 for sugarcane; 3.0 to 5.1 for sweet pepper; 1.9 to 3.0 for sweet potato; 1.6 to 2.8 for tomato; 1.2 to 2.3 for watermelon. These CU, NIR values were estimated with Kc from climatic areas similar to those of Trinidad. Experimental work to determine Kc in Trinidad is lacking.

20.4 SUMMARY

This chapter discusses climatic data of five weather stations in Trinidad and potential evapotranspiration with Hargreaves-Samani and modified Blaney-Criddle methods, and presents seasonal estimates of net irrigation requirements and total water con-sumption by 20 vegetable crops. Potential evapotranspiration range (mm/day) was 3.60 to 4.6 at the University of West Indies, 3.6 to 4.8 at Piarco, 3.6 to 4.7 at Hollis, 3.6 to 4.8 at Navet, and 4.1 to 5.3 at Penal. For the island of Trinidad ET varies from 3.6 to 5.3 mm/day.

Seasonal net irrigation requirements for various vegetable crops are useful in the design and management of irrigation systems in Trinidad. Based on net irrigation re-quirements, farmer/extensionist/investigator can determine gross irrigation require-ment for micro, sprinker – and gravity irrigation systems for each vegetable crop. This chapter provides basic ET information to the irrigator. Lysimeter studies should be conducted to determine crop coefficients in Trinidad, and to verify NIR values of this study. Blaney-Criddle coefficients for Trinidad are:

$$K_1 = 0.41776 + 0.024912 \times T \tag{10}$$

$$K_2 = 8.128 + 0.46 \times T \tag{11}$$

where, T is average temperature in °C; K_1, K_2 are the regression coefficients. Using an isohyetal map of Trinidad, one can find mean annual distribution of rainfall (inches) for any location in Trinidad. Similar results and data base can be developed for other countries in the world using methods presented in this chapter.

KEYWORDS

- agroclimatology
- agronomic crops
- Blaney Criddle method
- Blaney Criddle modified method
- crop water consumption
- evapotranspiration
- Hargreaves-Samani method
- Isohyetal map
- net irrigation requirement
- Puerto Rico
- temperature coefficients
- Trinidad
- United States Department of Agriculture [USDA]
- University of West Indies
- USDA Soil Conservation Service [USDA-SCS]
- vegetable crops

REFERENCES

1. Doorenbos, J.; Pruitt, W. O. **1977,** *Crop Water Requirements.* FAO Irrigation and Drainage Paper *24,* Food and Agriculture Organization of the United Nations, Rome.
2. Goyal, M. R.; **1989,** *Estimation of monthly water consumption of selected vegetable crops in the semiarid and humid regions of Puerto Rico,* Unpublished report by Agricultural Experiment Station, UPR Library of Congress Catalog #88-51661 and ISBN 0-962105-0-5. 454 pp.
3. Goyal, M. R.; González-Fuentes, E. A. **1989,** *Climatological Data for Seven Agricultural Experiment Substations in Puerto Rico.* Mimeographed Report, 75 pp.
4. Goyal, M. R.; González-Fuentes, E. A. Water requirements for vegetable production in Puerto Rico. Proceedings of National Irrigation and Drainage Symposium, Am. Soc. Civil Engineers, Lincoln-NE, **1988,** July 19–22.

5. Goyal, M. R. y González-Fuentes, E. A. Irrigation requirements for plantains in seven ecological regions of Puerto Rico [Spanish]. *J. Agric. Univ. P.R.,* **1989,** *72(4),* 599–608.
6. Goyal, M. R.; González-Fuentes, E. A. Monthly consumptive use of rice in semi-arid and humid regions of Puerto Rico. *J. Agric. Univ. of P.R.* **1989,** *73(1),* 31–44.
7. Hackbart, C. A. *Consumptive Use Computer Model and SCS Users Guide.* USDA-SCS, Forth Worth, TX; **1987,** 55 pp.
8. Hargreaves, G. H.; Samani, Z. A. Reference crop evapotranspiration from temperature. Appl. Eng. Agric.; ASAE **1985,** *1(2),* 96–9.
9. Rogers, J. S.; Harrison, D. S. **1977,** Irrigation water requirements for agronomic crops in Florida. Tech. Series WRC–5 by Agricultural Engineering Department, University of Florida, Gainesville-FL, Pages 1–2.
10. USDA-SCS, **1983,** *Technical Release No. 21.* United States Department of Agriculture-Soil Conservation Service, Washington D. C.

CHAPTER 21

CROP WATER STRESS INDEX FOR COMMON BEANS[1, 2]

VICTOR H. RAMIREZ BUILES, ERIC W. HARMSEN,
and TIMOTHY G. PORCH

CONTENTS

[1]Victor H. Ramirez Builes, Agroclimatology and Crop Science Researcher, National Coffee Research Center (Cenicafe), Chinchina (Caldas, Colombia). Email: <victor.ramirez@cafedecolombia.com>; Eric W. Harmsen. Professor, Department of Agricultural and Biosystems Engineering, University of Puerto Rico. Mayagüez, PR 00681, USA. Email: <harmsen1000@hotmail.com>; Timothy G. Porch, Genetics Researcher, Tropical Agricultural Research Station, USDA-ARS, Mayagüez - Puerto Rico, USA. <Timothy.Porch@ars.usda.gov>.
[2]Numbers in parentheses indicate the references in the bibliography.

21.1 INTRODUCTION

Common bean is highly susceptible to drought stress or water deficit, and the production of this crop in many places of the world is carried out under drought stress conditions, due to insufficient water supply by rainfall and/or irrigation. Drought stress influences several important plant processes, including plant water potential and stomatal behavior, which have direct effect on gas exchange. Changes in plant water status are directly related to the temperature crop canopy.

Permanent and intermittent drought stress adversely influences crop yield and growth. Methods for drought stress detection have been developed in a number of crops with different technology. Jones [19] described methods for drought detection and irrigation scheduling. The most popular methods are the thermal methods, which have been widely used with the aim of detecting drought stress and improving water management [4, 11, 13, 18, 20, 25, 26]. Measuring the canopy temperature by infrared thermometry is a popular technique because it is noninvasive, nondestructive and can be automated [6]. One of these methods that has been successfully applied since the 70's to detected drought stress uses the change in canopy temperature with respect to air temperature [12, 17].

One of the most widely used methods is the crop water stress index (CWSI) was proposed by Idso et al. [13]. The CWSI relates the difference between canopy and air temperature with the vapor pressures deficit. The index ranges from 0, (for nondrought stress conditions) to 1 (for maximum drought stress: a condition when water is not available for transpiration). The CWSI generates two baselines: an upper baseline for complete drought stress and a lower baseline for no drought stress, both curves being functions of the vapor pressure deficit (VPD). Jackson et al. [18] have shown that the CWSI is an index of transpiration reduction, as shown in Eq. (1), where, E is the actual evapotranspiration and E_p is the potential evapotranspiration or transpiration.

$$[(CWSI) = 1 - (E/E_p)] \tag{1}$$

The CWSI can be applied in the analysis of the irrigation scheduling. The CWSI has been increasing used recently due to the availability of the infrared thermometer, satellite thermal imaging and other remote sensing tools, that could be used in the detection of crop water stress at the macro and micro scales. Idso [14] determined the baseline model for several crops, including bean, as shown in Eq. (2). Erdem et al. [7] reported a lower baseline for a *P. vulgaris* L., cv., 'Sehirali 90,' as shown in Eq. (3). In Eqs. (2) and (3), T_c = Canopy temperature, T_a = Air temperature, and VPD = Vapor pressure deficit.

$$[T_c - T_a] = [(-2.35 \times VPD) + 2.91] \tag{2}$$

$$[T_c - T_a] = [(-2.69 \times VPD) + 3.53], \text{ lower baseline model} \tag{3}$$

Plant water status is a function of the available water in the soil. Water availability in plant tissues varies by cultivar, and genotype, which is directly related to water potential and stomatal control. This suggests that baseline models strongly depend on site location, crop species and variety [3, 8]. More research is needed in the development of baselines for the CWSI method.

The objectives of this chapter are to present the research results: (1) Develop baselines for different common bean with and without drought susceptibility in greenhouse and field environments; (2) Estimate the CWSI for common bean genotypes in greenhouse and field environments and relate the CSWI with yield components; (3) Relate the CWSI with the available soil water as a tool for crop water management; and (4) Detect the variability of these relationships among genotypes with and without drought stress.

21.2 METHODS AND MATERIALS

This experiment was conducted in a span of three years (2005, 2006 and 2007) and included several growing seasons in the greenhouse and field.

21.2.1 THE GREENHOUSE TRIALS

The greenhouse experiment was carried out at the USDA-TARS (Tropical Agric. Res. Station) in Mayagüez- Puerto Rico; coordinates 18° 12'22' N., 67° 8' 20" W at 18 meters of elevation above the sea level (m.a.s.l). Four trials were conducted during July September of 2005, and two during October–December of 2005 and 2006. Basic weather information was recorded during the study in the greenhouse (*see* Table 1). The common bean genotypes (*P. vulgaris* L.) evaluated in this research in the greenhouse were: Morales (white seed color), the most widely planted small white bean in Puerto Rico, SER 16 and SER 21 (red seed color), SEN 3 and SEN 21 (black seed color), and BAT 477 (cream seed color). BAT 477 has a plant architecture type III and the others are type II. Morales with unknown drought response and the other are drought tolerant. The experiments were conducted during 2005 and 2006 (Table 1). All pots were irrigated manually every day in the morning with desired amounts of water. Three water levels were used:

Level 1 – full water supply (no drought stress) using 80% of the daily available water (DAW) during the complete growing season;

Level 2 – stress 1 with 50% of the DAW before flowering and 60% of the DAW after flowering; and

Level 3 – stress 2 with 20% of the DAW before flowering and 40% of the DAW after flowering.

TABLE 1 Average weather conditions in the greenhouse during the 2005 and 2006 crop seasons, nd = no data was measured.

Weather Variables	July September	October–December
2005		
Air Temperature (°C)	27.55	26.06
Air Relative humidity (%)	84.29	82.53
Solar radiation (W.m²)	nd‡	nd
Wind speed (m.s⁻¹)	nd	nd
2006		
Air Temperature (°C)	26.90	26.58
Air Relative humidity (%)	78.75	77.35
Solar radiation (W.m²)	57.90	61.15
Wind speed (m.s⁻¹)	0.0088	0.0089

The DAW is defined as the total water required to keep the moisture at substrate field capacity (SFC). Volumetric moisture content was measured with a volumetric moisture sensor [theta probe soil moisture sensor, +/–0.01 $m^3.m^{-3}$ ML2X by Delta-T Devices Ltd.], and it was measured at different growth stages during the each season. A SFC test was previously carried out to estimate the total daily water needs for each pot and the measured value was 0.53 m^3 m^{-3} (+/– 0.010). At no time during the experiments did the soil moisture content reach the terminal drought stress level. In the greenhouse experiment, the total water applied per pot, was recorded daily.

21.2.2   THE FIELD TRIALS

The field experiments were conducted at the Agricultural Experimental Station of the University of Puerto Rico in Juana Diaz, PR, which is located in south coast of PR, at 18°0′N. latitude and 66°22′W longitude, elevation 21 m.a.s.l, within a semiarid climatic zone [9]. With an average annual rainfall of 838 mm, and the average rainfall is only 19.8 mm for January, 18.3 mm for February and 21.8 mm for March, respectively [USDA, 1979]. The annual average, minimum and maximum air temperatures are 26.22°C, 21.33°C, and 31.05°C, respectively. The daily average reference evapotranspiration values are 4.3, 3.4 and 5.5 mm per day [10]. The bean genotypes (*P. vulgaris* L.) evaluated in this research in greenhouse were: Morales and SER 16. The experiments were conducted during crop season of 2006 and 2007.

The field experiments were planted on February 15, 2006 and January 17, 2007. The soil at the UPR Agricultural Experiment is classified as a San Anton Clay Loam with 30% sand, 44% silt, 26% clay, and 1.28% of organic matter within the first 40 cm; with 0.30 $cm^3.cm^{-3}$ field capacity and 0.19 $cm^3.cm^{-3}$ wilting point [27]. One intermittent drought stress level was applied in both years at the beginning of the reproductive phase (R1: One blossom open at any node) to harvest. The drought stress was sufficient to allow the soil to dry to 75% of field capacity (FC), when the irrigation was applied. The stress level in 2006 was 18% corresponding to 387.3 mm of water applied as compared to the 472.5 mm total applied under the nondrought stress treatment; and in 2007, the stress level was 30.3% corresponding to 302.0 mm of water applied as compared to the 433.4 mm total applied under the nondrought stress treatment (Table 2).

TABLE 2 Irrigation dates and volumes of the various treatments at Juana Diaz- PR during 2006 and 2007.

Date	Growing Stage	Without Drought stress	With drought stress	Rainfall (mm)
		Irrigation (mm)		
Year 2006				
14 Feb	V_1	21.0	19.4	3.1
17 Feb	V_2	18.8	19.9	7.1
22 Feb	V_3	30.9	31.6	2.7
25 Feb	V_4	3.4	3.5	0.0
27 Feb	V_5	12.4	12.1	0.0

TABLE 2 *(Continued)*

Date	Growing Stage	Without Drought stress Irrigation (mm)	With drought stress	Rainfall (mm)
3 Mar	V^8	19.5	20.0	0.0
11 Mar	R_1*	15.3	0.0	56.1
14 Mar	R_2	24.1	6.4	0.0
16 Mar	R_2	0.0	5.1	34.0
25 Mar	R_4	22.3	0.0	37.3
29 Mar	R_5	32.8	16.6	2.6
8 Apr	R_8	8.4	0.0	106.2
11 Apr	R_9	14.5	3.6	0.0
Total	-----	**223.4**	**138.2**	**249.1**
Water deficit level	-----	-----	**18.0%**	----
Year 2007				
24 Jan	V_1	9.7	8.2	0.0
31 Jan	V_2	21.9	15.3	0.0
1 Feb	V_2	0.0	22.8	0.0
5 Feb	V_3	25.0	25.7	0.0
7 Feb	V_3	26.0	22.5	0.0
13 Feb	V_4	40.3	14.2	0.0
15 Feb	V_5	27.3	29.1	0.0
21 Feb	V_6	24.7	21.2	1.5
24 Feb	R_1*	10.8	0.0	0.0
26 Feb	R_2	12.7	0.0	0.0
1 Mar	R_3	29.9	10.1	0.0
5 Mar	R_4	34.2	22.5	0.0
6 Mar	R_4	0.0	9.3	0.0
9 Mar	R_5	60.2	19.6	0.0
12 Mar	R_6	27.3	13.2	0.0
15 Mar	R_6	31.9	0.0	0.7
20 Mar	R_7	15.4	15.4	0.4
23 Mar	R_8	14.5	8.6	19.7
28 Mar	R_8	0.0	0.0	13.9

TABLE 2   *(Continued)*

Date	Growing Stage	Without Drought stress Irrigation (mm)	With drought stress	Rainfall (mm)
30 Mar	R$_9$	0.0	0.0	17.5
Total	---	**411.8**	**257.7**	**53.7**
Water deficit level	----	------	**30.3%**	----

Foot Notes for table 2 [NDSU, 2003]:
*Drought stress beginning
V1: Completely unfolded leaves at the primary leaf node;
V2: First node above primary leaf node;
V3: Three nodes on the main stem including the primary leaf node. Secondary branching begins to show from branch of V1;
Vn: n-nodes on the main stem including the primary leaf node;
R1: One blossom open at any node;
R2: Pods at ½-long at the first blossom position.
R3: Pods at 1 inch long at first blossom position;
R4: Pods 2 inches long at first blossom position;
R5: Pods 3 plus inches long, seeds discernible by feel;
R6: Pods 4.5 inch long spurs (maximum length). Seeds at least ¼ inch long axis;
R7: Oldest pods have fully developed green seeds. Other parts of plant will have full-length Pods with seeds near same size;
R8: Leaves yellowing over half of plant, very few small new pod/blossom developing, small pods may be drying. Points of maximum production has been reached;
R9: Mature, at least 80% of the pods showing yellow and mostly ripe.

The volumetric moisture content was measured with a profile probe type PR2 sensor (Delta-T Devices, Ltd.). Two access tubes were installed in each main plot at 20 cm and 40 cm depths, and the irrigation was applied two times per week, using a drip irrigation system. Each main plot was divided into six subplots, which consisted of each genotype, two subplots (each with 10 rows) for SER 16 and three for Morales in 2006, and three for each one in 2007. The plant density was 13.5 plants-m^{-2} for Morales and 6.5 plants-m^{-2} for SER 16. The growing period was of 75 day in 2006 and 78 days in 2007. The other agronomic practices related to the crop were similar in the whole experiment and carried out at the same time, which include 560 pounds per hectare of NPK (16-4-4), weed control, and pest management.

The soil water balance was monitored daily to estimate the actual soil moisture (ASM) using Eq. (4), below:

$$ASM_{initial} = TAW - ET_c \qquad (4)$$

If [MR + ASM$_{initial}$] < TAW, then: ASM = MR + ASM$_{initial}$, and

If [MR + ASM$_{initial}$] > TAW, then: ASM = TAW

where, *TAW* is total available water; *ET$_c$* is crop evapotranspiration; and *MR* is the moisture recharge. The *TAW* was calculated using Eq. (5):

$$TAW = 1000(\theta_{FC} - \theta_{WP})Z_r \qquad (5)$$

where, θ_{FC} is a volumetric moisture content at field capacity and θ_{WP} is a wilting point, respectively, in $m^3.m^{-3}$; and Z_r is the root depth in meters. The moisture recharge was estimated from the following Eq. (6):

$$MR = ASM + R + I - RO - ET_c \qquad (6)$$

where, R is rainfall depth, I is irrigation depth, RO is runoff depth. The RO was measured in 12 drainage lysimeters and ET_c was estimated using the Penman-Monteith "one-step" model [1, 21, 22], with direct measurement of canopy and aerodynamic resistances during the whole growing season. For this purpose four automatic weather stations were located within the experimental plots as follows: genotype Morales with and without drought stress genotype SER 16 with and without drought-stress. Each weather station was equipped with Kipp and Zonen B.V. net radiometer (spectral range 0.2–100 µm), wind direction and wind speed with wind sensor-Met one 034B-L at 2.2 m; air temperature and relative humidity with HMP45C temperature and relative humidity probe at 2.0 m; soil temperature with TCAV averaging soil thermocouple probe at 0.08 m and 0.02 m depth, soil heat flux using soil heat flux plates at 0.06 m depth; and a volumetric soil moisture content with a CS616 water content reflectometer at 0.15 m depth. Six data points per minute were collected for each sensor and stored in a CR10X data logger (Campbell scientific, Inc.).

The stomatal resistance (r_L) was measured several times during the day from 7:00 to 17:00 in order to obtain a reasonable average value for each phenological growing phase, for each genotype and water level. Two leaf porometers were used: an AP4-UM-3 (Delta-T Devices Ltd.) during 2005 and a model SC-1 (Decagon Devices, Inc.) during 2006; and the readings were taken once per week. The leaf area index [LAI] was estimated using a nondestructive method described previously by Ramírez et al. [23]. Then according to the plant density, the LAI was estimated on a weekly basis. The aerodynamic resistance (r_a) was estimated with the Perrier equation [1, 5].

The soil in the lysimeter was encased in round polyethylene containers with an exposed soil surface of 0.22-m^2 and 0.8-m depth. The containers were sufficiently deep to accommodate the plant roots. The lysimeters were located within plots measuring 7-m wide by 61-m long, with the long dimension oriented in the direction of the prevailing wind. Daily rainfall was measured for each lysimeter with a manual rain gauge, and compared with an automated tipping bucket rain gauge (WatchDog™ Spectrum Technology, Inc.) located within the reference conditions. The irrigation was measured using a cumulative electronic digital flow meter (GPI, Inc.) and was recorded manually at the beginning and end of each irrigation event every three or four days. Two flow meters were placed on the irrigation supply lines, one on the well-watered treatment supply line and the second on the drought stress treatment water supply line. Water from RO and DP was removed from the collection containers periodically by means of a small vacuum pump (Shurflu-4UN26, 12 V, 4.5GPM). The depth of water in the soil profile was related to the soil moisture content using Eq. (7):

$$Si = \Sigma \ [(\theta v, i0 \ cm \ Z0 \ cm) + (\theta v, i10 \ cm \ Z10 \ cm) + ... + (\theta v, i60 \ cm \ Z60 \ cm)] \qquad (7)$$

where, S_i is the depth of soil water on day i [mm], $\theta_{v,i}$ is the volumetric soil moisture content on day i and Z is the thickness of the soil layer. Volumetric soil moisture was measured using a profile probe type PR2 sensor (Delta-T Devices, Ltd.) and measurements were obtained for each 10 cm depth interval.

To relate CWSI and yield components for both environments, an average value of CWSI estimated at 13:00 hour was used. This was the time of day when the water stress is likely to be highest, and when the need for irrigation using CWSI should be determined [16].

In the field experiment, the canopy temperatures were recorded on clear sky days, during the day of the year (DOY): 48 to 98 in 2006 and 31 to 71 in 2007. These days included vegetative and reproductive growing stages, similar to the periods evaluated in the greenhouse experiments. The lower (nonstressed) and upper (stressed) baselines (Fig. 1) were measured for each common bean genotype at different vapor pressure deficits and canopy temperature levels. Additionally, the lower base-line was estimated for each genotype for both environments from data for clear sky days for the treatments without drought stress.

For both environmental conditions: The leaf temperature was measured at different development stages and at different time interval during the day (7:00 a.m. to 6:00 p.m.). The canopy temperature (T_c) was measured using an infrared thermometer gun (MX4-TD +/–1°C, Raytek), a spectral range of 8 to 14 µm, and a resolution of 0.1°C. The measuring was made on a single leaf within the upper canopy structure. An automatic weather station (WatchDog-900ET, Spectrum Technologies, Inc.) was installed in the greenhouse and in the field. In addition to the weather station, the air temperature and absolute and relative humidity were measured in the greenhouse with a Hobo-Pro data-logger (Onset Computer Company, Pocassette, Maine). T_c was measured two times per replication in both environments. The crop water stress index (CWSI) was calculated using Eq. (8):

$$CWSI = \frac{\left[dT - dT_{low}\right]}{\left[dT_{up} - dT_{low}\right]} \tag{8}$$

where, dT is the measured difference between the temperatures of the crop canopy and air; dT_{low} is the measured difference between temperature of the canopy for well-watered crop and air temperature (lower baseline); and dT_{up} is the difference between the canopy temperature for nontranspiring crop and air temperature (upper baseline).

The data were subjected to the analysis of variance procedure for linear models to determine the relationship between (T_c-T_a) and VPD; and the relationship between CWSI and yield components, using Infostat statistical program version 3 and Sigma-Plot® program version 802, SPSS.

21.3   RESULTS AND DISCUSSION

21.3.1   THE BASELINES

The Fig. 1 represents the upper and lower baselines obtained for the genotype Morales in the field environment [Fig. 1a] and greenhouse [Fig. 1b]. The range of VPDs for the baseline were 0.8 to 3.5 kPa in the field and 0.1 to 3.5 in the greenhouse. The baseline developed by Erden et al. [7] was between 1.1 to 2.7 kPa.

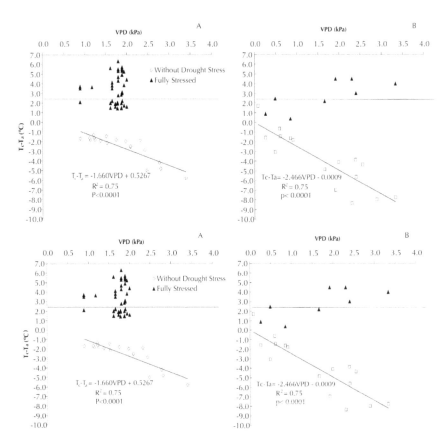

FIGURE 1 Canopy-air temperature differential (T_c-T_a) versus vapor pressure deficit (VPD) for full drought stressed and non-drought stressed common bean genotype *Morales* in (a) field environment and (b) Greenhouse environment.

The upper baseline represents the (T_c-T_a) for plants that are severely stressed, and were selected from the greenhouse and the field from plants under drought stress between 12:00 to 14:00 h. Then the average values of canopy temperature obtained from these plants were related with the average air temperature to obtain the upper baseline values. The values for upper baseline varied from 1.1 to 4.7°C, but differed among genotypes. For this study the upper baseline selected for each genotypes were:

Morales: 2.8°C (1 SD = 1.5°C); BAT 477: 3.1°C (1 SD = 1.5°C); SER 16: 3.1 °C (1 SD = 1.7°C); and SER 21: 2.9°C (1 SD = 1.5°C).

Erdem et al. [7] found 2.4°C as an upper baseline for bean in field environment (*P. vulgaris* L., cv., Sehirali 90).

Lower baselines are different among genotypes and environments (Table 3). The slope was over 2.17 for the greenhouse and was lower for the field. All the lower baseline models were statistically significant, and the coefficients of determination (R^2) were greater than 0.68. The correlation between (T_c-T_a) and VPD is affected by other micrometeorological variables such as clouds or wind, and equipment calibration [7]. For example, Ajayi and Olufayo [4] found that a higher correlation was obtained for low wind speeds in sorghum, with (T_c-T_a) from −2 to +8°C. Also, wind speed directly influences surface resistance and induces changes in the canopy temperature that are not necessarily indicative of drought stress (Table 3).

TABLE 3 Lower baseline functions for four common bean genotypes, in the greenhouse and field environment. **Notes:** †R² is the determination coefficient; ‡p-level is the probability level associated with the models; and §RMSE is the model root mean square error.

Genotype	Lower baseline functions	R²†	P-level²	RMSE§
	Greenhouse			
BAT 477	T_c-T_a=-2.17*VPD+0.12	0.64	0.0002	1.75
Morales	T_c-T_a=-2.47*VPD-0.0044	0.75	0.0001	1.50
SER 16	T_c-T_a=-2.29*VPD+0.17	0.77	0.0001	1.99
SER 21	T_c-T_a=-2.17*VPD-9.74	0.60	0.0001	1.99
	Field			
Morales	Tc-Ta=-1.66*VPD+0.5267	0.68	0.0001	0.68
SER 16	T_c-T_a=-1.33*VPD+0.1442	0.68	0.0001	0.42

Differences in CWSI between genotypes in the field environment were more evident in 2007, where the water deficit with respect to the control (well irrigated) was 30.3% as compared with 18% in 2006. The DOY 57, 64 and 71 during 2007 (See Figs. 2c and 2d) clearly showed the difference between genotypes SER 16 and Morales. The lowest CWSI in SER 16 in the 2007 season, can be attributed to a lower average stomatal resistance for those days, including 349, 690, and 187 sm⁻¹ respectively, compared with 769, 1747 and 449 sm⁻¹ for Morales (data not shown in the figure). Genotypic variations in CWSI and its relationship with stomatal conductance were also reported for seven winter wheat varieties by Alderfasi and Nilsen [3]. The CWSI for well irrigated treatments for both genotypes during 2006 were lower than 0.3 (Figs. 2a and 2b), similarly during 2007 for days with low wind speed.

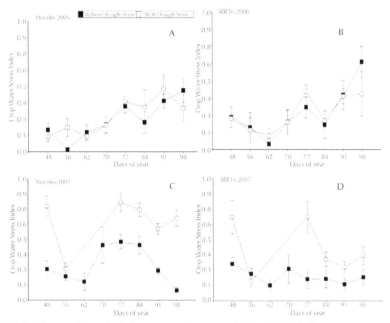

FIGURE 2   Seasonal trend of the crop water stress index (CWSI), for two common bean genotypes (Morales and SER16) and two growing seasons (2006–2007).

21.3.2   THE CWSI AND YIELD COMPONENTS

The drought stress treatments in the greenhouse were applied from the vegetative phase to maturity. The T_c was measured at 13:00 hour, when the drought stress was likely to be the highest (and when the maximum stress was in fact detected). The yield reduction in the four common bean genotypes were correlated with the CWSI, but with differences in magnitude (Fig. 3). For Morales, BAT 477 and SER 21, the linear models were statistically significant, while for SER 16 the relation was not linear (Fig. 3d).

The high slope in the genotype Morales relative to the others could be associated directly with the drought susceptibility of this genotype. Ten percent (10%) in yield reduction (RY = 0.9) is associated with CWSI value of 0.04 for BAT 477; 0.12 for SER16; 0.15 for SER 21; and 0.24 for Morales. Erdem et al. [7], working with bean (cv., *Sehirali* 90), demonstrated under field conditions that an average CWSI value of about 0.07 prior to irrigation produces the maximum yield. Albuquerque et al. [2] also working with bean, reported 0.15 as a CWSI limit for water management to avoid significant yield loss.

Similar results were observed in the field-environment, but regression functions were fitted, owing to the fact that the results were available only for the 2007 trial. These results indicate that the most susceptible genotypes were also Morales and BAT 477, which when reaching the same drought stress level produced the lowest seed yields and highest CWSI values (0.96 and 0.95, respectively). SER 16 showed the lowest CWSI of 0.82 under drought stress, but had a lower seed yield than SER 21 having a CWSI of 0.92. These results indicate that SER 21 under drought stress has a higher transpiration reduction than SER 16 (between 13:00 to 14:00 h), but can maintain a relatively high seed yield.

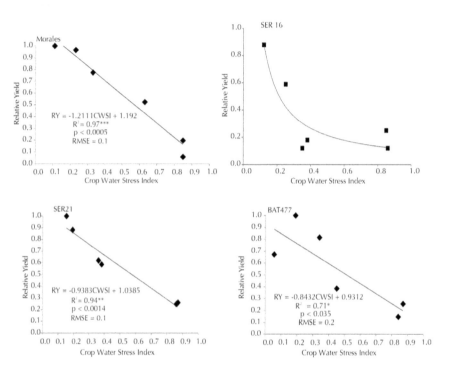

FIGURE 3 Relative yield (RY = $Y_{obs}/Y_{max.WS}$, Where Y_{obs}: is the yield observed an $Y_{max.WS}$ is the maximum yield observed without drought stress), as related to seasonal mean of crop water stress index (CWSI), under greenhouse environment for four common bean genotypes, during two years and three growing seasons.

21.3.3 CWSI AND SOIL MOISTURE IN FIELD

The crop water stress index was also well correlated with the water in the root zone. The linear regression models were fitted and these showed statistical significance for SER 16 and Morales (See Figs. 4a and 4b). If the water in the root zone is 50% of the total available water (TAW), the CWSI= 0.41 for SER 16 and 0.61 for Morales and if the water in the root zone is 75% of the TAW, then the CWSI of 0.55 for SER 16 and 0.79 for Morales were observed.

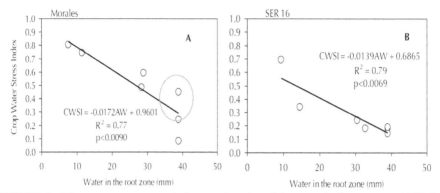

FIGURE 4   Mean crop water stress index as a function of the water in the root zone (AW) for two common bean genotypes: (a) Morales and (b) SER 16. The red open circle indicates low r_a (high wind conditions), which induced "physiological stress."

The CWSI is also affected by genotypic stomatal response, and the relationship between the aerodynamic and stomatal resistances. In the case of Morales (Fig. 4a), the red open circle indicated low value of r_a (windy conditions), which increased the CWSI. The red open circle corresponds to March 1 (DOY 60), where the mean wind speed during the canopy temperature reading was 5.8 ms^{-1}, and the mean daily r_a was 29.9 s.m^{-1}. The CWSI was 0.45 for Morales and 0.19 for SER 16 during the same day. Stomatal resistance (r_L) measured for the same period was 729 s.m^{-1} (1 SD = 236 s.m^{-1}) for Morales and 560 s.m^{-1} (1 SD = 717 s.m^{-1}) for SER16.

21.4   SUMMARY

The ability to detect and characterize the magnitude of drought stress has been an area of active research during the last three decades. With the development and increased popularity of the infrared thermometer, a thermal stress index has been proposed and applied. One of the most popular and useful is the crop water stress index (CWSI). The principal objective of this chapter was to develop baselines for CWSI for four common bean genotypes, and relate the index with yield components and soil available water under field and greenhouse environments. Three years of research (2005, 2006 and 2007) was conducted in two environments (greenhouse and field) in the west and south of Puerto Rico. Three water levels were applied in the greenhouse and two water levels in the field were used in randomized block experiments. Four common bean genotypes were studied: Morales, unknown drought response, and BAT477, SER16 and SER 21, drought tolerates. The CWSI was derived for a total of five growing seasons; including two field and three greenhouse experiments. The results indicate differences in drought tolerance between genotypes. The effect of wind induced additional "physiological stress" that was detected by the CWSI. The differences in the CWSI between genotypes were well correlated with the stomatal control, root available water, and yield components.

The CWSI was well correlated with yield components, but varied in magnitude among the different genotypes. The lower baselines derived from the greenhouse were different than those derived from the field, principally due to differences in atmospheric conditions, especially air temperature, and wind speed, with the field having windier and cooler conditions than the greenhouse. The high wind speeds, induced a physiological stress with increasing stomatal resistance and decreasing aerodynamic resistance. For the genotype Morales, the influence of wind speed was detected by the CWSI, which may indicate that this genotype is more stomatally susceptible under windy conditions than SER 16. The CWSI was well correlated with the water available in the root zone, indicating that this index is an excellent indicator of the plant-soil water status. The CWSI should, however, be used in combination with an analysis of wind speed and genotypic characteristics. The CWSI was also a good tool to characterize drought stress under greenhouse conditions. The upper and lower baselines developed as a part of the CWSI approach are highly genotype-dependent, and therefore, the applicability of the baselines developed in this study should be verified before they are used with other genotypes or varieties.

KEYWORDS

- air temperature, T_a
- canopy temperature, T_c
- common bean
- crop water stress index, CWSI
- crop yield
- drought stress
- leaf temperature
- physiological stress
- remote sensing
- soil available water
- standard deviation, SD
- vapor pressure deficit, VPD
- wind speed

REFERENCES

1. Allen G. R, Pereira, L. S.; Raes, D.; M.; Smith, **1998,** *Crop Evapotranspiration: Guidelines for Computing Crop Water Requirements.* Food and Agricultural Organization of the United Nations (FAO). Publication No. 56. Rome. 300p.
2. Albuquerque, P. E. P.; Gomide, R. L.; Klar, A. E.; Crop water stress index for bean obtained from temperature difference between canopy and air. *Water and Environment,* **1998,** *1,* 189–196.
3. Aldesfasi, A. A.; Nielsen, D. C.; Use of crop water stress index for monitoring water status and scheduling irrigation in wheat. *Agric. Water Manag.;* **2001,** *47,* 69–75.
4. Ajayi, A. E.; Olufayo, A. A.; Evaluation of two temperature stress indices to estimate grain sorghum yield and evapotranspiration. *Agron. J.;* **2004,** *96,* 1282–1287.

5. Alves, I.; Perrier, A.; Pereira, L. S.; Aerodynamic and surface resistance of complete cover crops: how good is the "Big Leaf"? *Trans. ASAE,* **1998,** *41(2),* 345–351.

6. Blom-Zandstra, M.; Metselaar, K.; Infrared thermometry for early detection of drought stress in *Chrysanthemum. HortScience,* **2006,** *41(1),* 136–142.

7. Erdem, Y.; Sehirali, S.; Erdem, T.; Kanar, D.; Determination of crop water stress index for irrigation scheduling of bean (*Phaseolus vulgaris* L.). *Turkish Journal of Agriculture and Forestry,* **2006,** *30,* 195–202.

8. Gardner, B. R.; Nielsen, D. C.; Shock, C. C.; Infrared thermometry and the crop water stress index. II. Sampling procedures and interpretation. *J. Prod. Agric.;* **1992,** *5,* 466–475.

9. Goyal, M. R.; Gonzalez, E. A.; Climatic data for agricultural experiment substations of Puerto Rico (Spanish). University of Puerto Rico – Mayaguez Campus, Agricultural Experiment Station. Publicación 88–70, Project C-411. **1989,** 87 pages.

10. Harmsen, E. W.; Goyal, M. R.; Torres, S.; Justiniano, Estimating evapotranspiration in Puerto Rico. *Journal Agricultural of the University of Puerto Rico,* **2002,** *86(1–2):* 35–54.

11. Howell, T. A.; Hatfield, J. L.; Yamada, H.; Davis, K. R.; Evaluation of cotton canopy temperature to detect crop water stress. Transaction of ASAE, **1984,** *27,* 84–88.

12. Idso, S. B.; Jackson, R. D.; Reginato, R. J. Remote sensing of crop yields. Science, **1977,** *196,* 19–25.

13. Idso, S. B.; Jackson, R. D.;, Pinter, P. J.; Hatfield, J. L. Normalizing the stress-degree-day parameter for environmental variability. Agricultural Meteorology, **1981,** *24,* 45–55.

14. Idso, S. B.; Non-water-stressed baselines: a key to measuring and interpreting plant water stress. Agricultural Meteorology, **1982,** *27,* 59–70.

15. InfoStat, **2003,** *InfoStat, version 3. User's Manual.* Grupo InfoStat, FCA, Universidad Nacional de Cordoba, Argentina. <http//www.infostat.com.ar>

16. Irmak. S, Hanan, D. Z.; Bastug, R.; Determination of crop water stress index for irrigation timing and yield estimation of corn. *Agron. J.;* **2000,** *92,* 1121–1227.

17. Jackson, R. D.; Reginato, R. J.; Idso, S. B.; Wheat canopy temperature: A practical tool for evaluating water requirements. Water Resources Research, **1977,** *13,* 651–656.

18. Jackson, R. D.; Idso, S. B.; Reginato, R. J.; Pinter Jr., P. J.; Canopy temperature as a crop water stress indicator. *Water Resources Research,* **1981,** *17,* 1133–1138.

19. Jones, G. H.; Irrigation scheduling: advantages and pitfalls of plant-based methods. *Journal Experimental Botany,* **2004,** *55(407),* 2427–2436.

20. Karamanos, A. J.; Papatheohari, A. Y.; Assessment of drought resistance of crop genotypes by means of the water potential index. *Crop Sci.;* **1999,** *39,* 1792–1797.

21. Kjelgaard, J. F.; Stokes, C. O.; Evaluating surface resistance for estimating corn and potato evapotranspiration with the Penman-Monteith model. Transaction of ASAE, **2001,** *44(4),* 797–805.

22. Monteith, J. L.; Unsworth, M. H.; *Principles of Environmental Physics.* Chapman and Hall Inc.; **1990,** 292 pp.

23. Ramirez, V. H.; Porch, G. T.; Harmsen, E. W.; Development of linear models for nondestructive leaflet area estimation in common bean (Phaseolus vulgaris L.) using direct leaf measurements. *J. Agric. Univ. PR.* **2008,** *92(3–4),* 171–182

24. SigmaPlot. **2005,** SigmaPlot.802 – SPSS. User manual. <http//www.sigmaplot.com>

25. Stökle, C. O.; Dugas, W. A.; Evaluating canopy temperature based index for irrigation scheduling. *Irrig. Sci.;* **1992,** *13(1),* 31–37.

26. Tanner, C. B.; Plant temperatures. *Agron. J.;* **1963,** *55,* 210–211.

27. U. S. Department of Agriculture (USDA), **1987,** Primary characterization data of San Anton soil, Guanica country, Puerto Rico. Soil Conservation Service, National Soil Survey laboratory, Lincoln, Nebraska.

28. Wanjura, D. F.; Upchurch, D. R.; Canopy temperature characterization of corn and cotton water status. *Transactions of ASAE,* **2000,** *43(4),* 867–875.

CHAPTER 22

EVAPOTRANSPIRATION: TEMPERATURE VERSUS ELEVATION RELATIONSHIPS[1]

MEGH R. GOYAL

CONTENTS

22.1 INTRODUCTION

Water consumptive use by crops in any particular location is probably affected more by temperature, which for long-time periods is a good measure of solar radiation, than any other factor. Abnormally low temperatures may retard plant growth and unusually high temperatures may produce dormancy. Temperature is one of the most dynamic properties of weather and is subject to daily and seasonal changes. Irrigation planning can be delayed or underestimated if temperature data for various locations in a particular locality is not available [6, 7].

This chapter discusses the research study: To establish relationships among monthly temperature (°C) and elevation (m) for Puerto Rico with available temperature data; To generate missing temperature data with these relationships; To present specific examples to estimate potential evapotranspiration using these results and Hargreaves – Samani method [7].

22.2 MATERIALS AND METHODS

For Puerto Rico and 30-four locations in Puerto Rico, Goyal et al. [3] determined relationships among monthly temperature (T) and elevation (X) for Puerto Rico with linear regression analysis. They used these relationships to estimate missing temperature data (°C) for January through December, for these 34 locations, such as: Mean maximum temperature (T_{MAX}), mean minimum temperature (T_{MIN}) and mean average temperature (T). They used available temperature data [10] and Publication No. 86–45 of the Weather Bureau, U.S. Department of Commerce [2]. Thirty-four locations [2] were Aceituna, Adjuntas, Barceloneta 2NNW, Bayamon Hato Tejas, Cabo Rojo, Calero Camp, Caonillas Villalba, Carite Plant I, Catano, Central San Francisco, Coamo Dam, Ensenada, Guajataca Dam, Guayabal Reservoir, Guayanilla, Gurabo, Indiera Baja, Jajome Alto, La Fe, Maricao, Maricao Fish Hatchery, Matrullas Dam, Maunabo 1SW, Mora Camp, Naguabo 6W, Paraiso, Peñuelas Salto Garzas, Puerto Real, Rio Blanco Lower, Rio Blanco Upper, San Lorenzo Espino, San Sebastian, Santa Isabel 3NW, Santa Rita, Toa Baja Constancia, Toro Negro Plant 2, Villalba, Yabucoa 1NNE, and Yauco 18. The climatic regions of Puerto Rico are shown in Fig. 1.

Goyal [4] estimated monthly ET_0 for Central Aguirre, Fortuna and Lajas Agricultural Experiment substations, and for Magueyes Island located on the south coast of Puerto Rico. Daily estimated values of ET_0 varied between 3.68 to 5.37 mm per day (0.15 to 0.21 in d^{-1}). The minimum estimated value of ET_0 occurred in December and the maximum occurred in July. The same procedure was applied for Vieques Island of Puerto Ric, with estimated monthly ET_0 ranging between 3.29 mm per day (0.13 in. d^{-1}) in December to 4.94 mm d^{-1} (0.19 in. d^{-1}) in July [5]. Goyal et al. [3 to 5] used the Hargreaves-Samani Eq. (1) for reference evapotranspiration [7]. In Eqs. (1) and (2): ET_0 (mm per day) is the reference evapotranspiration, R_a (mm d^{-1}) is the extraterrestrial radiation, R_s is the solar incident radiation, T is the mean daily average temperature (°C), and T_{min} and T_{max} are the mean daily minimum and maximum temperatures (°C), respectively.

$$ET_0 = [0.0023 \times R_a][(T + 17.8) \times (T_{max} - T_{min})^{0.5}] \tag{1}$$

$$ET_o = [0.0135 \times R_s][(T + 17.8)] \qquad (2)$$

22.3   RESULTS AND DISCUSSION

Goyal et al. [3] discusses and presents the results for 34 locations in Puerto Rico that also includes the tabulated values for [3]: mean daily maximum-, mean daily minimum-, mean daily average-, absolute highest- and lowest temperature (°C) versus elevation (m) relationships for January through December; estimates of maximum, minimum and average temperatures; observed and estimated temperatures for Juana Diaz Agricultural Experiment Substation; potential evapotranspiration estimations. They used the Hargreaves-Samani Eq. (1) [7] to estimate ET_o for 30-four locations within Puerto Rico. Average monthly minimum and maximum air temperatures were based on long-term measured data. Average monthly values of the extraterrestrial radiation (R_a) were based on the average latitude of Puerto Rico. They also developed a linear regression analysis in which several monthly climatic factors (mean daily minimum, maximum and average air temperature) were correlated with surface elevation in Puerto Rico using Eq. (3) below:

$$T = A + B \times X, \qquad (3)$$

where, T = temperature, °C; X = elevation above mean sea level, m; A and B are regression coefficients; and R^2 is the coefficient of determination.

It was found that the temperature was negatively correlated with elevation (Table 2). The relationships were linear. All regression coefficients were significant at P = 0.01. For various months, the coefficient of correlation varied from 0.89 to 0.96 for mean daily average temperature; 0.80 to 0.89 for mean daily maximum temperature; 0.67 to 0.79 for mean daily minimum temperature; 0.64 to 0.80 for highest temperature and 0.43 to 0.60 for lowest temperature. The coefficient of determination (R^2) varied from 0.78 to 0.92 for mean daily average temperature; 0.64 to 0.79 for mean daily maximum temperature; 0.44 to 0.63 for mean daily minimum temperature; 0.41 to 0.64 for highest temperature and 0.19 to 0.36 for lowest temperature. These coefficient of correlations imply that these relationships gave best estimates of mean daily temperature for January through December compared to highest and lowest monthly temperatures.

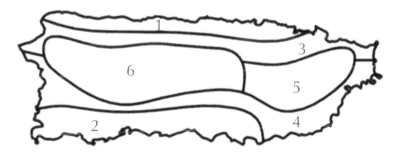

FIGURE 1   Climatic divisions of Puerto Rico: 1 North Coastal, 2 South Coastal, 3 Northern Slopes, 4 Southern Slopes, 5 Eastern Interior, and 6 Western Interior.

TABLE 1   Observed and estimated temperatures at Fruits Agricultural Experiment Substation, Juana Díaz, Puerto Rico.

MONTH	Temperature, °C									
	Mean						Absolute			
	Average		Maximum		Minimum		Highest		Lowest	
	OBS	EST	OBS	EST	OBS	EST	OBS	EST	OBS	EST
JAN	24.3	23.8	30.1	29.1	18.6	18.5	36.1	33.7	13.3	12.9
FEB	24.3	23.8	29.9	29.3	18.6	18.3	35.0	34.3	12.2	12.6
MAR	24.6	24.3	30.3	30.0	18.9	18.6	34.4	35.0	13.3	12.9
APR	25.4	25.1	30.7	30.5	20.3	19.8	34.4	35.5	14.4	14.3
MAY	26.4	26.1	31.2	31.1	21.8	21.2	35.6	35.6	16.1	15.5
JUN	27.1	26.8	31.7	31.7	22.6	21.9	36.7	35.9	14.4	16.7
JUL	27.1	26.8	31.7	31.7	22.6	21.9	36.7	35.9	14.4	16.7
AUG	27.6	27.1	32.5	32.0	22.6	22.1	ns	36.3	16.1	17.2
SEP	27.4	27.0	32.4	32.0	22.4	21.9	36.7	36.4	18.3	16.8
OCT	27.0	26.6	32.0	31.8	21.9	21.4	36.7	35.9	16.1	16.5
NOV	26.2	25.7	31.5	30.8	21.0	20.6	36.7	35.3	15.6	15.3
DEC	25.2	24.7	30.7	29.7	19.6	19.5	35.6	34.2	12.8	14.1
ANNUAL	26.1	25.7	81.3	30.8	20.9	20.5	36.0	35.3	14.8	15.1

Note: Weather station No. 7292 PONCE-4E at an elevation of 21 m.
Temperature values were estimated using relationships between temperature (°C) and elevation (m) given in Table 1 by Goyal et al. [3]. Observed temperatures are from climatography of the U.S. No. 86–45: Puerto Rico and U.S. Virgin Islands [2].

TABLE 2   Relationships among temperatures (T) and elevations (X) for Puerto Rico [3, 8].

| Month | Mean daily maximum temperatures, °C | | | Mean daily minimum temperatures, °C | | |
	A	B, $\times 10^{-5}$	R^2	A	B, $\times 10^{-5}$	R^2
Jan.	29.24	770	0.73	18.58	544	0.44
Feb.	29.37	752	0.72	18.37	558	0.46
Mar.	30.08	711	0.71	18.71	590	0.48
Apr.	30.59	687	0.71	19.90	686	0.63
May	31.16	707	0.76	21.23	608	0.63
June	31.76	686	0.73	21.92	577	0.59
July	32.07	717	0.64	22.14	591	0.58
Aug.	32.12	682	0.75	22.21	585	0.58
Sep.	32.12	696	0.79	21.95	586	0.62
Oct.	31.84	705	0.79	21.48	553	0.59
Nov.	30.89	706	0.75	20.68	562	0.55
Dec.	29.83	744	0.73	19.52	547	0.47

Linear regression equation: T = A + B×X, where, T = temperature, °C; X = elevation above mean sea level, m; A and B are regression coefficients; and R^2 is the coefficient of determination.

22.4   SOLVED EXAMPLES

22.4.1   EXAMPLE 1

The Fortuna Substation, Juana Diaz, P. R. is located at an elevation of 12 m and is identified as station No. 7292-Ponce 4E by U.S. Department of Commerce [2]. Estimate the monthly temperature for this location.

For Eq. (3), find values of regression coefficients (A and B from Table 1 given by Goyal et al., [3]) for weather station 7292-Ponce 4E. Using X = 12 meters in Eq. (3), calculate the temperature for each month. Estimated and observed temperature values are shown in Table 2.

22.4.2   EXAMPLE 2

Estimate potential evapotranspiration (PET or ET_o) using the Hargreaves and Samani method for January with the information given in Table 1 [Goyal et al., 3]. Use Eq. (1) in this chapter to estimate PET and Ra [mm per day] for Puerto Rico (18°N, Table 3) is 11.68 for January, 13.02 for February, 14.65 for March, 15.83 for April, 16.30 for May, 16.38 for June, 16.38 for July, 15.80 for August, 15.23 for September, 13.62 for October, 12.11 for November, 11.2 for December, respectively.

PET was calculated using Eqs. (1) and (2), values of R_a, T, T_{MAX} and T_{MIN}. PET values are given below:

Using Eq. (1): PET = 2.4 mm/day for January.
Using Eq. (2) and R_s = 6.9: PET = 3.4 mm/day for January.
Values for R_a are shown in Table 3 [8].

TABLE 3   Extraterrestrial radiation by month and latitude within Puerto Rico [8].

Month	Extraterrestrial radiation, R_a (Mega-Joules per m² per day)					
	Latitude (decimal degrees N)					
	17.90	18.00	18.10	18.20	18.30	18.40
Jan	27.90	27.85	27.80	27.74	27.69	27.64
Feb	31.36	31.32	31.27	31.23	31.19	31.14
Mar	35.33	35.30	35.28	35.25	35.23	35.20
Apr	38.03	38.02	38.02	38.02	38.01	38.01
May	39.02	39.03	39.04	39.06	39.07	39.09
June	39.07	39.09	39.12	39.14	39.16	39.19
July	38.91	38.93	38.95	38.97	38.99	39.01
Aug	38.30	38.31	38.31	38.32	38.32	38.33
Sep	36.38	36.36	36.35	36.33	36.32	36.31
Oct	32.91	32.88	32.84	32.81	32.77	32.74
Nov	29.10	29.05	29.01	28.96	28.91	28.86
Dec	26.89	26.84	26.78	26.73	26.67	26.61
July	38.91	38.93	38.95	38.97	38.99	39.01

22.4.3   EXAMPLE 3

Develop PET model for Puerto Rico as a function of incident solar radiation (R_s) and elevation (X) with temperature versus elevation relationships in Table 1 given by Goyal et al. [3]. The relationship between the temperature (T, °C) and elevation (X, m) is defined by the Eq. (3) in this chapter.

Combining Eqs. (2) and (3), we get:

$$PETNEW = [0.0135 \times R_s][(A + B \times X) + 17.8] \tag{4}$$

Rearranging the Eq. (4), we get:

$$PETNEW = R_s \times [(0.0135 \times A + 0.0135 \times B \times X) + (0.0135 \times 17.8)] \text{ or} \tag{5}$$

$$PETNEW = R_s \times [(0.2403 + 0.0135 \ x \ A) + \{(0.0135 \times B) \times X\}] \text{ or} \tag{6}$$

$$PETNEW = R_s \times [A_1 + B_1 \times X] \tag{7}$$

In Eq. (7): $A_1 = [0.2403 + 0.0135 \times A]$, and $B_1 = [0.0135 \times B]$. For January through December, Eq. (7) reduces to:

$$PETNEW_i = [(R_s \times A_{1,i}) + (B_{1,i} \times X)] \tag{8}$$

In Eq. (8), subscript $i = 1, 2, 3, 4, 5, 6, 7, 8, 9, 10, 11, 12$ for January, February, March, April, May, June, July, August, September, October, November, December, respectively. Introducing values of A_i and B_i from Table 1 given by Goyal et al. [3], we obtain values of $A_{1,i}$ and $B_{1,i}$. The Eq. (8) reduces to PET equations for Puerto Rico given in Table 4.

TABLE 4   Potential evapotranspiration equations for Puerto Rico using the temperature versus elevation relationships given by Goyal et al. [3]. Caution: The equations in this table are not valid for any other location in the world.

Month	PET model for Puerto Rico	Equation
Jan	$PETJAN = R_s \times [0.563085 - (8.8695 \times 10^{-5}) \times (X)]$	(9)
Feb	$PETFEB = R_s \times [0.56268 - (8.8425 \times 10^{-5}) \times (X)]$	(10)
Mar	$PETMAR = R_s \times [0.569295 - (8.721 \times 10^{-5}) \times (X)]$	(11)
April	$PETAPR = R_s \times [0.580635 - (8.7075 \times 10^{-5}) \times (X)]$	(12)
May	$PETMAY = R_s \times [0.594 - (8.8965 \times 10^{-5}) \times (X)]$	(13)
June	$PETJUN = R_s \times [0.602775 - (8.559 \times 10^{-5}) \times (X)]$	(14)
July	$PETJUL = R_s \times [0.605205 - (8.7075 \times 10^{-5}) \times (X)]$	(15)
Aug	$PETAUG = R_s \times [0.60723 - (8.5725 \times 10^{-5}) \times (X)]$	(16)
Sep	$PETSEP = R_s \times [0.60534 - (8.694 \times 10^{-5}) \times (X)]$	(17)
Oct	$PETOCT = R_s \times [0.600075 - (8.667 \times 10^{-5}) \times (X)]$	(18)
Nov	$PETNOV = R_s \times [0.588465 - 8.694 \times 10^{-5}) \times (X)]$	(19)
Dec	$PETDEC = R_s \times [0.57483 - 8.964 \times 10^{-5}) \times (X)]$	(20)
Annual	$PETANUAL = R_s \times [0.58779 - 8.748 \times 10^{-5}) \times (X)]$	(21)

Note: In Eqs. (9–21): $PETNEW_i$ = Monthly Potential evapotranspiration in mm per day, R_s = Incident solar radiation in mm per day and X = Elevation in m. The Eqs. (9–21) are only valid for Puerto Rico.

Eqs. (9–21) are only valid for Puerto Rico and for the elevation between 4.5 and 900 meters. The Hargreaves–Samani method (Eq. (1)) cannot be used to develop Eqs. (9–21) in Table 4, because the coefficient of correlation between (T_{MAX}–T_{MIN}) versus elevation is very poor (Table 1 by Goyal et al. (3)) for all months. This is explained by the fact that Rs varies at all locations in Puerto Rico.

When the temperature data are not available, Eqs. (9–21) are useful because elevation of a location is readily available from property records, U.S. Weather Bureau, U.S. Geological Survey, USDA-SCS, or another local government agency in Puerto Rico. It is not possible to combine Eqs. (9–21) into one equation as the relationships among temperature versus elevation are different for January through December.

22.5 SUMMARY

Relationships among mean daily maximum-, mean daily minimum-, mean daily average-, absolute highest-, absolute lowest temperature (°C) versus elevation (m) were determined for January through December for Puerto Rico in the Caribbean region. These relationships were found to be linear: (Y = A + B×X), where, A and B are linear regression coefficients, X = Elevation above mean sea level in meters, and Y = Temperature in °C. The coefficient of correlation varied from (–0.43) to (–0.96). Examples are presented to estimate PET and to develop PET model as a function of incident solar radiation and elevation. The procedure presented here can be used to develop simplified potential evapotranspiration equations for any other location in the world in the tropical climate.

KEYWORDS

- Caribbean region
- elevation
- evapotranspiration
- evapotranspiration, crop
- evapotranspiration, potential
- Hargreaves – Samani
- Puerto Rico
- regression analysis
- solar radiation
- temperature, average
- temperature, maximum
- temperature, mean
- temperature, minimum
- tropical region

REFERENCES

1. Allen, R. G.; Pereira, L. S.; Dirk Raes and M.; Smith, **1998,** *Crop Evapotranspiration Guidelines for Computing Crop Water Requirements.* FAO Irrigation and Drainage Paper 56, Food and Agriculture Organization of the United Nations, Rome. 300 pp.

2. Climatography of the United States No. 86–45 of Puerto Rico and US Virgin Islands. In: *Climatic Summary of the US Supplement for 1951 through 1960,* US Department of Commerce, Washington, D.C. USA.

3. Goyal, Megh R.; Gonzalez E. A.; Chao de Baez, C.; **1988,** Temperature versus elevation relationships for Puerto Rico. *J. Agric. UPR, 72(3),* 449–467.

4. Goyal, M. R. **1988a.** Research Note. Potential evapotranspiration for the south coast of Puerto Rico with the Hargreaves-Samani technique. *J. Agric. UPR, 72(1),* 57–63.

5. Goyal, M. R. **1988b.** Research note: Potential Evapotranspiration for Vieques Island, Puerto Rico, With the Hargreaves and Samani Model. *J. Agric. UPR, 72(1),* 177–78.

6. Hargreaves, G. H.; Z. A . Samani. Simplified irrigation scheduling and crop selection for El Salvador. Utah State University, Logan-Utah, USA. Personal communication.

7. Hargreaves, G. H.; Samani, Z. A.; Reference Crop Evapotranspiration from Temperature. Appl. Eng. Agric. of ASAE, **1985,** *1(2),* 96–99.

8. Harmsen, E. W.; Goyal, M. R.; S.; Torres-Justiniano, Estimating Evapotranspiration in Puerto Rico. *J. Agric. Univ. P.R.,* **2002,** *86(1–2),* 35–54.

9. Jensen, M. E.; Burman, R. D.; Allen, R. G.; Evapotranspiration and irrigation water requirements. ASCE Manuals and Reports on Engineering Practice No. 70, **1990,** 332 pp.

10. Temperature data: Puerto Rico and US Virgin Islands, updated **1984,** National Climatic Data Center, Asheville, NC 28801.

11. University of Puerto Rico-Mayaguez Campus, *J. Agric. Univ. P.R.,* **1986,** *70(4),* 267–275.

CHAPTER 23

GENERATION OF MISSING CLIMATIC DATA: PUERTO RICO[1]

MEGH R. GOYAL and S. F. SHIH

CONTENTS

[1] This chapter has been summarized from the report by Dr. S. F. Shih on, "Irrigation Requirement Estimation in Puerto Rico: Evapotranspiration", that was submitted to Dr. Megh R Goyal, Principal Investigator at Agricultural Experiment Station of University of Puerto Rico – Mayaguez Campus as part of Caribbean Basin Agricultural Development, USDA-ARS – 23 project. We thank the support by Agricultural Experiment Stations at the University of Puerto Rico, Mayaguez Campus and the University of Florida; and the Administrative staff of CBAG USDA–ARS, Washington, DC. We also thank S. Cheng, Research Assistant for his help and guidance in the analysis of data. © Copyright RevistaInvestigaciones de UNISARC Columbia.

23.1   INTRODUCTION

Evapotranspiration (ET) is an important process of the hydrologic cycle. Approximately 75% of the total rainfall on a continental basis is returned to the atmosphere by evaporation and transpiration [4]. The ET represents the quantity of water which must be supplied at all times to maintain a soil water balance. Knowledge of ET is necessary in planning and implementing soil water management systems. Numerous approaches can be used to estimate ET-, such as mass (water vapor) transfer, energy budget, water budget, soil moisture budget, groundwater fluctuations, empirical formulae, and combination (of energy budget and mass transfer) methods. These different techniques have been developed partly in response to the availability of different types of data for estimating ET. Each method has certain advantages and limitations (*see* Chapter 5). Availability of specific types of data is often a limiting factor in the choice of calculation technique for practical applications. The choice of calculation technique also depends on the intended use [6, 8] and on the time scale required by the problem. For example, irrigation management requires daily estimates of ET to allow producers to make rational decisions concerning the timing and amount of irrigation. Different types of vegetation and stages of growth must be considered because they have a considerable influence on the daily rate of ET. In contrast, basin-level planning may require monthly estimates of ET to project changes in water supplies and requirements during the year. The combination energy-balance mass-transport method [9] has displayed the best overall fit among the empirical methods for ET estimation [6, 7, 13]. However, the combination method [9, 10] requires a variety of climatological data such as net radiation, air temperature, relative humidity, and wind velocity. If sufficient climatological data are not available for using the Penman method, an alternative method should be chosen. Shih [14] suggested that in choosing an ideal ET equation, one should minimize input of climatological data without affecting the accuracy of estimation, so that not only the multi – collinearnity problem among the data can be eliminated but also the data availability can be improved considerably. For example, Shih et al. [13] used the modified Blaney-Criddle method [12] as follows to predict ET:

$$ET = k*f'$$

$$f' = \{[2.54*PR] * [1.8*T_{avg} + 32]\} / [100] \qquad (1)$$

where, k = Coefficient for modified Blaney – Criddle method; T_{avg} = Mean monthly temperature in °C; PR = Percent of annual solar radiation during the month; ET = Evapotranspiration; and f' = Monthly Et factor in mm.

The use of Eq. (1) to predict the irrigation requirement in south Florida gave results close to those estimated by the Penman and water-budget methods. However, it needs to be noted that Eq. (1) uses only temperature and solar radiation instead of using temperature, solar radiation, relative humidity, and wind speed as in the Penman method. Unfortunately, both the Penman method and Shih's modified Blaney-Criddle method involves the solar radiation parameter. Due to errors in radiation measurement which tend toward the low side in the Hargreaves method, Shih [14] indicated the need to develop a more reliable method for a specific location. Unfortunately, solar radiation data often comprize one of the most incomplete records in weather stations. For instance, the University of Florida cooperated with the University of Puerto Rico in 1985 for using climatological data to estimate the ET requirement in Puerto Rico during the 10-year period of 1975–1984. It was found that both solar radiation and relative humidity data were missing in part of the records [15]. Since solar radiation data often present cyclic patterns, the interpolation of missing data using sources either from adjacent stations or from the historical records of the station itself are not always applicable; thus an alternative technique was developed by Dr. S. F. Shih [15].

Therefore, the objectives of this research were to: (1) Introduce some common methods used to interpolate missing data; (2) Develop an alternative method for generating missing data which undergo cyclic patterns in nature; and 3. Demonstrate the practical application of this alternative method for the case of missing solar radiation.

23.2 MATERIALS AND METHODS

23.2.1 CLIMATOLOGICAL DATA

Climatological data gathered from stations located in Puerto Rico at Lajas, Gurabo, Corozal, Isabela, Fortuna, Adjuntas, and Río Piedras (Fig. 1) were used in this study. The index number, division, county, latitude, longitude, and elevation of each station are given in Table 1. The climatological data (1975–1984) for these seven stations were obtained from the National Oceanic and Atmospheric Administration (NOAA) – National Climatic Center at Asheville, North Carolina [2]. The climatological data used in this study were the temperature, solar radiation, relative humidity, and wind speed which are the principal data required for estimation of potential evapotranspiration [1, 3, 9, 10, 12–14, 16]. The available records of solar radiation during 1975–1984 indicate the data before 5/76 and after 4/82 were not available for any of the seven stations, and many records were missing during the 5/76–4/82 period. The unit of solar radiation is Langleys/day (one Langley/day = 41.84 KJ/m²-day), which is the same unit as reported by NOAA data source for solar radiation [2].

Some relative humidity records were also missing at Isabela (8/75–4/76, 11–12/76, 11/78–1/79, 6/79–5/80; 12/80, 1/81, 5/81–12/81, 2/82–12/84), Fortuna (8/75–4/76, 11–12/76, 5/80–11/81, 9/82–8/83), Adjuntas (8/75–4/76, 11–12/76, 7/79–12/84), and Rio Piedras (8/75–4/76, 11–12/76, 1/80, 5/80–12/81, 2/82–12/84). The unit of percentage was used in the relative humidity data [2].

23.2.2 FOUR METHODS OF GENERATION OF MISSING DATA [12–15]

23.2.2.1 CORRELATION TECHNIQUE

This technique is commonly used to interpret missing data. However, this technique has some inherent limitations: (1) A high correlation among the stations must exist. (2) Some historical data in each station must be available not only for establishing the regression coefficient but also to serve as a source of input data for interpolation of the missing data. A stepwize regression technique is used to find the best correlation between the station with missing data and the stations without data missing. The general model used is shown in Eq. (2), where, S_m = Dependent variable: the station with missing data to be generated; S_n = Independent variable: the nth station without data missing; N = Total number of stations used as independent variables; and a_0, a_1, ..., and a_N = Regression coefficients.

$$S_m = [a_0] + [\textstyle\sum_{n=1}^{N}(a_n) * (S_n)], \; m?n \qquad (2)$$

23.2.2.2   HISTORICAL DATA INTERPOLATION

This method uses the historical data available from the station to interpolate missing data for the same station. This method is simple to use but the accuracy needs further testing. It is recommended that this method should be implemented when the correlation technique is not applicable. The mathematical formulation is expressed in Eq. (3), where, S_{mj} = the missing jth month data to be interpolated; for station S_m; S_{mk} = the kth month data that is available from station S_m; and M = Total number of monthly data records available from station S_m.

$$S_{mj} = [\textstyle\sum_{k=1}^{M} \{(S_{mk}) \div (M)\}], \; j?k \qquad (3)$$

23.2.2.3   AVERAGE APPROACH

This method uses the data available from other stations to interpolate missing data for a single station. This method is simple to use and can be implemented even under the following conditions: (1) No historical data are available for the station; (2) The number of records is not large enough to establish a relationship through other methods. The mathematical expression is shown in Eq. (4), where, S_{mj} = The station S_m with missing jth month data to be interpolated; S_{nj} = The nth station possessing jth month data; and N = Total number of stations used to interpolate the missing data.

$$S_{mj} = [\textstyle\sum_{n=1}^{N} \{(S_{nj}) \div (N)\}], \; m?n \qquad (4)$$

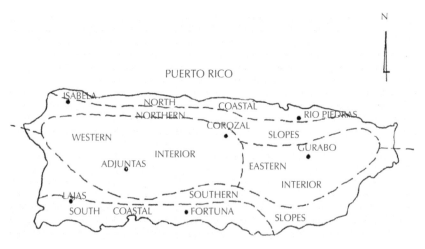

FIGURE 1   Seven weather stations at the Agricultural Experiment Substations, University of Puerto Rico – Mayagüez Campus. Climatic zones are shown with the dotted lines.

TABLE 1   Location of seven weather stations in Puerto Rico.

Location	Weather station number	Division	County	Latitude	Longitude	Elevation (m.s.L.), m
Adjuntas	0061	06	Ponce	18°11' N	66° 48' W	558
Corozal	2934	06	Arecibo	18°20' N	66° 48' W	198
Gurabo	4276	05	Humacao	18°15' N	66° 48' W	49
Isabela	4702	03	Mayagüez	18°28' N	66° 48' W	128
Lajas	5097	02	Mayagüez	18°03' N	66° 48' W	27
Rio Piedras	8306	01	San Juan	18°24' N	66° 48' W	26
Fortuna	---	---	Juana Díaz	18°01' N	66° 48' W	21

23.2.2.4   SPECTRAL ANALYSIS APPROACH

This method uses the consecutive historical data available at a station S_m to interpolate missing data for the same station. It assumes that the data variation with time has a base period of 12 months. The mathematical model is expressed in Eq. (5), where, $S_m(t)$ = The consecutive data available at the station with missing data; A_0 = A constant term; A_i, B_i = Coefficients of the ith harmonic component; $W = [2\pi/12]$; i = index of harmonic components; t = index of months, $t = 1$ for the first data point; $N1$ = Total number of harmonic components to be used in the model ($N_i = 6$ was used in this study); and $R(t)$ = residuals of original data.

$$S_m (t) = [A_0] + \{\Sigma_{i=1}^{N1}[(A_i \cos(Wit) + B_i(Wit)]\} + R(t), \tag{5}$$

In order to implement the model described in Eq. (5), following steps are involved:

Step 1. Removal of cyclic movement from the original data by performing the least squares linear regression analysis with the model shown in Eq. (4). Only the harmonic components with at least one significant coefficient are retained in the model.

Step 2. Bartlett's white noise test [Priestley 1981,] is applied to the residuals, $R(t)$. If the residuals are white noise, the final form of the model has been attained. Otherwize, **Step 3** is followed.

Step 3. Spectral analysis is performed on the residuals $R(t)$ to find the embedded frequency components in the residuals and to define the existence of frequency peaks in the periodogram.

Step 4. A mathematical model is established for $R(t)$ in Eq. (6), where, a_0 = Constant term; a_{pi} and b_{pi} = Coefficients of the ith harmonic component in the residuals $R(t)$; W_p = Frequency of the pth peak in the periodogram of $R(t)$, [from **Step 3**]; i = Index of harmonic components; t = index of months; p = Index of frequency peaks in the periodogram; N_2 = Total number of harmonic components in the model, ($N_2 = 3$ was used in this study); N_p = Total number of peaks in the periodogram of $R(t)$; and $RR(t)$ = residuals of $R(t)$.

$$R(t) = [a_0] + \{\Sigma_{p=1}^{N_p} \Sigma_{i=1}^{N_2} \ [(a_{pi} \cos(W_p it) + b_{pi} \sin(W_p it)]\} + RR(t), \tag{6}$$

Least squares linear regression is again applied to $R(t)$. Only those harmonic components with at least one significant coefficient are retained, and the N2 in Eq. (6) is replaced by N(P), which is the number of harmonic components for the p-th frequency with significant coefficients. The frequencies "W from **Step 3**" and "coefficients a_p and b_p from Eq. (6)" are used as initial values in **Step 5**.

Step 5. The coefficients of Eq. (6) are finalized using nonlinear regression analysis.

Step 6. The Bartletts white noise test is performed on the residuals RR(t). If RR(t) is white noise, the final form of the model has been reached. Otherwize, **Steps 3 through 6** are repeated until either the white noise is reached or the R2 value has reached a relatively high value (e.g., 0.7). Several factors need to be considered before implementing this method:

The data must have a cyclic movement in nature.

1. The data must have a relatively long period of consecutive records.
2. This method is particularly useful for generating missing data for the time period which has no data available.

23.3 RESULTS AND DISCUSSION

23.3.1 GENERATION OF MISSING SOLAR RADIATION DATA

The Eqs. (7) to (17) indicate the final results for this study. The historical solar radiation data not only exhibited a seasonal variation in nature but were also unavailable before 5/75 and after 4/82. However, weather stations at Corozal, Isabela and Fortuna had some consecutive records that were used for generating the missing data by the spectral analysis method.

23.3.1.1 ISABELA WEATHER STATION

The consecutive data between April 1977 and October 1980 (no missing data in this period) were chosen as the basis for model development.

Step 1. Least squares linear regression was used to estimate the coefficients A1 and B1 by the method given in Eq. (5). The results showed that the R2 value was 0.53 and only the B1 value was significant at the 0.05 level. Thus, only one harmonic component of the original data was retained in the model of solar radiation, SR(t), as shown in Eq. (7).

Step 2. The white noise test showed that the R(t) was not the white noise. Thus, the following steps 3 to 5 were taken.

Step 3. Spectral analysis was applied to the R(t) data. After examining the periodogram of R(t), only one peak was found, occurring at frequency W = 0.15. Least squares linear regression analysis was applied to estimate the a and b in Eq. (6), and only the a = 10.87 and b = 35.29 coefficients were significant at the 0.05 level. Thus, only one cyclical movement of the residual data was retained in the initial model as indicated in Eq. (8).

Step 4. The initial R(t) model in Eq. (8) was subjected to nonlinear regression analysis. The nonlinear regression estimates were W = 0.114, a = 32.17 and b = 18.60; and the Eq. (8) was rewritten as given shown in Eq. (9).

Step 5. The white noise test was applied to the RR(t) data. The results showed that the RR(t) was white noise at the 0.05 significance level. The final form of the equation for generating missing solar radiation data for the Isabela Station is described in Eq. (10). The R^2 value was 0.76, and the simulated results for six years (1977–1982) and available observation data are plotted in Fig. 2. This Eq. (9) model appears to be able to generate missing data for the Isabela Station.

Isabela Experiment Station, Puerto Rico: (7)

$$SR(t) = [368.96 + 4.71* \cos(0.524t) + 58.67* \sin(0.524t) + R(t)]$$

$$R(t),\text{initial} = [10.87* \cos(0.15t) + 35.29* \sin(0.15t) + RR(t)]$$ (8)

$$R(t) = [32.17* \cos(0.114t) + 18.60* \sin(0.114t) + RR(t)]$$ (9)

$$SR(t) = [\{368.96 + 4.71* \cos(0.524t) + 58.67* \sin(0.524t)\}$$
$$+ \{32.17* \cos(0.114t) + 18.60* \sin(0.114t)\}]$$ (10)

23.3.1.2   COROZAL WEATHER STATION

The data between December 1978 and April 1982 were used to develop the model. The procedures were similar to the procedures used for the Isabela Station except that the A_1, A_2, A_3, B_1, B_2, and B_3 of original data as given in Eq. (5) were significant at the 0.05 level. The spectral analysis of $R(t)$ showed that there were two peaks W_1 and W_2 present in the $R(t)$ data. The coefficients of a_{11}, a_{21}, b_{11}, and b_{21} were significant at the 0.05 level. The final form of the model for the Corozal weather Station is given Eq. (11). The R^2 value was 0.70. The simulated results for six years (1977–1982) and available observation data are plotted in Fig. 3. This Eq. (11) model appears to be able to synthesize missing data for the Corozal Station.

Corozal: (11)

$$SR(t) = [341.81 - \{24.92* \cos(0.524t) + 19.03* \sin(0.524t) +$$
$$23.82* \cos(1.047t) + 1.34* \sin(1.047t)\} + \{36.24* \cos(1.571t) -$$
$$11.15* \sin(1.571t) - 36.09* \cos(0.674t) + 16.92* \sin(0.674t)\} +$$
$$\{29.63* \cos(0.923) + 6.14* \sin(0.923)\}]$$

Furtuna: (12)

$$SR(t) = [413.80 - \{46.22* \cos(0.524t) - 7.03* \sin(0.524t) + 35.09*$$
$$\cos(1.047t) - 5.21* \sin(1.047t)\} - \{2.38* \cos(0.299t) - 18.08* \sin(0.299t) +$$
$$31.00* \cos(2.275t) + 4.09* \sin(2.275t)\}]$$

Lajas: (13)

$$SR_1 = [116.42 + 0.60* SR_5], \quad R^2 = 0.88$$

Rio Piedras: (14)

$$SR_7 = [175.59 + 0.77* SR_4], \quad R^2 = 0.61$$

Gurabo: (15)

$$SR_2 = [444.74 + 0.44* SR_4], \quad R^2 = 0.62$$

23.3.1.3   FORTUNA WEATHER STATION

The data between January 1979 and August 1980 were chosen for developing the model. The procedures used for the Fortuna Station were similar to the one used for the Corozal Station except that the A_3 and B_3 coefficients for the original data were not significant. The final form of the model for the Fortuna weather station is described in Eq. (12). The R^2 value was 0.88 and the simulated results for six years (1977–1982) and available observation data are plotted in Fig. 4. This Eq. (12) appears to be able to generate missing data for the Fortuna Station.

23.3.1.4 LAJAS STATION

The consecutive data between December 1978 and May 1980 were used to attempt the development of a spectral analysis model. The results, however, indicated that the R^2 value was not high enough to draw a meaningful model. This could be due to the consecutive record consisting of only 18 months of data, which may be too short period for establishing a reliable coefficient for harmonic components. Thus, the correlation technique as presented (*see* Eq. (2)) was used for the Lajas Station. After the stepwize linear regression analysis was implemented, the results showed that the solar radiation data from the Lajas Station (SR_1) were highly correlated with those from the Fortuna Station (SR_5), and can be expressed by Eq. (13). This Eq. (13) was used to generate missing data for the Lajas Station.

23.3.1.5 RIO PIEDRAS STATION

The solar radiation data from Isabela station (SR_4) were used to establish a model for interpolating the missing data for the Rio Piedras Station (SR_7) by using the correlation technique as presented (Eq. (2)). The results are expressed as in Eq. (14). The R^2 value of 0.61 is considered to be acceptable in this study, mainly because the historical data from the Rio Piedras Station are too sparse to advance a further formulation.

23.3.1.6 GURABO STATION

The solar radiation data for Gurabo station were used to formulate a model for generating missing data for the Gurabo Station by using the correlation technique (Eq. (2)). The results showed that only the Isabela Station data (SR_4) are correlated with the Gurabo Station data (SR_2), as described in Eq. (15).

23.3.1.7 ADIUNTAS STATION

There were no historical solar radiation data available from the Adjuntas Station. Therefore, only the average approach method (Eq. (4)) and the solar radiation data available from the other six stations were used to generate the solar radiation for the Adjuntas Station.

23.3.2 GENERATION OF MISSING RELATIVE HUMIDITY DATA

Relative humidity data were partially missing from the Isabela, Fortuna, Adjuntas, and Rio Piedras Stations. The relative humidity data were not varied with a base period of 12 months as was the case for the solar radiation data. Thus, the spectral analysis approach was not applicable to generate missing relative humidity data.

Isabela: (16)
$$RH_4 = [67.90 + 0.17* RH_1 - 0.37* RH_3 + 0.27* RH_7], \ R^2 = 0.63$$

Adjuntas: (17)
$$RH_6 = [30.49 + 0.49* RH_1 + 0.09* RH_3], \ R^2 = 0.58$$

Furthermore, there were some recorded relative humidity data available from each station. Thus, the average approach method (Eq. (4)) was also not applicable for simulating the missing relative humidity data. Therefore, only the correlation technique (Eq. (2)) and the historical data interpolation method (Eq. 3) were used to generate missing relative humidity data. The correlation technique was used to generate missing data for the Isabela and Adjuntas Stations. The results are expressed as in Eqs. (16) and (17), where, RH_1 = relative humidity at Lajas Station; RH_3 = relative humidity at Corozal Station; RH_4 = relative humidity at Isabela Station; RH_6 = relative humidity at Adjuntas Station; and RH_7 = relative humidity at Rio Piedras Station.

The historical data interpolation method (Eq. (3)) was used to generate missing relative humidity data for the Fortuna and Rio Piedras stations.

23.4 SUMMARY

Climatic data such as temperature, solar radiation, relative humidity and wind speed have been widely used to estimate evapotranspiration. Most of the solar radiation data and portions of the relative humidity data are missing from the historic records in Puerto Rico. Depending upon the availability and characteristics of records, four methods (Correlation technique, historical data interpolation, average approach and spectral analysis) were used for generating missing climatic data. The limitations of each method are discussed. The results showed that the spectral analysis approach, correlation technique and average approach have been successfully applied to generate missing solar radiation data for Puerto Rico. The R^2 values varied from 0.61 to 0.88. The correlation technique and historical data interpolation methods have been satisfactorily used to interpolate missing relative humidity data. The R^2 varied from 0.58 to 0.63.

Depending upon the availability and characteristics of records, four methods (Correlation technique, historical data interpolation, average approach and spectral analysis) can be successfully used for generating missing climatic data for any location in the world.

KEYWORDS

- averaging historical data
- climatic data
- correlation technique
- energy budget
- evapotranspiration
- groundwater fluctuations
- historical data interpolation
- irrigation
- Periodogram
- Puerto Rico
- regional average
- regression technique
- relative humidity
- soil moisture budget
- solar radiation

- **spectral analysis**
- **synthetic data**
- **time series analysis**
- **water budget**
- **water vapor transfer**

REFERENCES

1. Blaney, H. F.; Criddle, W. D.; **1950,** Determining water requirements in irrigated area from climatological and irrigation data. Dept, Agr. US Soil Conservation Services, SCS-TP-96, pp. 48.
2. Climatological data for Puerto Rico: 1975–1984. NOAA-USDC, Asheville-NC-USA.
3. Doorenbos, J. Pruitt, W. D.; **1977,** *Guidelines for Predicting Crop Water Requirements.* Food and Agriculture Organization of the United Nations, Irrigation and Drainage Paper No. 24, pages 144.
4. Gray, D. M, **1970,** *Handbook on the Principles of Hydrology.* Water Information Center, Inc.; Port Washington, New York.
5. Goyal, M. R.; Shih, S. F.; Generation of missing climatic data for Puerto Rico. Corporación Universitaria Santa Rosa de Cabal [UNISARC], Colombia. Boletín: Investigaciones de UNISARC, **2006,** *5(1),* January): 1–11.
6. Jensen, M. E.; **1974,** *Consumptive Use of Water and Irrigation Water Requirements.* Am. Soc. Civil Engr.; New York, pages 215.
7. Ligon, J. T.; Wilson, T. V.; **1975,** Water balance computations based on evapotranspiration estimates. ASAE Paper No. 75-2027.
8. Linsley, R. K.; Kohier, M. A.; J. Paulhus, L. H.; **1975,** *Hydrology for Engineers.* McGraw-Hill Book Co. Inc.; New York.
9. Penman, H. L.; Natural evaporation from open water, bare soil and grass. Proceedings of the Royal Society, Series A.; **1948,** *193,* 120–145.
10. Penman, H. L.; **1956,** Estimating evaporation. Trans. Am. Geophys. Union, *37,* 43–50.
11. Priestley, M. B.; **1981,** *Spectral Analysis and Time Series.* Academic Press, Inc.; Orlando, Florida.
12. Shih, S. F.; Myhre, D. L.; J. W.; Mishoe and G.; Kidder, Water management for sugarcane production in Florida Everglades. Proc. International Soc. of Sugar Cane Technologists, 16th Congress, San Paulo-Brazil, **1977,** *2,* 995–1010.
13. Shih, S. F.; Allen, L. H.; Jr.; Hanimond, L. C.; Jones, J. U.; Rogers, J. S.; Smajstrla, A. G.; Basin wide water requirement estimation in southern Florida. ASAE Transactions, **1983,** *26(3),* 760–766.
14. Shih, S. F.; Data requirement for evapotranspiration estimation. J. Irrig. Drain. Eng., Am. Soc. Civil Engr.; **1985,** *110(3),* 263–274.
15. Shih, S. F.; Keng, K. S.; **1987,** Generation of missing climatic data for Puerto Rico. Technical meeting paper #1987–2502 by American Society of Agricultural Engineers, St Joseph-MI.
16. Thornthwaite, C. W.; Report of the committee on transpiration and evaporation: 1943–1944. Trans. Am. Geophys. Union, **1944,** *25,* 683–693.

CHAPTER 24

ESTIMATION OF PAN EVAPORATION COEFFICIENTS[1]

ERIC W. HARMSEN, ANTONIO GONZÁLEZ-PÉREZ,
and AMOS WINTER

CONTENTS

[1] This research is supported by University of Puerto Rico, Agricultural Experiment Station Grant SP-368, NOAA-CREST and NASA-EPSCoR. Modified and reprinted with permission from: "Harmsen, Eric W., Antonio González-Pérez and Amos Winter, 2004. Re-evaluation of pan evaporation coefficients at seven locations in Puerto Rico. J. Agric. Univ. P.R., 88(3–4):109–122." © Copyright Journal of Agriculture Univ. P.R.

24.1 INTRODUCTION

The pan evaporation method is widely used to schedule irrigation because it is easy and inexpensive. The University of Puerto Rico Agricultural Experiment Station (UPR-AES) is currently promoting this method in the "Technological Package (Spanish)" guidance publications for various crops [1]. A number of studies have been performed to determine optimal irrigation rates based on pan evaporation data in Puerto Rico for several crops, such as: Tanier [2], bananas under mountain conditions [3], bananas under semiarid conditions [4], plantains under semiarid conditions [5], watermelon under semiarid conditions [6], and sweet peppers under humid conditions [7, 8]. Harmsen [9] presented a summary of these studies.

The evapotranspiration (ET) can be estimated using pan evaporation method with the following equations:

$$ET_{pan} = K_c \times Et_{o\text{-}pan} \tag{1}$$

$$Et_{o\text{-}pan} = K_p \times E_{pan} \tag{2}$$

where, ET_{pan} = Actual crop ET, based on the pan-derived reference ET, $Et_{o\text{-}pan}$; K_p = Pan coefficient; E_{pan} = Class A pan evaporation; and K_c = Crop coefficient. According to Allen et al. [10], estimates of ET from pan data are generally recommended for periods of 10 days or longer. It is recommended that the Eqs. (1) and (2) should be used for ET estimations usually for periods of four to seven days in Puerto Rico. Most of the studies have recommended applying water to plants at a rate equal to 1 to 1.5 times the pan-estimated ET rate to maximize crop yield. Because this approach is easy and inexpensive, these studies represent valuable contributions to agricultural production in the tropics.

Problems, however, may result from this approach because of the inherent differences in water loss from an open water surface and a crop [10]. Another potential limitation is that only a single value of crop coefficient is commonly used, and by definition the crop coefficient varies throughout the season. The magnitude of the crop coefficient depends on crop height, leaf area, crop color, stomatal resistance, and crop maturity.

Although recommended irrigation application rates by this method may maximize crop yields, the method may also result in the over application of water early in the crop season, leading to the degradation of groundwater resources from leaching of agricultural chemicals.

In Puerto Rico, the K_p values commonly used were derived from a study by Goyal and González [11] using data from the seven agricultural substations located at Adjuntas, Corozal, Juana Díaz (Fortuna), Gurabo, Isabela, Lajas, and Río Piedras. Figure 1 shows the location of the substations and the Climate Divisions established by the National Oceanic and Atmospheric Administration (NOAA). These data were developed on the basis of the ratio of long-term monthly average reference evapotranspiration (estimated from an equation) to pan evaporation:

$$K_p = ET_0 / E_{pan} \tag{3}$$

where, K_p is the pan coefficient; ET_0 is reference or potential ET; and E_{pan} is the pan evaporation rate. Mean daily values of pan evaporation were derived from a University of Puerto Rico Agricultural Experiment Station document *Climatological Data from the Experimental Substations of Puerto Rico* [12]. Goyal and González [11] estimated the potential ET by using the Soil Conservation Service (SCS) Blaney-Criddle method [13]. In a recent study by the American Society of Civil Engineers (ASCE), the SCS Blaney-Criddle method was found to produce large errors relative to weighing lysimeter data indicating overestimation on average by 17% in humid regions and underestimation on average by 16% in arid regions [14].

In a study that compared seasonal consumptive use for pumpkin and onion at two locations in Puerto Rico, Harmsen et al. [15] reported large differences between the SCS Blaney-Criddle method [11] and the Penman-Monteith method. The Penman-Monteith approach used crop coefficients as determined by the FAO procedure [10]. Crop stage durations, used to construct the crop coefficient curves, were based on crop growth curve data presented by Goyal [16]. The maximum observed differences in the estimated seasonal consumptive use were on the order of 100 mm per season. The study concluded that large potential differences can be expected between the SCS Blaney-Criddle and the Penman-Monteith methods, with under-estimations in some months and overestimations in other months.

Because of inherent errors associated with the SCS Blaney-Criddle method, the published values of K_p for Puerto Rico may not be accurate. The United Nations Food and Agriculture Organization (FAO) currently recommends using the ratio of pan evaporation divided by the Penman-Monteith-estimated reference ET for calculating the pan coefficient [10]. The Penman-Monteith based reference ET was found to have a high degree of accuracy in the above-mentioned ASCE study [14], with errors not exceeding ±4 percent.

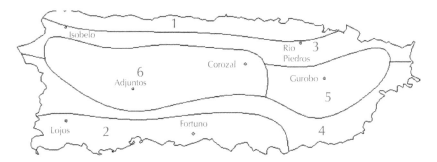

FIGURE 1 UPR Agricultural Experiment Substation locations and NOAA climate divisions of Puerto Rico: 1, North Coastal; 2, South Coastal; 3, Northern Slopes; 4, Southern Slopes; 5, Eastern Interior; and 6, Western Interior.

This chapter indicates how to update pan coefficient values for the seven substations in Puerto Rico using the Penman-Monteith reference ET, and to incorporate 20 years of additional pan evaporation data. As part of the study, long-term trends in pan evaporation data were evaluated.

24.2 MATERIALS AND METHODS

24.2.1 PAN EVAPORATION DATA

Historical pan evaporation data were evaluated to determine whether decreasing or increasing trends existed in the data. Roderick and Farquhar [17] and Ohmura and Wild [18] have reported that pan evaporation rates have been decreasing globally. The cause of the decrease has been attributed to the decrease in solar irradiance (during the last decade) and changes in diurnal temperature range and vapor pressure deficit [17]. If in fact pan evaporation is changing in Puerto Rico, then the more recent data (e.g., for the last 20 years) may provide better estimates of the pan coefficient than would longer term average data. Updated pan evaporation data were obtained from NOAA's Climatological Data Sheets. To evaluate possible trends, pan evaporation data were plotted graphically, and regression analysis was used to determine whether the regression coefficient (i.e., the slope) of the best-fit linear model was significantly different from zero. All statistical analyzes were performed by using the statistical software package StatMost Version 3.6 [19].

24.2.2 REFERENCE ET

The long-term monthly reference ET was estimated by using the Penman-Monteith equation [10]:

$$ET_0 = \frac{0.408\Delta(R_n - G) + \gamma\left(\dfrac{900}{T + 273}\right)u_2\left(e_s - e_a\right)}{\Delta + \gamma\left(1 + 0.34\,u_2\right)} \tag{4}$$

where :

ET_0	=	ET [mm day^{-1}]
Δ	=	Slope of the vapor pressure curve [kPa°C^{-1}]
R_n	=	Net radiation at the crop surface [MJ m^{-2}day^{-1}]
G	=	Soil heat flux density [MJ m^{-2} day^{-1}]
ρa	=	Mean air density at constant pressure [kg m^{-2}]
Cp	=	Specific heat at constant pressure [MJ kg^{-1}C^{-1}]
e_s-e_a	=	Vapor pressure deficit [kPa]
e_s	=	Saturation vapor pressure [kPa]
e_a	=	Actual vapor pressure [kPa]
r_a	=	Aerodynamic resistance [s m^{-2}]
r_s	=	The bulk surface resistance [s m^{-2}]
λ	=	Latent heat of vaporization [MJ kg^{-1}]
γ	=	The psychrometric constant [kPa°C^{-1}]

Eq. (4) applies specifically to a hypothetical grass reference crop with an assumed crop height of 0.12 m, a fixed surface resistance of 70 sec/m and a solar reflectivity of 0.23. The FAO recommends using the Penman-Monteith method over all other methods even when local data are missing. Studies have shown that using estimation procedures for missing data with the Penman – Monteith equation will generally provide more accurate estimates of ET_0 than will other available methods requiring less data input [10].

Of the various climate parameters needed to calculate ET_0 with Eq. (4), only air temperature (T) and wind speed (u) were available for all seven experimental substations in Puerto Rico; however, wind speed was not measured consistently. For example, in the case of Lajas, wind speed data were available only during the following years: 1963, 1966 to 1969, 1971 to 1978, 1983 to 1985 and 1987 to 1990. Wind speeds were measured at 0.33 m above the ground and therefore needed to be adjusted to the two-meter value (u_2) using the logarithmic adjustment equation presented by Allen et al. [10].

Relative humidity (needed to estimate actual vapor pressure) is measured at the substations by using a sling psychrometer, but only once in 24 hours; thus, these data do not represent daily average values. Therefore, the actual vapor pressure was derived from the dew point temperature (T_{dew}). Long-term average dew point temperature was estimated from the minimum air temperature plus or minus a temperature correction factor. Temperature correction factors, developed for the six NOAA Climate Divisions for Puerto Rico (*see* Fig. 1), were obtained previously from Harmsen et al. [20]. Net radiation was estimated from solar radiation (R_s) by using the method presented by Allen et al. [10] involving the use of a simple equation for island settings (elevations <100 m) or by the Hargreaves radiation equation (elevations >100 m), based on air temperature differences. Pan coefficients were estimated from Eq. (3). Statistical comparisons were made between K_p from average pan evaporation data collected between 1960 and 1980 and K_p from data collected between 1981 and 2000.

24.3 RESULTS AND DISCUSSION

Figure 2 shows the monthly average pan evaporation for the seven experimental substations, based on approximately 40 years of pan evaporation data. Note that pan evaporation was highest for Fortuna and lowest for Adjuntas for most months of the year. Figures 3, 4 and 5 show the average monthly pan evaporation with time. Figure 3 shows the sites that had significant decreasing pan evaporation with time; Fig. 4 shows the sites that had significant increasing pan evaporation with time; and Fig. 5 shows the sites that had no significant increase or decrease in pan evaporation with time. Increases and decreases, as expressed by the linear regression coefficients, associated with Figs. 3 and 4, were significant at or below the 5% probability level.

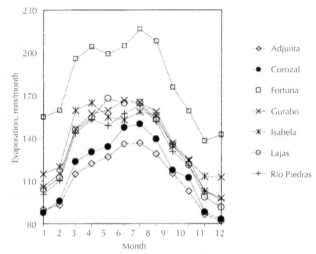

FIGURE 2   Long-term average monthly pan evaporation for the seven substations in Puerto Rico.

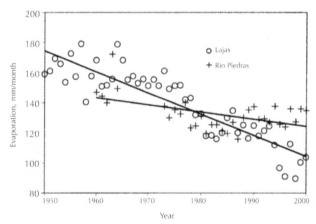

FIGURE 3   Average monthly pan evaporation with time at Lajas and Río Piedras, Puerto Rico.

Regression coefficients associated with the linear regression lines shown in Fig. 5 were not statistically significant. The linear regression results are summarized in Table 1. Following noteworthy results can be summarized from Figs. 3 to 5 and Table 1:

- Lajas had the greatest decrease in the average monthly pan evaporation: 1.4 mm per month (average) per year. This amount is equivalent to a drop of 56 mm per month in the pan evaporation in 40 years. This is a very significant reduction considering that the average pan evaporation in 2002 was only 103.9 mm per month in Lajas. It will be interesting to see whether this trend continues in the future or whether it begins to level off.

- The decreasing pan evaporation observed at Lajas and Río Piedras is consistent with the observed decreasing trend globally.
- Pan evaporation at two sites (Adjuntas and Gurabo) increased. These results are contrary to the observed global decrease in pan evaporation. Both sites are located in humid areas. It is interesting to note that Adjuntas is at a relatively high elevation (549 m), whereas Gurabo is at a relatively low elevation (48 m).

Figure 6 shows the estimated long-term average monthly reference ET for each substation. As with pan evaporation (Fig. 2), Fortuna shows the highest ET_o, and Adjuntas shows the lowest values during most of the year. However, ET_o for Lajas was essentially identical to that of Fortuna, whereas the Lajas pan evaporation (Fig. 2) was lower than that of Fortuna. There are two possible explanations for this:

1. The local environment may have gradually changed in the vicinity of the evaporation tank in Lajas. For example, installation of new structures, establishment of trees, or relocation of the evaporation tank. Development of the Lajas Valley may also have influenced a change in pan evaporation at the substation.
2. Pan evaporation and reference ET may not be directly comparable. Allen et al. [10] list the following factors that may cause significant differences in loss of water from a water surface and from a cropped surface:

- Reflection of solar radiation from the water surface might be different from the assumed 23% for the grass reference surface.
- Storage of heat within the pan can be appreciable and may cause significant evaporation during the night while most crops transpire only during the daytime.
- There are differences in turbulence, temperature and humidity of the air immediately above the respective surfaces.
- Heat transfer occurring through the sides of the pan can affect the energy balance.

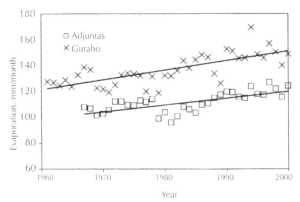

FIGURE 4   Average monthly pan evaporation with time at Adjuntas and Gurabo, Puerto Rico.

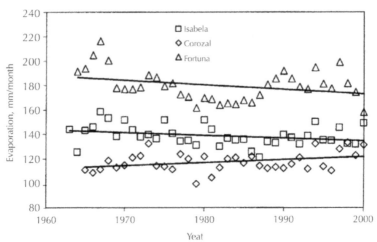

FIGURE 5 Average monthly pan evaporation with time at Corozal, Isabela and Fortuna, Puerto Rico.

Monthly average pan coefficients were estimated for each month at each of the seven experimental substations on the basis of pan evaporation data from 1960 (approximate) to 1980 and from 1981 to 2000. (For convenience, hereafter the earlier period will be referred to as 1960 to 1980 and the latter period as 1981 to 2000.

A Student t-Test analysis indicated that the difference between the mean K_p based on the two time periods was highly significant. Table 2 presents the results of the t-Test. The difference in the mean K_p for all locations for the two time periods was 0.15. The average K_p equaled 0.75 for 1960 to 1980 and 0.91 for 1981 to 2000. A comparison was also made between the K_p values of Goyal and González [11] and the 1981 to 2000 K_p values from this study (*see* Table 5).

A significant difference was observed between the two datasets at the 0.01% probability level, with a difference in the mean K_p of 0.08. The average value of the K_p of Goyal and González [11] was 0.82.

To understand whether the difference in the mean pan evaporation between the two periods (1960 to 1980 and 1981 to 2000) is significant on a practical level (independent of statistical significance), Eq. (2) was used to estimate the difference in the reference ET for a given amount of pan evaporation. Suppose the annual pan evaporation for a certain location was 1500 mm; then the K_p difference of 0.15 is equivalent to $[0.15 \times 1500 \text{ mm}] = 225$ mm in the annual reference ET. For an average farm size of 18 hectares in Puerto Rico [21], this is equivalent to 40,500 m^3 of water (or 10.7 million gallons).

Because there was a significant difference between the mean K_p for the last 20 years and that of the subsequent 20-year period, this study recommends that crop water use estimates use K_p values from the most recent 20 years. Tables 3–5 give the average monthly reference ET, pan evaporation and pan coefficients, respectively.

The methodologies used in this paper can be considered sufficiently general and therefore could be applied at other locations throughout the world.

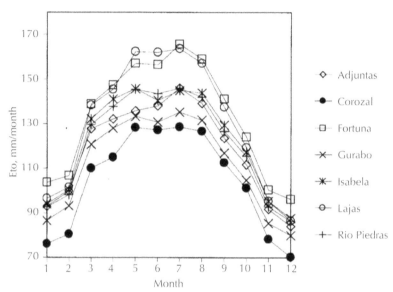

FIGURE 6   Long-term average monthly reference ET for the seven substations.

TABLE 1   Linear regression results for the pan evaporation data from the seven substations.

Station	Latitude	Ele-vation (m)	NOAA Climate Division	Regression Coeffi-cient (slopeofline)	r^2	Significant at the 5% level	Trend
Gurabo	18°15′N	48	5	0.029	0.55	Yes	Increasing
Adjuntas	18°11′N	549	6	0.021	0.47	Yes	Increasing
Corozal	18°20′N	195	6	0.010	0.11	No	Increasing
Isabela	18°28′N	126	3	-0.008	0.08	No	Decreasing
Fortuna	18°01′N	21	2	-0.015	0.10	No	Decreasing
RíoPiedras	18°24′N	100	3	-0.019	0.28	Yes	Decreasing
Lajas	18°03′N	27	2	-0.055	0.81	Yes	Decreasing

24.4   SUMMARY

The objective of the research in this chapter was to update pan evaporation coefficient (K_p) values for the seven University of Puerto Rico, Agricultural Experimental Substations, based on updated pan evaporation data and the Penman-Monteith reference ET. Therefore, historical pan evaporation data for seven experimental substations. Were evaluated to determine whether increasing or decreasing trends existed.

Significant decreasing pan evaporation was observed at Lajas and Río Piedras. Significant increasing pan evaporation was observed at Gurabo and Adjuntas. No significant trends were observed at Fortuna, Isabela or Corozal. A significant difference was found to exist between the mean K_p calculated with pan evaporation data from 1960 to 1980 and that with data from 1981 to 2000. An updated table of monthly average pan coefficients is provided (Table 5) that can be used to estimate ET_{pan} for the seven substations.

TABLE 2 Results of a Student t-test comparing monthly pan coefficients based on pan evaporation data from 1960 (approximate) to 1980 versus 1981 to 2000.

t-Test Analysis Results			
Confidence Level = 0.95 [Two Tail Test]			
	1960 to 1981	1981 to 2000	
Sample Size	84	84	
Number of Missing	0	0	
Minimum	0.5694	0.6732	
Maximum	1.1579	1.2473	
Standard Deviation	0.1398	0.1319	
Standard Error	0.0153	0.0144	
Coeff of Variation	18.5924	14.5441	
Mean	0.7520	0.9067	Difference = 0.1547
Variance	0.0196	0.0174	Ratio = 1.1242
Paired	t-Value	Probability	DF Critical t-Value
	9.5097	6.24516E-015	83 1.9890

Co-Variance = 0.0074; Std Deviation = 0.0163

TABLE 3 Long-term average reference ET (ET_o, mm/month) for the seven experimental substations.

	Jan	Feb	Mar	Apr	May	Jun	Jul	Aug	Sep	Oct	Nov	Dec
Adjuntas	93	100	128	132	136	138	146	139	124	112	92	84
Corozal	76	80	110	115	128	127	129	127	112	101	78	70
Fortuna	104	107	139	147	157	156	166	159	141	124	100	96
Gurabo	87	93	121	128	133	131	135	132	117	105	85	80
Isabela	94	100	132	141	146	141	145	144	129	118	95	88
Lajas	97	102	138	145	162	162	164	157	137	120	95	87
Río Piedras	93	98	130	138	145	143	146	142	127	116	93	86
Average	93	100	128	132	136	138	146	139	124	112	92	84

TABLE 4. Average monthly pan evaporation (E_{pan}, mm per month) based on 1981 through 2000 pan evaporation data for seven experimental substations.

	Jan	Feb	Mar	Apr	May	Jun	Jul	Aug	Sep	Oct	Nov	Dec
Adjuntas	92	95	122	129	131	143	141	136	124	112	93	86
Corozal	91	100	130	136	136	148	152	146	122	114	90	87
Fortuna	152	159	193	199	194	202	212	210	176	161	139	139
Gurabo	117	121	155	167	168	177	176	172	147	133	110	105
Isabela	112	120	154	164	153	153	163	158	134	126	108	108
Lajas	89	92	120	127	139	133	131	135	135	108	89	81
Río Piedras	97	109	141	153	145	152	160	147	130	122	100	94

TABLE 5 Pan Coefficients (K_p, no units) based on 1981 through 2000 pan evaporation data, for seven experimental substations.

	Jan	Feb	Mar	Apr	May	Jun	Jul	Aug	Sep	Oct	Nov	Dec
Adjuntas	1.02	1.05	1.05	1.03	1.04	0.97	1.04	1.02	1.00	1.00	0.99	0.98
Corozal	0.84	0.80	0.85	0.85	0.94	0.86	0.85	0.87	0.92	0.88	0.87	0.81
Fortuna	0.68	0.67	0.72	0.74	0.81	0.78	0.78	0.76	0.80	0.77	0.72	0.69
Gurabo	0.74	0.77	0.78	0.77	0.80	0.74	0.77	0.77	0.80	0.79	0.77	0.76
Isabela	0.84	0.84	0.86	0.86	0.95	0.92	0.89	0.91	0.97	0.93	0.88	0.82
Lajas	1.08	1.10	1.15	1.14	1.17	1.22	1.25	1.16	1.02	1.10	1.08	1.07
Río Piedras	0.95	0.90	0.92	0.90	1.00	0.95	0.91	0.96	0.97	0.95	0.93	0.92

Additional research is needed to help explain the significant reduction in the pan evaporation observed at Lajas as compared to that of other locations. The K_p data presented in Table 5 are valid for data obtained from the pan located at the Lajas Experiment Station. However, if pan evaporation is obtained from another source in the vicinity of Lajas, these data should be compared with the experiment station evaporation data to verify consistency between the two data sources. If large differences exist, then an adjustment should be made in the Lajas K_p values presented in Table 5. Further research is also needed to investigate the reason for the observed variations in the trends in pan evaporation (i.e., increasing at some locations and decreasing at other locations).

KEYWORDS

- **Blaney-Criddle method**
- **crop water use**
- **reference evapotranspiration**
- **evapotranspiration**
- **Food and Agriculture Organization, FAO**
- **pan coefficient**
- **pan evaporation**
- **Penman – Monteith method**
- **Puerto Rico**
- **regression analysis**
- **water management**

REFERENCES

1. Rivera, Rivera, L. E.; **2004,** Personal communication. UPR-Agricultural Experiment Station.
2. Goenaga, R.; Growth, nutrient uptake and yield of tanier (Xanthosoma spp.) grown under semi-arid conditions. *J. Agric. Univ. P.R.,* **1994,** *78(3–4),* 87–98.
3. Goenaga, R.; H.; Irizarry, Yield of banana grown with supplemental drip irrigation on an Ultisol. Expl. Agric.; **1998,** *34,* 439–448.
4. Goenaga, R.; H.; Irizarry, **1995,** Yield performance of banana irrigated with fractions of Class A pan evaporation in a semiarid environment. *Agron. J. 87,* 172–167.
5. Goenaga, R.; H.; Irizarry and E.; González, **1993,** Water requirement of plantains (Musa acuminate x Musa balbisiana AAB) grown under semiarid conditions. Trop. Agric. (Trinidad), *70(1),* 3–7.
6. Santana-Vargas, J. **2000,** Patrones de humedecimiento bajo el sistema de micro riego subterráneo. Thesis, M. S.; Department of Agronomy and Soils. University of Puerto Rico-Mayagu□ez Campus.
7. Harmsen, E. W.; J. Colón-Trinidad, C.; Arcelay and E.; Sarmiento-Esparra, 2002a. Evaluation of the Pan Evaporation Method for Scheduling Irrigation on an Oxisol in Puerto Rico. Proceedings of the Caribbean Food Crops Society, Thirty Eighth Annual Meeting, **2002,** Martinique. 38.
8. Harmsen, E. W.; J. Colón-Trinidad, C. L.; Arcelay and D.; Cádiz-Rodríguez, **2003,** Evaluation of percolation and nitrogen leaching from a sweet pepper crop grown on an Oxisol soil in Northwest Puerto Rico. Proceedings of the 39th Annual Meeting of the Caribbean Food Crops Society, **2003,** Grenada. 39.
9. Harmsen, E. W.; Fifty years of crop evapotranspiration studies in Puerto Rico. J. Soil Water Conserv.; **2003,** *58(4),* 214–223.
10. Allen, R. G.; Pereira, L. S.; D.; Raes and M.; Smith, **1998,** Crop Evapotranspiration Guidelines for Computing Crop Water Requirements. FAO Irrigation and Drainage Paper *56,* Food and Agriculture Organization of the United Nations, Rome.
11. Goyal, M.; González-Fuentes, E. A.; **1989a.** Evaporation Pan coefficients for Puerto Rico [Spanish]. *J. Agric. Univ. P.R., 73(1),* 89–92.
12. Goyal, M. R.; González, E. A.; **1989b.** Climatic data for agricultural substations of Puerto Rico [Spanish]. Agricutural Experiment Station-University of Puerto Rico, Rio Piedras. 87 pages.
13. SCS, **1970,** Irrigation Water Requirements. Technical Release No. 21. USDA-Soil Conservation Service, Engineering Division.

14. Jensen, M. E.; Burman R. D.; Allen, R. G.; **1990,** Evapotranspiration and irrigation water re-quirements. ASCE Manuals and Reports on Engineering Practice No. 70. 332

15. Harmsen, E. W.; Caldero J.; Goyal, M. R.; **2001,** Consumptive Water Use Estimates for Pump-kin and Onion at Two Locations in Puerto Rico. In: Proceedings of the Sixth Caribbean Islands Water Resources Congress. Silva Araya, W. F. (ed). University of Puerto Rico, Mayagu□ez, PR 00680.

16. Goyal, M. R.; **1989,** Estimation of Monthly Water Consumption by Selected Vegetable Crops in the Semiarid and Humid Regions of Puerto Rico. AES Monograph 1999–00, Agricultural Experiment Station, University of Puerto Rico, Río Piedras, PR.

17. Roderick, M. L.; Farquhar, G. D.; The cause of decreased pan evaporation over the past 50 years. **2002,** *298,* 1410–1411.

18. Ohmura, A.; M.; Wild, Is the hydrologic cycle accelerating? Science, **2002,** *298,* 1345–1346.

19. Dat@xiom Software, Inc.; **2001,** User's Guide StatMost Statistical Analysis and Graph–ics. Fourth Edition. Dataxiom Software, Inc.; 3700, Wilshire Blvd. Suite 1000, Los Angeles, CA 90010. (http://www.dataxiom.com).

20. Harmsen, E. W.; Goyal, M. R.; Torres-Justiniano, S.; Estimating evapotranspiration in Puerto Rico. *J. Agric. Univ. P.R.,* **2002b,** *86(1–2),* 35–54.

21. Census of Agriculture for Puerto Rico, **1998,** US Department of Agriculture, National Agricul-tural Statistics Service, AC97-A-52.

CHAPTER 25

DAILY EVAPOTRANSPIRATION ESTIMATIONS USING SATELLITE REMOTE SENSING[1]

ERIC W. HARMSEN, JOHN R. MECIKALSKI, MELVIN J. CARDONA-SOTO, ALEJANDRA ROJAS-GONZALEZ, and RAMON E. VASQUEZ

CONTENTS

[1]Modified and reprinted with permission: Harmsen, E. W.; J. Mecikalski, M. J.; Cardona-Soto, A.; Rojas Gonzalez; Vasquez, R. Estimating daily evapotranspiration in Puerto Rico using satellite remote sensing. WSEAS Transactions on Environment and Development, **2009**, *6(5)*, 456–465. © Copyright   WSEAS Transactions on Environment and Development

25.1 INTRODUCTION

Determination of evapotranspiration (ET) is important for evaluation of hydrologic resources of a region, and evaluating irrigation requirements. Because of the inter-relation between components of the hydrologic cycle, ET is important in the evaluation of soil water content, surface runoff, and aquifer recharge. ET is defined as the combination of evaporation from soil and plant surfaces, and transpiration from plant leaves. Evaporation is the process whereby liquid water is converted to water vapor and removed from the evaporating surface [3]. Transpiration is the vaporization of liquid water contained in plant tissues and its subsequent removal to the atmosphere. Crops predominately lose water through small openings in their leaves called stomata. ET can be expressed in units of mm/day (or in/day), or as an energy flux in units of MJ m^{-2} day^{-1} [3]. ET is important because it is often the largest component of the hydrologic cycle after rainfall. Under arid conditions, potential ET can easily exceed rainfall. Remote sensing methods for estimating ET are needed for tropical conditions. Various techniques have been developed based on radiation methods [e.g., 20] and surface energy budgets [e.g., 1 and 5].

The objective of this chapter is to develop an algorithm for estimating daily, high resolution (1-km), crop reference ET (ET$_0$) over Puerto Rico. In this chapter, three radiation-based ET$_0$ methods will be tested and compared.

25.2 TECHNICAL APPROACH

In this study, the ET flux was estimated using the Penman-Monteith [3], Priestly Taylor [18] and Hargreaves-Samani [7] methods, in combination with a solar radiation product of the GOES-12 satellite. Solar radiation was derived using the radiative transfer model of Diak et al. [19]. Input required for the Penman-Monteith was based on procedures developed for Puerto Rico by Harmsen et al. [11].

Of the three methods, the Penman-Monteith (PM) method is generally regarded as superior because it takes into account the major variables, which control ET [3], and the method has been rigorously validated under diverse conditions throughout the world [18]. The Penman-Monteith method is given by Allen et al. [3, 12] as described below in Eq. (1):

$$ET_0 = \frac{0.408\Delta\left(R_n - G\right) + \left(\dfrac{900}{T+273}\right)u_2\left(e_s - e_a\right)}{\Delta + \gamma\left(1 + 0.34\,u_2\right)} \tag{1}$$

where: Δ is a slope of the vapor pressure curve [k Pa°C^{-1}], R_n is net radiation [MJ.m^{-2}.day^{-1}], G is soil heat flux density [MJ.m^{-2}.day^{-1}], γ is psychrometric constant [k Pa °C^{-1}], T is mean daily air temperature at 2 m height [°C], u_2 is wind speed at 2 m height [m s^{-1}], e_s is the saturated vapor pressure and e_a is the actual vapor pressure [kPa]. Eq. (1) applies specifically to a hypothetical reference crop with an assumed crop height of 0.12 m, a fixed surface resistance of 70 sec m^{-1} and an albedo of 0.23.

The Priestly Taylor equation (PT) represents a simplification of the Penman equation [16, 17,] and is valid for humid conditions:

$$ET_0 = \alpha \frac{\Delta(R_n - G)}{(\Delta + \gamma)} \qquad (2)$$

where: α is the Priestly-Taylor constant equal to 1.26, and the other variables/parameters were previously defined in this chapter. A value of 1.32 has been recommended for estimates from vegetated areas as a result of the increase in surface roughness [14]. In this chapter a value of 1.3 is used. The Hargreaves-Samani (HS) reference ET [7] is described in Eq. (3):

$$ET_o = [0.0135]\,R_s[(T + 17.8)] \qquad (3)$$

In Eq. (3), ET_o and solar radiation (insolation, R_s) are in the same equivalent units of water evaporation [$L\,T^{-1}$], and T is mean temperature in °C. Harmsen et al. [9] reported good agreement between the PM and HS methods for 34 locations in Puerto Rico.

Daily average temperature was estimated using the regression equations of Goyal et al. [6], which relate temperature to elevation in Puerto Rico. The equations provide values of daily mean temperature for each month of the year. The monthly data were regressed to obtain a polynomial equation relating the day of the year with air temperature. The average daily air temperature was "nudged" based on the actual average daily temperature measured from the Natural Resource Conservation Service (NRSC) Soil Climate Analysis Network (SCAN) sites in western Puerto Rico. These sites include coastal and mountainous conditions.

An average value of 1.9 m/s was used for wind speed in the PM model based on the published average winds speeds for the six NOAA Climate Divisions for Puerto Rico [9]. This value is close to the worldwide average value of 2 m/s recommended by the FAO [3] in the absence of observed data.

Saturated and actual vapor pressures are estimated based on the average and dew point temperatures, respectively. For convenience, in this study the dew point temperature was assumed to be equal to the minimum temperature based on the regression method for minimum temperature of Goyal et al. [6] and nudged using actual air temperature data from the seven SCAN stations. For humid conditions, use of minimum temperature for dew point temperature is generally a valid assumption. For the drier south and south-west part of Puerto Rico, however, the assumption may lead to errors in the ET_o calculation.

Solar radiation (R_s) was estimated with the radiative transfer model of Diak et al. [4] using data from the visible-channel of the GOES satellite. More information on this R_s product can be found in Sumner et al. [20]. The methods presented in Allen et al. [12] were used to calculate extraterrestrial radiation (R_a), R_n and G.

25.3 RESULTS AND DISCUSSION

In this section, the ET_o estimates are presented based on the PM, PT and HS methods for March 5th, 2009. Table 1 shows the weather information for the seven SCAN stations in Puerto Rico for this day. Figure 1 shows a visible satellite (GOES) image at 15:15 local time (19:15 UTC), indicating large-scale cloud bans covering the region. The National Weather Service in San Juan reported haze, fog and light rain during the day. The National Weather Service (NWS) reported severe rain in Vaga Alta, Puerto Rico with flooding reported at 15:38 local time (19:38 UTC). However, other locations in Puerto Rico experienced little or no rainfall during the day (Table 1). Figure 2 shows the NEXRAD radar total storm rainfall at 15:26 local time (19:26 UTC), indicating rain bands extending across a significant portion of the island.

Figure 3 shows the estimated average air temperature distribution in Puerto Rico on March 5th, 2009. The average air temperature was based on the regression method of [6] which relates temperature with surface elevation (Fig. 4). The estimated versus observed average air temperature are shown in Fig. 5. The regression equation was used to estimate the average air temperature in Fig. 3.

TABLE 1   Weather information from the seven SCAN stations on March 5th, 2009.

Site	Isabela	Maricao	Guilarte	Fortuna	Combate	Mayaguez	Bosque Seco
	Elevation (m)						
	15	746	1019	28	10	14	165
Rainfall (mm)	2.8	1.8	14.7	0	0	1.5	0
Average Temperature (C)	23.1	17.6	15.8	23.7	23.8	23.1	22.5
Minimum Temperature (C)	21.9	15.9	14.0	20.6	21.3	21.4	19.5
Maximum Temperature (C)	24.4	18.8	16.9	27.0	27.4	25.4	26.7
Relative Humidity (%)	77.4	96.6	97.1	75.7	68.5	79.6	78.8
Wind Speed (m/s)	4.8	2.3	0.8	2.4	0.9	0.8	0.05
Solar Radiation (W/m²)	255	215	92	304	332	304	211

FIGURE 1   Visible satellite image of Caribbean region at 15:15 local time (19:15 UTC).

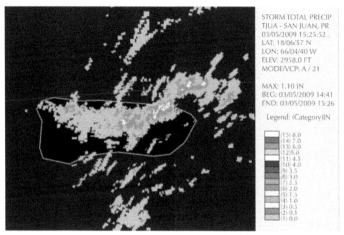

FIGURE 2   NEXRAD radar storm total precipitation in inches over Puerto Rico at 15:26 local time (19:26 UTC).

FIGURE 3   Estimated average air temperature on March 5, 2009.

FIGURE 4   Surface elevation in Puerto Rico.

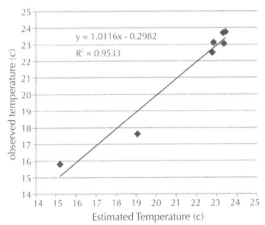

FIGURE 5 Estimated versus observed daily average temperature at the seven SCAN stations in Puerto Rico. The regression equation was used to estimate air temperature in Fig. 3.

Figure 6 shows the distribution of solar insolation across Puerto Rico on March 5, 2009. The figure indicates that the west and southwest parts of the island received significantly more solar insolation than central, northern and northeastern Puerto Rico. The southeast received an intermediate level of solar insolation. This spatial pattern of the solar insolation is apparent in the NEXRAD radar storm total precipitation distribution (Fig. 2).

Figures 7–9 show the daily ET_o estimated using the PM, PT and HS methods, respectively. The ET_o spatial distributions closely match the solar insolation pattern (Fig. 6). In general the three methods are in good agreement. The PM method produced the lowest ET_o values, as compared to the PT and HS methods (*see* differences in the figure color bars). The lowest ET_o values occur in the mountain areas associated with the lowest air temperatures (Fig. 3), and where solar insolation was the lowest.

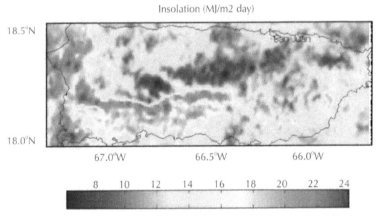

FIGURE 6 Integrated daily solar insolation for Puerto Rico on March 5, 2009.

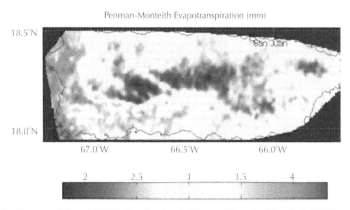

FIGURE 7 Penman-Monteith ET$_o$ distribution in Puerto Rico for March 5, 2009.

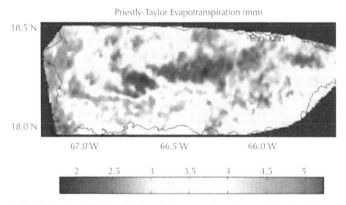

FIGURE 8 Priestly Taylor ET$_o$ distribution in Puerto Rico for March 5, 2009.

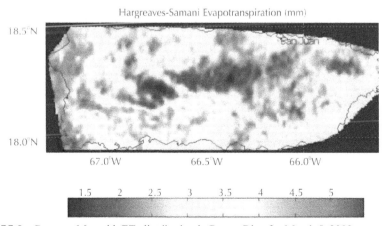

FIGURE 9 Penman-Monteith ET$_o$ distribution in Puerto Rico for March 5, 2009.

TABLE 2 ET_0 estimated by PM, PT and HS methods for March 5th, 2009 compared with the long-term average ET_0 calculated with the Puerto Rico ET (PRET) computer program [8].

Station	Elevation (m)	Latitude	PRET	PM	PT	HS
Isabela	15	18.28	4.7	3.8	3.6	3.7
Maricao	746	18.15	3.9	3.8	4.1	4.3
Guilarte	1019	18.15	3.7	2.4	2.3	1.9
Fortuna	28	18.03	5.0	3.9	3.7	3.9
Combate	10	17.98	5.0	3.8	3.6	3.8
Mayaguez	14	18.22	4.5	3.9	3.8	4.0
Bosque	165	17.97	5.1	3.4	3.1	3.0

Table 2 compares the PM, PT and HS ET_0 values at the SCAN stations with the long-term average ET_0 as calculated by the computer program PRET [8]. All values for March 5th were lower than the long-term average values (PRET). The lowest value of ET_0 was associated with the Guilarte site where the observed and estimated solar radiation were 92 and 118 W/m², respectively, and observed and estimated average daily temperatures were 15.8°C and 15.2°C, respectively.

25.3.1 LIMITATIONS OF METHODS

Theoretically, the PM method is the most accurate of the three; however, numerous assumptions were made in developing the input for the PM method. For example, the wind speed was assumed to be 1.9 m/s over the entire island. Table 1 indicates that average daily wind speeds at the SCAN stations varied between 0.05 to 4.8 m/s, with an average of 1.2 m/s. Future efforts need to incorporate spatially varied wind speed. Air temperature was estimated as described in the Methods section. As can be seen from Figure 5, there was excellent agreement between the estimated and observed temperatures at the seven SCAN sites. However, these stations are limited to locations in western Puerto Rico. Central, northern and northeastern Puerto Rico received relatively low levels of insolation as compared to west and southern Puerto Rico, and consequently the estimated air temperatures in those areas may be overestimated.

25.4 SUMMARY

A technique is presented in which satellite solar insolation estimates are used to predict daily reference evapotranspiration (ET_0) using the Penman-Monteith (PM), Preistly Taylor (PT) and Hargreaves-Samani (HS) methods for Puerto Rico. For this approach, average, minimum and maximum daily air temperatures were obtained from a regression procedure that depends on surface elevation and day of the year. The air temperature was adjusted using actual daily temperatures from several locations in PR. Dew point temperature was assumed to be equal to the daily minimum temperature, and a value of 1.9 m/s was assumed for wind speed. As an example, ET_0 was estimated for March 5, 2009 using the three methods, with the Penman-Monteith method producing the lowest values. This research represents a preliminary step in the development of an ET_0 product for PR. This product is a potentially valuable tool for conducting water resource studies and for supporting irrigation scheduling efforts, not only for Puerto Rico/Carribbean region but also worldwide tropical regions.

KEYWORDS

- **actual vapor pressure**
- **elevation**
- **evapotranspiration**
- **evapotranspiration, crop**
- **evapotranspiration, reference**
- **GOES**
- **Hargreaves-Samani**
- **net radiation**
- **Penman-Monteith**
- **Priestly Taylor**
- **psychrometric constant**
- **Puerto Rico**
- **regression analysis**
- **remote sensing**
- **satellite**
- **saturated vapor pressure**
- **soil heat flux density**
- **solar radiation**
- **temperature**
- **tropical region**
- **vapor pressure curve**
- **wind speed**

REFERENCES

1. Allen, R. G.; M.; Tasumi, R.; Trezza, Robison, C. W.; M.; Garcia, D.; Toll, K.; Arsenault, J. Hendrickx, M. H.; Kjaersgaard, J.; Comparison of Evapotranspiration Images Derived from MODIS Landsat along the Middle Rio Grande. **2008,** Proceedings of the ASCE World Environmental and Water Resources Congress, Ahupua'a.

2. Allen, R. G.; Walter, I. A.; Elliott, R.; Howell, R.; Itenfisu, D.; Jensen, M.; Snyder, R. L.; The *ASCE Standardized Reference Evapotranspiration Equation.* Environmental and Water Resources Institute of the American Society of Civil Engineers. **2005,** 57.

3. Allen, R. G.; Pereira, L. S.; Dirk Raes, Smith, M.; *Crop Evapotranspiration Guidelines for Computing Crop Water Requirements.* FAO Irrigation and Drainage Paper 56,Food and Agriculture Organization of the United Nations, Rome. **1998,** 300.

4. Diak, G. R.; Bland, W. L.; Mecikalski, J. R.; A note on first estimates of surface insolation from GOES-8 visible satellite data. Agric. For. Meteor.; **1996,** *82,* 219–226.

5. Gowda, P. H.; Chávez, J. L.; Colaizzi, P. D.; Evett, S. R.; Howell, T. A.; Tolk, J. A.; Remote sensing based energy balance algorithms for mapping ET: Current status and future challenges. Transactions of the American Society of Agricultural and Biological Engineers. **2007**, *50(5),* 1639-1644.

6. Goyal, M. R.; González, E. A.; C.; Chao de Báez, Temperature versus elevation relationships for Puerto Rico. J.; Agric. UPR, **1988**, *72(3),* 449-467.

7. Hargreaves, G. H.; Samani, Z. A.; Reference Crop Evapotranspiration from Temperature. Appl. Eng. Agric. of ASAE, **1985**, *1(2),* 96-99.

8. Harmsen, E. W.; Gonzaléz, A.; Technical Note: A Computer Program for Estimating Crop Evapotranspiration in Puerto Rico. J.; Agric. UPR. **2005**, *89(1–2),* 107–113.

9. Harmsen, E. W.; Goyal, M. R.; Torres Justiniano, S.; Estimating Evapotranspiration in Puerto Rico. *J. Agric. Univ. P. R.* **2002**, *86(1–2),* 35–54.

10. Harmsen, E. W.; Miller, N. L.; Schlegel, N. J.; Gonzalez, J. E.; Seasonal Climate Change Impacts on Evapotranspiration, Precipitation Deficit and Crop Yield in Puerto Rico. J.; Agricultural Water Management, **2009**, *96,* 1085–1095.

11. Harmsen, E. W.; Gomez, S. E.; Cabassa, E.; Ramirez, N. D.; Pol, S. C.; Kuligowski, R. J.; Vasquez, R.; Satellite Sub- Pixel Rainfall Variability. Wseas Transactions on Signal Processing, **2008**, *8(7),* 504–513.

12. Harmsen, E. W.; Ramirez Builes, V. H.; Dukes, M. D.; Jia, X.; Gonzalez, J. E.; Pérez Alegía, L. R.; A Ground-Based Method for Calibrating Remotely Sensed Surface Temperature for use in Estimating Evapotranspiration. Wseas Transactions on Environment and Development, **2009**, *5(1),* 13-23.

13. Jensen, M. E.; Burman, R. D.; Allen, R. G.; *Evapotranspiration and irrigation water requirements.* ASCE Manuals and Reports on Engineering Practice No. 70, **1990**, 332

14. Morton, F. I.; Operational estimates of areal evapotranspiration and their significance to the science and practice of hydrology. *J. Hydrology,* **1983**, *66,* 1-76.

15. Paech, S. J.; Mecikalski, J. R.; Sumner, D. M.; Pathak, C. S.; Wu, Q.; Islam, S.; Sangoyomi, T. A calibrated, high-resolution OES satellite solar insolation product for a climatology of Florida evapotranspiration. *J. Am. Water Res. Assoc.,* **2009**, *45(6),* 1328–1342.

16. Penman, H. L.; Natural evaporation from open water, bare soil, and grass. Proc. Royal Soc.; London, A **1948**, *193,* 120–145.

17. Penman, H. L.; **1963**, *Vegetation and Hydrology.* Tech. Commission 53. Commonwealth Bureau of Soils, Harpenden, England.

18. Priestly, C. H. B.; Taylor, R. J.; On the assessment of surface heat flux and evaporation using large scale parameters. *Monthly Weather Review,* **1972**, *100,* 81-92.

19. Ramírez-Beltran, N. D.; Kuligowski, R. J.; Harmsen, E. W.; Castro, J. M.; Cruz-Pol, S.; Cardona-Soto, M.; Rainfall Estimation from Convective Storms Using the Hydro-Estimator and NEXRAD.; *Wseas Transactions on Systems,* **2008**, *7(10),* 1016-1027.

20. Sumner, D. M.; Pathak, C. S.; Mecikalski, J. R.; Paech, S. J.; Wu, Q.; Sangoyomi, T.; **2008**, Calibration of GOES-derived Solar Radiation Data Using Network of Surface Measurements in Florida, USA. *Proceedings of the ASCE World Environmental and Water Resources Congress 2008, Ahupua'a.*

CHAPTER 26

VAPOR FLUX MEASUREMENT SYSTEM[1]

ERIC W. HARMSEN, VICTOR H. RAMIREZ BUILES,
MICHAEL D. DUKES, XINHUA JIA, LUIS R. PEREZ ALEGRIA,
and RAMON E. VASQUEZ

CONTENTS

[1]Authors appreciate financial support from NOAA-CREST (NA17AE1625), NASA-EPSCoR (NCC5–595), USDA-TSTAR-100, USDA Hatch (H-402), NASA-URC, and UPRM-TCESS. We would like to thank the following students for their contributions to this paper: Javier Chaparro, Antonio Gonzalez, and Richard Diaz. Modified and reprinted with permission: "Harmsen, E.W., V. H. Ramirez Builes, M. D. Dukes, X. Jia, L. R. Pérez Alegía, R. Vasquez, 2008. An Inexpensive Method for Validating Remotely Sensed Evapotranspiration. Proceedings of the 4th WSEAS International Conference on Advances in Remote Sensing, Venice -Italy. Nov. 20–23. pp 118–123." <World Scientific and Engineering Academy and Society Press, support@ wseas.org>" © Copyright WSEAS 4th 2008 Conference on Advances in Remote Sensing

26.1 INTRODUCTION

Accurate estimates of actual evapotranspiration (ET) are costly to obtain. An inexpensive alternative is to estimate actual evapotranspiration by multiplying a potential or reference evapotranspiration [1, 2, 3] by a crop coefficient (Kc) [4, 5]. This approach has been promoted by the United Nations Food and Agriculture Organization (FAO) for more than 30 years through their Irrigation and Drainage Paper No. 24 [1] and more recently in Paper No. 56 [6]. Even though they have reported values for Kc for numerous crops, many crops grown in the world are not included in their lists, and coefficients for mixed natural vegetation are generally not available. Although crop coefficients derived in other parts of the world can be used to provide approximate estimates of evapotranspiration, the crop coefficient in fact depends upon the specific crop variety and other local conditions [7].

To avoid the need for using crop coefficients, a direct approach can be used to estimate actual evapotranspiration. Current methods for estimating actual evapotranspiration include weighing lysimeter, eddy covariance, scintillometer, and Bowen ratio methods. Each of these methods has certain limitations. A meteorological method is described in this paper which provides an estimate of the actual ET from short natural vegetation or agricultural crops and is less expensive than the other methods mentioned above. The specific objectives of this chapter are to describe a relatively inexpensive method for estimating hourly actual evapotranspiration; and to present results from validation studies conducted in Florida and Puerto Rico (PR).

26.2 TECHNICAL APPROACH

26.2.1 THEORETICAL ANALYSIS

The method used in this study consisted of equating the ET flux equations based on the generalized Penman-Monteith (GPM) combination method [6] with a humidity gradient (HG) method [8]. In the procedure, the value of one of the resistance factors (either the aerodynamic resistance, r_a, or the bulk surface resistance, r_s) is adjusted in the two equations until their ET time series curves approximately coincide. A similar approach was used by Alves et al. [9] in which an independent estimate of ET was derived from the Bowen ratio method, r_a was obtained from a theoretical equation, and r_s was obtained by inversion of the Penman-Monteith equation. The GPM combination equation is described below [6]:

$$ET = \frac{\Delta\left(R_n - G\right) + \rho_a c_p \left(\dfrac{e_s - e_a}{r_a}\right)}{\lambda\left[\Delta + \gamma\left(a + \dfrac{r_s}{r_a}\right)\right]} \tag{1}$$

where: Δ is the slope of the vapor pressure curve [k.Pa°.C⁻¹], R_n is the net radiation[MJ.m⁻².day⁻¹], G is the soil heat flux density[MJ.m⁻².day⁻¹], ρ_a is air density [kg/m³], c_p is specific heat of air [MJ.kg⁻¹°C⁻¹], latent heat of vaporization [MJ kg⁻¹],γ is a psychrometric constant [k.Pa.°C⁻¹], T is the air temperature at 2 m height [°C], u_2 is the wind speed at 2 m height [m.s⁻¹], e_s is the saturated vapor pressure [kPa] and e_a is the actual vapor pressure [kPa], r_a is the aerodynamic resistance [s.m⁻¹] and r_s is the bulk surface resistance [s.m⁻¹]. Eq. (1) applies specifically to a hypothetical reference crop with an assumed crop height of 0.12 m, a fixed surface resistance of 70 sec.m⁻¹ and albedo of 0.23.

Evapotranspiration (ET) can also be estimated by means of a humidity gradient Eq. (2):

$$ET = \left(\frac{\rho_a c_p}{\gamma \rho_w}\right)\left(\frac{\rho v L - \rho v H}{r_a - r_s}\right) \tag{2}$$

In eq. /2/: ρ_a is the mean air density of dry air [kg.m⁻1], c_p specific heat [MJ.Kg⁻¹.⁰C⁻¹], ρ_w is the density of water [kg.m⁻¹], ρ_v is the water vapor density of the air [kg.m⁻¹], and L and H are vertical positions above the ground [m], r_a is the aerodynamic resistance [s.m⁻¹] and r_s is the bulk surface resistance [s.m⁻¹]. The water vapor densities were calculated from the actual vapor pressures and air temperatures using the ideal gas equation. In this study L and H were 0.3 m and 2 m above the ground, respectively. Eq. (2) is essentially identical to the latent heat flux equation presented by Monteith and Unsworth [8], eq. (17.3)]. The method, which effectively combines Eqs. (1) and (2), allows for the solution of r_s. In this study, the value of the aerodynamic resistance (r_a) is estimated using the following Eq. [6]:

$$r_a = \frac{\ln\left[\dfrac{z_m - d}{z_{0m}}\right]\ln\left[\dfrac{z_h - d}{z_{0h}}\right]}{k^2 u_2} = \frac{\varsigma}{u_2} \tag{3}$$

where: z_m is the height of wind measurement [m], z_h is the height of humidity measurement [m], d is zero plane displacement height equal to 0.67 h [m], h is crop height [m], z_{0m} is roughness length governing momentum transfer equal to 0.123 h [m], z_{0h} is roughness length governing transfer of heat and vapor equal to 0.1 z_{0m} [m], and k is the von Karman's constant (0.41). The variable ς is defined as $r_a u_2$. Allen et al. [6] reported a value of equal to 208 for a theoretical reference grass with a height h = 0.12 m. Eq. (3) and the associated estimates of d, z_{0m} and z_{0h} are applicable for a wide range of crops [6]. A study of surface and aerodynamic resistance performed by [10] determined that eq. (3) will produce reliable estimates of r_a for small crops.

In this study, the value of r_s is obtained by a graphical procedure in which successive adjustments are made to r_s until the time-series plots of ET (during the daylight hours) from Eqs. (1) and (2) approximately coincide. Adjustment of the average daily r_s value is considered acceptable when the values of the integrated daily total ET from the two equations are within 0.01 mm.

26.2.2   FIELD DATA ANALYSIS

Climatological data were saved on a Campbell Scientific, Inc. (CSI) CRX10 data logger every 10 seconds. Net radiation was measured using a CSI NR Lite Net Radiometer. Wind speed was measured 3 m above the ground using a CSI MET One 034B wind speed and direction sensor. The wind speed at 3 m was adjusted to the 2 m height using the logarithmic relation presented by [3]. Soil water content was measured using a CSI CS616 Water Content Reflectometer. Soil temperature was measured using two

CSI TCAV Averaging Soil Temperature probes, and the soil heat flux at 8 cm below the surface was measured using a CSI HFT3 Soil Heat Flux Plate.

An initial test using two temperature/relative humidity (Temp/RH) sensors simultaneously, positioned at the same height in close proximity revealed nonconstant differences in RH between the two sensors. Differences in RH ranged from −5% to +8.5% (Fig. 1). Errors of this magnitude were unacceptable for use in estimating the vertical humidity gradient. Therefore, to obtain accurate estimates of the humidity gradient, a single Temp/RH sensor (Vaisala HMP45C) was used, which was automatically moved between two vertical positions (0.3 m and 2 m) every 2 minutes.

An automated elevator device was developed for moving the Temp/RH sensor between the two vertical positions [11]. The device consisted of a PVC plastic frame with a 12 volt DC motor (1/30 hp) mounted on the base of the frame. One end of a 2-m long chain was attached to a sprocket on a shaft on the motor and the other end to a sprocket at the top of the frame. Waterproof limit switches were located at the top and bottom of the frame to limit the range of vertical movement.

The new method was verified by comparing ET results for April 5th and 6th, 2005, with an eddy covariance system at the University of Florida (UF) Plant Science Research and Education Unit (PSREU) near Citra, Florida. The eddy station was located in the center of a 23 ha bahia grass field and the shortest distance from the station to the edge of the field was 230 m.

A second validation was conducted in grass and sweet corn fields located at the University of PR Agricultural Experiment Station at Lajas, PR. Comparisons for the grass were made on December 21, 22, and 23, 2006, and on January 3, 9, 10 and 11, 2007. Comparisons for the sweet corn were made on June 6, 7, 8, 9, 10 and 11, 2007.

A CSI CSAT3-3D Sonic Anemometer and CSI KH_2O krypton Hygrometer are the major instruments used in the eddy covariance systems. The anemometer measured wind speeds and the speed of sound using three pairs of nonorthogonal sonic transducers to detect any vertical wind speed fluctuations. The anemometer was set up facing the prevailing wind to minimize the negative effect by the anemometer arms and other supporting structures. The frequency of the CSAT3 is 10 Hz with an output averaged every 30-minutes. The KH_2O Krypton Hygrometer was mounted 10 cm away from the center of the CSAT3, with the source tube (the longer tube) on the top and the detector tube (the shorter tube) on the bottom. The output voltage of the hygrometer is proportional to the attenuated radiation, which is in turn related to vapor density. The frequency of the hygrometer is 10 Hz with an average output every 30-minute.

The eddy covariance-derived 30-minute latent heat fluxes were corrected for temperature induced fluctuations in air density [12], for the hygrometer sensitivity to oxygen [13], and for energy balance closure. Sensible heat fluxes were corrected for differences between the sonic temperature and the actual air temperature [14]. Both the sensible and latent heat fluxes were corrected for misalignment with respect to the natural wind coordinate system [15]. The Bowen-ratio method was used to close the surface energy balance relationship [16]. Flux and atmospheric measurements were logged using a CSI CR23X data logger. During certain periods, such as early mornings and after precipitation, the hygrometer measurements were not available due to the moisture obscuring the lens. The data analysis was conducted for daytime measurements, based on the available energy for evapotranspiration.

26.3   RESULTS AND DISCUSSION

For convenience, the equipment used in this study involving a standard weather station and an elevator device for obtaining the temperature and humidity gradients, will be referred to as the "ET station." To estimate the ET using data from the ET station the following steps were used:

1.   The data were read into a spreadsheet macro which, among other things, separated the "up" and "down" humidity and temperature data, and calculated actual vapor pressures.
2.   The aerodynamic resistance (r_a) was estimated using Eq. (3).
3.   The ET estimates from Eqs. (1) and (2) were plotted together on the same graph, and the value of r_s was adjusted until the two datasets approximately coincided. The two datasets were considered to be in agreement when their total daily ET was within 0.01 mm of each other.

Table 1 lists the estimated daily ET (expressed in mm) from the eddy covariance systems and the ET station for 15 dates at locations in Florida and Puerto Rico. The ET estimates by the two methods were in reasonably good agreement. The average ratio of the ET from the eddy system and the ET station was 1.03. Figures 1 through 3 show ET from the eddy systems plotted against ET from the ET station. The average coefficient of determination (r^2) was 0.87.

TABLE 1   Comparison of daily ET determined from eddy covariance system and the ET station for 15 dates from 2005 to 2006 in Florida, PR.

Date	Vegetation	Location	Eddy Covariance ET(mm)	ET Station (mm)	$[ET_{eddy}]/[ET_{station}]$
4/5/2005	Grass	Florida	3.92	4.11	0.95
4/6/2005	Grass	Florida	3.78	3.66	1.03
12/21/2006	Grass	Puerto Rico	2.89	2.85	1.01
12/22/2006	Grass	Puerto Rico	5.14	4.60	1.12
12/23/2006	Grass	Puerto Rico	3.80	3.40	1.12
1/3/2007	Grass	Puerto Rico	3.09	2.82	1.10
1/9/2007	Grass	Puerto Rico	2.00	2.10	0.95
1/10/2007	Grass	Puerto Rico	2.90	2.50	1.16
1/11/2007	Grass	Puerto Rico	2.20	2.20	1.00
6/6/2007	Corn	Puerto Rico	5.40	5.60	0.96
6/7/2007	Corn	Puerto Rico	5.97	6.00	1.00
6/8/2007	Corn	Puerto Rico	6.39	6.32	1.01
6/9/2007	Corn	Puerto Rico	7.00	7.60	0.92
6/10/2007	Corn	Puerto Rico	6.90	6.10	1.13
6/11/2007	Corn	Puerto Rico	5.70	6.10	0.93
Average	--	--	--	--	**1.03**

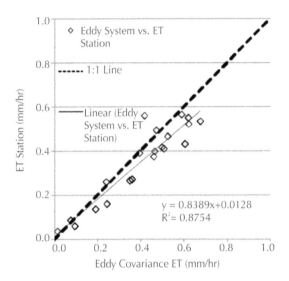

FIGURE 1 Half-hour values of grass evapotranspiration (expressed in mm/hr) estimated using the eddy covariance system and ET station on April 5th and 6th, 2005 at the University of Florida Plant Science Research and Education Center near Citra, FL.

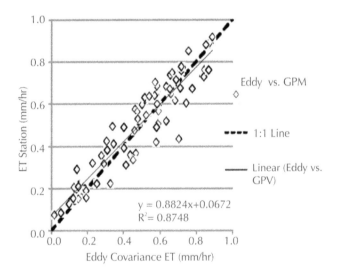

FIGURE 2 Half-hour values of corn evapotranspiration (expressed in mm/hr) estimated using the eddy covariance system and ET station for June 6, 7, 8, 9, 10 and (11–2007) at the University of PR Agricultural Experiment Station, Lajas, PR.

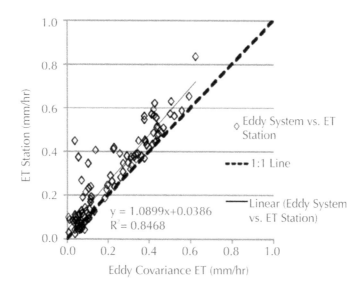

FIGURE 3   Half-hour values of grass evapotranspiration (expressed in mm hr^{-1}) estimated using the eddy covariance system and ET station for December 21, 22, 23 (2006) and January 3, 9, 10, 11 (2007) at the University of PR Agricultural Experiment Station, Lajas, PR.

26.4   SUMMARY

The ET estimates from the ET station and eddy covariance methods were in reasonably good agreement. Because of the relatively low cost of the method described in this chapter, numerous stations could be deployed over a region with the purpose of validating or calibrating remote sensing estimates of ET. The system described in this chapter is approximately 20 to 7 times less expensive than the weighing lysimeter and eddy covariance methods, respectively. The Bowen Ratio method, although relatively inexpensive, nevertheless is about twice the cost of the system described in this chapter. The procedure can be adopted for other locations worldwide.

KEYWORDS

- aerodynamic resistance
- ATLAS
- Bowen ratio
- corn
- crop water requirement
- eddy covariance
- evapotranspiration
- FAO
- Florida

- **Generalized Penman-Monteith combination method**
- **grass**
- **humidity gradient**
- **irrigation**
- **lysimeter**
- **NASA**
- **neural network**
- **Penman-Monteith**
- **Puerto Rico**
- **remote sensing**
- **surface resistance**
- **surface temperature**

REFERENCES

1. Doorenbos, J. Pruitt, W. O.; **1977**, *Guidelines for Predicting Crop Water Requirements*. FAO Irrigation and Drainge Paper 24, Revised. United Nations, Rome.
2. Wang, Y. M.; Traore S.; Kerh, T.; Neural network approach for estimating reference evapotranspiration from limited climatic data in Burkina Faso. WSEAS Transactions on Computers, **2008,** *7(6),* 704–713.
3. Teixeira, J. Shahidian S.; J. Rolim. Regional analysis and calibration for the South of Portugal of a simple evapotranspiration model for use in an autonomous landscape irrigation controller. WSEAS Transactions on Environment and Development, **2008,** *4(8),* 676–686.
4. Jensen, M. E.; Burman, R. D.; Allen, R. G.; **1990,** *Evapotranspiration and Irrigation Water Requirements*. ASCE Manuals and Reports on Engineering Practice No. 70. 332
5. Hajare, H. V.; Raman N. S.; Dharkar, E. J. New technique for evaluation of crop water requirement. WSEAS Transactions on Environment and Development, **2008,** *4(5),* 436–446.
6. Allen, R. G.; Pereira, S. L.; D.; Raes, M.; Smith, M. **1998,** *Crop Evapotranspiration: Guidelines for Computing Crop Water Requirements*. Food and Agricultural Organization of the United Nations (FAO) 56. Rome. 300 pages.
7. Harmsen, E. W. A Review of evapotranspiration studies in Puerto Rico. J. Soil and Water Conservation, **2003,** *58(4),* 214–223.
8. Monteith, J. L.; Unsworth, M. H.; **2008,** *Principles of Environmental Physics*. 3rd ed.; Academic Press. 418 pages.
9. Alves, I.; A.; Perrier, Pereira, L. S.; Aerodynamic and surface resistances of complete cover crops: How good is the "big leaf"? Transactions of the ASAE, **1998,** *41(2),* 345–351.
10. Kjelgaard, J. F.; Stokes, C. O.; Evaluating surface resistance for estimating corn and potato evapotranspiration with the Penman-Monteith model. Trans. American Society of Agricultural Engineers, **2001,** *44(4),* 797–805.
11. Harmsen, E. W.; R.; Díaz and J. Chaparro. **2004,** A ground-based procedure for estimating latent heat energy fluxes. Proceedings of the NOAA Educational Partnership Program, Education and Science Forum, New York City, NY, October 21st–23rd.
12. Webb, E. K.; Pearman, G. I.; Leuning. R.; Correlation of flux measurements for density effects due to heat and water vapor transfer. Quarterly Journal of the Royal Meteorological Society, **1980,** *106,* 85–100.

13. Tanner, B. D.; Greene, J. P.; **1989,** Measurement of sensible heat and water vapor fluxes using eddy correlation methods. Final report prepared for US Army Dugway Proving Grounds, Dugway, Utah, 17 p.

14. Schotanus, P.; F. Nieuwstadt, T. M.; H. A. R. De Bruin. **1983,** Temperature measurement with a sonic anemometer and its application to heat and moisture fluxes. Boundary-Layer Meteorol, *26,* 81–93.

15. Baldocchi, D. D.; Hicks, B. B.; Meyers, T. P.; **1988,** Measuring biosphere atmosphere exchanges of biologically related gases with micrometeorological methods. Ecology, *69(5),* 1331–1340.

16. Twine, T. E.; Kustas, W. P.; Norman, J. M.; Cook, D. R.; Houser, P. R.; Meyers, T. P.; Prueger, J. H.; Starks, P. J. Wesely, M. L.; **2000,** Correcting eddy covariance flux underestimates over a grassland. Agricultural and Forest Meteorology, *103,* 279–300.

17. Ramírez-Builes, Víctor H.; **2007,** Plant water relationships for several common bean genotypes (Phaseolus vulgaris L.) with and Without drought stress conditions. MS Thesis. Department of Agronomy and Soils, University of Puerto Rico-Mayaguez, Puerto Rico, December 2007.

18. González, J. E.; Luvall, J. C.; D.; Rickman, Comarazamy, D. E.; Picón, A. J. Harmsen, E. W.; H.; Parsiani, N.; Ramírez, R.; Vázquez, R.; Williams, Waide, R. B.; Tepley, C. A.; Urban heat islands developing in coastal tropical cities. EOS Transactions, A. G. U.; **2005,** *86, 42,* 397 and 403.

19. Jarvis, P. G.; **1981,** Stomatal conductance gaseous exchange and transpiration. Pages 175–204, In: *Plants and their Atmospheric Environment* by J. Grace E. D.; Ford and Jarvis, P. G.; editors. Blackwell, Oxford U. K.

20. Luvall, J. C.; Lieberman, D.; Lieberman M.; Hartshorn G. S.; Peralta. R.; Estimation of tropical forest canopy temperatures, thermal response numbers, and evapotranspiration using an aircraft-based thermal sensor. Photogrammetric Engineering and Remote Sensing, **1990,** *56(10),* 1393–1401.

21. Holbo, H. R.; Luvall, J. C.; Modeling surface temperature distributions in forest landscapes. Remote Sensing of Environment, **1989,** *27,* 11–24.

22. Quattrochi, D. A.; Luvall, J. C.; Thermal infrared remote sensing for analysis of landscape ecologic processes: methods and applications. Landscape Ecology, **1990,** *14,* 577–598.

23. Turner M. G.; Gardner, R. H.; **1991,** *Quantitative Methods in Landscape Ecology: the Analysis and Interpretation of Landscape Heterogeneity.* Springer-Verlag Inc.; New York.

CHAPTER 27

CLIMATE CHANGE IMPACTS ON AGRICULTURAL WATER RESOURCES: 2090[1]

ERIC W. HARMSEN, NORMAN L. MILLER, NICOLE J. SCHLEGEL, and JORGE E. GONZALEZ

CONTENTS

[1] Modified and reprinted with permission from: Harmsen, E. W.; Miller, N. L.; Schlegel, N. J.; Gonzalez, J. E. Seasonal Climate Change Impacts on Evapotranspiration, Precipitation Deficit and Crop Yield in Puerto Rico. *J. Agric. Water Manage.* **2009,** *96(7),* 1085–1095. © Copyright Elsevier 2009 Journal Agricultural Water Management

27.1 INTRODUCTION

In recent years great emphasis has been given to the potential impact that human in-duced increases in atmospheric carbon dioxide (CO_2) will have on the global climate during the next 50 to 100 years [1, 2]. Significant changes are expected to occur in the air temperature, sea surface temperature, sea level, and the magnitude and frequency of extreme weather events. Potential impacts on water resources in rain-dominated catchments, such as those found in the Caribbean Region [3] include: higher precipita-tion extremes, increase in streamflow seasonal variability, with higher flows during the wet season and lower flows during the dry season; increase in extended dry period probabilities; and a greater risk of droughts and flood. Extended dry periods and the potential for greater evaporation will have a negative impact on lake levels used for freshwater supply. Groundwater use will likely be increased in the future due to in-creasing demand, and because groundwater may be needed to offset declining surface sources during the drier months. Extended dry periods will also reduce soil moisture and therefore increase water demand by irrigated agricultural.

This chapter addresses the global warming-temperature dependent changes in ref-erence evapotranspiration (ET_o) and precipitation deficit (or precipitation excess) for the twenty-first century at three locations on the Island of Puerto Rico. In this chapter, future values of reference evapotranspiration and precipitation deficit were estimated, based on data from a general circulation model (GCM). This study is the first of its kind in Puerto Rico and provides potentially important information for water resource planners.

Numerous other studies have been conducted using GCM output for hydrologic model forcing. Bouraoui et al. [4] coupled the hydrologic model ANSWERS [5] with a GCM showing that although large-scale GCM output data could be one of the best available techniques to estimate the effects of increasing greenhouse gases on pre-cipitation and evapotranspiration, their coarse spatial resolution was not compatible with watershed hydrologic models. Bouraoui et al. [4] proposed a general methodol-ogy to disaggregate large-scale GCM output directly to hydrologic models and il-lustrated by predicting possible impacts of CO_2 doubling on water resources for an agricultural catchment close to Grenoble, France. The results showed that the doubling atmospheric CO_2 would likely reduce aquifer recharge causing a negative impact on groundwater resources in the study area. However, the authors warned the results were obtained from only one GCM and since many uncertainties still exist among different models, they must be used with caution. The disparate spatial scales between GCMs and hydrologic models requires that statistical or dynamic downscaling techniques be used [6].

Miller et al. [7] analyzed the sensitivity of California streamflow's timing and amount using two GCM projections and the U.S. National Weather Service – Rive Forecast Center's Sacramento-Snow model and found that regardless of the GCM projection, the hydrologic response will lead to decreased snowpack, early runoff, and increased flood likelihoods, with a shift in streamflow to earlier in the season. Maurer and Duffy [8] evaluated the impact of climate change on streamflow in California based on downscaled data from ten GCMs. They observed significant detection of decreasing summer flows and increasing winter flows, despite the relatively large intermodel variability between the 10 GCMs. Brekke et al. [9] evaluated water resources for the San Joaquin Valley in California using two GCMs (HadCM2 and PCM). They predicted impacts on reservoir inflow, storage, releases for deliveries, and streamflow. They concluded that the results were too broad to provide a guide for selection of mitigation projects. Most of the impact uncertainty was attributed to differences in projected precipitation type (rain, snow), amount, and timing by the two GCMs. Dettinger et al. [10] applied a component resampling technique to derive streamflow probability distribution functions (PDFs) for climate change scenarios using six GCMs. The results indicated that although the total amount of total streamflow per water year in California did not change significantly, the mean 30-year (1961–1990) climatological peak streamflow shifted 15 to 25 days earlier under the climate projection scenario, as was observed initially in 1987 [11]. The results were consistent with Stewart et al. [12] who evaluated 302 western North American gauges for their trends in steamflow timing across western North America.

Regional or mesoscale models have also been used to evaluate potential future impacts on water resources. For example, Pan et al. [13] coupled the National Center for Atmospheric Research (NCAR)/Penn State University mesoscale model version 5 (MM5), the U.S. Department of Agriculture (USDA) Soil Water Assessment Tools (SWAT), and the California Environmental Resources Evaluation System (CERES) together to form a two-way coupled soil-plant-atmosphere agroecozystem model. The purpose of this coupled model approach was to predict seasonal crop-available water, thereby allowing evaluation of alternative cropping systems.

The water cycle of tropical islands in the Caribbean Region is determined by a unique set of external and local factors. Although the general characteristics of the hydrological cycle are well understood, little information is available on the sensitivity of flux rates and therefore, relative importance of the various components of the hydrologic cycle, especially under different global climate change scenarios and local land use practices in tropical regions. Furthermore, there is a lack of understanding relative to the linkage between mesoscale weather processes and the hydrologic cycle at the basin scale. Improving our understanding of these processes is crucial for managing risks in the future related to climate and land use change. This study presents a methodology that can be used to evaluate reference evapotranspiration and precipitation deficit (as defined by Ref. [14]), and can potentially be applied at other locations throughout the world. Other components of the hydrologic water balance and relative crop yield reduction are also evaluated.

27.2 MATERIALS AND METHODS

The objective of this study was to analyze future precipitation, reference evapotranspiration, precipitation deficit and relative crop yield reduction at three locations in western PR. Although the temperature and precipitation data were downscaled to specific locations (Adjuntas, Mayaguez and Lajas, PR), generic values were assumed for other parameters required in the analysis. For example, soil texture was assumed to be clay, as this is the dominant soil texture in all three areas. This assumption affects the values of the soil field capacity and wilting point. Average values of evapotranspiration crop coefficients and yield response factors were used for the generic crop, and average monthly runoff coefficients were used based on values derived from the two principal watersheds in the study area (Añasco and Guanajibo Watersheds).

Near-surface air temperature and precipitation were statistically downscaled to the three sites matching historical distributions (1960 to 2000) using the method of Miller et al. [15, 16]. Historical near surface air temperatures were obtained at 2-m height above the ground surface. The site locations were selected because they represent a relatively wide range of conditions within the region (Fig. 1, Table 1). Adjuntas is humid, receives a large amount of precipitation, is at a relatively high elevation, the topography is mountainous and is located relatively far from the coast. Mayagüez is humid, receives a large amount of precipitation, is located immediately adjacent to the Mayagüez Bay, the elevation is close to sea level, topography is relatively flat near the ocean but rises in elevation away from the ocean. Lajas is less humid than the other two locations, receives less precipitation, is located in a flat valley, and is about half the distance to the ocean as Adjuntas. The Lajas Valley, designated by the Commonwealth of Puerto Rico as an Agricultural Reserve, is well-known for its elaborate irrigation and drainage system. Irrigation water is derived from the Lago Loco reservoir located at the eastern end of the Lajas Valley [17].

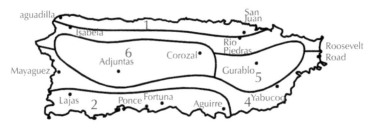

FIGURE 1 Map of Puerto Rico showing the locations of Adjuntas (A), Mayagüez (M) and Lajas (L). Numbers indicate National Oceanographic and Atmospheric Administration (NOAA) Climatic Divisions: 1, North Coastal; 2, South coastal; 3, Northern Slopes; 4, Southern Slopes; 5, Eastern Interior; and 6; Western Interior.

TABLE 1 Latitude, elevation, average precipitation, average temperature, NOAA Climate Division and distance to the coast for the three study locations in Puerto Rico.

Location	Units	Adjuntas	Mayaguez	Lajas
Latitude	Decimal degrees	18.18	18.33	18.00
Elevation from m.s.l.	m	549	20	27
Annual Rainfall	mm	1871	1744	1143
Tmean	°C	21.6	25.7	25.3
Tmin	°C	15.2	19.8	18.8
Tmax	°C	27.9	30.5	31.7
NOAA Climate Division	---	6	4	2
Distance to Ocean	km	22	3	10

TABLE 2 Temperature correction factor K_o used in Eq. 2 for NOAA Climatic Divisions 2, 4 and 6 within Puerto Rico [Harmsen et al., 2002].

NOAA Climate Division	2	4 and 6
K_o (°C)	-2.9	0

[a] See Fig. 1 for climatic divisions 2, 4 and 6. Note that climatic divisions 1, 3 and 5 in Fig. 1 are not relevant in this chapter.

The GCM data were obtained from the Department of Energy (DOE)/National Center for Atmospheric Research (NCAR) Parallel Climate Model (PCM) [18]. The emission scenarios considered are from the Intergovernmental Panel on Climate Change Special Report on Emission Scenarios (IPCC SRES) B1 (low) A2 (mid-high) and A1fi (high) [19].

Reference evapotranspiration (ET_0) was estimated using the Penman-Monteith (PM) method [20]:

$$ET_0 = \frac{0.408\Delta(R_n - G) + \gamma\left(\frac{900}{T+273}\right)u_2(e_s - e_a)}{\Delta + \gamma(1 + 0.34\,u_2)} \tag{1}$$

where: Δ is a slope of the vapor pressure curve [kPa.°C^{-1}], R_n is net radiation (MJ.m^{-2} .day^{-1}), G is soil heat flux density (MJm^{-2}day^{-1}), γ is psychrometric constant (kPa°C^{-1}), T is mean daily air temperature at 2 m height (°C), u_2 is wind speed at 2 m height (m s^{-1}), e_s is the saturated vapor pressure and e_a is the actual vapor pressure (kPa). Eq. (1) applies specifically to a hypothetical reference crop with an assumed crop height of 0.12 m, a fixed surface resistance of 70 sec m^{-1} and an albedo of 0.23. Vapor pressure was calculated using the following equation:

$$e(T) = 0.6108 \exp\left(\frac{17.27T}{T + 237.3}\right) \tag{2}$$

where: e(T) is vapor pressure (KPa) evaluated at temperature T (°C). Saturated and actual vapor pressures were estimated using eq. (2) with the mean monthly air temperature (T_{mean}, °C) and mean monthly dew point temperature (T_{dew}, °C), respectively.

In this study T_{dew} was assumed to be equal to T_{min} for the Adjuntas and Mayaguez sites, a valid assumption for reference conditions or regions characterized as humid and subhumid [21]. For nonreference sites or arid and semiarid climates, the T_{dew} can be estimated from the following relation [20]: $T_{dew} = T_{min} + K_o$, where, K_o is a temperature correction factor ($K_o < 0$). Lajas is located in the National Oceanographic and Atmospheric Administration's (NOAA) Climate Division 2 for PR (Fig. 1), which is classified as semiarid. Harmsen et al. [22] determined a value of $K_o = -2.9$°C for this climate division, which was used in this study for the Lajas site (Table 2).

The FAO recommends that wind speed be estimated from nearby weather stations, or as a preliminary first approximation, the worldwide average of 2 m/s can be used. In this study we used the wind speed values presented by Harmsen et al. [22], which were based on average station data within the Climatic Divisions established by the NOAA (Fig. 1), and are presented in Table 3. The data in Table 3 were derived from wind speed sensors located at airports and university experiment stations. Average wind speeds were based on San Juan and Aguadilla for Div. 1; Ponce, Aguirre, Fortuna and Lajas, for Div. 2; Isabela and Rio Piedras for Div. 3; Mayagüez, Roosevelt Rd. and Yabucoa for Div. 4; Gurabo for Div. 5; and Corozal and Adjuntas for Div. 6. The sensor heights were 10 m and 0.58 m above the ground for the airports and experiment stations, respectively. The experiment station wind speed sensors were the standard agricultural cup-type anemometer which measures the daily distance in miles. Information about the airport wind speed sensors was not available. Harmsen et al. [22] obtained the wind speed data from the International Station Meteorological Climate Summary [23]. Measured wind speeds were adjusted to the wind speed at 2 m above the ground using the following equation [20]: $u_2 = (4.87\ u_z) / (\ln(67.8\ z - 5.42))$, where u_z is the wind speed at height z above the ground.

TABLE 3 Average daily wind speeds [at 2 m height above the ground] during 12-months and NOAA Climatic Divisions within Puerto Rico [Harmsen et al. 2002, 22].

Month	Average daily wind speed (m/s)[b]					
	NOAA Climate Division[a]					
	1	2	3	4	5	6
Jan	2.7	1.8	2.2	1.8	1.1	1.3
Feb	2.8	2.0	2.4	2.0	1.3	1.5
Mar	3.0	2.2	2.6	2.1	1.4	1.5
Apr	2.9	2.1	2.4	2.1	1.5	1.5
May	2.6	2.2	2.2	2.0	1.6	1.6
Jun	2.6	2.4	2.4	2.0	1.7	1.8
Jul	2.9	2.4	2.7	2.0	1.6	1.8
Aug	2.7	2.1	2.5	1.8	1.3	1.5
Sep	2.1	1.7	2.0	1.6	1.1	1.2
Oct	1.9	1.5	1.8	1.6	0.9	1.1
Nov	2.2	1.4	2.0	1.6	0.9	1.0
Dec	2.6	1.5	2.3	1.6	0.9	1.0

[a] See Fig. 1 for NOAA Climate Divisions
[b] Averages are based on:
San Juan and Aguadilla for Division 1;
Ponce, Aguirre, Fortuna and Lajas, for Division 2;
Isabela and Rio Piedras for Division 3;
Mayaguez, Roosevelt Road and Yabucoa for Division 4;
Gurabo for Division 5;
and Corozal and Adjuntas for Division 6.

Solar radiation (R_s) was estimated using the Hargreaves' radiation formula [24]:

$$R_s = k_{Rs}(T_{max} - T_{min})^{1/2} R_a \qquad (3)$$

where: k_{Rs} is an adjustment factor equal to 0.16 for interior locations (Adjuntas) and 0.19 for coastal locations (Mayagüez and Lajas); T_{max} and T_{min} are the mean monthly maximum and minimum air temperature (°C), respectively; and R_a is the extraterrestrial radiation (Wm^{-2}). The various formulas used to calculate R_a (Eq. (2)), R_n and G (Eq. (1)) are presented by Allen et al. [20].

The precipitation deficit (PD) was estimated by subtracting the monthly cumulative ET_o from the monthly cumulative precipitation (P). This approach has been used previously by De Pauw [14] in an agroecological study of the Arabian Peninsula. A positive value indicates water in excess of crop water requirements and a negative value indicates a deficit in terms of crop water requirements. It should be noted that we estimated PD using the reference evapotranspiration and not the actual crop evapotranspiration.

Relative crop yield reduction was estimated from the expression presented by Allen et al. [20]:

$$YR = Ky\,(1-Ks) \tag{4}$$

where: YR is relative crop yield reduction, K_y is a yield response factor, K_s is a water stress coefficient defined as the ratio of ET_{cadj} to ET_c.

$$ET_{cadj} = K_s\,ET_c \tag{5}$$

and

$$ET_c = K_c ET_o, \tag{6}$$

where, ET_{cadj} is the adjusted crop evapotranspiration accounting for limited water availability, ET_c is the crop evapotranspiration under well watered conditions, ET_o is crop reference evapotranspiration, and K_c is the evapotranspiration crop coefficient.

In this study a generic crop with K_c, and K_y values equal to 1 is considered. The assumption of a K_c equal to 1 is especially applicable for long season crops such as banana, pineapple, sugar cane, and citrus, in which the mid season lengths are 120–180 day, 600 days, 135–210 days, and 120 days, respectively. For these same crops, average mid season K_c values are 1.15, 0.5, 1.25 and 0.8 (average 0.94). Here we assume the generic crop has a seasonal yield response factor K_y equal to 1. Allen et al. [20] reported K_y values for 24 crops with an average value of 1.04. Considering the evapotranspiration crop coefficient values for just these 24 crops, the average K_c rounds to 1.1. However, crops are within the "mid" growth stage only a portion of the time, and during other periods the K_c would be lower; therefore a lower value of 1.0 is justified.

The crop stress coefficient, K_s, was determined as follows: for soil moisture values between the soil field capacity (θ_{FC}) and the threshold moisture content (θ_t), equal to the θ_{FC} minus the readily available water (RAW), K_s was equal to 1. Between the θ_t and the soil wilting point (θ_{WP}), K_s varied linearly between 1 (at θ_t) and 0 (at θ_{WP}). RAW is defined as p TAW, where p is the average fraction of the total available water (TAW) that can be depleted from the root zone before moisture stress occurs and ET is reduced. In this study we used a value of p equal to 0.5, a recommended value for forage crops, grain crops and deep rooted row crops [25].

The volumetric soil moisture content is needed to estimate K_s and YR. In this analysis a generic vertical one meter clay soil profile was assumed (predominant soil texture in Puerto Rico) with the following characteristics [26]: soil porosity (φ) = 530 mm, field capacity (FC) = 440 mm and wilting point (WP) = 210 mm. The mean-monthly soil moisture content was derived from the following water balance:

$$S_{i+1} = P_i - ET_{cadj,\, i} - RO_i - Rech_i + S_i,$$ (7)

where, S_{i+1} is the depth of soil water at the beginning of month $i+1$ (mm), S_i is the depth of soil water in the profile at the beginning of month i(mm), P_i is precipitation during month i(mm), $ET_{cadj,\, i}$ is actual evapotranspiration during month i(mm), RO_i is surface runoff during month i(mm) and $Rech_i$ is percolation or aquifer recharge during month i(mm).

Surface runoff was determined based on the following simple monthly runoff equation: $RO = C\ P$, where P is monthly precipitation and C is monthly runoff coefficient. The monthly values of C were derived from the ratio of runoff (stream flow) and precipitation data from the two principal watersheds in the study area (Añasco and the Guanajibo watersheds). The 12 monthly C values (January through December) were 0.40, 0.29, 0.30, 0.31, 0.51, 0.38, 0.30, 0.29, 0.52, 0.52, 0.60, and 0.64, respectively. Historical average streamflow was obtained from the USGS for Water Year 2002 [27]. Average monthly watershed precipitation was derived from interpolated rain gauge data obtained from the USGS, covering the period between 1990 to 2000.

Aquifer recharge was estimated from the following relations:

$$S_{i+1} = Pi - ET_{cadj,\, i} - RO_i + S_i$$ (8a)

If $S_{i+1} \leq FC$ then $Rech_i = 0$ (8b)

If $S_{i+1} > FC$ then $Rech_i = S_{i+1} - FC$, and $S_{i+1} = FC$ (8c)

27.3   RESULTS AND DISCUSSION

Figure 2 shows the average daily air temperatures for the three locations derived from historical records. The slopes of the trend lines were 9×10^{-5} °C/day, 8×10^{-5} °C/day and 5×10^{-6} °C/day, respectively, for Adjuntas, Mayagüez and Lajas. The slopes for the Adjuntas and Mayagüez data were statistically significant at the 95% confidence level. However, the slope for the Lajas data was not significant. From 1970 to 2000 the average temperature at Adjuntas increased by 0.99°C. From 1961 to 2000 the average temperature for Mayagüez increased by 1.17°C. These increases in temperature are significantly greater than the global average increase of 0.6 ± 0.2°C during the last century.

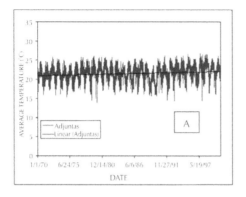

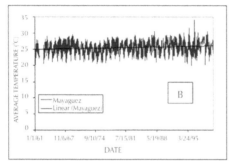

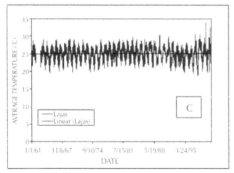

FIGURE 2   Historic daily mean air temperatures at Adjuntas (a), Mayaguez (b) and Lajas ©, PR. Linear trend lines and associated equations have been included.

Since the slope associated with the Lajas regression equation was not significant, an estimate of the increase in temperature based on the slope is not appropriate. It should be noted that the nonsignificant increase in air temperature for Lajas is anomalous when compared with the data presented by Ramirez-Beltran et al. [28], who indicated an average trend in air temperature in Puerto Rico, based on data from 53 stations collected between 1950 and 2006, similar to those shown in Figs. 2a and 2b, for Adjuntas and Mayagüez, respectively. Similar increasing air temperature trends have been observed in the Dominican Republic, Haiti, Jamaica and Cuba [28].

Whatever caused the Lajas historical air temperature data to respond differently than the other two sites (possibly moved instrument, change of instruments, station proximity to paved road, and/or land use change), the temperature increase predicted by the statistical downscaling procedure preserved the increase in temperature for Lajas for the next 100 years, as shown in Figure 3 (Scenario A2). Figure 3 also shows predicted minimum and maximum air temperatures. Figs. 4, 5 and 6 show the air temperature difference ($T_{max}-T_{min}$), vapor pressure deficit (VPD), and reference evapotranspiration (ET_o) for the A2 scenario for Lajas during the next 100 years. Increasing variance can be observed in the $T_{max}-T_{min}$, VPD and ET_o data, which is probably due to the increasing variance evident in the mean air temperature (Fig. 3). Interestingly, the variance in the minimum temperature can be seen to decrease with time.

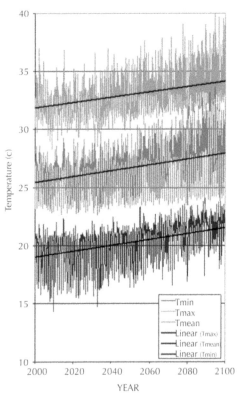

FIGURE 3 Monthly minimum, mean and maximum air temperature for the A2 scenario at Lajas. Linear regression trend lines are shown.

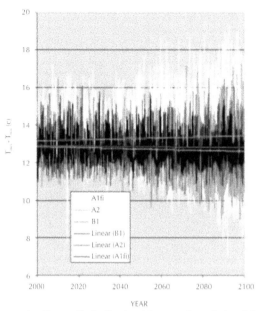

FIGURE 4   Mean monthy Tmax–Tmin for the A2 scenario at Lajas. Linear regression trend lines are shown.

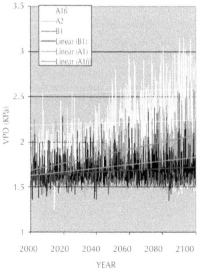

FIGURE 5   Mean monthly vapor pressure deficit (VPD) for the A2 scenario at Lajas. Linear regression trend lines are shown.

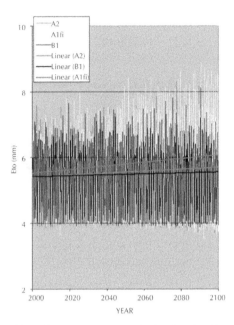

FIGURE 6   Mean monthly reference evapotranspiration for the A2 scenario at Lajas. Linear regression trend lines are shown.

For the wettest (September) and driest (February) months, respectively, Figure 7 shows increasing precipitation during September (i.e., positive slope in the linear regression trend line) and a slight decrease in precipitation during February (i.e., negative slope). Figure 8 shows the predicted monthly average precipitation for each month of the year for the years 2000 and 2090 for the three climate change scenarios for the Lajas location. The predicted precipitation values are based on 20-year averages, for example, the average monthly precipitation for 2000 was based on the average of the monthly precipitation from 1990 through 2010. Note that the 2000 precipitation results vary slightly between scenarios. This is due to the influence of the climate change scenario during the period between 2000 and 2010. Slight variations in the 2000 results for other predicted variables between scenarios will also be observed.

Figure 9 indicates that the B1 scenario average monthly precipitation does not change significantly between 2000 and 2090 for the months of November through July. However, for the A1fi scenario, the 2090 monthly precipitation dropped markedly (−50 mm on average) during these months, relative to 2000. In all scenarios the 2090 rainfall increased significantly during September (150 mm on average), relative to 2000. The results are consistent with other studies indicating the rainy season in the Caribbean will become wetter and the dry season will become drier [e.g., 2, 3, 29, 30].

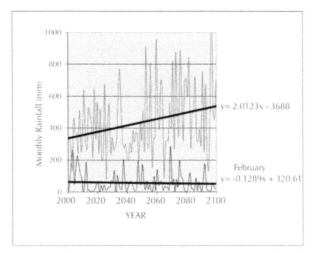

FIGURE 7 Estimated precipitation at Lajas for climate change scenario A2 for February and September.

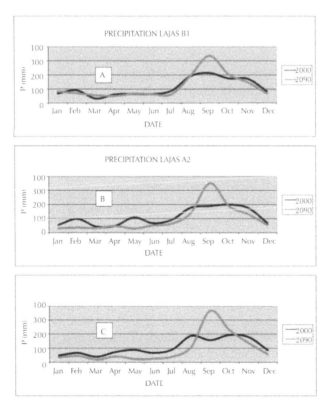

FIGURE 8 Precipitation (P) for Lajas for scenario B1 (a), A2 (b) and A1fi © by month for 2000 and 2090 for Lajas, PR.

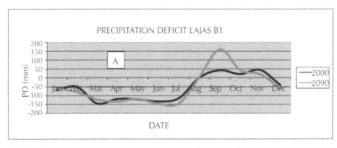

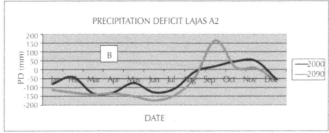

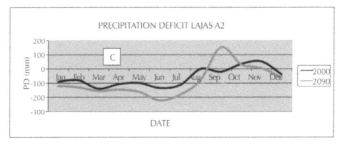

FIGURE 9 Precipitation deficit (PD) for Lajas for scenario B1 (a), A2 (b) and A1fi © by month for 2000 and 2090 for Lajas, PR. A negative value indicates a deficit and a positive value indicates an excess precipitation with respect to crop water requirements.

Table 4a presents PD for the three locations and the three climate change scenarios for the months of February and September, for the years 2000, 2050 and 2090. Note that all of the values for February are negative indicating a deficit in terms of crop water requirements and all but one value for September are positive indicating an excess in terms of crop water requirements. Table 4b presents the difference in the PD relative to the year 2000.

Table 4a shows increasing deficits in February at all locations for the A1fi and A2 scenarios. Although there was an increase in the deficit for the B1 scenario in February, the trend is not as clear. Interestingly the largest deficit occurred for the A2 scenario (−130.8 mm), not the A1fi scenario, which was expected since the A1fi scenario produces higher air temperatures. However the deficit associated with the A1fi scenario for Adjuntas for 2090 was essentially identical (−130.5 mm). The higher (or equal) value of PD for the A2 scenario relative to the A1fi scenario was likely caused by the fact that the A2 scenario produced slightly lower rainfall (35.89 mm) as compared to the rainfall (42.52 mm) produced under the A1fi scenerio.

TABLE 4 Estimated September precipitation deficit (A) and change in precipitation deficit (PD) relative to the year 2000 (B) for Adjuntas, Mayaguez and Lajas, for 2000, 2050 and 2090. Values represent 20-year averages. A negative value indicates a precipitation deficit and a positive value indicates a precipitation excess relative to crop water requirements.

4A		Precipitation deficit (mm)					
		February			September		
Scenario	Year	Adjuntas	Mayaguez	Lajas	Adjuntas	Mayaguez	Lajas
	2000	−6.2	−52.7	−80.3	169.1	100.5	−21.5
A1fi	2050	−25.6	−70.3	−105.2	250.4	178.0	9.7
	2090	−35.8	−84.5	−130.5	480.7	377.4	150.4
	2000	36.9	−22.2	−42.3	222.2	144.0	17.6
A2	2050	−28.6	−77.2	−103.0	339.3	241.4	84.0
	2090	−41.2	−113.9	−130.8	467.1	344.8	162.7
	2000	14.4	−38.2	−51.5	249.0	168.1	40.6
B1	2050	−18.9	−72.5	−92.1	301.9	206.5	74.4
	2090	−3.7	−72.1	−80.1	437.2	305.3	159.0

4B		Change in precipitation deficit (PD, mm) relative to year 2000					
		February			September		
Scenario	Year	Adjuntas	Mayaguez	Lajas	Adjuntas	Mayaguez	Lajas
	2000	0.0	0.0	0.0	0.0	0.0	0.0
A1fi	2050	−19.3	−17.6	−24.9	81.3	77.5	31.2
	2090	−29.6	−31.8	−50.2	311.5	276.9	171.9
	2000	0.0	0.0	0.0	0.0	0.0	0.0
A2	2050	−65.5	−55.0	−60.7	117.1	97.5	66.4
	2090	−78.1	−91.7	−88.5	244.9	200.9	145.1
	2000	0.0	0.0	0.0	0.0	0.0	0.0
B1	2050	−33.2	−34.3	−40.6	52.8	38.4	33.9
	2090	−18.1	−33.9	−28.5	188.1	137.2	118.4

Notes: Table 4A presents PD (mm) for the three locations and the three climate change scenarios for the months of February and September, and for the years 2000, 2050 and 2090. Note that all of the values for February are negative indicating a deficit in terms of crop water requirements, and all but one value for September is positive indicating an excess in terms of crop water requirements.
Table 4B presents the difference in the PD relative to the year 2000.

Tables 5a and 5b, respectively, present the February and September average components of the hydrologic water balance for the three study areas for years 2000, 2050 and 2090 under climate change scenarios B1, A2 and A1fi. The predicted components of the hydrologic water balance are based on 20-year averages.

From Table 5a (February), the following observations can be made: in general, ET_{cadj}, surface runoff and soil moisture content decreased with time. An exception to this is scenario B1 for Adjuntas, in which the ET_{cadj} more or less remained the same.

Aquifer recharge in February generally decreased, and in all cases dropped to zero by 2090, except for scenario B1 at Adjuntas. It is noted that the current condition (2000) aquifer recharge in most cases is negligible. From Table 5b (September), the following observations can be made: precipitation, surface runoff and aquifer recharge increased with time.

TABLE 5A   February components of the hydrologic water balance for the three study areas for years 2000, 2050 and 2090 under climate change scenarios B1, A2 and A1fi.

SITE	SCENARIO	YEAR	P	ET$_o$	ET$_{cadj}$	RO	RECH	MC	YR
					mm				%
Adjuntas	A1FI	2000	97	103	67	28	8	32	35
		2050	76	102	54	22	0	30	46
		2090	60	96	45	17	0	28	53
	A2	2000	139	102	75	40	14	34	26
		2050	72	100	53	21	2	29	47
		2090	56	97	52	17	0	29	46
	B1	2000	116	102	69	33	23	32	32
		2050	83	102	67	28	0	31	34
		2090	98	101	72	27	6	33	28
Lajas	A1FI	2000	66	146	43	19	0	26	69
		2050	54	159	38	16	0	25	76
		2090	43	173	33	12	0	24	81
	A2	2000	98	140	65	27	0	28	53
		2050	48	151	37	14	0	25	75
		2090	36	167	35	10	0	25	79
	B1	2000	88	140	51	25	8	28	62
		2050	54	146	42	15	0	25	70
		2090	68	148	42	19	0	26	70
Mayaguez	A1FI	2000	72	124	49	20	0	27	60
		2050	57	127	42	16	0	26	67
		2090	45	129	31	13	0	25	76
	A2	2000	72	124	49	30	0	27	60
		2050	53	130	39	15	0	26	70
		2090	41	136	33	12	0	25	76
	B1	2000	87	126	50	25	5	28	59
		2050	60	133	44	17	0	26	67
		2090	72	144	47	21	0	26	67

NOTE: P is precipitation; ET$_o$ is reference evapotranspiration; ET$_{cadj}$ is the actual evapotranspiration adjusted for soil moisture availability; RO is surface runoff; RECH is aquifer recharge; and SM is soil moisture; and YR is relative crop yield reduction.

TABLE 5B   September components of the hydrologic water balance for the three study areas for years 2000, 2050 and 2090 under climate change scenarios B1, A2 and A1fi.

SITE	SCENARIO	YEAR	P	ET_o	ET_{cadj}	RO	RECH	MC	YR
					mm				%
Adjuntas	A1FI	2000	297	128	125	153	56	41	3
		2050	376	125	119	194	88	41	5
		2090	599	118	118	309	224	43	0
	A2	2000	348	126	125	180	75	42	1
		2050	462	123	123	239	142	43	0
		2090	583	116	116	309	219	44	0
	B1	2000	374	125	125	188	94	42	0
		2050	426	124	121	211	136	41	2
		2090	560	123	123	270	219	43	0
Lajas	A1FI	2000	157	178	110	81	8	31	37
		2050	210	200	130	108	8	32	35
		2090	366	216	171	189	74	36	20
	A2	2000	189	171	136	105	6	34	20
		2050	262	178	146	119	35	35	18
		2090	351	188	161	182	59	38	14
	B1	2000	211	171	128	109	15	34	25
		2050	250	175	131	129	33	35	25
		2090	335	176	145	173	74	37	17
Mayaguez	A1FI	2000	254	154	143	131	32	39	7
		2050	323	145	133	167	65	40	9
		2090	523	146	145	270	171	42	1
	A2	2000	254	154	143	155	32	39	7
		2050	401	159	157	207	94	42	2
		2090	509	164	164	263	163	42	0
	B1	2000	323	155	145	167	61	40	6
		2050	370	163	147	191	94	40	9
		2090	488	183	177	252	153	40	3

P **is** precipitation; ET_o is reference evapotranspiration; ET_{cadj} is the actual evapotranspiration adjusted for soil moisture availability; RO is surface runoff; Rech is aquifer recharge; and SM is soil moisture; and YR is relative crop yield reduction.

Table 5b indicates that surface runoff is predicted to increase during September for all scenarios and locations. This is a positive result with respect to irrigation water supply; however, surface water may suffer due to increased soil erosion and may lead to accelerated filling of reservoirs by sedimentation. There was no clear trend with ET_{cadj}; Adjuntas ET_{cadj} decreased, while Lajas and Mayaguez increased. Soil moisture content in general increased, except for the B1 scenario for Mayaguez, in which the moisture content remained constant at 40 percent. In February the ET_{cadj} was markedly lower than the ET_o (Table 5a), whereas for September the ET_{cadj} was similar in magnitude to ET_o.

Little research has been done on the impacts of climate change on aquifer recharge [3]. However, for the three scenarios considered in this study, aquifer recharge increased at all locations by 108 mm (overall average) between 2000 and 2090 in September, the season when the majority of the island's aquifer recharge occurs. The February overall average aquifer recharge decreased by only by 5 mm. Since the drier months do not contribute significantly to aquifer recharge, a large increase during the wet season will likely produce a net increase in the annual aquifer recharge. This is good news from a groundwater production standpoint. Increasing aquifer recharge also suggests that groundwater levels may increase and this may help to minimize saltwater intrusion near the coasts as sea levels rize, provided that groundwater use is not oversubscribed. Saltwater intrusion has already been observed at coastal locations in Puerto Rico (e.g., [31]).

The relative crop yield reduction (YR) increased in February (Table 5a), and decreased or remained essentially unchanged in September (Table 5b). A note is in order relative to the interpretation of YR. Typically, YR is used to estimate the seasonal crop yield reduction. In this study, we are applying the index on a monthly basis. Crop seasons in Puerto Rico for typical agricultural crops are three to four months in duration, or longer. Therefore, an estimated YR value for a single month should not be taken as a seasonal relative crop yield reduction. Rather, the monthly YR should only be viewed as a contributor toward the overall seasonal yield reduction.

Figure 10 shows the average monthly variation in the relative crop yield reduction for Lajas for 2000 and 2090 for the three climate change scenarios. Under current conditions, without irrigation, crops grown in Lajas will experience a significant yield reduction. This can be seen from the results for the current (Year 2000) period in the Figure10 (a, b, c). Under the B1 scenario for Lajas(Figure 10A), the relative crop yield reduction did not change significantly in the future. However, under the A1fi scenario (Figure 10C), the relative crop yield reduction increased significantly in the future during the May/June period (greater than 20%). The relative crop yield reduction decreased for all scenarios during September owing to higher soil moisture conditions.

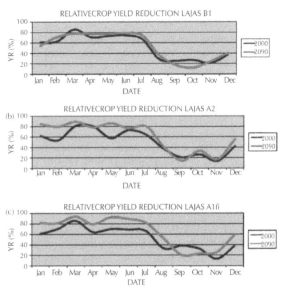

FIGURE 10   Relative crop yield reduction (YR) for Lajas for scenario B1 (a), A2 (b) and A1 (c) by month for 2000 and 2090 for Lajas, PR.

27.4   LIMITATIONS IN THE RESULTS PRESENTED

The results presented in this chapter should necessarily be viewed with caution since they are based in part on coarse resolution GCM data downscaled to single sites. As Pielke et al. [32] rightly point out, future "agricultural impacts extend far beyond a global mean temperature and include other anthropogenic climate forcings." Some of these forcings include land-use change, atmospheric aerosols, and complex nonlinear feedbacks, not accounted for in present-day, and likely next-generation, GCMs. Statistical downscaling itself assumes that the predictor – predictand relationship remains constant in time with stationary dynamic conditions under future climate change [33]. Furthermore, this study was based on only one GCM and since many uncertainties still exist among different models, the results need to be used with caution [4].

Several simplifying assumptions were made with respect to parameters used in the analysis, which may also contribute to uncertainty in the results of this study. However, it is quite possible that the uncertainties in the assumptions made relative to the parameters are less than the uncertainties associated with the future climate predictions, and therefore, a more precise parameterization may be unwarranted.

27.5   SUMMARY

The purpose of this study was to estimate reference evapotranspiration, precipitation deficit and relative crop yield reduction for a generic crop under climate change conditions for three locations in western Puerto Rico: Adjuntas, Mayagüez, and Lajas. Precipitation and temperature data from the DOE/NCAR PCM global circulation model was statistically downscaled to the three study locations. The 100 year (2000 to 2100) climate change/hydrologic analysis focused on the driest and wettest months of the year (i.e., February and September, respectively). The results from this study are consistent with other studies which indicate that the rainy season will become wetter and the dry season will become drier. This has important implications on agricultural water management. With increasing precipitation deficits during the dry months, the agricultural sector's demand for water will increase, which may lead to conflicts in water use.

The analysis revealed that lower soil moisture and increases in the relative crop yield reduction were associated with increasing precipitation deficits during the dry season. Relative crop yield reduction decreased during September, and was associated with increasing precipitation excess. Runoff and aquifer recharge can be expected to increase in the future during the wet season. The additional surface runoff can possibly be captured in newly constructed reservoirs to offset the higher irrigation requirements during the drier months, however, increased surface runoff may be associated with increased soil erosion and degradation of reservoirs. Increased aquifer recharge during the wet season will help offset potential increased demand for water and may increase groundwater water levels in coastal areas, which will help to counter the growing threat of saltwater intrusion in Puerto Rico's coastal aquifers.

KEYWORDS

- aquifer
- aquifer recharge
- climate change
- crop coefficient
- crop yield
- Dominican Republic
- downscaling
- evapotranspiration
- gcm
- general circulation model, GCM
- Haiti
- humid zone
- IPCC
- irrigation requirement
- precipitation
- precipitation excess
- precipitation deficit
- Puerto Rico
- rainfall
- reference evapotranspiration
- regression analysis
- runoff
- semiarid zone
- soil moisture
- soil moisture storage
- water balance
- water resource
- water shed

REFERENCES

1. IPCC, **2001**, Climate Change 2001, In: *The Scientific Basis. Contribution of Working Group I to the Third Assessment Report of the Intergovernmental Panel on Climate* Change, Houghton, J. T.; Ding, Y.; Griggs, D. J.; Noguer, M.; van der Linden, P. J.; Dai, X.; Maskell, K.; Johnson, C. A. (eds.). Cambridge University Press, 881 pp.
2. IPPC, 2007a. Climate Change 2007, In: *The Physical Science Basis – Summary for Policy Makers.* Contribution of Working Group *I* to the Fourth Assessment Report of the Intergovernmental Panel on Climate Change. Released February, **2007**, 21 pp.

3. IPPC, 2007b. Fourth Assessment Report, Working Group II Report, Impacts, Adaptation and Vulnerability. Chapter 3, 174–210. In: *Freshwater resources and their management.*

4. Bouraoui F.; Vachaud, G.; Haverkamp, B.; A distributed physical approach for surface subsurface water transport modeling in agricultural watersheds. *J. Hydrol.* **1997,** *203(4),* December), 79–92.

5. Beasley, D. B.; Huggins, L. F.; Monke, E. J.; Answers: A model for watershed planning. Trans. ASAE, **1980,** *23(4),* 938–944.

6. Charles, S. P.; Bates, B. C.; Whetton, P. H.; Hughes, J. P.; Validation of downscaling models for changed climate conditions: case study of south-western Australia. *Clim Res.;* **1999,** *12,* 1–14.

7. Miller, N. L.; Bashford, K. E.; Strem, E.; Potential Impacts of Climate Change on California Hydrology. *J. Amer. Water Resources Assoc.;* **2003,** 771–784.

8. Maurer, E. P.; Duffy, P. B.; Uncertainty in projections of streamflow changes due to climate change in California. Geophysical Resarch Letters, **2005,** 32(LO3704): 1–5.

9. Brekke, L. D.; Miller, N. L.; Bashford, K. E.; N. Quinn, W. T.; Dracup, J. A.; Climate change impacts uncertainty for water resources in the San Joaquin River Basin, California. Journal of the American Water Resources Association. February, **2004,** 149–164.

10. Dettinger, M. D.; Cayan, D. R.; Large-scale atmospheric forcing of recent trends toward early snowmelt runoff in California. J.; Clim.; **1995,** *8,* 606–623.

11. Roos, M.; Possible change in California snowmelt patterns. Proc.; Fourth Pacific Climate Workshop, Pacific Grove, CA, **1987,** 22–31.

12. Pan, A.; E.; Takle, Horton, R.; Segal, M.; Warm-seasonal soil moisture prediction using a coupled regional climate model. 16th Conference on Hydrology, **2002,** January 13–18. Orlando, Florida.

13. Stewart, R. I.; Cyan, D. R.;, Dettinger, M. D.; Changes towards earlier stream flow timing across Western North America. Journal of Climate, **2005,** *18,* 1136–1155.

14. De Pauw, E.; An agroecological exploration of the Arabian Peninsula. ICARDA, Aleppo, Syria, **2002,** 77.

15. Miller, N. L.; Jin, J.; Hayhoe, K.; Projected Extreme Heat and Energy Demand under Future Climate Scenarios. Proceedings of the Caribbean Climate Symposium, University of Puerto Rico – Mayagüez Campus, **2006,** April 24–25.

16. Miller, N. L.; Hayhoe, K.; Jin, J.; Auffhammer, M.; **2007,** Climate, Extreme Heat, and Electricity Demand in California, JAMC.

17. Molina-Rivera, W.; **2005,** USGS, Estimated Water Use in Puerto Rico, 2000, USGS Open-File Report 2005–1201.

18. Washington, W. M.; Weatherly, J. W.; Meehl, G. A.; Semtner, A. J.; Bettge, T. W.; Craig, A. P.; Strand, W. G.; Arblaster, J.; Wayland, V. B.; James, R.; Zhang, Y.; Parallel climate model (PCM) control and 1% per year CO_2 simulations with a 2/3 degree ocean model and 27 km dynamical sea ice model. Clim. Dyn.; **2000,** *16,* 755–774.

19. Nakicenovic, N.; Alcamo, J.; Davis, B. de Vries, G.; Fenham, J.; Gaffin, S.; Gregory, K.; Grubler, A.; Jung, T. Y.; **2000,** Intergovernmental Panel on Climate Change Special Report on Emission Scenarios. Cambridge University Press.

20. Allen, R. G.; Pereira, L. S.; Dirk Raes; Smith, M.; **1998,** *Crop Evapotranspiration Guidelines for Computing Crop Water Requirements.*FAO Irrigation and Drainage Paper *56,*Food and Agriculture Organization of the United Nations, Rome.

21. Allen, R. G.; Assessing integrity of weather data for reference evapotranspiration estimation.ASCE J. of Irr.and Drainage Eng.; March/April, **1996,** *122(2),* 97–106.

22. Harmsen, E. W.; Goyal, M. R.; Torres Justiniano, S.; Estimating Evapotranspiration in Puerto Rico. J.; Agric. Univ.; P. R.; **2002,** *86(1–2),* 35–54.

23. National Climate Data Center, 1992. International Station Meteorological Climate Summary (ISMCS), Version 2.

24. Hargreaves, G. H.; Samani, Z. A.; Reference crop evapotranspiration from temperature. *J. Appl. Eng. Agric.;* **1985,** *1(2),* 96–9.

25. Keller, J.; Bliesner, R.; **1990,** *Sprinkler and Trickle Irrigation.*Van Nostrand Reinhold Publisher.

26. Schwab, G. O.; Fangmeier, D. D.; Elliot, W. J.; **1996,** *Soil and Water Management Engineering.*4th edition.John Wiley and Sons Publisher.

27. USGS, *Water Resource Data, Puerto Rico and the US Virgin Islands, Water Year 2004,* Water Data Report PR-0401. **2004,** 578.

28. Ramirez-Beltran, N. D.; Julca, O.; **2006,** Detection of a Local Climate Change, The 18th Conference on Climate Variability and Change, at the 86th American Meteorological Society Annual Meeting, Atlanta Georgia.

29. Pulwarty, R. S.; **2006,** Climate Change in the Caribbean: Water, Agriculture, Forestry Mainstreaming Adaptation to Climate Change (MACC). Issue Paper (DRAFT). University of Colorado and NOAA/Climate Diagnostics Center Boulder CO 80305.

30. Scatena, F. N.; An assessment of climate change in the Luquillo Mountains of Puerto Rico. Proceedings of the Third International Symposium on Water Resources, Fifth Caribbean Island Water Resources Congress, American Water Resources Association. July, San Juan, PR, **1998,** 193–198.

31. Rodríguez-Martínez, J.; L.; Santiago-Rivera, Rodríguez, J. M.; Gómez-Gómez, F.; **2005,** Surface-Water, Water-Quality, and Ground-Water Assessment of the Municipio of Ponce, Puerto Rico, 2002–2004. Scientific Investigations Report 2005–5243. U. S.; Department of the Interior, US; Geological Survey. p. 96.

32. Pielke, R. A.; Sr. Adegoke, J. O.; Chase, T. N.; Marshall, C. H.; Matsui, T.; Niyogi, D.; **2006,** A new paradigm for assessing the role of agriculture in climate system and in climate change. *Agric. for Meteorol.* In press.

33. Mearns, L. O.; Giorgi, F.; Whetton, P.; Pabon, D.; Hulme, M.; Lal, M.; **2003,** Guidelines for Use of Climate Scenarios Developed from Regional Climate Model Experiments. The Intergovernmental Panel on Climate Change Data Distribution Center. Final Version, 10/30/03.

CHAPTER 28

EVAPOTRANSPIRATION USING SATELLITE REMOTE SENSING FOR THE TROPICAL CLIMATE[1]

ERIC W. HARMSEN, JOHN MECIKALSKI, ARIEL MERCADO VARGAS, and PEDRO TOSADO

CONTENTS

[1]Modified and reprinted with permission: "Harmsen, E. W., J. Mecikalski, A. Mercado, and P. Tosado, 2010. Estimating evapotranspiration in the Caribbean region using satellite remote sensing. Presented at 2010 Summer Specialty Conference by American Water Resources Association (AWRA, http://www.awra.org/), San Juan, Puerto Rico, August 30 to September 1, 2010. Pages 1–6." © Copyright 2010 Conference AWRA

28.1　INTRODUCTION

Remote sensing methods for estimating evapotranspiration (ET) are needed for tropical conditions. Various techniques have been developed based on radiation [17] and surface energy budget methods [1, 5]. In this study a solar radiation based methodology is used for estimating reference and actual ET for the Caribbean region. Remote sensing of solar radiation has several important advantages over the use of pyranometers networks including large spatial coverage, relatively high spatial resolution, and the availability of data in remote, inaccessible regions and countries that may not have the means to install a ground-based pyranometers network [17]. This chapter introduces a remote sensing methodology to estimate reference ET for Puerto Rico, Haiti and the Dominican Republic. Actual ET is estimated for Puerto Rico using an energy balance technique and applied in a water balance calculation. Results for surface runoff, aquifer recharge and soil moisture are presented. The development of the methodology has advanced more quickly in Puerto Rico; therefore, the information presented here can be considered a prototype of what is being developed for the other two countries (i.e., Haiti and the Dominican Republic). This methodology can also be used for other tropical regions of the world.

28.2　THEORETICAL METHODS FOR EVAPOTRANSPIRATION ESTIMATIONS

Reference ET is estimated with the original radiation-based Hargreaves formula [10, 11] is given in Eq. (1). The daily actual evapotranspiration (ET_a) was obtained by converting the latent heat flux (LE) for each 1 km^2 pixel to an equivalent depth of water using the latent heat of vaporization. The latent heat, LE, was estimated using the Eq. (2) described by Monteith and Unsworth [13].

$$ET_0 = 0.0135 \, [(0.408 \, R_s) \, (T+17.8)] \tag{1}$$

$$LE = \frac{\rho C_p \left(e_0(T_s) - (T_a) \right)}{\gamma \left(r_a + r_s \right)} \tag{2}$$

$$e = 0.6018 \left(\exp\left[\frac{17.27T}{T+237.3} \right] \right) \tag{3}$$

In Eq. (1): R_s is solar radiation and T_a is the average daily air temperature. The 0.408 value converts the solar radiation from units of MJ/m^2 day to mm of water per day. In Eq. (2): ρ is the density of dry air; C_p is the specific heat of air; γ is psychrometric constant; $e_0(T_s)$ is the saturated vapor pressure at the effective surface temperature T_s; $e(T_a)$ is the actual vapor pressure at the air temperature T_a; and ra and rs are aerodynamic and surface resistances, respectively.

Air temperature (Ta) was obtained from lapse rates calibrated for Puerto Rico by Goyal et al. [7] with regression equations relating average air temperature with surface elevation. To verify the appropriateness of using the estimated air temperature, comparisons were made with measured air temperature at two coastal locations (Isabela and Fortuna – Juana Diaz) and two mountain locations (Guilarte and Maricao) in Puerto Rico during the ten day period of analysis (June 20 to June 29, 2010). A high correlation between measured and estimated air temperature was obtained using regression analysis: $T_{a, measured} = [(1.062 *T_{a, estimated}) - 0.78]$, with a coefficient of determination, $r^2 = 0.94$. The vapor pressures were calculated using Eq. (3) described by Allen et al. [1], where T is temperature.

The effective surface temperature is difficult to obtain from remote sensing, since the satellite brightness temperature obtained may be that of the cloud top and not the ground surface, if clouds are present. Therefore, T_s was obtained by an implicit approach similar to that described by Lascano and van Bavel [12]. In this study, a surface energy balance was employed to obtain actual ET [18]:

$$[Rn–LE (Ts)–H(Ts)–G] = 0 \tag{4}$$

where, R_n is the net radiation, G is the soil heat flux, assumed to be zero for the daily analysis, and H is the sensible heat flux, Eq. (5):

$$H = \frac{\rho C_p (T_s - T_a)}{r_a} \tag{5}$$

The single unknown variable (T_s) was obtained using the recursive root function **fzero** in MatLab [Version 10, <http://www.mathworks.com>]. After obtaining the value of T_s for each pixel, the values of LE and H are calculated, the form of the surface energy balance used in this study is shown in Eq. (4).

Net radiation is obtained from the calculation procedure presented by Allen et al. [1], which requires solar radiation (R_s) and albedo (α), which in this study are derived from the radiative transfer model of Diak et al. [4] using half-hourly data from the visible channel of NOAA's GOES-13 [Geostationary Operational Environment Satellite] satellite. More information about the R_s and α remote sensing products can be found in Sumner et al. [17].

Aerodynamic resistance (r_a) was calculated with Eq. (6) described by Yunhao et al. [18]:

$$r_a = r_{ao} \cdot \Phi + r_{bh} \tag{6}$$

In Eq. (6), r_{ao} is the aerodynamic resistance under conditions of neutral atmospheric stability [1] and is described Eq. (7).

$$r_{ao} = \frac{\ln\left[\frac{(z - z_{disp})}{z_0}\right] \ln\left[\frac{z - z_{disp}}{(0.1) z_0}\right]}{k^2 u} \tag{7}$$

where, z is the virtual height at which meteorological measurements are taken, in this study assumed to be within the inertial sublayer and $= [1.5(z_0/0.13)]$ as described by Monteith and Unsworth [13]. The roughness length (z_0) and the zero plane displacement (z_{disp}) for various land use/vegetation categories were obtained from ATMET [2]. The term ($z_0/0.13$) is equivalent to the canopy height (h). The parameter k is Von Karman's constant (k = 0.41). Six-hour values of wind speed for Puerto Rico, obtained from the National Weather Service's National Digital Forecast Database [14], were averaged to obtain the daily average wind speed (u). Although the wind speed is a model forecast, it is the best source of spatially distributed wind speed over the island. The atmospheric stability coefficient is estimated using Eq. (8) described by Yunhao et al. [18]:

$$\phi = \left[1 - \frac{\eta\left(z - z_{disp}\right)g\left(T_s - T_a\right)}{T_0 u_2} \right] \tag{8}$$

where, g is acceleration of gravity, and the coefficient η is commonly taken as 5 [18]. The temperature, T_0, is the average of the values of T_s and T_a. The excess resistance is calculated using: $r_{bh} = 4/U^*$, where, the friction velocity is $U^* = [k\,u] / \ln [(z-z_{disp})/z_0)]$. Surface or canopy resistance (r_s) was estimated using the Eq. (9) described by Ortega-Farias et al. [15]:

$$r_s = \frac{\rho C_p VPD}{\Delta\left(R_n - G\right)C_f}\left(\frac{\theta - \theta_{WP}}{\theta_{FC} - \theta_{WP}}\right)^{-1} \tag{9}$$

where, Δ is the slope of the saturation vapor pressure curve at the mean temperature; VPD is the vapor pressure deficit; C_f is a calibration coefficient equal to 1 in this study; θ is volumetric soil moisture content in the root zone; and θ_{FC} and θ_{WP} are the volumetric soil moisture content at field capacity and wilting point, respectively. Values of field capacity and wilting point were obtained from regression equations based on percent sand, silt and clay presented by Cemek et al. [3]. Percentages of sand, silt and clay for Puerto Rico were obtained from the weblink [http://soils.usda.gov/survey/geography/ssurgo/] of Soil Survey Geographic (SSURGO) Database of the USDA Natural Resource Conservation Service.

A water balance was performed in each of the 1 km² pixels with Eq. (10), where, SMD1 and SMD2 are the depths of water in the soil profile at the beginning and end of the day (24 hours), respectively; PRECIP is rainfall; RO is surface run-off; ET_a is the actual evapotranspiration (described above); and DP is deep percolation or the soil water that passes below the root zone. The water balance analysis is performed over the soil profile depth equal to the root depth (R_{depth}). Root depth for various land use/vegetation categories were obtained from ATMET [2]. The 24-hour rainfall is obtained from NOAA's Advanced Hydrologic Prediction Service (AHPS) website [http://water.weather.gov/precip/]. In Puerto Rico, the source of the AHPS rainfall is NEXRAD radar and rain gauge data. Runoff is estimated [Eq. (11)] using the Curve Number method of the USDA – Soil Conservation Service [1972]. The maximum potential difference between rainfall and runoff at the moment of rainfall initiation is calculated with Eq. (12). CN is the runoff curve number and is adjusted for antecedent soil moisture conditions [6].

$$SMD2 = PRECIP - ET_0 - RO - DP + SMD1 \tag{10}$$

$$RO = [(PRECIP - 0.2S)^2] / [(PRECIP + 0.8S] \tag{11}$$

$$S = [(25400 / CN) - 254] \qquad (12)$$

An initial value of SMD2 is calculated with a modification of Eq. (10): $SMD2_i$ = PRECIP – ET_o – RO – DP + SMD1. If the value of $SMD2_i$ is larger than the depth of water in the soil profile at field capacity (FCD), then DP = $[SMD2_i - FCD]$ and the value of SMD2 is equal to FCD. If however, $SMD2_i$ < FCD, then DP = 0, and SMD2 = $SMD2_i$.

28.3   RESULTS AND DISCUSSION

Figures 1 and 2 show the estimated reference evapotranspiration (ETo) for Puerto Rico, Haiti and the Dominican Republic, respectively, on June 29, 2010. Figure 3 shows the estimated actual evapotranspiration for Puerto Rico for the same day. By taking the ratio of ET_a to ET_o, the "crop" coefficient (K_c) was obtained (Fig. 4). Estimated K_c values are between 0.3 and 0.95, which are quite reasonable. Figure 5 shows the total rainfall that occurred in Puerto Rico on June 29, 2010. The figure indicates a relatively large amount of rain fell in the north-western portion [humid region] of the island. In this study, CN values were not available, therefore a new CN map was developed for Puerto Rico based on land cover and soil hydrologic group obtained from the SSURGO data (Fig. 6).

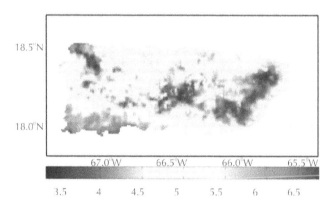

FIGURE 1   Estimated reference evapotranspiration (ET_o) for Puerto Rico on June 29, 2010.

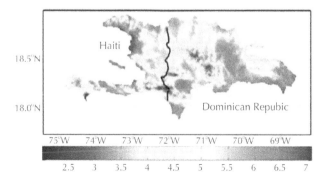

FIGURE 2   Estimated reference evapotranspiration (ET_o) for Haiti and the Dominican Republic on June 29, 2010.

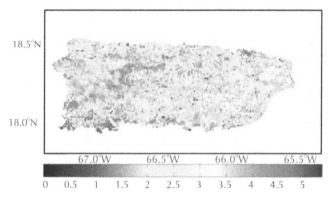

FIGURE 3   Estimated actual evapotranspiration (ET$_a$) for Puerto Rico on June 29, 2010.

FIGURE 4   Estimate "crop" coefficient (Kc) over Puerto Rico on June 29, 2010.

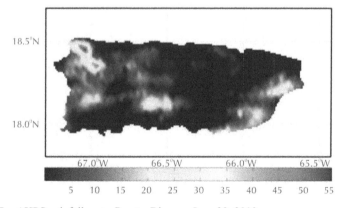

FIGURE 5   AHPS rainfall over Puerto Rico on June 29, 2010.

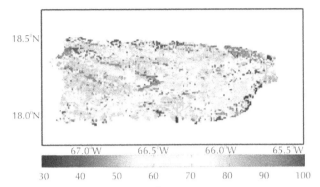

FIGURE 6 Curve Number map for Puerto Rico.

Figures 7–9 show the estimated surface runoff, deep percolation and soil moisture content on June 29, 2010. To initialize the analysis, a value of the soil moisture equal to the field capacity was set for each pixel on June 19, 2010. Subsequently, the soil moisture was updated daily as described in the previous section in this chapter. Due to abundant rainfall during the 10-day analysis period, the soil moisture was still very close to the field capacity throughout most of the island on June 29.

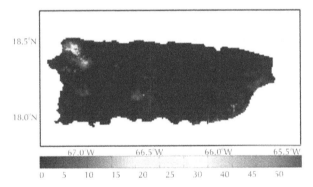

FIGURE 7 Estimated surface runoff in Puerto Rico on June 29, 2010.

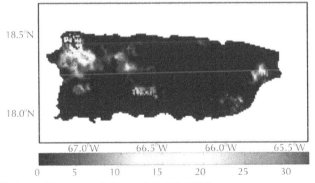

FIGURE 8 Estimated deep percolation on June 29, 2010.

FIGURE 9   Estimated soil moisture in Puerto Rico on June 29, 2010.

28.4   SUMMARY

In early 2009, collaboration between the University of Puerto RicoMayagüez Campus and the University of Alabama in Huntsville resulted in the availability of a solar radiation satellite remote sensing product for Puerto Rico, the Dominican Republic, Haiti and Cuba. The half-hourly and daily integrated data are available at 1 km resolution for Puerto Rico and 2 km resolution for the other islands. These data are extremely valuable for the purpose of analyzing water resource-related problems.

This chapter describes the technical approach for estimating reference evapotranspiration (ET), actual ET, surface runoff, deep percolation and soil moisture content. Results of estimated reference ET are presented for Puerto Rico, Haiti and the Dominican Republic for June 29, 2010. A method for performing a water balance analysis over Puerto Rico is also described. Estimates of actual evapotranspiration, surface runoff, deep percolation and soil moisture content for Puerto Rico on the same day were also made. Future efforts should investigate the quality of the pyranometers used in this study and possible improvement of the satellite algorithm.

This research represents a preliminary step in the development of a suite of remote sensing products that are potentially valuable tools for conducting water resource studies for the Caribbean region. The method described in this chapter is potentially useful for estimating evapotranspiration using satellite remote sensing in other tropical regions of the world.

KEYWORDS

- aerodynamic resistance
- alabama
- canopy resistance
- crop coefficient
- Cuba
- Dominican Republic
- evapotranspiration

- **evapotranspiration, actual**
- **evapotranspiration, reference**
- **field capacity**
- **Florida, USA**
- **GOES satellite**
- **Haiti**
- **Hargreaves method**
- **latent heat**
- **pan evaporation**
- **Puerto Rico**
- **pyranometer**
- **rainfall**
- **runoff**
- **satellite remote sensing**
- **satellite remote sensing**
- **solar radiation**
- **surface resistance**
- **vapor pressure deficit**
- **water balance**
- **wilting percentage**

REFERENCES

1. Allen, R. G.; Pereira, L. S.; Dirk Raes; Smith, M.; **1998,** *Crop Evapotranspiration Guidelines for Computing Crop Water Requirements.* FAO Irrigation and Drainage Paper 56, Food and Agriculture Organization of the United Nations, Rome. Pp. 300.
2. ATMET, **2005,** ATMET Technical Note, Number 1, Modifications for the Transition From LEAF–2 to LEAF–[<http: //www.atmet.com/html/docs/rams/RT1□leaf2□3.pdf>].
3. Cemek, B.; Meral, R.; Apan M.; Merdun, H.; Pedo transfer functions for the estimation of field capacity and permanent wilting point. Pakistan J. Biological Sciences, **2004,** *7(4),* 535–541.
4. Diak, G. R.; Bland W. L.; Mecikalski, J. R.; A note on first estimates of surface insolation from GOES–8 visible satellite data. *Agric. For. Meteor.* **1996,** *82,* 219–226.
5. Evett, T.; Howell, T. A.; Tolk, J. A.; Remote sensing based energy balance algorithms for mapping ET: Current status and future challenges. Transactions of the American Society of Agricultural and Biological Engineers, **2007,** *50(5),* 1639–1644.
6. Fangmeier, D. D.; Elliot, W. J. Workman, S. R.; Huffman, R. L.; Schwab, G. O.; **2006,** *Soil and Water Conservation Engineering.* Fifth Edition. John Wiley & Sons.
7. Goyal, M. R.; González, E. A.; Chao de Báez, C.; Temperature versus elevation relationships for Puerto Rico. *J. Agric. UPR,* **1988,** *72(3),* 449–467.
8. Gowda, P. H.; Chávez, J. L.; Colaizzi, P. D.; Evett, S. R.; Howell, T. A.; Tolk, J. A.; Remote Sensing based Energy Balance Algorithms for Mapping ET: Current Status and Future Challenges. Transactions of the ASABE, **2007,** *50(5),* 1639–1644.

9. Hargreaves, G. H.; Moisture availability and crop production. Transactions of the ASAE, **1975,** *18(5),* 980–984.

10. Hargreaves, G. H.; Samani, Z. A.; Estimating potential evapotranspiration. J. Irrig. Drain. Division, Proceedings of the ASCE, **1982,** *108,* No. IR3, 223–230.

11. Harmsen, E. W.; Mecikalski, J.; Cardona-Soto, M. J.; Rojas Gonzalez A.; Vasquez, R.; Estimating daily evapotranspiration in Puerto Rico using satellite remote sensing. WSEAS Transactions on Environment and Development, **2009,** *6(5),* 456–465.

12. Lascano, R. J. C. H. M. van Bavel, **2007,** Explicit and recursive calculation of potential and actual evapotranspiration. *Agron. J. 99,* 585–590.

13. Monteith, J. L.; Unsworth, M. H.; **2007,** *Principles of Environmental Physics.* 3rd Edition, 440 pages. Academic Press.

14. DFD, **2010,** National Weather Service National Digital Forecast Database. [<http: //www. weather.gov/forecasts/graphical/sectors/puertorico.php>].

15. Ortega-Farias, S.; Poblete C.; Zuñiga, M.; **2008,** Evaluation of a two-layer model to estimate vine transpiration and soil evaporation for vineyards. Proceedings of the ASCE World Environmental and Water Resources Congress **2008,** Ahupua'a.

16. Shuttleworth, W. J.; Wallace, J. S.; Evaporation from sparse crops–an energy combination theory, Quarterly Journal of the Royal Meteorological Society, **1985,** *111,* 839–855.

17. Sumner, D. M.; Pathak, C. S.; Mecikalski, J. R.; Paech, S. J. Wu, Q.; Sangoyomi, T.; **2008,** Calibration of GOES-derived solar radiation data using network of surface measurements in Florida, USA. Proceedings of the ASCE World Environmental and Water Resources Congress 2008.

18. Yunhao, C.; Xiaobing, L.; Peijun, S.; Estimation of regional evapotranspiration over Northwest China by using remotely sensed data. *J. Geo. Sci.,* **2001,** *11(2),* 140–148.

CHAPTER 29

IRRIGATION SCHEDULING FOR SWEET PEPPER[1]

ERIC W. HARMSEN, JOEL TRINIDAD-COLON, CARMEN L. ARCELAY, and DIONEL C. RODRIGUEZ

CONTENTS

29.1   INTRODUCTION

A study was conducted to evaluate the influence of agricultural lime (CaCO$_3$) on the movement and uptake of inorganic nitrogen for a sweet pepper crop (*Capsicum annuum*) grown on an Oxisol soil (Coto clay) in north-west Puerto Rico. The Coto clay soil, which contains the 1:1 kaolinite mineral, has a low pH (4 to 4.5). The 1:1 type clays are known to possess a net positive charge at low pH, resulting in the adsorption of negatively charged ions such as nitrate. From an environmental standpoint this characteristic of the 1:1 clay is favorable, since nitrate leaching, a major cause of groundwater pollution in many areas, is reduced relative to soils with net negative charge. However, agricultural plants, such as sweet peppers, favor a higher soil pH (approximately 6.5), which can be obtained by the application of agricultural lime. This, however, may have the negative effect of increasing the potential for nitrate leaching, as the net charge on the soil particles becomes negative with increasing pH.

This chapter describes the results of a nitrogen leaching analysis for two sweet pepper crop seasons. The analysis was based on multiplying the daily percolation flux through the soil profile by the measured concentration of nitrogen below the root zone. Irrigations were scheduled using the pan evaporation method for estimating crop water requirements. No significant difference in nitrogen leaching was observed for the lime and no-lime treatments. This was attributed to the low nitrate retention capacity of this soil, even at low pH. The average percent of nitrogen leached during the 1st and 2nd season, relative to the amounts applied, were 26% and 15%, respectively. Leaching events were associated with large rainstorms, suggesting that leaching of N. would have occurred regardless of the irrigation scheduling method used.

29.2   TECHNICAL APPROACH

29.2.1   EXPERIMENTAL SITE

Sweet pepper crops were planted at the UPR Experiment Station at Isabela in northwest PR (Fig. 1) during March 2002, and January 2003. Harmsen et al. [1] provided a detailed description of the experimental layout of the field site. The soil at the Isabela Experiment Station belongs to the Coto series. It is a very fine kaolinitic, isohyperthermicTypicEutrustox. These are very deep, well drained, moderately permeable soils formed in sediments weathered from limestone. The available water capacity is moderate, and the reaction is strongly acidic throughout the whole profile. Consistence is slightly sticky and slightly plastic in the Oxic horizons. A strong, stable granular structure provides these soils with a very rapid drainage, despite their high clay content [2]. Average values of hydraulic properties published for the Coto clay soil near the study area are as follows: air dry bulk density 1.39 g/cm^3, porosity 48%, field capacity 30%, wilting point 23%, available water holding capacity (AWHC) 9% [12]. The AWHC of this soil is low for clay. Typical values for clay are 15 to 20% [3]. A small value of AWHC means that there is a greater potential for leaching since the soil moisture content associated with the field capacity is more easily exceeded.

FIGURE 1   Location of field site at Isabela, PR.

The experimental site of 0.1 ha was divided into four blocks, each block divided into four plots, one for each treatment, for a total of 16 plots. The plots measure 67 m². The treatments included two lime levels (lime and no lime) and two fertigation frequencies (F1 and F2). Each plot had four beds covered with plastic (silver side exposed) with two rows of sweet pepper plants per bed. The transplanted sweet peppers were grown in rows 91 cm apart, 30 cm apart along rows, with beds 1.83 meter on center. This gave a plant population of approximately 37,000 plants per hectare. There was an initial granular application of triple super-phosphate of 224 Kg/ha and 80 Kg/ha of 10–10–10 fertilizer. Peppers were planted from March 11th through March 13th, 2002 and January 27 through January 31th, 2003. KNO3 and urea were injected through the drip irrigation system throughout the season at different frequencies (weekly (F1) or bi-weekly (F2)). The total nitrogen applied during the season was 225 Kg/ha. After transplanting, soil samples were taken bi-weekly at 20 cm increments, down to an 80 cm depth from each plot to be analyzed for moisture content and nitrogen concentration. Each date in which soil samples were collected, whole plants were harvested for growth data. Periodic pesticide applications were made to control weeds and insects affecting crop growth.

29.2.2   WATER BALANCE

A water balance approach [Eq. (1)] was used in this study to estimate percolation past the root zone.

$$\text{PERC} = R - RO + \text{IRR} - ET_c + \Delta S \tag{1}$$

where, PERC is percolation below the root zone, R is rainfall, IRR is irrigation, RO is surface runoff, ET_c is crop evapotranspiration, and $\Delta S = S1 - S2$, where S1 and S2 are the water stored in the soil profile at times 1 and 2, respectively. The units of each term in Eq. (1) are in mm of water per day. Rainfall was obtained from a tipping bucket-type rain gauge located on the Isabela Experiment Station property. The rain gauge was located within a weather station complex located approximately 0.4 km from the study area. The weather station consisted of a 10 meter (high wind resistant) tower with lighting protection, data logger and radio communication system, and sensors to measure the following parameters: wind direction and speed, temperature, relative humidity, barometric pressure, cumulative rainfall, and solar radiation [4]. Irrigation (IRR) was applied through a drip irrigation system. The inline-type emitters produced a flow of 1.9 liters per hour per emitter at a design pressure of 10 pounds per square inch (psi). Emitters were spaced every 30 cm. Irrigations (IRR) were scheduled based on the estimated evapotranspiration rate as determined with Eq. (2), where ET_{pan} is the pan evaporation-derived evapotranspiration, K_c is the evapotranspiration crop coefficient for sweet peppers [5], which varied daily; K_p is the average annual value of the pan coefficient equal to 0.78 for Isabela, PR [6]. A cumulative water meter was used to control the gallons of irrigation water applied. The evapotranspiration term in Eq. (1) was estimated with Eq. (3), where: K_c is the crop coefficient (dimensionless) and ET_0 (mm/day) is the reference evapotranspiration obtained using the Penman-Monteith method [5], as described in Eq. (4).

$$\text{IRR} = ET_{pan} = (K_c \times K_p \times E_{pan}) \tag{2}$$

$$\text{ETc} = K_c * ET_0 \tag{3}$$

$$ET_0 = \frac{\left[0.408\Delta\left(R_n - G\right)\right]\gamma\left[\left(\dfrac{900}{T+273}\right)u_2\left(e_s - e_a\right)\right]}{\left[\Delta + \gamma\left(1 + 0.34\, u_2\right)\right]} \tag{4}$$

In Eq. (4): Δ is the slope of the vapor pressure curve [kPa °C^{-1}], R_n is net radiation [MJ m^{-2} day^{-1}], G is soil heat flux density [MJ m^{-2} day^{-1}], γ is psychrometric constant [kPa °C^{-1}], T is mean daily air temperature at 2 m height [°C], u_2 is wind speed at 2 m height [m s^{-1}], e_s is the saturated vapor pressure and e_a is the actual vapor pressure [kPa]. Eq. (4) applies specifically to a hypothetical reference crop with an assumed crop height of 0.12 m, a fixed surface resistance of 70 sec m^{-1} and an albedo of 0.23.

Data required by Eq. (4) were obtained from the weather station located near the study area. Wind speeds obtained from the 10 m high tower were adjusted to the 2 m wind speed, required by the Penman-Monteith method, by means of an exponential relationship. Initial values of the crop coefficient were obtained from the literature for sweet pepper for the initial, mature and end crop stages (FAO Paper No. 56). Adjustments of K_c were made during the calibration of Eq. (1) as described later in this section. ET_0 was estimated on a daily basis using a spreadsheet program. The calculation methodology is described by Allen, et al. [5]. The values of S in Eq. (1) and (2) were obtained from Eq. (5).

$$S = \theta_v * Z \tag{5}$$

where, θ_v is the vertically averaged volumetric soil moisture content over the depth Z, obtained by multiplying the moisture content, mass-basis (θ_m), by the soil bulk density and dividing by the density of water. The soil bulk densities were obtained from undisturbed soil cores.

Between sampling dates when measured values of θ_v were not available, daily values were estimated using Eq. (1) along with information about the moisture holding capacity of the soil. In this method, if the water added to the profile by rainfall or irrigation exceeds the soil moisture holding capacity (or field capacity), then the excess water was assumed to be equal to PERC and the moisture content was set equal to the field capacity on that day. This approach has previously been used for irrigation scheduling [7], waste landfill leachate estimation [8] and estimation of aquifer recharge rates [9, 10]. In this study, the effective field capacity of the soil was determined in-situ by saturating the soil and obtaining the soil moisture content within 48 hours.

Calibration of the water balance equation was accomplished by adjusting the ratio of runoff to rainfall (RO/R) within reasonable limits, until the measured and estimated soil moisture content were in reasonable agreement. [1 – (RO/R)] represents the fraction of rainfall that infiltrates into the soil bed. This contribution of water can occur in several ways for the plastic covered bed-type system used in this study. Rainfall may enter directly through the holes in the plastic made for the plants. Rainfall that runs off of the plastic into the furrow or that falls directly into the furrow may also be absorbed into the beds. Under flood conditions, which occurred on several occasions during the two crop seasons, water could have entered the beds under a positive water pressure. For nonflooding rainfall events, soil water may move from the furrows into the beds by means of unsaturated flow, which is controlled by the pore water pressure gradient between the furrow and the bed.

29.2.3 NITROGEN LEACHING

Nitrogen leaching (nitrate and ammonium) was estimated by multiplying the daily value of PERC by the concentration of nitrogen within the 60 to 80 cm depth of soil. This vertical interval was considered to be below the root zone, since plant roots were not observed within this interval any time throughout the two seasons. The following Eq. (6) was used to estimate nitrate and ammonium leaching, respectively:

$$LNO_3 = [0.01 \times (\rho_b) \times (NO_3) \times (PERC)] / [\theta_{vol}] \qquad (6a)$$

$$LNH_4 = [0.01 \times (\rho_b) \times (NO_4) \times (PERC)] / [\theta_{vol}] \qquad (6b)$$

where, LNO_3 and LNH_4 are the kg of nitrate and ammonium leached below the root zone per hectare, NO_3 and NH_4 are the nitrate and ammonium soil concentration in mg/kg in the 60 to 80 cm depth interval, PERC is the percolation rate in mm, ρ_b is the bulk density (gm/cm^3), and θ_{vol} is a volumetric moisture content (cm^3/cm^3) in the 60 to 80 cm depth interval. Equations (6a) and (6b) were used on a daily basis. Each measured value of soil concentration used in Eqs. (6a) and (6b) were based on the average of four replications. Values of NO_3 and NH_4 between sampling dates were linearly interpolated.

29.3 RESULTS AND DISCUSSION

The Coto clay soil was analyzed for various physical and hydraulic properties (Table 1). The soil has a relatively high sand content and high hydraulic conductivity in the 0–20 cm interval, which accounts for it high water intake capacity. We observed on several occasions the rapid infiltration of water after large rainfall events. In fact, the value of hydraulic conductivity for the 0–20 cm interval is similar to sand, which averages

900 cm/day [11]. Bulk density, porosity, hydraulic conductivity, moisture content at 0.33 and 15 bars pressure, and AWHC were obtained from undisturbed cores in the laboratory.

TABLE 1 Physical and hydraulic properties of Coto clay in the 0–20, 20–40, 40–60 and 60–80 cm depth intervals.

Soil depth cm	% Sand[1]	% Silt[1]	% Clay[1]	Soil Classi- fication	Bulk density	Porosity
0-20	35.10	19.35	45.55	silty clay	1.36	0.49
20-40	28.72	1.85	69.43	clay	1.36	0.49
40-60	22.50	5.00	72.50	clay	1.31	0.51
60-80	20.00	5.80	74.20	clay	1.29	0.51

Soil depth cm	Hydraulic conductivity (cm/day)	In-Situ field capacity year-1 site	In-Situ field capacity year-2 site	Moisture content at 0.33 bar pressure (FC)	Moisture content at 15 bar pressure (PW)	Available water holding capacity (AWHC)
0-20	1210.06	0.33	0.44	0.44	0.39	0.05
20-40	316.99	0.33	0.37	0.37	0.27	0.10
40-60	70.10	0.37	0.36	0.36	0.31	0.05
60-80	12.19	0.37	0.38	0.38	0.3	0.08

[1] Soil texture data for the 40-60 cm and 60-80 cm were obtained from Soil Conservation Service [12]. All other data were measured during the project.

Measured soil pH soil was between 4 and 5. Laboratory incubation tests were performed to determine the proper amount of lime needed to be applied to the soil to increase the pH to around 6.5 in the limed treatments; this amount was 7.4 tons lime/ha. The first year the pH did not respond as expected in the limed plots, and therefore, this may have contributed to there being no significant difference observed in the estimated nitrate losses by leaching between the lime and no-lime treatments. The second year the amount of lime applied to the limed treatments was doubled (14.8 tons lime/ha) and pH levels rose as expected.

Figure 2 shows a comparison of the evapotranspiration derived from pan and Penman-Monteith methods during Year 2. ET_{pan} was observed to have higher variability than ET_c. For reference, Figure 2 also shows the ET_c based on long-term average climate data for Isabela, PR. The seasonal ET [mm per season] was 447 for the methods of pan, 402 for Penman-Monteith method based on weather station data and 511 for Penman-Monteith based on long-term data, respectively.

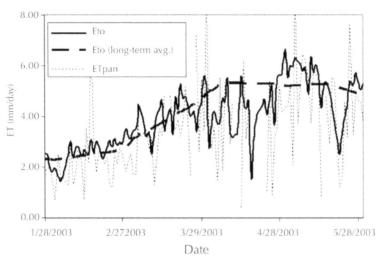

FIGURE 2 Daily values of evapotranspiration for a sweet pepper crop between January 27 to June 12, 2003 at Isabela, PR. Evapotranspiration was derived from the pan evaporation and Penman-Monteith methods.

The water balance, Eq. (1), was calibrated for the site conditions. Figure 3 shows the simulated and measured average soil moisture content for Year 1 and Year 2. The measured moisture contents shown in Fig. 3 represent the vertically averaged moisture content over all 16 plots. The minimum and maximum measured soil moisture content is also shown in Fig. 3. Vertically averaged values of the in-situ-measured field capacity equal to 0.39 and 0.35 were used in the Year 1 and Year 2 analyzes, respectively (averages from Table 1). It was necessary to use a value of RO/R = 0.25, reasonable agreement between the estimated and measured soil moisture content. During Year 1, the beginning of the season was quite wet. On April 6, 2002, a 176 mm rainfall occurred, which caused severe flooding of the study area. During Year 2, a rainy period occurred during April 5th through April 18th with flooding observed in the field plots. The largest rainfall of the season occurred on April 10, 2003 equal to 97 mm.

According to the procedure described above, percolation occurred on those days when the estimated moisture content exceeded the field capacity moisture content (0.39 for Year 1 and 0.35 during Year 2). On those days, the water in excess of the field capacity was assigned to PERC and the moisture content set equal to the field capacity. This can be seen in Figure 3 for those days in which the moisture content curve touched the dashed horizontal line associated with the field capacity moisture content. Figure 4 shows the estimated percolation during the Year 1 and Year 2. During the April 6, 2002 rainfall event of 175 mm, 43 mm were converted to percolation. During the April 10th, 2003 rainfall event of 97 mm, 31 mm were converted to percolation. Recall that only 25 percent of the rainfall was allowed to infiltrate, which was equal to 44 mm on April 6, 2002 and 24 mm on April 10, 2003. In the latter case 18 mm of irrigation was also applied, which together (24 mm + 18 mm) equaled 42 mm. In this case 31 mm was lost to percolation and 11 mm was stored in the root zone. Table 2 shows the Year 1 and Year 2 seasonal components of the water balance.

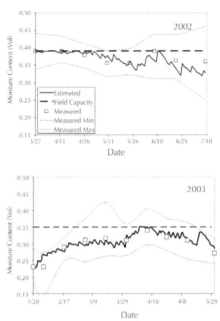

FIGURE 3   Estimated and measured volumetric soil moisture content between March 27 and
July 9, 2002 and January 27 and June 12, 2003.

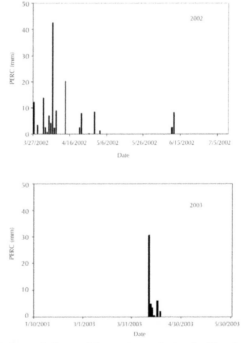

FIGURE 4   Estimated percolation past the root zone during the Year 1 and Year 2 seasons.

TABLE 2   Components of the seasonal water balance for years-1 (2002) and year-2 (2003).

	Year 1 (2002)	Year 2 (2003)
R-RO	175	136
IRR	350	411
ETc	416	441
ΔS	50	-52
PERC	159	54

Table 3 compares the Year 1 and Year 2 results of the nitrogen leaching analysis. The leached nitrate and ammonium estimates were obtained from Eqs. (5a) and (5b), respectively. Figure 5 shows the nitrate concentrations in the 60–80 cm depth interval during the Year 1 season. During Year 1 the range of estimated nitrogen leached was between 36 and 67 **kg**/ha. During Year 2, the range of estimated nitrogen leached was between 27 and 36 **kg**/ha. Interestingly, the amount of nitrate lost (average of all treatments) on April 6, 2002 and April 10, 2003 was 19.6 **kg**/ha and 20.1 **kg**/ha, respectively. For years 1 and 2 this represented 34% and 60% of the total N. lost by leaching during the two seasons, respectively. Figure 6 shows the estimated percent of nitrogen (i.e., nitrate plus ammonium) leached relative to N. applied (225 **kg**/ha) during the Year 1 and Year 2 seasons for the four experimental treatments.

TABLE 3   Nitrate, ammonium and nitrate plus ammonium (total) leached during Year 1 and 2 for the four experimental treatments.

		Year-1 (2002)				Year-2 (2003)			
	Units	LF1	LF2	NLF1	NLF2	LF1	LF2	NLF1	NLF2
NO3	kg/ha	36	50	47	42	34	32	34	24
NH4	kg/ha	10	13	21	11	2	3	2	3
Total	kg/ha	46	63	67	54	36	35	36	27
Total	%	21	28	30	24	16	16	16	12

The smallest amount of nitrogen leaching occurred in the LF1 treatment in 2002 and the NLF2 treatment during the second year. There is no clear difference between either the lime or fertigation treatments. Ammonium leaching was typically much lower than nitrate leaching (Table 3) except in the case of treatment NLF1 in 2002, in which 21 **kg**/ha ammonium was leached as compared to 47 **kg**/ha nitrate. The fact that no clear difference was observed between nitrogen leaching for the two lime treatments is consistent with laboratory studies currently being conducted on the Coto clay soil at the University of Puerto Rico Mayaguez Campus, which indicates that the pH at which this soil will possess a net positive charge (< 4) is below the native pH measured in the field (around 4.3).

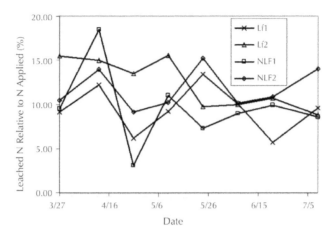

FIGURE 5 Year 1 Soil nitrate concentrations in the 60–80 cm depth interval. Values between the sampling dates were obtained by linear interpolation.

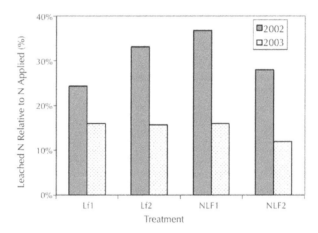

FIGURE 6 Estimated nitrogen leached during the Year 1 and Year 2. LF1 is the Lime-Fertigation 1 treatment, LF2 is the Lime-Fertigation 2 treatment, NLF1 is the No-Lime-Fertigation 1 treatment, NFL2 is the No-Lime-Fertigation 2 treatment.

29.4 LIMITATIONS OF NITROGEN LEACHING

There are several sources of uncertainty in the estimates of nitrogen leaching, which includes:

- Between sampling dates, soil nitrogen concentrations were derived by linear interpolation. Nitrogen concentrations were measured every two weeks. In some cases, the average nitrate concentration was observed to change as much as 15 mg/kg in the 60–80 cm depth interval. The estimated nitrogen leaching would be in error if these concentrations did not change linearly between sampling dates.

- The method of estimating percolation in this study does not account for the leaching that can potentially occur by unsaturated flow. All leaching was assumed to occur when the moisture content of the soil exceeded the soil field capacity. However, significant downward gradients can exist which would result in unsaturated flow. Although not presented in this paper, continuous soil pressure data obtained from vertically spaced tensiometers indicated downward hydraulic gradients throughout most of the season.

29.5   SUMMARY

This chapter described the results of a nitrogen leaching analysis for two sweet pepper crop seasons. The study was conducted on an Oxisol soil in NW Puerto Rico. The analysis was based on multiplying the daily percolation flux through the soil profile by the measured concentration of nitrogen below the root zone. Irrigations were scheduled using the pan evaporation method for estimating crop water requirements. Estimated percolation in 2002 was three times greater than occurred in 2003, whereas the nitrogen leached during 2002 was only slightly greater than two times the nitrogen leached during 2003.

No clear difference in nitrogen leaching was observed for the lime and no-lime treatments. This result is consistent with on-going studies of the Coto clay, which indicate that this soil has little to no capacity to retain nitrate. The average percent of nitrogen (nitrate plus ammonium) leached during the 1st and 2nd season, relative to the amounts applied, were 26% and 15%, respectively. Leaching events were associated with large rainstorms, suggesting that leaching of N would have occurred regardless of the irrigation scheduling method used. During the first and second seasons, respectively, 34% and 60% of the total N lost by leaching occurred during a single day (April 6 in 2002 and April 10 in 2003) when flooding was observed in the study areas.

KEYWORDS

- agricultural lime
- aquifer recharge
- available water holding capacity
- deep percolation
- emitter
- evaporation
- evapotranspiration
- fertigation
- fertilization
- field capacity
- humid region
- irrigation
- irrigation scheduling

- **lime fertigation**
- **linear interpretation**
- **nitrogen leaching**
- **oxisol**
- **pan evaporation**
- **Penman-Monteith**
- **percolation**
- **Puerto rico**
- **rainfall**
- **runoff**
- **soil moisture content**
- **surface runoff**
- **sweet pepper**
- **vegetable crops**
- **water balance**
- **wilting percentage**

REFERENCES

1. Harmsen, E. W.; Goyal, M. R.; Torres Justiniano, S.; Estimating evapotranspiration in Puerto Rico. *J. Agric. Univ. P.R.* **2002,** *86(1–2)*, 35–54.
2. Keng, J. C. W.; Scott T. W.; Lugo-López, M. A.; Fertilizer for sweet pepper under drip irrigation in an Oxisol in north-western Puerto Rico. *J. Agric. Univ. P.R.* **1981,***65(2),* 123–128.
3. Keller J.; Bliesner, R.; **1990,** *Sprinkler and Trickle Irrigation.* Van Nostr and Reinhold Publisher.
4. Zapata, R.; López, R. R.; Vázquez, D.; **2001,** Collection of meteorological data for improvement of building codes for Puerto Rico. Proceedings of the Sixth Caribbean Islands Water Resources Congress. Editor: Walter F.; Silva Araya. University of Puerto Rico, Mayagüez, PR 00680.
5. Allen, R. G.; Pereira, S. L.; Raes, D.; Smith, M. M.; *Crop Evapotranspiration: Guidelines for Computing Crop Water Requirements.* Food and Agricultural Organization of the United Nations (FAO) Report #56, Rome. **1998,** 300 pp.
6. Goyal, M.; Gonzalez Fuentes, E. A.; Pan coefficients for Puerto Rico [Spanish]. *J. Agric. Univ. P.R.* **1989,** *73(1).*
7. Shayya, W. H.; Bralts, V. F.; **1994,** *Guide to SCS-Microcomputer irrigation scheduling package.* SCS Scheduler Version 3.00. Developed under joint agreement 68–5D21–5-008 between Michigan State University and the USDA Soil Conservation Service.
8. Fenn, D. G.; Hanley, K. J.; DeGeare, T. V. **1975,** *Use of the Water Balance Method for Predicting Leachate Generation from Solid Waste Disposal Sites.* U. S.; Environmental Protection Agency. EPA/530/SW-168, October, 1975.
9. Thornthwaite, C. W.; Mather, J. R.; The water balance. Drexel Institute of Technology, *Publ. Climatology,* **1955,** *8(1),* 104
10. Papadopulos & Associates, Inc.; MathSoft, Inc. **1994,** Aquifer Data Evaluation for Pumping Tests (ADEPT).
11. Freeze, A. R.; Cherry, J. A.; **1979,** *Groundwater.* Prentice Hall Publisher.
12. Soil Conservation Service, *Soil Survey Laboratory Data and Descriptions for Some Soils of Puerto Rico and the Virgin Islands.* U. S.; Department of Agriculture Soil Survey Investigations Report No. 12. August, **1967,** 191 p.

CHAPTER 30

WEB-BASED IRRIGATION SCHEDULING[1,2]

ERIC W. HARMSEN

CONTENTS

[1]Eric W. Harmsen, PhD, P.E., Professor, Department of Agricultural and Biosystems Engineering, University of Puerto Rico Mayagüez, PR 00681, USA. Email: <eric.harmsen@upr.edu>. This research was conducted with a the financial support from NOAA-CREST (grant NA06OAR4810162) and USDA Hatch Project (Hatch-402) – Agricultural Experiment Station, University of Puerto Rico.
[2]Modified and printed with permission from, "Harmsen, E.W., 2012. Web based irrigation scheduling. J. Agric. Univ. P.R., 96(3-4)". © Copyright J. Agric. Univ. P.R.

30.1   INTRODUCTION

There is antidotal evidence that many farmers in Puerto Rico and the developing countries do not employ scientific methods for scheduling irrigation for their crops. Instead, the pump is turned on for an arbitrary amount of time without knowing whether the amount of water applied is too much or too little. Over application of water can lead to the waste of water, energy, chemicals and money, and also may lead to the contamination of ground and surface waters. Under application of irrigation can lead to reduced crop yields and a loss of revenue to the grower.

There are various approaches for scheduling irrigation. One approach is to supplement rainfall with enough irrigation so that the cumulative rainfall and irrigation, over a specific period of time (e.g., one day, one week, one season), matches the estimated potential evapotranspiration, which is equivalent to the crop water requirement. Potential evapotranspiration (ET_c) can be estimated by the product of a crop coefficient (K_c) and the reference evapotranspiration (ET_o). Traditionally, potential evapotranspiration is derived from pan evaporation data or meteorological data from weather stations. Another approach involves monitoring the soil moisture and applying irrigation sufficient to maintain the soil moisture content within a predetermined range. In this paper we present an approach based on applying irrigation to the crop to meet the crop water requirements (i.e., potential evapotranspiration), but instead of using pan evaporation or meteorological data, we use a remote sensing technique. The advantage of the method is that reference evapotranspiration can be estimated at a 1 km resolution for the entire island each day. If the relatively simple approach presented in this chapter is used it can potentially lead to increased efficiency of water and energy use, and help to reduce crop water stress and losses in crop yields.

In this chapter, a set of steps are provided to estimate the irrigation requirement for locations within Puerto Rico. A detailed example problem is then given to use of the method.

30.2   METHODS FOR IRRIGATION SCHEDULING

Potential crop evapotranspiration is estimated using Eq. (1):

$$ET_c = K_c ET_o \tag{1}$$

$$ET_0 = \frac{\left[0.408\Delta\left(R_n - G\right)\right]\gamma\left[\left(\dfrac{900}{T+273}\right)u_2\left(e_s - e_a\right)\right]}{\left[\Delta + \gamma\left(1 + 0.34\,u_2\right)\right]} \tag{2}$$

where: ET_o is reference evapotranspiration, Δ is slope of the vapor pressure curve (kPa °C^{-1}), R_n is net radiation (MJ m^{-2} day^{-1}), G is soil heat flux density (MJ m^{-2} day^{-1}), γ is psychrometric constant (kPa °C^{-1}), T is mean daily air temperature at 2 m height (°C), u_2 is wind speed at 2 m height (m s^{-1}), e_s is the saturated vapor pressure and e_a is the actual vapor pressure (kPa). Eq. (1) applies specifically to a hypothetical reference crop with an assumed crop height of 0.12 m, a fixed surface resistance of 70 sec m^{-1} and an albedo of 0.23. The crop coefficient, which changes throughout the crop season are shown in Fig. 1. During the initial crop growth stage, the value of the crop coefficient is $K_{c\,ini}$. During the mid season the crop coefficient is $K_{c\,mid}$, and at the end of the late season the crop coefficient is $K_{c\,end}$. The values of $K_{c\,ini}$, $K_{c\,mid}$ and $K_{c\,end}$ can be obtained from published tables [1].

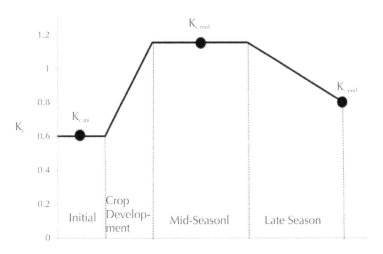

FIGURE 1   Crop Coefficient Curve [1].

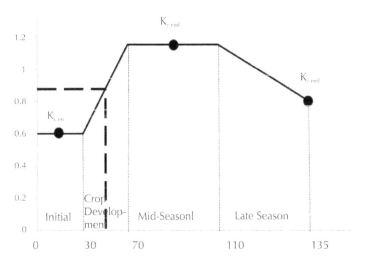

FIGURE 2   Crop coefficient curve with an example problem in this chapter. The heavy dashed line applies to the example problem with day of season 46–50 (horizontal axis) corresponding to an approximate crop coefficient of 0.85 (vertical axis).

30.2.1 STEPS TO ESTIMATE IRRIGATION REQUIREMENT

Step 1: Create an evapotranspiration crop coefficient curve for the crop. The following link to the Food and Agriculture Organization (FAO) Document No. 56 [1] provides tables of crop stage growth (Table 11) and K_c values (Table 12) for a large number of crops: http://www.fao.org/docrep/X0490E/x0490e00.htm. The K_c curve should look like Fig. 1 when step 1 is finished. Note that crop coefficient curves can also be created by using computer programs such as PRET [8] or CropWat [3].

Step 2: Go to the following website address to obtain the appropriate ET_o map(s) for the specific location:http://academic.uprm.edu/hdc/GOES-PRWEB_RESULTS/reference_ET/. Note that if you are irrigating every day, then you need only to obtain the ET_o for yesterday's date. If, however, one is irrigating once per week, for example, then one will need to obtain the ET_o values from the maps for the previous week. In this latter example, one will need to sum up the daily values of ET_o to obtain a value of the weekly ET_o.

Step 3: From the K_c curve obtained in Step 1, determine a representative value of K_c for the current growth stage of the crop.

Step 4: Estimate the crop water requirement (crop evapotranspiration) using Eq. (1): $ET_c = K_c \times ET_o$.

Step 5: Estimate the required amount of irrigation in depth units: Irrigation = (ET$_c$ – Rainfall). If the estimated Irrigation is negative, then one does not need to irrigate.

It is recommended that rainfall be measured on the farm with a rain gauge, however, if measured rainfall is not available, the approximate value of the rainfall (derived from NEXRAD radar) can be obtained at the following website: http://academic.uprm.edu/hdc/GOES-PRWEB_RESULTS/rainfall/. It will also be necessary to measure the irrigation volume. A digital or mechanical flow meter which measures the cumulative volume in gallons is recommended.

The irrigation scheduling approach described above is based on various simplifying assumptions (e.g., surface runoff and deep percolation are ignored). The FAO [1] has suggested corrections to K_{cini} for time interval between wetting events, evaporative power of the atmosphere and magnitude of the wetting events, and corrections to K_c mid and K_c end for air humidity, crop height and wind speed; however, these corrections have been ignored in order to preserve simplicity in the approach presented above. Despite the simplifying assumptions, the approach should significantly improve water management on a farm if currently there is no irrigation scheduling method being used.

30.3 EXAMPLE PROBLEM

A detailed example problem is presented here to illustrate the use of the proposed methodology. In this problem, we will determine the irrigation requirement for the 5 day period [February 15–19, 2012] for a tomato crop being grown in Juana Diaz, Puerto Rico, USA. Table 1 summarizes the information used in the example problem. Table 2 provides the important web addresses necessary for obtaining data for use in the example problem.

TABLE 1   Information for the example problem in this chapter.

Location	Juana Diaz, Puerto Rico
Site Latitude	18.02 degrees N.
Site Longitude	66.52 degrees W
Site Elevation above sea level	21 m
Crop	Tomato
Planting Date	1-Jan-12
Rainfall	A rain gauge is not available
Information	on or near the farm
Type of irrigation	Drip
Approximate wetted area of the field	50%
Irrigation system efficiency	85%
Field Size	10 acres
Pump capacity	300 gpm

TABLE 2   Web addresses used to obtain information for solving the example problem in this chapter.

Length of Growth Stages (Table 11) and Crop Coefficients (Table 12)	http://www.fao.org/docrep/X0490E/x0490e00.htm
Daily Reference ET Results for Puerto Rico[1]	http://academic.uprm.edu/hdc/GOES-PRWEB_RE-SULTS/reference_ET/
Daily NEXRAD Rainfall For Puerto Rico	http://academic.uprm.edu/hdc/GOES-PRWEB_RE-SULTS/rainfall/

[1]The web subdirectory contains Penman-Monteith, Hargreaves-Samani and Priestly Taylor ET_o data.

TABLE 3   Crop growth stage lengths and crop coefficient data for the example problem in this chapter (Fig. 2).

Initial crop growth stage	30 days
Crop development growth stage	40 days
Mid-season growth stage	40 days
Late-season growth stage	25 days
Total length of season	135 days
$K_{c ini}$	0.6
$K_{c mid}$	1.15
$K_{c end}$	0.8

Step 1: With the information in Table 1 it is now possible to construct the crop coefficient curve by consulting the FAO Document No. 56, Table 11. (Lengths of crop development stages for various planting periods and climatic regions) and Table 12 (Single time-averaged crop coefficients...). FAO Document No. 56 is available online at the web address given in Table 2. Table 3 summarizes the crop stage and crop coefficient information. The crop coefficient curve constructed from the data in Table 2 is

shown in Fig. 2. The approximate average crop coefficient for February 15–19 (day of season 46–50) is approximately 0.85.

Step 2: Determine the reference evapotranspiration for the five-day period. Figure 3 shows the estimated reference evapotranspiration for Puerto Rico on February 15, 2012 obtained from the web address provided in Table 2. Note that the preferred reference evapotranspiration method is used (i.e., Penman-Monteith method). The estimated ET_o for the site location on Feb. 15, 2012 is 2.95 mm. Using a similar procedure, the ET_o for Feb. 16, 17, 18 and 19 is 2.8 mm, 3.1 mm, 3.5 mm and 3.7 mm, respectively. Summing up the ET_o values comes to a total crop water requirement (for the five days) of 16.1 mm.

Step 3: A rain gauge is not available on or near the farm for the example problem; therefore it is necessary to obtain rainfall information from the NEXRAD radar. Figure 4 shows the NEXRAD rainfall for Puerto Rico for February 15, 2012. At the site location no rainfall was estimated from the NEXRAD radar. Checking the other maps for the other days reveals that no significant rainfall occurred at the site. Therefore all of the crop water requirement will have to be satisfied with irrigation.

Step 4: The crop water requirement for the time period can now be estimated as follows:

$ET_c = K_c ET_o = (0.85)(16.1 \text{ mm}) = 13.7$ mm (slightly greater than one-half of an inch).

Step 5: Determine the number of hours that the pump should be run to apply the 13.7 mm of water. A form of the well-known irrigation Eq. (3) [2] can be used which is shown below:

$$T = 17.817 \times [D \times A]/[Q \times \text{eff}] \tag{3}$$

where T is time in hours, D is depth of irrigation water in mm, A is effective field area in acres, Q is flow rate in gallons per minute and eff is irrigation system efficiency. Using D = 13.7 mm, A = 10 acres, Q = 300 gallons per minute and eff = 0.85, yields:

T = 17.817 × [13.7 × 10] / [300 × 0.85] = 9.6 hours

REFERENCE ET (mm) Penman-Monteith 15-Feb- 2012

FIGURE 3 Estimated reference evapotranspiration (ET_o) for Feb. 15, 2012. The approximate ET_o at the site locationis 2.95 mm.

FIGURE 4 Estimated NEXRAD rainfall for Feb. 15, 2012.

30.4 IRRIGATION MANAGEMENT

To evaluate the irrigation management with the approach described in this chapter, construction of a graph similar to the one shown in Fig. 5 is recommended. The graph shows the cumulative depth of irrigation and ET_c plotted with time. The goal of irrigation scheduling is to try to match the applied irrigation with the ET_c. By the end of the season, the cumulative irrigation (plus rainfall) should more or less equal the cumulative ET_c. If these two curves stay close together, this is an indication that good irrigation management is being achieved. Note that the graph shown in Fig. 5 is not related to the example problem given above.

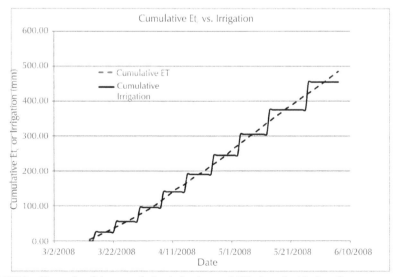

FIGURE 5 Example of the cumulative irrigation and ET_c plotted with time for a crop season. *Note:* This graph is not related to the example problem in this chapter.

30.5 SUMMARY

Irrigation scheduling is critically important to avoid the loss of water, fuel and chemicals by over application of water, or a reduction in crop yield if too little water is applied. In this chapter a web-based irrigation scheduling approach is described. The approach is based on applying irrigation water at the rate of the estimated potential evapotranspiration, which is equivalent to the crop water requirement. Reference evapotranspiration is obtained from an operational water and energy balance algorithm (GOES-PREWEB) which produces a suite of hydro-climate variables on a daily basis for Puerto Rico. The algorithm produces daily estimates of the Penman-Monteith, Priestly Taylor and Hargreaves-Samani reference evapotranspiration. The crop coefficient curve is constructed per the methodology recommended by the United Nations Food and Agriculture Organization (FAO). Daily rainfall can be obtained from radar (NEXRAD) if rain gauge data is not available for the farm. A detailed example is provided for a farm growing tomato in Juana Diaz, PR. The approach is relatively simple and the near-real time data is available to any farmer in Puerto Rico with internet access. Using the procedures described in this chapter, the approach could be developed at any location throughout the world.

KEYWORDS

- **crop coefficient**
- **crop water use**
- **evapotranspiration**
- **fao**
- **goes-prweb**
- **Hargreaves-Samani**
- **irrigation**
- **irrigation management**
- **irrigation scheduling**
- **NEXRAD**
- **Penman-Monteith**
- **Priestley-Taylor**
- **Puerto Rico**
- **radar**
- **web-based**

REFERENCES

1. Allen, R. G.; Pereira, L. S.; Raes, D.; Smith, M.; **1998,** *Crop Evapotranspiration Guidelines for Computing Crop Water Requirements*. FAO Irrigation and Drainage Paper *56,* Food and Agriculture Organization of the United Nations, Rome.

2. Fangmeier, D. D.; Elliot, W. J.; Workman, S. R.; Huffman, R. L.; Schwab, G. O.; *Soil and Water Conservation Engineering*. Fifth Edition.Delmar Cengage Learning, **2005,** 502 pp.

3. FAO, **2012,** CROPWAT 8.0. Decision support tool developed by the Land and Water Development Division of the United Nation's Food and Agriculture Organization. http: //www.fao.org/nr/water/infores_databases_cropwat.html

4. Hargreaves, G. H.; Moisture availability and crop production. TRANSACTIONS of the ASAE **1975,** *18(5),* 980–984.

5. Hargreaves, G. H.; Samini, Z. A. Reference Crop evapotranspiration from temperature. Appl. Eng. Agri.; ASAE **1985,***1(2),* 96–99.

6. Harmsen, E. W.; Mecikalski, J.; Mercado, A.; Tosado Cruz, P.; Estimating evapotranspiration in the Caribbean Region using satellite remote sensing. Proceedings of the AWRA Summer Specialty Conference, Tropical Hydrology and Sustainable Water Resources in a Changing Climate. San Juan, Puerto Rico. August 30–September 1, **2010,** 42–47.

7. Harmsen, E. W.; Mecikalski, J.; Cardona-Soto, M. J.; Rojas González, A.; Vásquez, R.; Estimating daily evapotranspiration in Puerto Rico using satellite remote sensing. WSEAS Transactions on Environment and Development, **2009,** *6(5),* 456–465.

8. Harmsen, E. W.; Gonzaléz, A.; Technical Note: A computer program for estimating crop evapotranspiration in Puerto Rico. *J. Agric. Univ. P.R.* **2005,** *89(1–2)*, 107–113. [http: //pragwater.com/crop-water-use/]

9. Jensen, M. E.; Burman, R. D.; Allen, R. G.; *Evapotranspiration and Irrigation Water Requirements*. ASCE Manuals and Reports on Engineering Practice No. 70. 332 pages. Jensen, M. E.; Burman, R. D.; Allen, R. G.; Evapotranspiration and irrigation water requirements. ASCE Manuals and Reports on Engineering Practice No. 70. **1990,** 332.

10. Priestly, C. H. B.; Taylor, R. J.; On the assessment of surface heat flux and evaporation using large scale parameters. *Mon. Weath. Rev.* **1972,** *100,* 81–92.

APPENDIX A

[Modified and reprinted with permission from: Megh R. Goyal, 2012. Appendices, Pages 317–332. In: *Management of Drip/Trickle or Micro Irrigation* edited by Megh R. Goyal, New Jersey, USA: Apple Academic Press Inc.]

APPENDIX A
CONVERSION SI AND NON SI UNITS

To convert the column 1 in the Column 2, Multiply by	Column 1 Unit SI	Column 2 Unit Non-SI	To convert the column 2 in the column 1 Multiply by
		LINEAR	
0.621	kilometer, km (10^3m)	miles, mi	1.609
1.094	meter, m	yard, yd	0.914
3.28	meter, m	feet, ft	0.304
3.94×10^{-2}	millimeter, mm (10^{-3})	inch, in	25.4
		SQUARES	
2.47	hectare, he	acre	0.405
2.47	square kilometer, km^2	acre	4.05×10^{-3}
0.386	square kilometer, km^2	square mile, mi^2	2.590
2.47×10^{-4}	square meter, m^2	acre	4.05×10^{-3}
10.76	square meter, m^2	square feet, ft^2	9.29×10^{-2}
1.55×10^{-3}	mm^2	square inch, in^2	645
		CUBICS	
9.73×10^{-3}	cubic meter, m^3	inch-acre	102.8
35.3	cubic meter, m^3	cubic-feet, ft^3	2.83×10^{-2}
6.10×10^4	cubic meter, m^3	cubic inch, in^3	1.64×10^{-5}
2.84×10^{-2}	liter, L (10^{-3} m^3)	bushel, bu	35.24
1.057	liter, L	liquid quarts, qt	0.946
3.53×10^{-2}	liter, L	cubic feet, ft^3	28.3
0.265	liter, L	gallon	3.78
33.78	liter, L	fluid ounce, oz	2.96×10^{-2}
2.11	liter, L	fluid dot, dt	0.473
		WEIGHT	
2.20×10^{-3}	gram, g (10^{-3} kg)	pound,	454
3.52×10^{-2}	gram, g (10^{-3} kg)	ounce, oz	28.4
2.205	kilogram, kg	pound, lb	0.454
10^{-2}	kilogram, kg	quintal (metric), q	100
1.10×10^{-3}	kilogram, kg	ton (2000 lbs), ton	907
1.102	mega gram, mg	ton (US), ton	0.907
1.102	metric ton, t	ton (US), ton	0.907
		YIELD AND RATE	
0.893	kilogram per hectare	pound per acre	1.12
7.77×10^{-2}	kilogram per cubic meter	pound per fanega	12.87
1.49×10^{-2}	kilogram per hectare	pound per acre, 60 lb	67.19
1.59×10^{-2}	kilogram per hectare	pound per acre, 56 lb	62.71
1.86×10^{-2}	kilogram per hectare	pound per acre, 48 lb	53.75
0.107	liter per hectare	galloon per acre	9.35
893	ton per hectare	pound per acre	1.12×10^{-3}
893	mega gram per hectare	pound per acre	1.12×10^{-3}
0.446	ton per hectare	ton (2000 lb) per acre	2.24
2.24	meter per second	mile per hour	0.447
		SPECIFIC SURFACE	
10	square meter per kilogram	square centimeter per gram	0.1
10^3	square meter per kilogram	square millimeter per gram	10^{-3}
		PRESSURE	
9.90	megapascal, MPa	atmosphere	0.101
10	megapascal	bar	0.1
1.0	megagram per cubic meter	gram per cubic centimeter	1.00
2.09×10^{-2}	pascal, Pa	pound per square feet	47.9
1.45×10^{-4}	pascal, Pa	pound per square inch	6.90×10^3

To convert the column 1 in the Column 2, Multiply by	Column 1 Unit SI	Column 2 Unit Non-SI	To convert the column 2 in the column 1 Multiply by
		TEMPERATURE	
$1.00\,(K-273)$ --	Kelvin, K	centigrade, °C ------------	$1.00\,(C+273)$
$(1.8\,C+32)$ --	centigrade, °C	Fahrenheit, ° F ------------	$(F-32)/1.8$
		ENERGY	
9.52×10^{-4} ----	Joule J	BTU -------------------------	1.05×10^{3}
0.239 ----------	Joule, J	calories, cal ------------------	4.19
0.735 ----------	Joule, J	feet – pound -----------------	1.36
2.387×10^{5} ---	Joule per square meter	calories per square centimeter ---	4.19×10^{4}
10^{5} -------------	Newton, N	dynes -----------------------	10^{-5}
		WATER REQUIREMENTS	
9.73×10^{-3} -----	cubic meter	inch acre --------------------	102.8
9.81×10^{-3} -----	cubic meter per hour	cubic feet per second --------	101.9
4.40 -------------	cubic meter per hour	galloon (US) per minute ----	0.227
8.11 -------------	hectare-meter	acre-feet ---------------------	0.123
97.28 -----------	hectare-meter	acre-inch --------------------	1.03×10^{-2}
8.1×10^{-2} ------	hectare centimeter	acre-feet --------------------	12.33
		CONCENTRATION	
1 ----------------	centimol per kilogram	milliequivalents per 100 grams --------------------	1
0.1 -------------	gram per kilogram	percents ---------------------	10
1 ----------------	milligram per kilogram	parts per million -------------	1
		NUTRIENTS FOR PLANTS	
2.29 ----------- P		P_2O_5 -------------------------	0.437
1.20 ----------- K		K_2O --------------------------	0.830
1.39 ----------- Ca		CaO -------------------------	0.715
1.66 ----------- Mg		MgO -------------------------	0.602

NUTRIENT EQUIVALENTS

Column A	Column B	Conversion A to B	Equivalent B to A
N	NH_3	1.216	0.822
	NO_3	4.429	0.226
	KNO_3	7.221	0.1385
	$Ca(NO_3)_2$	5.861	0.171
	$(NH_4)_2SO_4$	4.721	0.212
	NH_4NO_3	5.718	0.175
	$(NH_4)_2HPO_4$	4.718	0.212
P	P_2O_5	2.292	0.436
	PO_4	3.066	0.326
	KH_2PO_4	4.394	0.228
	$(NH_4)_2HPO_4$	4.255	0.235
	H_3PO_4	3.164	0.316
K	K_2O	1.205	0.83
	KNO_3	2.586	0.387
	KH_2PO_4	3.481	0.287
	Kcl	1.907	0.524
	K_2SO4	2.229	0.449
Ca	CaO	1.399	0.715
	$Ca(NO_3)_2$	4.094	0.244
	$CaCl_2 \cdot 6H_2O$	5.467	0.183
	$CaSO_4 \cdot 2H_2O$	4.296	0.233
Mg	MgO	1.658	0.603
	$MgSO_4 \cdot 7H_2O$	1.014	0.0986
S	H_2SO_4	3.059	0.327
	$(NH_4)_2SO_4$	4.124	0.2425
	K_2SO_4	5.437	0.184
	$MgSO_4 \cdot 7H_2O$	7.689	0.13
	$CaSO_4 \cdot 2H_2O$	5.371	0.186

INDEX